THE SOCIOLOGY OF SCIENCE

THE SOCIOLOGY OF SCIENCE

Edited by BERNARD BARBER
BARNARD COLLEGE, COLUMBIA UNIVERSITY
and WALTER HIRSCH
PURDUE UNIVERSITY

GREENWOOD PRESS, PUBLISHERS
WESTPORT, CONNECTICUT

Library of Congress Cataloging in Publication Data

Barber, Bernard, ed.
The sociology of science.

Reprint of the ed. published by Free Press of Glencoe, New York.
Bibliography: p.
Includes index.
1. Science--Social aspects. I. Hirsch, Walter, 1919- II. Title.
[Q175.5.B37 1978] 500 78-5937
ISBN 0-313-20403-9

Reprinted with the permission of The Free Press, A Division of Macmillan Publishing Co., Inc.

Reprinted in 1978 by Greenwood Press, Inc.
51 Riverside Avenue, Westport, CT. 06880

Printed in the United States of America

10 9 8 7 6 5 4 3 2 1

Contents

4. The Organization of Scientific Work and Communication among Scientists

5. The Social Process of Scientific Discovery

THE SOCIOLOGY OF SCIENCE

Introduction

THIS READER *in the sociology of science has two main purposes. The first is to provide instructive and readable material on science as a social phenomenon: its essentially social character, its sociohistorical development, its patterns of organization, the social images of science, social influences on the processes of discovery, and the social responsibilities of science. The second purpose is to contribute to the further development of the sociology of science as a specialized field of study of great intellectual and practical importance.*

We hope that our first purpose, to provide readable and instructive material on science as a social phenomenon, will meet the current needs and arouse the new interest of at least two different groups. One group consists of college students, of all college students no matter what their major field of academic interest. Because science is of such central social and cultural importance in our society, no one is properly educated today if he does not have an understanding of science equal to his understanding of other major social institutions. Recently C. P. Snow has lamented the harm that may occur because even many educated men do not understand such substantive components of science as the second law of thermodynamics, to use Snow's example. Furthermore, we need to lament a second kind of ignorance of science, an ignorance that we take even more for granted than the substantive one that Sir Charles has pointed out—science as a social phenomenon. It should be one of the key aims of any college education nowadays to see to it that no student persists in this second kind of ignorance.

The other group for whom we are trying to provide readable and instructive material in the sociology of science consists of what are often ambiguously called "general readers." For our part we take this term to refer to those college-educated adults who are continuing to educate themselves by reading on their own. For the same reasons that all college students should know something of the sociology of science, general readers with a wide variety of concerns should too. We are, however, especially hopeful that our general readers will include many practicing scientists, for whom some of our selections may bring the pleasure and profit of seeing themselves, their products, and their problems in a new way. We also hope for the attention of those who make practical decisions concerning science—executives in the universities, in government, and in many areas of business.

Our second main purpose, we have said, is to contribute to the further development of the sociology of science as a specialized field of research. It is perhaps all too easy to say of almost any part of society that it has not yet received sufficient scientific study. But whatever may be true in fact of various other parts of society, certainly there is a minimum of rhetorical and moral exaggeration in pointing to the neglect of the sociology of science. Colleges and universities with courses in this subject or with professors who devote a major part of their research efforts to it are a handful at the most, despite the fact that among the few scholars who do seriously cultivate the sociology of science are included men like Robert K. Merton and Talcott Parsons, men who are outstanding for their contributions to sociology as a whole as well as to this particular field. Their theoretical and research interests in the sociology of science are a sterling guarantee of its importance. We hope that this Reader will become a focus for further teaching at the graduate level and for independent research by young sociologists. We think that they will find in these selections much work that would be distinguished for either theoretical sophistication or methodological rigor in any collection of sociological writings. Though it is not yet a large accumulation, the sociology of science can offer the young sociologist some of the finest models of work that his science now possesses. Such models should provide an inspiration for specialization in this field.

Although we have sought to provide technical, theoretical, and methodological models for the would-be sociologist of science, we have not neglected the excellent work by those who are not professional specialists in this field, nor have we failed to include some selections that are implicitly sociological even when they are not cast in the technical language of our specialty. As the contents of this Reader and our introductions to its several parts will show, the sociology of science can build on good foundations erected by men from several different scholarly specialties. It is of first importance for the sociologist of science to recognize that he must exchange ideas with many neighboring professionals in the historical and social science fields, even though his own basic concern is always for the advancement of a system of sociological analysis in its own right, however varied its sources.

Whether technically professional or not, our selections also seek to educate the novice in the sociology of science in the variety of different methods for research that are now available to him. Thus, our selections exemplify, among others, the methods of historical research, content analysis, participant observation, public opinion polling, and analysis of census data. Some of our selections, though not so many as we should like, pay careful attention to problems of sampling and statistical analysis. Since the methodology of research is no less important than its concepts and theories, we wish to demonstrate in this Reader that the sociology of science can use all the most recently developed sociological research tools; it can flourish in the field as well as in the library.

To attract new students and researchers to a field of scientific work is not enough. Their data must be accessible. Up to now data about science as a social phenomenon have been difficult to find—not so much the historical data, of course, as the contemporary data. This difficulty is not at all surprising. Many social objects of study resist those who would scrutinize them in a scientific fashion. There are always good social reasons for such resistance. Therefore, it only seems ironic that natural scientists should be reluctant to be studied by social scientists. In this respect, with regard to the sources of their resistance to being studied scientifically, scientists are like men in a variety of other social roles. Some of our selections show the social sources of resistance by scientists to the sociology of science. Resistance can be overcome, however, when men see that what they feared is in some good measure useful to them in the very activities that are being studied and that they value. We hope, then, that this Reader will prove to some practicing scientists that the sociology of science is sufficiently useful for the practice of science and for its values to earn it a more cordial welcome than it has sometimes had. We hope that scientists, by making themselves and their activities more accessible to sociologists of science, will furnish one of the indispensable ingredients of a further-developing research specialty.

As an aid to additional reading and research, we have appended to our readings a Selected Bibliography. This Bibliography will bring up to date Bernard Barber's Sociology of Science: A Trend Report and Bibliography (UNESCO, Paris: Current Sociology, V, no. 2, 1956). There are about 150 items in this new bibliography; they are not, however, annotated, as is the UNESCO bibliography.

Finally, a few words should be said about the relations between this Reader and Bernard Barber's Science and the Social Order (The Free Press of Glencoe, 1952; reprinted as a paperback, with a new introduction, by Collier Books, 1962), the only existing attempt to provide a theoretically systematic and factually comprehensive account of the sociology of science. This Reader is neither so systematic nor so inclusive as Science and the Social Order, but at many points it goes much deeper into factual detail and into some specific points of theoretical analysis as well. Furthermore this Reader is an important supplement to the other book, providing a good deal of work that has been done since Science and the Social Order was written. Some thirty of the approximately forty selections reprinted in this Reader have been published since 1952. One section in the Reader, "The Social Image of the Scientist and His Self-Conceptions," containing six selections, represents a kind of work on which practically nothing was available ten years ago. In sum, while the two books can be used entirely separately, each will add something to the other if they are used together.

The Social Nature of Science and the Scientific Role

1

THE two selections in this first part of the reader concern themselves with an analysis of some fundamental aspects of the social nature of scientific ideas and the scientific role. Parsons, whose interest in science stems from the most general theoretical concerns for sociology as a whole, begins by showing why "science" of some sort must exist in every society, why there can be no social action without a certain amount of rational empirical knowledge of the physical, the biological, and the social worlds. He then defines the basic characteristics of scientific ideas as a structure in their own right and indicates how these several characteristics vary in different societies. The characteristics of systematization and generalization, for example, are very much greater in "modern" science than in the science of primitive or ancient civilized societies. For social reasons, which he tries to specify, greater systematization and generality in science are hard to achieve; but this achievement is aided by certain kinds of social norms and by the development of specialized, professional, full-time roles for "the scientist." It is of interest, in this connection, to recall that the term "scientist" itself was coined only in the early nineteenth century by the Reverend William Whewell, Professor of Moral Philosophy in Cambridge University. Moreover, in his Life of Lord Kelvin, *Thompson remarks that this distinguished scientist, as well as such men as Huxley, Lord Rayleigh, and Sir John Lubbock, even late in the nineteenth century, refrained from using the term "scientist" because of their dislike of it. The specialized role of scientist, which we take so much for granted, is,* as we can see from this and other evidence, a recent development in the history of human society. In brief form, which selections in other parts of the reader develop in greater detail, Parsons has described the norms and other elements of the Western cultural tradition that have been particularly powerful supports for the development of modern science. He has also touched upon the organizational supports for the scientific role, stressing especially the great importance of the university; and he has not neglected some of the problems of the professionalization of the scientific role, what he calls "the communication gap," and the disturbing impact of scientific discoveries on a variety of social, economic, and religious vested interests.

The selection by Merton, though illustrating its analysis chiefly with the case of the Nazi destruction of German science, is oriented to the general sociological problem of the sources of support for science in modern society and hostility to it. These elements of support and hostility can be found in all modern societies, though in degrees of difference in the different countries that are of the utmost importance for the development of science in each country. Merton's discussion serves to remind us that support and hostility for science may stand in a precarious balance; we can alter the balance in the direction we desire by better understanding its social components. In Part 6, for example, Shils's discussion of the dangers of politically-imposed secrecy in American science supplements Merton's analysis very nicely. So also do selections in Part 2 by Joravsky and Labedz, which analyze past and present conditions in Russian science.

No more than any other social institution, can science be taken for granted. Merton and Parsons, in the two selections of Part 1, provide an introduction to the basic social elements that define, support, and oppose the fullest development of science.

1 *The Institutionalization of Scientific Investigation*

TALCOTT PARSONS

THE DIFFERENCE between science and empirical lore, and the corresponding difference between scientific investigation and empirically cognitive problem-solving on a common sense level, are gradual and in a sense relative. What was the technical science of yesterday has in certain respects become the common sense of today—as in the case of the heliocentric theory of the solar system or the germ theory of disease. But though the borderline is indistinct, in fundamental pattern there is a sharp difference. The generality of science far transcends the boundaries of particular practical fields of instrumental interest, and cuts across many of them. The role of the scientist becomes technical and his specialized interests and procedures are of "no use" except for his own specialized purposes. The knowledge he possesses is only with difficulty if at all accessible to the untrained layman. The ultimate judgment of it must lie with his professionally qualified peers. Thus, the status of any given item of knowledge, as belonging to science or to common sense, may be doubtful. And the variations between these two types is a matter of degree. But the distinction is nonetheless vital.

The basic norms of scientific knowledge are perhaps four, empirical validity, logical clarity or precision of the particular proposition, logical consistency of the mutual implications of propositions, and generality of the "principles" involved, which may perhaps be interpreted to mean range of mutually verified implications.[1] Very specific propositions

---

Reprinted from The Social System, *by Talcott Parsons (New York: The Free Press of Glencoe, 1951), pp. 335–345.*

1. The reader may for purposes of orientation to the general nature of science and its processes of development, be referred to James B. Conant, *On Understanding Sci-*

of particular fact may be held to be verified with a certainty approaching absoluteness. The more general the proposition the less that order of approach to certainty in the sense that it is inconceivable that it should ever have to be modified, is possible. But science "progresses" in proportion as it is possible to relate very particular facts to generalized systems of implication. Hence it is not possible to use only the one criterion. It would be generally admitted that analytical mechanics before the relativity and quantum era was in some sense a more "advanced" science than botanical taxonomy, even though the "meaning" of many of the generalized propositions of mechanics was in certain respects seriously in doubt, and taxonomy ordered enormous numbers of facts, with very little in its logical structure which was questionable or controversial.

There is, therefore, not merely the question of whether or not a given item of knowledge "belongs" to science, but implied in the term used above, "basic norms," is the implication that there are levels of scientific advancement. Moreover this advancement does not consist only of discrete additions to existing knowledge of facts, but of the relation of this knowledge of fact to systematization and generalized theoretical analysis. This gives us the sense in which science, specifically on the cultural level, is a dynamic thing. Its inherent structure is one of variant levels of advancement. Such a type of culture element contains, in its relation to action, an inherent element of instability. There is always the possibility that someone will make a new discovery. This may be merely a specific addition to knowledge of fact, in which case it will simply be fitted in with the rest in its proper place. But it may be something which necessitates the *reorganization* to a greater or less degree, of the systematized body of knowledge.

This growth-oriented dimension of scientific knowledge as a part of culture is of particular interest here. For this ties in with action; scientific innovation is not a culturally automatic process, but is an action process, and as such involves all the fundamental elements which are relevant to the analysis of action-processes.

There is, however, as Kroeber has brought out with particular clarity,[2] an inherent element of "cultural structure" which provides a partial but very important set of determinants of this process. For precisely as a cognitive system the body of scientific knowledge, in any given field at a given time, is definitely structured. Advance does not and cannot take place in random fashion in all directions at once, i.e., unselectively. It is structured by the intrinsic cultural features of that

ence. Some of the best treatments of sociological problems relative to science are to be found in R. K. Merton, *Social Theory and Social Structure,* Part IV. Also the author's *Structure of Social Action,* especially Chapters I and XIX contains some relevant discussions. Cf. also "The Institutionalization of Social Science and the Problems of the knowledge. Certain problems are inherent in this structure. Facts which are discovered may be more or less relevant to these problems. Even if discovered by chance, the consequences of a discovery are thus a function of the way in which it fits into the structure of existing knowledge, and its problem-structure. The possibilities inherent in any given knowledge-system and related problem-structure are not random and infinite, but finite and specifically structured. There will then be a determinate process of working out the possibilities inherent in a knowledge structure the building of which has once begun, until these possibilities have finally become exhausted. Kroeber uses this type of analysis most convincingly to show that creativity in scientific advance, as well as for example in the arts, is not a simple function of the supply of biologically gifted individuals, but depends on the job there is for them to do. By exactly the same token it cannot be simply a function of favorable states of the social structure. This is authentically a *cultural* factor.

Empirical knowledge is an essential part of all action, particularly when the instrumental aspect is highly developed. There is hence an inherent interest not only in the application of such knowledge, but in its further development. But at the same time we have seen that there are strongly counteracting factors of such a character that unless investigation becomes the primary technical function of specialized roles, the advancement of knowledge is often very slow and halting. Perhaps the most fundamental reason is that for the "practical man" the primary focus is on the attainment of the immediate goal itself, and knowledge constitutes simply *one* of the available resources for achieving it. But, furthermore, practical action tends, for a variety of reasons, to be imbedded in a matrix of non-rational orientation patterns (including the functional equivalents of magic) which, because they are not directly empirically grounded, can only be stabilized by being traditionalized. Indeed the general pressure toward stabilization of a system of action militates against the advancement of knowledge because this obviously has many repercussions besides making the effective attainment of the particular goal more feasible. Also the practical man does not have a direct interest in the further ramifications of a scientific body of knowledge beyond his immediate sphere of practical interest. All this is attested by a considerable amount of tension between scientists and practical men even in the fields where science has been most successfully applied.

These reasons why the practical man develops resistances to him-

Conference," Chapter XIV of *Perspectives on a Troubled Decade.* Published by the Conference on Science, Philosophy and Religion, 1950.

2. See A. L. Kroeber, *Configurations of Culture Growth,* and *Anthropology,* 1948 Edition.

self pushing the advancement of empirical knowledge forward are also in turn connected with the fact that beyond certain points this advancement only becomes possible through the kinds of technical means which involve specialization of roles. Knowledge itself becomes technical, and takes long training to master. Furthermore, investigation is a process which requires special skills which in certain respects go beyond the mastery of the bases in established knowledge from which any given phase of the process starts. Hence the above considerations about the way in which changes in empirical knowledge impinge on any system of social interaction, become even more cogent when the place in a differentiated instrumental complex of the specialized role of scientific investigation is taken into account.

First, the technical nature of specialized science means that there develops what may be called a communication gap. The scientist is inevitably dependent on "laymen" for support and for the provision of facilities. But in circumstantial detail the layman is not technically competent to judge what the scientist is doing, he has to take it "on authority." This general situation is accentuated by the fact that there is often a large gap between the frontiers of scientific investigation and the practical results which the practical man can most readily appreciate, understand and use. This is particularly because the cognitive structure of science is such that the ramifications of scientific problems cannot be restricted to the solution of the kind of applied problem area in which a practical man is interested. He, therefore, from his own perspective often does not have the basis for seeing that what the scientist is doing is of "any use."

Such a communication gap between roles always creates a problem of control. Besides the question of why an activity should be supported, which doesn't seem to be of any particular use, there are two types of foci of anxiety. In the first place the scientist must do a good many things which impinge upon others. Partly these things are just "queer" and their motives seem to be unfathomable. But sometimes they are potentially dangerous to some laymen, e.g., explosions in the chemical laboratory, and partly they impinge on touchy areas of sentiment. Thus, the dissection of cadavers by anatomists and medical students long had to be carried out surreptitiously and today, with all the prestige of medical science, some religious groups permit autopsies only in the few cases where it is legally required. The activities of the Society for Prevention of Cruelty to Animals in trying to limit the use of animals in scientific research are well known. Scientists may also impose burdens on the layman if only by even asking him to give some of his time for a purpose which as a layman he does not understand. This is frequently the case in social science research, e.g., in interviewing subjects.

Along with having to secure facilities from others in ways which impinge on their interests and sentiments, the scientist is faced with a good deal of anxiety about the implications of the results of his work. This is particularly true because the crucial significance of generality of implication of science means that it is not possible to limit these implications to the solution of clear-cut and limited practical problems. This is perhaps particularly important in relation to the ideological, philosophical and religious fields. The motives of adherence to ideological and religious ideas are usually differently structured from those of belief in the simpler bits of empirical lore; the very fact that the elements of the situation involved present such extremely difficult cognitive problems and yet the affective interests in a clear-cut definition of the situation are so strong, makes that clear. But certainly no large-scale development of science is possible without some important impingement and hence in part an upsetting effect, on ideological and religious positions which play an important part in the cultural tradition.

In general the practical man is hedged in by considerations which make only relatively ad hoc and limited resources available for him to make use of. Where the operations of scientific investigation may involve far-reaching repercussions on the sentiment system of a society, it is unlikely that the major impetus for such a development will come alone or even primarily[3] from practical "interests." At the least practical interests will have to be combined with a cultural situation which lends prestige to the relevant types of activity on bases other than their promise of improvements of practical efficiency.

This appears very definitely to be the case with science. The place of science in Western society is part of the ascendancy of a cultural tradition which involves a high valuation of certain types of rationality of understanding of the empirical world, on grounds apart from the promise of practical applicability of the results of that understanding. Once such valuation is established and built into the institutional system, it comes to be strongly reinforced by the practical fruits, once science has been permitted to develop far enough so that these fruits have become relatively impressive. In all probability only when such a combination has become firmly established does it become possible for scientific investigation to acquire the level of prestige which it has enjoyed in the modern Western world. But even here there are many elements of ambivalence in public attitudes toward science and the scientist, which are expressed in much irrational and some relatively rational opposition to his role.[4]

3. The wielder of political power is the most obvious exception, since he often cannot avoid such repercussions.

4. Thus an article in the Vienna *Presse*, which the author happened to see while

The obverse of this is that there is a strong non-rational element in the popular support of the scientist. He is the modern magician, the "miracle man" who can do incredible things. Along with this in turn goes a penumbra of belief in pseudo-science. Scientists themselves are, like other people, far from being purely and completely rational beings. Their judgment, particularly toward and beyond the fringes of their technical competence, is often highly fallible, and things are said in the name of science which are very far from meeting the standards of scientific demonstrability, or short of that, showing the degree of tentativeness and suspension of judgment which is indicated in the light of the deficiencies of the evidence. From this professionally internal penumbra there is a shading off into the ideas about scientific matters current among the lay public, where pseudo-science is much more prolific. It is important to note that the conditions under which science could have a well-established place in the social system are such that the presence of this pseudo-scientific penumbra seems to be inevitable. It can, of course, to a degree be limited and controlled, and in fact is, but it seems unlikely that it can be eliminated.

In the broadest possible sense, the most important feature of the Western cultural tradition as a bulwark of science is its strong universalistic trend. This means in the first instance a strong emphasis on the importance of knowledge, an emphasis evidenced for instance in the stress on rational theology in all the most important branches of Christianity. That the hospitality to science was greatly increased in the "ascetic" branches of Protestantism, as compared with the Catholic tradition, is shown by Merton's analysis of the religious situation in England at the time of the great scientific developments of the seventeenth century.[5]

The valuation of knowledge in a secular direction greatly increased in post-mediaeval times, connecting with the revival of interest in and prestige of the traditions of classical antiquity. Eventually, in the Western world the doctrine that a gentleman ought to be an educated man, first as part of his ascribed aristocratic role, gradually shifted until it has at least become true in more recent times that an educated man was to be considered a gentleman, that is, knowledge became the most important single mark of generalized superiority. This is, of course, a highly schematic statement of a very complex development.

The primary core of the Western tradition of higher education as the mark of the gentleman was in its earlier phases humanistic rather than scientific, though the place of mathematics is of considerable im-

in Austria in the summer of 1948, argued with the utmost seriousness that it would have been a good thing for civilization had the Church in the long run won out against Galileo, that is, had the development of modern science been suppressed. The argument was that science had opened a Pandora's box with the contents of which portance. But the great tradition of humanistic learning shares many features with science, above all the respect for impartial objectivity and hence for evidence; in the first instance shown in the concern for the authenticity of historical and literary texts, which is by no means so prominent in many other great literate traditions. The humanistic scholar in this sense was in many respects the precursor of the scientist and is of course today his colleague in the most highly educated sector of the population.

In the most modern era this cultural tradition has above all become embodied in the university as its principal institutionalized frame. Not least important of the facts about the modern university is that it combines the highest levels of development of the functions of pure scientific investigation in the same basic organizational complex with the humanistic branches of learning which have formed the primary core of the most highly rationalized part of the great expressive cultural tradition of the West. This includes, of course, theology as the rational foundation of religious beliefs. Furthermore, of course, it is highly significant that a large part of the fundamental training function of the major branches of applied science, especially in medicine and engineering, has become an integral part of the university.

Apart from the institutionalization of the specific role of the scientist as such, which will be commented upon presently, this situation has the great importance of directly integrating the role of the scientist with those of the other principal "experts" in major branches of the cultural tradition. The scientist has the support of being considered part of the same cultural complex which includes the humanities. Not least important, he shares with them the function of educating the primary elite elements of the oncoming generation in the society. In so far as the doctrine is upheld that in general the "leading men" of the society should be educated men in the modern sense, their elite status carries with it commitment to a value-system of which the values of the scientist, and the valuation of his activities and their results, form an integral part. This integration of science, both with the wider cultural tradition of the society, and with its institutional structure, constitutes the *primary* basis of the institutionalization of scientific investigation as part of the social structure. It means that the scientist shares the status in the universities with the other key groups who are primary culture bearers and on terms such that the values of science come to be inculcated in the value-system of society generally through the education of its primary elite elements. Without this it is highly

humanity was unable to cope, and a kind of intellectual authoritarianism which would limit investigation to fields known to be "safe" was the only solution. Who is to say that there is no force in such an argument?

5. Robert K. Merton, *Science and Society in Seventeenth Century England.*

doubtful whether even at its most recent stages of development, the interest of practical men in the fruits of science could alone long sustain scientific investigation as the function of a major type of specialized social role.

In more specific terms, it is first important to note that status on the faculty of a university gives the scientist a clearly institutionalized role with all its concomitants. In terms of modern organization in the occupational field, it gives him both a source of remuneration for himself and of course his family, and a "market" for his products, through putting him in relation to students and professional colleagues and providing or encouraging publication channels for his work. It provides him by and large with the increasingly necessary but expensive facilities for his work, and the framework of the all-important co-operative relationships with colleagues and others. By giving him an "orthodox" occupational role, it gives not only him, but members of his family, an acceptable status in the society, e.g., he "earns a living." Moreover, the university, both through its general prestige and sometimes through specific administrative action, serves to protect his freedom to carry out his function in the face of forces in the society which tend to interfere with it.[6]

The occupational role which the scientist occupies, with its center of gravity in the university, is an integral part of the general occupational system. Moreover it is of the special type we have above called a professional role.[7] The fact that it shares the pattern elements of universalism, affective neutrality, specificity and achievement orientation with the occupational system in general does not require special comment here. But it is worth while to call attention to the fact that as a professional role it is institutionalized predominantly in terms of collectivity—rather than self-orientation.

There seem to be two primary contexts in which self-orientation in the scientific world would tend to be seriously dysfunctional. One is the implication of the saying "knowledge is power." It is indeed true that in a sufficiently large number and strategically important type of cases the discoveries of the scientist would, if uncontrolled, put him in a position to interfere with the interests and sentiments of others.

6. The fact that mechanisms of social control sometimes fail to operate successfully is no evidence that they do not exist, or are not effective in other connections.

7. There are, of course, many possibilities of dysfunctional phenomena developing when scientific investigation is thus institutionalized. Thus exposure to the criticism of colleagues may be associated with a tendency to sterile pedantry and perfectionism in detail which sacrifices the importance of bold ideas. In general the minimization of refined competitive ratings in university faculties—the treatment of the "company of scholars" as a "company of equals"—may be interpreted to be an adaptive structure with the function of counteracting some of these dysfunctional tendencies. Cf. Logan Wilson, *The Academic Man.*

These others, the "public," are in need of protection in the whole field of the uses of science. A major factor in this need lies in the gap in communication between expert and layman occasioned by the technical character of science. The layman is unable to protect his own interests in a "market situation." Thus, in a certain sense, the scientist is institutionally endowed with authority, he is recognized as "an authority" in his field, and the general analysis of the functional reasons for the association of other-orientation with authority applies.

The second dysfunctional possibility is that of the "monopolization" of knowledge in its bearing on the process of scientific advancement itself. Such monopolization would not only restrict the rate and spread of scientific advance, by making it more difficult to build on what others had done, but it would also seriously interfere with the social control mechanisms of science internal to itself. It is a cardinal fact that the scientist is, through discussion and publication, exposed to the criticism of his professional colleagues to an unusual degree, including the checking of his results through replication on the part of others. The idea that a "scientist's theory is his castle" which must not be trespassed upon, except on terms laid down by himself, would be incompatible with this discipline which is so important to the maintenance of standards of objectivity.[8] Finally, it should not be forgotten that the scientist requires "privileges" from his fellow men. Collectivity-orientation certainly does much to validate his claims to them. Thus, when the social scientist requests information in an interview the subject may very likely ask, "why do you want to know this?" The legitimation, which must be explicit or implicit in the answer, is that it is in the interests of the advancement of knowledge, not the personal "axe to grind" that the interviewer may conceivably have. Often explicit safeguards against misuse of information, i.e., generally "personal" or "partisan" rather than scientific use, have to be given.

8. This is what Merton, "Science and Democratic Social Structure," *op. cit.*, Chapter XII, calls the "communism" of science. He gives an admirable functional analysis of its significance.

2 *Science and the Social Order*

ROBERT K. MERTON

ABOUT THE TURN OF THE CENTURY, Max Weber observed that "the belief in the value of scientific truth is not derived from nature but is a product of definite cultures."[1] We may now add: and this belief is readily transmuted into doubt or disbelief. The persistent development of science occurs only in societies of a certain order, subject to a peculiar complex of tacit presuppositions and institutional constraints. What is for us a normal phenomenon which demands no explanation and secures many self-evident cultural values, has been in other times and still is in many places abnormal and infrequent. The continuity of science requires the active participation of interested and capable persons in scientific pursuits. This support of science is assured only by appropriate cultural conditions. It is, then, important to examine those controls which motivate scientific careers, which select and give prestige to certain scientific disciplines and reject or blur others. It will become evident that changes in institutional structure may curtail, modify or possibly prevent the pursuit of science.[2]

Sources of Hostility toward Science

HOSTILITY toward science may arise under at least two sets of conditions, although the concrete systems of values—humanitarian,

---

Reprinted from Social Theory and Social Structure, *by Robert K. Merton (New York: The Free Press of Glencoe, 1957, revised edition), Chapter XV, pp. 537–561.*

Read at the American Sociological Society Conference, December, 1937. The writer is indebted to Professor Read Bain, Professor Talcott Parsons, Dr. E. Y. Hartshorne, and Dr. E. P. Hutchinson for their helpful suggestions.

1. Max Weber, *Gesammelte Aufsätze zur Wissenschaftslehre*, 213; *cf.* Sorokin, *Social and Cultural Dynamics*, esp. II, Chap. 2.

2. *Cf.* Merton, *Science, Technology, and Society in Seventeenth Century England*, Chap. XI.

economic, political, religious—upon which it is based may vary considerably. The first involves the logical, though not necessarily correct, conclusion that the results or methods of science are inimical to the satisfaction of important values. The second consists largely of non-logical elements. It rests upon the feeling of incompatibility between the sentiments embodied in the scientific ethos and those found in other institutions. Whenever this feeling is challenged, it is rationalized. Both sets of conditions underlie, in varying degrees, current revolts against science. It might be added that such reasoning and affective responses are also involved in the social approval of science. But in these instances science is thought to facilitate the achievement of approved ends, and basic cultural values are felt to be congruent with those of science rather than emotionally inconsistent with them. The position of science in the modern world may be analyzed, then, as a resultant of two sets of contrary forces, approving and opposing science as a large-scale social activity.

We restrict our examination to a few conspicuous instances of certain revaluation of the social role of science, without implying that the antiscience movement is in any sense thus localized. Much of what is said here can probably be applied to the cases of other times and places.[3]

The situation in Nazi Germany since 1933 illustrates the ways in which logical and non-logical processes converge to modify or curtail scientific activity. In part, the hampering of science is an unintended by-product of changes in political structure and nationalistic credo. In accordance with the dogma of race purity, practically all persons who do not meet the politically imposed criteria of "Aryan" ancestry and of avowed sympathy with Nazi aims have been eliminated from universities and scientific institutes.[4] Since these outcasts include a considerable number of eminent scientists, one indirect consequence of the racialist purge is the weakening of science in Germany.

Implicit in this racialism is a belief in race defilement through actual or symbolic contact.[5] Scientific research by those of unimpeachable "Aryan" ancestry who collaborate with non-Aryans or who even accept their scientific theories is either restricted or proscribed. A new racial-political category has been introduced to include these incorrigible Aryans: the category of "White Jews." The most prominent member of this new race is the Nobel Prize physicist, Werner Heisenberg, who has

3. The premature death of E. Y. Hartshorne halted a proposed study of science in the modern world in terms of the analysis introduced in this chapter.

4. See Chapter III of E. Y. Hartshorne, *The German Universities and National Socialism* (Cambridge: Harvard University Press, 1937), on the purge of the universities; *cf. Volk und Werden,* 5, 1937, 320–1, which refers to some of the new requirements for the doctorate.

5. This is one of many phases of the introduction of a caste system in Germany. As R. M. MacIver has observed, "The idea of defilement is common in every caste system." *Society,* 172.

persisted in his declaration that Einstein's theory of relativity constitutes an "obvious basis for further research."[6]

In these instances, the sentiments of national and racial purity have clearly prevailed over utilitarian rationality. The application of such criteria has led to a greater proportionate loss to the natural science and medical faculties in German universities than to the theological and juristic faculties, as E. Y. Hartshorne has found.[7] In contrast, utilitarian considerations are foremost when it comes to official policies concerning the directions to be followed by scientific research. Scientific work which promises direct practical benefit to the Nazi party or the Third Reich is to be fostered above all, and research funds are to be re-allocated in accordance with this policy.[8] The rector of Heidelberg University announces that "the question of the scientific significance [*Wissenschaftlichkeit*] of any knowledge is of quite secondary importance when compared with the question of its utility."[9]

The general tone of anti-intellectualism, with its depreciation of the theorist and its glorification of the man of action,[10] may have long-run rather than immediate bearing upon the place of science in Germany. For should these attitudes become fixed, the most gifted elements of the population may be expected to shun those intellectual disciplines which have thus become disreputable. By the late 30's, effects of this antitheoretical attitude could be detected in the allocation of academic interests in the German universities.[11]

It would be misleading to suggest that the Nazi government has completely repudiated science and intellect. The official attitudes

6. *Cf.* the official organ of the SS, the *Schwarze Korps*, July 15, 1937, 2. In this issue Johannes Stark, the president of the Physikalisch-Technischen Reichsanstalt, urges elimination of such collaborations which still continue and protests the appointment of three university professors who have been "disciples" of non-Aryans. See also Hartshorne, *op. cit.*, 112–3; Alfred Rosenberg, *Wesen Grundsätze und Ziele der Nationalsozialistischen Deutschen Arbeiterpartei* (München: E. Boepple, 1933), 45 ff.; J. Stark, "Philipp Lenard als deutscher Naturforscher," *Nationalsozialistische Monatshefte*, 1936, 71, 106–11, where Heisenberg, Schrödinger, von Laue, and Planck are castigated for not having divorced themselves from the "Jewish physics" of Einstein.

7. The data upon which this statement is based are from an unpublished study by E. Y. Hartshorne.

8. *Cf. Wissenschaft und Vierjahresplan*, Reden anlässlich der Kundgebung des NSD-Dozentenbundes, January 18, 1937; Hartshorne, *op. cit.*, 110 ff.; E. R. Jaensch, *Zur Neugestaltung des deutschen Studententums und der Hochschule* (Leipzig, J. A. Bart, 1937), esp. 57 ff. In the field of history, for example, Walter Frank, the director of the Reichsinstitut für Geschichte des neuen Deutschlands, "the first German scientific organization which has been created by the spirit of the national-socialistic revolution," testifies that he is the last person to forego sympathy for the study of ancient history, "even that of foreign peoples," but also points out that the funds previously granted the Archaeological Institute must be re-allocated to this new historical body which will "have the honor of writing the history of the National Socialist Revolution." See his *Zukunft und Nation* (Hamburg, Hanseatische Verlagsanstalt, 1935), esp. 30 ff.

toward science are clearly ambivalent and unstable. (For this reason, any statements concerning science in Nazi Germany are made under correction.) On the one hand, the challenging scepticism of science interferes with the imposition of a new set of values which demand an unquestioning acquiescence. But the new dictatorships must recognize, as did Hobbes who also argued that the State must be all or nothing, that science is power. For military, economic and political reasons, theoretical science—to say nothing of its more respectable sibling, technology—cannot be safely discarded. Experience has shown that the most esoteric researches have found important applications. Unless utility and rationality are dismissed beyond recall, it cannot be forgotten that Clerk Maxwell's speculations on the ether led Hertz to the discovery that culminated in the wireless. And indeed one Nazi spokesman remarks: "As the practice of today rests on the science of yesterday, so is the research of today the practice of tomorrow."[12] Emphasis on unity requires an unbanishable minimum of interest in science which can be enlisted in the service of the State and industry.[13] At the same time, this emphasis leads to a limitation of research in pure science.

Social Pressures on Autonomy of Science

AN ANALYSIS of the role of science in the Nazi state uncovers the following elements and processes. The spread of domination by one segment of the social structure—the State—involves a demand for primary loyalty to it. Scientists, as well as all others, are called upon to relinquish adherence to all institutional norms which, in the opinion of political authorities, conflict with those of the State.[14] The norms

9. Ernst Krieck, *Nationalpolitische Erziehung* (Leipzig, Armanen Verlag, 1935), (19th Printing), 8.

10. The Nazi theoretician, Alfred Baeumler, writes: "Wenn ein Student heute es ablehnt, sich der politischen Norm zu unterstellen, es x. B'ablehnt, an einem Arbeits- oder Wehrsportlager teilzunehmen, weil er damit Zeit für sein Studium versäume, dann zeigt er damit, dass er nichts von dem begriffen hat, was um ihn geschieht. Seine Zeit kann er nur bei einem abstrakten, richtungslosen Studium versäumen." *Männerbund und Wissenschaft* (Berlin, Junker & Dünnhaupt, 1934), 153.

11. Hartshorne, *op. cit.*, 106 ff.; *cf. Wissenschaft und Vierjahresplan, op. cit.*, 25–6, where it is stated that the present "breathing-spell in scientific productivity" is partly due to the fact that a considerable number of those who might have received scientific training have been recruited by the army. Although this is a dubious explanation of that particular situation, a prolonged deflection of interest from theoretical science will probably produce a decline in scientific achievements.

12. Professor Thiessen in *Wissenschaft und Vierjahresplan, op. cit.*, 12.

13. For example, chemistry is highly prized because of its practical importance. As Hitler put it, "we will carry on because we have the fanatic will to help ourselves and because in Germany we have the chemists and inventors who will fulfill our needs." Quoted in *Wissenschaft und Vierjahresplan, op. cit.*, 6; *et passim.*

14. This is clearly put by Reichswissenschaftsminister Bernhard Rust, *Das nationalsozialistische Deutschland und die Wissenschaft* (Hamburg, Hanseatische Verlagsanstalt, 1936), 1–22, esp. 21.

of the scientific ethos must be sacrificed, in so far as they demand a repudiation of the politically imposed criteria of scientific validity or of scientific worth. The expansion of political control thus introduces conflicting loyalties. In this respect, the reactions of devout Catholics who resist the efforts of the political authority to redefine the social structure, to encroach upon the preserves which are traditionally those of religion, are of the same order as the resistance of the scientist. From the sociological point of view, the place of science in the totalitarian world is largely the same as that of all other institutions except the newly-dominant State. The basic change consists in placing science in a new social context where it appears to compete at times with loyalty to the state. Thus, coöperation with non-Aryans is redefined as a symbol of political disloyalty. In a liberal order, the limitation of science does not arise in this fashion. For in such structures, a substantial sphere of autonomy—varying in extent, to be sure—is enjoyed by non-political institutions.

The conflict between the totalitarian state and the scientist derives in part, then, from an incompatibility beween the ethic of science and the new political code which is imposed upon all, irrespective of occupational creed. The ethos of science[15] involves the functionally necessary demand that theories or generalizations be evaluated in terms of their logical consistency and consonance with facts. The political ethic would introduce the hitherto irrelevant criteria of the race or political creed of the theorist.[16] Modern science has considered the personal equation as a potential source of error and has evolved impersonal criteria for checking such error. It is now called upon to assert that certain scientists, because of their extra-scientific affiliations, are *a priori* incapable of anything but spurious and false theories. In some instances, scientists are required to accept the judgments of scientifically incompetent political leaders concerning *matters of science*. But such politically advisable tactics run counter to the institutionalized norms of science. These, however, are dismissed by the totalitarian state as "liberalistic" or "cosmopolitan" or "bourgeois" prejudices,[17] inasmuch

15. The ethos of science refers to an emotionally toned complex of rules, prescriptions, mores, beliefs, values and presuppositions which are held to be binding upon the scientist. Some phases of this complex may be methodologically desirable, but observance of the rules is not dictated solely by methodological considerations. This ethos, as social codes generally, is sustained by the sentiments of those to whom it applies. Transgression is curbed by internalized prohibitions and by disapproving emotional reactions which are mobilized by the supporters of the ethos. Once given an effective ethos of this type, resentment, scorn and other attitudes of antipathy operate almost automatically to stabilize the existing structure. This may be seen in the current resistance of scientists in Germany to marked modifications in the content of this ethos. The ethos may be thought of as the "cultural" as distinct from the "civilizational" component of science. *Cf.* R. K. Merton, "Civilization and culture," *Sociology and Social Research*, 1936, 21, 103–113.

16. *Cf.* Baeumler, *op. cit.*, 145. Also Krieck (*op. cit.*), who states: "Nicht alles,

as they cannot be readily integrated with the campaign for an unquestioned political creed.

From a broader perspective, the conflict is a phase of institutional dynamics. Science, which has acquired a considerable degree of autonomy and has evolved an institutional complex which engages the allegiance of scientists, now has both its traditional autonomy and its rules of the game—its ethos, in short—challenged by an external authority. The sentiments embodied in the ethos of science—characterized by such terms as intellectual honesty, integrity, organized scepticism, disinterestedness, impersonality—are outraged by the set of new sentiments which the State would impose in the sphere of scientific research. With a shift from the previous structure where limited loci of power are vested in the several fields of human activity to a structure where there is one centralized locus of authority over all phases of behavior, the representatives of each sphere act to resist such changes and to preserve the original structure of pluralistic authority. Although it is customary to think of the scientist as a dispassionate, impersonal individual—and this may not be inaccurate as far as his technical activity is concerned—it must be remembered that the scientist, in company with all other professional workers, has a large emotional investment in his way of life, defined by the institutional norms which govern his activity. The social stability of science can be ensured only if adequate defences are set up against changes imposed from outside the scientific fraternity itself.

This process of preserving institutional integrity and resisting new definitions of social structure which may interfere with the autonomy of science finds expression in yet another direction. It is a basic assumption of modern science that scientific propositions "are invariant with respect to the individual" and group.[18] But in a completely politicized society—where as one Nazi theorist put it, "the universal meaning of the political is recognized"[19]—this assumption is impugned. Scientific findings are held to be merely the expression of race or class or nation.[20]

was den Anspruch auf Wissenschaftlichkeit erheben darf, liegt auf der gleichen Rang- und Wertebene; protestantische und katholische, französische und deutsche, germanische und jüdische, humanistische oder rassische Wissenschaft sind zunächst nur Möglichkeiten, noch nicht erfüllte oder gar gleichrangige Werte. Die Entscheidung über den Wert der Wissenschaft fällt aus ihrer 'Gegenwärtigkeit,' aus dem Grad ihrer Fruchtbarkeit, ihrer geschichtsbildenden Kraft. . . ."

17. Thus, says Ernst Krieck: "In the future, one will no more adopt the fiction of an enfeebled neutrality in science than in law, economy, the State or public life generally. The method of science is indeed only a reflection of the method of government." *Nationalpolitische Erziehung,* 6. *Cf.* Baeumler, *op. cit.,* 152; Frank, *Zukunft und Nation,* 10; and contrast with Max Weber's "prejudice" that "Politik gehört nicht in den Hörsaal."

18. H. Levy, *The Universe of Science* (New York, Century Co., 1933), 189.

19. Baeumler, *Männerbund und Wissenschaft,* 152.

20. It is of considerable interest that totalitarian theorists have adopted the radical

As such doctrines percolate to the laity, they invite a general distrust of science and a depreciation of the prestige of the scientist, whose discoveries appear arbitrary and fickle. This variety of anti-intellectualism which threatens his social position is characteristically enough resisted by the scientist. On the ideological front as well, totalitarianism entails a conflict with the traditional assumptions of modern science.

Functions of Norms of Pure Science

ONE sentiment which is assimilated by the scientist from the very outset of his training pertains to the purity of science. Science must not suffer itself to become the handmaiden of theology or economy or state. The function of this sentiment is likewise to preserve the autonomy of science. For if such extra-scientific criteria of the value of science as presumable consonance with religious doctrines or economic utility or political appropriateness are adopted, science becomes acceptable only in so far as it meets these criteria. In other words, as the pure science sentiment is eliminated, science becomes subject to the direct control of other institutional agencies and its place in society becomes increasingly uncertain. The persistent repudiation by scientists of the application of utilitarian norms to their work has as its chief function the avoidance of this danger, which is particularly marked at the present time. A tacit recognition of this function may be the source of that possibly apocryphal toast at a dinner for scientists in Cambridge: To pure mathematics, and may it never be of any use to anybody!

The exaltation of pure science is thus seen to be a defence against the invasion of norms which limit directions of potential advance and threaten the stability and continuance of scientific research as a valued social activity. Of course, the technological criterion of scientific achievement has also a positive social function for science. The increasing comforts and conveniences deriving from technology and ultimately from science invite the social support of scientific research. They also testify to the integrity of the scientist, since abstract and difficult theories which cannot be understood or evaluated by the laity are presumably

relativistic doctrines of *Wissenssoziologie* as a political expedient for discrediting "liberal" or "bourgeois" or "non-Aryan" science. An exit from this cul-de-sac is provided by positing an Archimedean point: the infallibility of *der Führer* and his *Volk* (*cf.* General Hermann Goering, *Germany Reborn*, London, Mathews & Marrot, 1934, 79). Politically effective variations of the "relationism" of Karl Mannheim (e.g. *Ideology and Utopia*) have been used for propagandistic purposes by such Nazi theorists as Walter Frank, Krieck, Rust, and Rosenberg.

21. For example, Pareto writes: "The quest for experimental uniformities is an end in itself." See a typical statement by George A. Lundberg. "It is not the business of a chemist who invents a high explosive to be influenced in his task by considerations as to whether his product will be used to blow up cathedrals or to build tunnels

proved in a fashion which can be understood by all, *i.e.*, through their technological applications. Readiness to accept the authority of science rests, to a considerable extent, upon its daily demonstration of power. Were it not for such indirect demonstrations, the continued social support of that science which is intellectually incomprehensible to the public would hardly be nourished on faith alone.

At the same time, this stress upon the purity of science has had other consequences which threaten rather than preserve the social esteem of science. It is repeatedly urged that scientists should in their research ignore all considerations other than the advance of knowledge.[21] Attention is to be focused exclusively on the scientific significance of their work with no concern for the practical uses to which it may be put or for its social repercussions generally. The customary justification of this tenet—which is partly rooted in fact[22] and which, in any event, has definite social functions, as we have just seen—holds that failure to adhere to this injunction will encumber research by increasing the possibility of bias and error. But this *methodological* view overlooks the *social* results of such an attitude. The objective consequences of this attitude have furnished a further basis of revolt against science; an incipient revolt which is found in virtually every society where science has reached a high stage of development. Since the scientist does not or cannot control the direction in which his discoveries are applied, he becomes the subject of reproach and of more violent reactions in so far as these applications are disapproved by the agents of authority or by pressure groups. The antipathy toward the technological products is projected toward science itself. Thus, when newly discovered gases or explosives are applied as military instruments, chemistry as a whole is censured by those whose humanitarian sentiments are outraged. Science is held largely responsible for endowing those engines of human destruction which, it is said, may plunge our civilization into everlasting night and confusion. Or to take another prominent instance, the rapid development of science and related technology has led to an implicitly anti-science movement by vested interests and by those whose sense of economic justice is offended. The eminent Sir Josiah Stamp

through the mountains. Nor is it the business of the social scientist in arriving at laws of group behavior to permit himself to be influenced by considerations of how his conclusions will coincide with existing notions, or what the effect of his findings on the social order will be." *Trends in American Sociology* (edited by G. A. Lundberg, R. Bain, and N. Anderson, New York, Harper, 1929), 404–5. Compare the remarks of Read Bain on the "Scientist as Citizen," *Social Forces*, 1933, 11, 412–15.

22. A neurological justification of this view is to be found in E. D. Adrian's essay in *Factors Determining Human Behavior* (Harvard Tercentenary Publications, Cambridge, 1937), 9. "For discriminative behavior . . . there must be some interest: yet if there is too much the behavior will cease to be discriminative. Under intense emotional stress the behavior tends to conform to one of several stereotyped patterns."

and a host of less illustrious folk have proposed a moratorium on invention and discover,[23] in order that man may have a breathing spell in which to adjust his social and economic structure to the constantly changing environment with which he is presented by the "embarrassing fecundity of technology." These proposals have received wide publicity in the press and have been urged with unslackened insistence before scientific bodies and governmental agencies.[24] The opposition comes equally from those representatives of labor who fear the loss of investment in skills which become obsolete before the flood of new technologies and from the ranks of those capitalists who object to the premature obsolescence of their machinery. Although these proposals probably will not be translated into action within the immediate future, they constitute one possible nucleus about which a revolt against science in general may materialize. It is largely immaterial whether these opinions which make science ultimately responsible for undesirable situations are valid or not. W. I. Thomas' sociological theorem—"If men define situations as real, they are real in their consequences"—has been repeatedly verified.

In short, this basis for the re-valuation of science derives from what I have called elsewhere the "imperious immediacy of interest."[25] Concern with the primary goal, the furtherance of knowledge, is coupled with a disregard of those consequences which lie outside the area of immediate interest, but these social results react so as to interfere with the original pursuits. Such behavior may be rational in the sense that it may be expected to lead to the satisfaction of the immediate interest. But it is irrational in the sense that it defeats other values which are not, at the moment, paramount but which are none the less an integral part of the social scale of values. Precisely because scientific research is not conducted in a social vacuum, its effects ramify into other spheres of value and interest. In so far as these effects are deemed socially undesirable, science is charged with responsibility. The goods of science are

23. Of course, this does not constitute a movement opposed to science as such. Moreover, the destruction of machinery by labor and the suppression of inventions by capital have also occurred in the past. *Cf.* R. K. Merton, "Fluctuations in the rate of industrial invention," *Quarterly Journal of Economics*, 1935, 49, 464 ff. But this movement mobilizes the opinion that science is to be held strictly accountable for its social effects. Sir Josiah Stamp's suggestion may be found in his address to the British Association for the Advancement of Science, Aberdeen, 6 Sept. 1934. Such moratoria have also been proposed by M. Caillaux (*cf.* John Strachey, *The Coming Struggle for Power*, New York: 1935, 183), by H. W. Sumners in the U.S. House of Representatives, and by many others. In terms of current humanitarian, social and economic criteria, some of the products of science are more pernicious than beneficial. This evaluation may destroy the rationale of scientific work. As one scientist pathetically put it: if the man of science must be apologetic for his work, I have wasted my life. *Cf. The Frustration of Science* (ed. by F. Soddy), (New York: Norton, 1935), 42 *et passim.*

24. English scientists have especially reacted against the "prostitution of scientific

no longer considered an unqualified blessing. Examined from this perspective, the tenet of pure science and disinterestedness has helped to prepare its own epitaph.

Battle lines are drawn in terms of the question: can a good tree bring forth evil fruit? Those who would cut down or stunt the tree of knowledge because of its accursed fruit are met with the claim that the evil fruit has been grafted on the good tree by the agents of state and economy. It may salve the conscience of the individual man of science to hold that an inadequate social structure has led to the perversion of his discoveries. But this will hardly satisfy an embittered opposition. Just as the *motives* of scientists may range from a passionate desire in the furtherance of knowledge to a profound interest in achieving personal fame and just as the *functions* of scientific research may vary from providing prestige-laden rationalizations of the existing order to enhancing our control of nature, so may other social *effects* of science be considered pernicious to society or result in the modification of the scientific ethos itself. There is a tendency for scientists to assume that the social effects of science *must* be beneficial in the long run. This article of faith performs the function of providing a rationale for scientific research, but it is manifestly not a statement of fact. It involves the confusion of truth and social utility which is characteristically found in the non-logical penumbra of science.

Esoteric Science as Popular Mysticism

ANOTHER relevant phase of the connections between science and the social order has seldom been recognized. With the increasing complexity of scientific research, a long program of rigorous training is necessary to test or even to understand the new scientific findings. The modern scientist has necessarily subscribed to a cult of unintelligibility. There results an increasing gap between the scientist and the laity. The

effort to war purposes." Presidential addresses at annual meetings of the British Association for the Advancement of Science, frequent editorials and letters in *Nature* attest to this movement for "a new awareness of social responsibility among the rising generation of scientific workers." Sir Frederick Gowland Hopkins, Sir John Orr, Professor Soddy, Sir Daniel Hall, Dr. Julian Huxley, J. B. S. Haldane, and Professor L. Hogben are among the leaders of the movement. See, for example, the letter signed by twenty-two scientists of Cambridge University urging a program for dissociating science from warfare (*Nature*, 137, 1936, 829). These attempts for concerted action by English scientists contrast sharply with the apathy of scientists in this country toward these questions. (This observation holds for the period prior to the development of atomic weapons.) The basis of this contrast might profitably be investigated. In any event, although this movement may possibly derive from the sentiments, it may serve the function of eliminating one source of hostility toward science in democratic regimes.

25. Merton, "The unanticipated consequences of purposive social action," *op. cit.*

layman must take on faith the publicized statements about relativity or quanta or other such esoteric subjects. This he has readily done in as much as he has been repeatedly assured that the technologic achievements from which he has presumably benefited ultimately derive from such research. Nonetheless, he retains a certain suspicion of these bizarre theories. Popularized and frequently garbled versions of the new science stress those theories which seem to run counter to common sense. To the public mind, science and esoteric terminology become indissolubly linked. The presumably scientific pronouncements of totalitarian spokesmen on race or economy or history are for the uninstructed laity of the same order as announcements concerning an expanding universe or wave mechanics. In both instances, the laity is in no position to understand these conceptions or to check their scientific validity and in both instances they may not be consistent with common sense. If anything, the myths of totalitarian theorists will seem more plausible and are certainly more comprehensible to the general public than accredited scientific theories, since they are closer to common-sense experience and cultural bias. Partly as a result of scientific advance, therefore, the population at large has become ripe for new mysticisms clothed in apparently scientific jargon. This promotes the success of propaganda generally. The borrowed authority of science becomes a powerful prestige symbol for unscientific doctrines.

Public Hostility toward Organized Scepticism

ANOTHER feature of the scientific attitude is organized scepticism, which becomes, often enough, iconoclasm.[26] Science may seem to challenge the "comfortable power assumptions" of other institutions,[27] simply by subjecting them to detached scrutiny. Organized scepticism involves a latent questioning of certain bases of established routine, authority, vested procedures and the realm of the "sacred" generally. It is true that, *logically*, to establish the empirical genesis of beliefs and values is not to deny their validity, but this is often the psychological effect on the naïve mind. Institutionalized symbols and values demand attitudes of loyalty, adherence and respect. Science which asks questions of fact concerning every phase of nature and society comes into psychological, not logical, conflict with other attitudes toward these same data which have been crystallized and frequently ritualized by other institutions. Most institutions demand unqualified faith; but the institution of science makes scepticism a virtue. Every institution involves, in this sense, a sacred area, which is resistant to profane examination in

26. Frank H. Knight, "Economic psychology and the value problem," *Quarterly Journal of Economics*, 1925, 39, 372–409. The unsophisticated scientist, forgetting that scepticism is primarily a methodological canon, permits his scepticism to spill over into the area of value generally. The social functions of symbols are ignored and

terms of scientific observation and logic. The institution of science itself involves emotional adherence to certain values. But whether it be the sacred sphere of political convictions or religious faith or economic rights, the scientific investigator does not conduct himself in the prescribed uncritical and ritualistic fashion. He does not preserve the cleavage between the sacred and the profane, between that which requires uncritical respect and that which can be objectively analyzed.[28]

It is this which in part lies at the root of revolts against the so-called intrusion of science into other spheres. In the past, this resistance has come for the most part from the church which restrains the scientific examination of sanctified doctrines. Textual criticism of the Bible is still suspect. This resistance on the part of organized religion has become less significant as the locus of social power has shifted to economic and political institutions which in their turn evidence an undisguised antagonism toward that generalized scepticism which is felt to challenge the bases of institutional stability. This opposition may exist quite apart from the introduction of certain scientific discoveries which appear to invalidate particular dogmas of church, economy and state. It is rather a diffuse, frequently vague, recognition that scepticism threatens the *status quo*. It must be emphasized again that there is no logical necessity for a conflict between scepticism, within the sphere of science, and the emotional adherences demanded by other institutions. But as a psychological derivative, this conflict invariably appears whenever science extends its research to new fields toward which there are institutionalized attitudes or whenever other institutions extend their area of control. In the totalitarian society, the centralization of institutional control is the major source of opposition to science; in other structures, the extension of scientific research is of greater importance. Dictatorship organizes, centralizes and hence intensifies sources of revolt against science which in a liberal structure remain unorganized, diffuse, and often latent.

In a liberal society, integration derives primarily from the body of cultural norms toward which human activity is oriented. In a dictatorial structure, integration is effected primarily by formal organization and centralization of social control. Readiness to accept this control is instilled by speeding up the process of infusing the body politic with new cultural values, by substituting high-pressure propaganda for the slower process of the diffuse inculcation of social standards. These differences in the mechanisms through which integration is typically effected permit a greater latitude for self-determination and autonomy to various

they are impugned as "untrue." Thus, social utility and truth are once again confused.

27. Charles E. Merriam, *Political Power* (New York, McGraw-Hill, 1934), 82–3.

28. For a general discussion of the sacred in these terms, see Durkheim, *The Elementary Forms of the Religious Life*, 37 ff., *et passim*.

institutions, including science, in the liberal than in the totalitarian structure. Through such rigorous organization, the dictatorial state so intensifies its control over non-political institutions as to lead to a situation which is different in kind as well as degree. For example, reprisals against science can more easily find expression in the Nazi state than in America, where interests are not so organized as to enforce limitations upon science, when these are deemed necessary. Incompatible sentiments must be insulated from one another or integrated with each other if there is to be social stability. But such insulation becomes virtually impossible when there exists centralized control under the aegis of any one sector of social life which imposes, and attempts to enforce, the obligation of adherence to its values and sentiments as a condition of continued existence. In liberal structures the absence of such centralization permits the necessary degree of insulation by guaranteeing to each sphere restricted rights of autonomy and thus enables the gradual integration of temporarily inconsistent elements.

Conclusions

THE main conclusions of this paper may be briefly summarized. There exists a latent and active hostility toward science in many societies, although the extent of this antagonism cannot yet be established. The prestige which science has acquired within the last three centuries is so great that actions curtailing its scope or repudiating it in part are usually coupled with affirmation of the undisturbed integrity of science or "the rebirth of true science." These verbal respects to the pro-science sentiment are frequently at variance with the behavior of those who pay them. In part, the anti-science movement derives from the conflict between the ethos of science and of other social institutions. A corollary of this proposition is that contemporary revolts against science are *formally* similar to previous revolts, although the *concrete* sources are different. Conflict arises when the social effects of applying scientific knowledge are deemed undesirable, when the scientist's scepticism is directed toward the basic values of other institutions, when the expansion of political or religious or economic authority limits the autonomy of the scientist, when anti-intellectualism questions the value and integrity of science and when non-scientific criteria of eligibility for scientific research are introduced.

This paper does not present a program for action in order to withstand threats to the development and autonomy of science. It may be suggested, however, that as long as the locus of social power resides in any one institution other than science and as long as scientists themselves are uncertain of their primary loyalty, their position becomes tenuous and uncertain.

Science and Society: Reciprocal Relations

2

THE eleven selections in Part 2 of the Reader have been chosen to illustrate a variety of ways in which science and society influence each other. A variety of societies and historical periods have also been purposely included to show both that there are constant patterns of interaction between science and society in different places and times and that the whole historical development of science has been socially conditioned. Finally, these selections exemplify a considerable number of different research methods, each of them useful for its special task of analysis.

The first two selections, both by Merton, are outstanding classics in the sociology of science, as important and impressive today as they were when first published some twenty-five years ago. The first is a demonstration of the way in which a certain set of religious ideas and values, what Merton calls the "Puritan ethos," was a support for the development of science in seventeenth-century England and also in later historical periods and other places right down to the present time. Until quite recently, and perhaps still in some quarters, it was an accepted idea that science and religion were always and necessarily at warfare with each other. This belief was a product of certain nineteenth-century misconceptions about the relations between science and religion. Merton's research, as well as other historical and sociological work, has recently shown that science and religion in Western society have strongly supported each other in many ways, though also incompatible in some other ways. Of particular value in Merton's first essay is the "Postscript," written in 1957 and bringing down to that date the scholarly discussion of his theory that had been published in the previous twenty years. It should be noted especially that, in dealing with this discussion, Merton tries to show how the problem should be redefined in order to deal with it more effectively. That is, his comments bring out clearly that progress in solving a scientific problem is often made by redefining it, not just by answering the old statement of the problem in either-or, yes-no terms.

The second paper by Merton makes an important point that is often ignored in discussions of whether or not religion had an influence on the rise of science and capitalism. This point is that other social factors, in this case socioeconomic factors, contributed to the rise of science along with the Puritan ethos. In trying to demonstrate the importance of religious ideas and values, one does not deny the additional influence of other factors, economic or otherwise. Again, by redefining the problem, by asking what several factors contributed to the rise of science in seventeenth-century England, Merton has advanced our understanding. Still further, by analyzing the various "pure" and "applied" influences upon the selection of scientific problems by members of the Royal Society in the seventeenth century, Merton has cast light on another general problem in the sociology of science, the relative importance of "pure" and "applied" sources of scientific research. His data show that both are important, and so it has remained in science down to our own day. Either-or discussions of this problem too are not so much properly scientific as ideological, seeking to damn one or the other of two continually important sources of scientific research and discovery.

The paper by Gillispie on "Science in the French Revolution" is an abstract, in effect, of his book on that subject that will shortly be published. Gillispie's paper properly stresses both the diverse attitudes toward science during the Revolution and the diverse consequences, favorable and unfavorable. Especially important, on the favorable side, was the establishment of a whole new set of scientific schools and research institutions.

Coming down to the nineteenth century, in America, Shryock's discussion of the social sources of American indifference to basic science is an excellent analysis of some of the factors that produced a situation lasting, indeed, nearly up to our own times. The kind of understanding that Shryock contributes has been helpful in knowing what improves support for basic science, a factor now recognized by everyone to be of fundamental importance. It should perhaps be pointed out that since Shryock wrote this paper some very important work on nineteenth-century American science, by Hunter Dupree of the University of California and Edward Lurie of the Wayne State University (see Selected Bibliography) has produced evidence that would modify the description and analysis presented by Shryock.

The two selections on science in Russian society notably improve our knowledge about a subject on which most writing has been either merely ideological partisanship or superficial journalism. Joravsky's paper especially, which contains some material from his recent book, is an outstanding contribution to scholarship in this field. Both papers indicate, again, that science and society are involved in a set of complex relationships, some mutually favorable, others not. Certainly they show, in a better way even than do sputniks and missiles, why science is not "dead" in Russia, a misconception that was held in some scientific circles in the West as recently as ten years ago.

Because of the complex and reciprocal relations between science and society and because of our meager understanding of these relations, the accuracy with which predictions can be made about the occurrence of scientific discovcrics or about the effects of these discoveries on society is not very great. As the sociology of science makes progress, the accuracy and specificity of our predictions will increase. Lilley's paper is an incisive discussion of the problem of prediction of scientific discoveries and of their social consequences. He evaluates the success of past attempts at prediction and, by examining the sources of success and failure, shows what we need to do to make our predictions better.

How people in a society understand what science is, how much they know of its substance, the kinds and degree of interest they have in science, and whether or not they approve of science—all these are social elements that affect science very much. Evidence on how matters stand in these respects in recent American society is summed up in Withey's paper, which is based on a series of public opinion poll surveys, several of them made by the Survey Research Center of the University of Michigan, with which Withey is connected. This paper also indicates, apart from its substance, how valuable a research tool the public opinion survey has become. Other selections in this reader give further evidence of the value of the public opinion survey for the sociology of science.

The next two selections, by Goodrich, Knapp, and their associates, represent the central findings of two studies on the social origins of American scientists reported in detail in two books and sponsored by Wesleyan University, of which Goodrich and Knapp are members of the faculty. The social factors that favor the "production" of scientists are obviously important in modern society, and these two studies throw a good deal of light on the relevance of such factors as religion, size, and "social atmosphere" of undergraduate colleges; parental occupations and social class position; economic resources; scholarship programs; and "good teaching." The two studies also show how the impact of these factors changes through even relatively short historical periods.

The last selection in this part of the reader is a happy example of how the sociology of science is, in some of its parts, becoming cumulative. Holland's study of the "Undergraduate Origins of American Scientists" builds directly on the work by Knapp and Goodrich. Holland's data, based on studies of the high-school students who do or do not go to college and who go to one college and not to another, suggest that the colleges that Knapp and Goodrich found to be "more productive" of scientists are so, in part, because they are selected by the better high-school students with the higher motivations for academic and scientific achievement. It is of interest to note that Holland, who is director of research for the National Merit Scholarship Corporation, has been led to his study by the immediately practical interests of the scholarship program with which he is associated. The National Merit Scholarship Corporation wants to distribute its scholarships to the students and colleges that are most likely to be involved in the production of scientists and scholars. His study, along with those by Knapp and Goodrich, provides some of the information needed to carry out the program in the most effective way.

One final remark should be made about the subject of this part of the Reader. Although it may seem that the eleven selections in this part cover a very large area of the reciprocal relations between science and society, it must be stressed that this area is in fact so large and complex that these eleven selections offer only an introduction. Other important materials, including those in the Selected Bibliography at the end of this Reader and those referred to in the selections we have printed, can profitably be consulted.

3 *Puritanism, Pietism, and Science*

ROBERT K. MERTON

In his prolegomena to a cultural sociology Alfred Weber has discriminated between the processes of society, culture, and civilization.[1] Since his primary interest lay in differentiating these categories of sociological phenomena, Weber in large measure ignored their specific interrelationships, a field of study which is fundamental for the sociologist. It is precisely this interaction between certain elements of culture and civilization, with especial reference to seventeenth-century England, which constitutes the object-matter of the present essay.

The Puritan Ethos

The first section of this paper outlines the Puritan value-complex in so far as it was related to the notable increase of interest in science during the latter part of the seventeenth century, while the second presents the relevant empirical materials concerning the differential cultivation of natural science by Protestants and other religious affiliates.

It is the thesis of this study that the Puritan ethic, as an ideal-typical expression of the value-attitudes basic to ascetic Protestantism generally, so canalized the interests of seventeenth-century Englishmen as to constitute one important *element* in the enhanced cultivation of science.

Reprinted from Social Theory and Social Structure, *by Robert K. Merton (New York: The Free Press of Glencoe, 1957, revised edition), Chapter* XVIII, *pp.* 574–606.

1. Alfred Weber, "Prinzipielles zur Kultursoziologie: Gesellschaftsprozess, Zivilisationsprozess und Kulturbewegung," *Archiv für Sozialwissenschaft und Sozialpolitik*, xlvii, 1920, 47, 1–49. See the similar classification by R. M. MacIver, *Society: Its Structure and Changes*, chap. xii; and the discussion of these studies by Morris Ginsberg, *Sociology* (London, 1934), 45–52.

The deep-rooted religious *interests*[2] of the day demanded in their forceful implications the systematic, rational, and empirical study of Nature for the glorification of God in His works and for the control of the corrupt world.

It is possible to determine the extent to which the values of the Puritan ethic stimulated interest in science by surveying the attitudes of the contemporary scientists. Of course, there is a marked possibility that in studying the avowed motives of scientists we are dealing with rationalizations, with derivations, rather than with accurate statements of the actual motives. In such instances, although they may refer to isolated specific cases, the value of our study is by no means vitiated, for these conceivable rationalizations themselves are evidence (Weber's *Erkenntnismitteln*) of the motives which were regarded as socially acceptable, since, as Kenneth Burke puts it, "a terminology of motives is moulded to fit our general orientation as to purposes, instrumentalities, the good life, etc."

Robert Boyle was one of the scientists who attempted explicitly to link the place of science in social life with other cultural values, particularly in his *Usefulness of Experimental Natural Philosophy*. Such attempts were likewise made by John Ray, whose work in natural history was path-breaking and who was characterized by Haller as the greatest botanist in the history of man; Francis Willughby, who was perhaps as eminent in zoology as was Ray in botany; John Wilkins, one of the leading spirits in the "invisible College" which developed into the Royal Society; Oughtred, Wallis, and others. For additional evidence we can turn to the scientific body which, arising about the middle of the century, provoked and stimulated scientific advance more than any other immediate agency: the Royal Society. In this instance we are particularly fortunate in possessing a contemporary account written under the constant supervision of the members of the Society so that it might be representative of their views of the motives and aims of that association. This is Thomas Sprat's widely read *History of the Royal-Society of London*, published in 1667, after it had been examined by Wilkins and other representatives of the Society.[3]

Even a cursory examination of these writings suffices to disclose one

2. "Nicht die ethische Theorie theologischer Kompendien, die nur als ein (unter Umständen allerdings wichtiges) Erkenntnismittel dient, sondern die in den psychologischen und pragmatischen Zusammenhängen der Religionen gegründeten praktischen Antriebe zum Handeln sind das, was in Betracht kommt [unter 'Wirtschaftsethik' einer Religion]." Max Weber, *Gesammelte Aufsätze zur Religionssoziologie* (Tübingen, 1920), 1, 238. As Weber justly indicates, one freely recognizes the fact that religion is but *one* element in the determination of the religious ethic, but none the less it is at present an insuperable, and for our purposes, unnecesary task to determine *all* the component elements of this ethic. That problem awaits further analysis and falls outside the scope of this study.

3. *Cf.* C. L. Sonnichsen, *The Life and Works of Thomas Sprat* (Harvard Uni-

outstanding fact: certain elements of the Protestant ethic had pervaded the realm of scientific endeavour and had left their indelible stamp upon the attitudes of scientists toward their work. Discussions of the why and wherefore of science bore a point-to-point correlation with the Puritan teachings on the same subject. Such a dominant force as was religion in those days was not and perhaps could not be compartmentalized and delimited. Thus, in Boyle's highly commended apologia for science it is maintained that the study of Nature is to the greater glory of God and the Good of Man.[4] This is the motif which recurs in constant measure. The juxtaposition of the spiritual and the material is characteristic. This culture rested securely on a substratum of utilitarian norms which constituted the measuring-rod of the desirability of various activities. The definition of action designed for the greater glory of God was tenuous and vague, but utilitarian standards could easily be applied.

Earlier in the century, this keynote had been sounded in the resonant eloquence of that "veritable apostle of the learned societies," Francis Bacon. Himself the initiator of no scientific discoveries, unable to appreciate the importance of his great contemporaries, Gilbert, Kepler, and Galileo, naïvely believing in the possibility of a scientific method which "places all wits and understandings nearly on a level," a radical empiricist holding mathematics to be of no use in science, he was, nevertheless, highly successful as one of the principal protagonists of a positive social evaluation of science and of the disclaim of a sterile scholasticism. As one would expect from the son of a "learned, eloquent, and religious woman, full of puritanic fervour" who was admittedly influenced by his mother's attitudes, he speaks in the *Advancement of Learning* of the true end of scientific activity as the "glory of the Creator and the relief of man's estate." Since, as is quite clear from many official and private documents, the Baconian teachings constituted the basic principles on which the Royal Society was patterned, it is not strange that the same sentiment is expressed in the charter of the Society.

versity, unpublished doctoral dissertation, 1931), 131 ff., where substantial evidence of the fact that the *Historian's* representative of the views of the Society is presented. It is of further interest that the statements in Sprat's book concerning the aims of the Society bear a distinct similarity on every score to Boyle's characterizations of the motives and aims of scientists in general. This similarity is evidence of the dominance of the ethos which included these attitudes.

4. Robert Boyle, *Some Considerations Touching the Usefulness of Experimental Natural Philosophy* (Oxford, 1664), 22 ff. See, also, the letters of William Oughtred in *Correspondence of Scientific Men of the Seventeenth Century*, edited by S. J. Rigaud (Oxford, 1841), xxxiv, *et passim;* or the letters of John Ray in the *Correspondence of John Ray*, edited by Edwin Lankester (London, 1848), 389, 395, 402, *et passim.*

In his last will and testament, Boyle echoes the same attitude, petitioning the Fellows of the Society in this wise: "Wishing them also a happy success in their laudable attempts, to discover the true Nature of the Works of God; and praying that they and all other Searchers into Physical Truths, may cordially refer their Attainments to the Glory of the Great Author of Nature, and to the Comfort of Mankind."[5] John Wilkins proclaimed the experimental study of Nature to be a most effective means of begetting in men a veneration for God.[6] Francis Willughby was prevailed upon to publish his works—which he had deemed unworthy of publication—only when Ray insisted that it was a means of glorifying God.[7] Ray's *Wisdom of God*, which was so well received that five large editions were issued in some twenty years, is a panegyric of those who glorify Him by studying His works.[8]

To a modern, comparatively untouched by religious forces, and noting the almost complete separation, if not opposition, between science and religion today, the recurrence of these pious phrases is apt to signify merely customary usage, and nothing of deep-rooted motivating convictions. To him these excerpts would seem to be a case of *qui nimium probat nihil probat*. But such an interpretation is possible only if one neglects to translate oneself within the framework of seventeenth-century values. Surely such a man as Boyle, who spent considerable sums to have the Bible translated into foreign tongues, was not simply rendering lip service. As G. N. Clark very properly notes in this connection:

> There is . . . always a difficulty in estimating the degree to which what we call religion enters into anything which was said in the seventeenth century in religious language. It is not solved by discounting all theological terms and treating them merely as common form. On the contrary, it is more often necessary to remind ourselves that these words were then seldom used without their accompaniment of meaning, and that their use did generally imply a heightened intensity of feeling.[9]

The second dominant tenet in the Puritan ethos designated social welfare, the good of the many, as a goal ever to be held in mind. Here again the contemporary scientists adopted an objective prescribed by

5. Quoted by Gilbert, Lord Bishop of Sarum, *A Sermon Preached at the Funeral of the Hon. Robert Boyle* (London, 1692), 25.
6. *Principles and Duties of Natural Religion* (London, 1710—sixth edition), 236 *et passim*.
7. *Memorials of John Ray*, 14 f.
8. *Wisdom of God* (London, 1691), 126–129, *et passim*.
9. G. N. Clark, *The Seventeenth Century* (Oxford, 1929), 323.
10. Thomas Sprat, *History of the Royal Society*, 78–79.
11. *Ibid.*, 341–2.
12. Ray, *Wisdom of God*, 125.
13. Sprat, *op. cit.*, 344–5.
14. Richard Baxter, *Christian Directory* (London, 1825—first published in 1664),

the current values. Science was to be fostered and nurtured as leading to the domination of Nature by technological invention. The Royal Society, we are told by its worthy historian, "does not intend to stop at some particular benefit, but goes to the root of all noble inventions."[10] But those experiments which do not bring with them immediate gain are not to be condemned, for as the noble Bacon has declared, experiments of Light ultimately conduce to a whole troop of inventions useful to the life and state of man. This power of science to better the material condition of man, he continues, is, apart from its purely mundane value, a good in the light of the Evangelical Doctrine of Salvation by Jesus Christ.

And so on through the principles of Puritanism there was the same point-to-point correlation between them and the attributes, goals, and results of science. Such was the contention of the protagonists of science at that time. Puritanism simply made articulate the basic values of the period. If Puritanism demands systematic, methodic labour, constant diligence in one's calling, what, asks Sprat, more active and industrious and systematic than the Art of Experiment, which "can never be finish'd by the perpetual labours of any one man, nay, scarce by the successive force of the greatest Assembly?"[11] Here is employment enough for the most indefatigable industry, since even those hidden treasures of Nature which are farthest from view may be uncovered by pains and paticncc.[12]

Does the Puritan eschew idleness because it conduces to sinful thoughts (or interferes with the pursuit of one's vocation)? "What room can there be for low, and little things in a mind so usefully and successfully employ'd [as in natural philosophy]?"[13] Are plays and playbooks pernicious and flesh-pleasing (and subversive of more serious pursuits)?[14] Then it is the "fittest season for experiments to arise, to teach us a Wisdome, which springs from the depths of Knowledge, to shake off the shadows, and to scatter the mists [of the spiritual distractions brought on by the Theatre]."[15] And finally, is a life of earnest activity within the world to be preferred to monastic asceticism? Then recognize the fact that the study of natural philosophy "fits us not so well for the secrecy of a Closet: It makes us serviceable to the World."[16] In short,

I, 152; II, 167. *Cf.* Robert Barclay, the Quaker apologist, who specifically suggests "geometrical and mathematical experiments" as innocent divertissements to be sought instead of pernicious plays. *An Apology for the True Christian Divinity* (Phila., 1805 —first written in 1675), 554–5.

15. Sprat, *op. cit.*, 362.

16. *Ibid.*, 365–6. Sprat perspicaciously suggests that monastic asceticism induced by religious scruples was partially responsible for the lack of empiricism of the Schoolmen. "But what sorry kinds of Philosophy must the Schoolmen needs produce, when it was part of their Religion, to separate themselves, as much as they could, from the converse of mankind? When they were so far from being able to discover the secrets of Nature, that they scarce had opportunity to behold enough of its common works," *Ibid.*, 19.

science embodies two highly prized values: utilitarianism and empiricism.

In a sense this explicit coincidence between Puritan tenets and the qualities of science as a calling is casuistry. It is an express attempt to fit the scientist *qua* pious layman into the framework of the prevailing social values. It is a bid for religious and social sanction, since both the constitutional position and the personal authority of the clergy were much more important then than now. But this is not the entire explanation. The justificatory efforts of Sprat, Wilkins, Boyle, or Ray do not simply present opportunistic obsequiousness, but rather an earnest attempt to justify the ways of science to God. The Reformation had transferred the burden of individual salvation from the Church to the individual, and it is this "overwhelming and crushing sense of the responsibility for his own soul" which explains the acute religious interest. If science were not demonstrably a lawful and desirable calling, it dare not claim the attention of those who felt themselves "ever in the Great Taskmaster's eye." It is to this intensity of feeling that such apologias were due.

The exaltation of the faculty of reason in the Puritan ethos—based partly on the conception of rationality as a curbing device of the passions—inevitably led to a sympathetic attitude toward those activities which demand the constant application of rigorous reasoning. But again, in contrast to medieval rationalism, reason is deemed subservient and auxiliary to empiricism. Sprat is quick to indicate the pre-eminent adequacy of science in this respect.[17] It is on this point probably that Puritanism and the scientific temper are in most salient agreement, for the combination of *rationalism and empiricism* which is so pronounced in the Puritan ethic forms the essence of the spirit of modern science. Puritanism was suffused with the rationalism of neo-Platonism, derived largely through an appropriate modification of Augustine's teachings. But it did not stop there. Associated with the designated necessity of dealing successfully with the practical affairs of life within this world—a derivation from the peculiar twist afforded largely by the Calvinist doctrine of predestination and *certitudo salutis* through successful worldly activity—was an emphasis upon empiricism. These two currents brought to convergence through the logic of an inherently consistent

17. Sprat, *op. cit.*, 361. Baxter in a fashion representative of the Puritans decried the invasion of "enthusiasm" into religion. Reason must "maintain its authority in the command and government of your thoughts." *CD.*, ii, 199. In like spirit, those who at Wilkins' lodgings laid the foundation of the Royal Society "were invincibly arm'd against all the inchantments of Enthusiasm." Sprat, *op. cit.*, 53.

18. On the basis of this analysis, it is surprising to note the statement *accredited* to Max Weber that the opposition of the Reformers is sufficient reason for not coupling Protestantism with scientific interests. See *Wirtschaftsgeschichte* (München, 1924), 314. This remark is especially unanticipated since it does not at all accord with Weber's discussion of the same point in his other works. *Cf. Religionssoziologie*,

system of values were so associated with the other values of the time as to prepare the way for the acceptance of a similar coalescence in natural science.

Empiricism and rationalism were canonized, beatified, so to speak. It may very well be that the Puritan ethos did not directly influence the method of science and that this was simply a parallel development in the internal history of science, but it is evident that through the psychological compulsion toward certain modes of thought and conduct this value-complex made an empirically-founded science commendable rather than, as in the medieval period, reprehensible or at best acceptable on sufferance. This could not but have directed some talents into scientific fields which otherwise would have engaged in more highly esteemed professions. The fact that science to-day is largely if not completely divorced from religious sanctions is itself of interest as an example of the process of secularization.

The beginnings of such secularization, faintly perceptible in the latter Middle Ages, are manifest in the Puritan ethos. It was in this system of values that reason and experience were first markedly considered as independent means of ascertaining even religious truths. Faith which is unquestioning and not "rationally weighed," says Baxter, is not faith, but a dream or fancy or opinion. In effect, this grants to science a power which may ultimately limit that of theology.

Thus, once these processes are clearly understood, it is not surprising or inconsistent that Luther particularly, and Melanchthon less strongly, execrated the cosmology of Copernicus and that Calvin frowned upon the acceptance of many scientific discoveries of his day, while the religious ethic which stemmed from these leaders invited the pursuit of natural science.[18] In so far as the attitudes of the theologians dominate over the, in effect, subversive religious ethic—as did Calvin's authority in Geneva until the early eighteenth century—science may be greatly impeded. But with the relaxation of this hostile influence and with the development of an ethic, stemming from it and yet differing significantly, science takes on a new life, as was indeed the case in Geneva.

Perhaps the most directly effective element of the Protestant ethic for the sanction of natural science was that which held that the study of nature enables a fuller appreciation of His works and thus leads us

I, 141, 564; *Wissenschaft als Beruf* (München, 1921), 19–20. The probable explanation is that the first is not Weber's statement, since the *Wirtschaftsgeschichte* was compiled from classroom notes by two of his students who may have neglected to make the requisite distinctions. It is unlikely that Weber would have made the elementary error of confusing the Reformers' opposition to certain scientific discoveries with the unforeseen consequences of the Protestant ethic, particularly since he expressly warns against the failure to make such discriminations in his *Religionssoziologie*. For perceptive but vague adumbrations of Weber's hypothesis, see Auguste Comte, *Cours de philosophie positive* (Paris, 1864), IV, 127–130.

to admire the Power, Wisdom, and Goodness of God manifested in His creation. Though this conception was not unknown to medieval thought, the consequences deduced from it were entirely different. Thus Arnaldus of Villanova, in studying the products of the Divine Workshop, adheres strictly to the medieval ideal of determining properties of phenomena from *tables* (in which all combinations are set forth according to the canons of logic). But in the seventeenth century, the contemporary emphasis upon empiricism led to investigating nature primarily through observation.[19] This difference in interpretation of substantially the same doctrine can only be understood in the light of the different values permeating the two cultures.

For a Barrow, Boyle, or Wilkins, a Ray or Grew, science found its rationale in the end of all of existence: glorification of God. Thus, from Boyle:[20]

> . . . God loving, as He deserves, to be honour'd in all our Faculties, and consequently to be glorified and acknowledg'd by the acts of Reason, as well as by those of Faith, there must be sure a great Disparity betwixt that general, confus'd and lazy Idea we commonly have of His Power and Wisdom, and the Distinct, rational and affecting notions of those Attributes which are form'd by an attentive Inspection of those Creatures in which they are most legible, and which were made chiefly for that very end.

Ray carries this conception to its logical conclusion, for if Nature is the manifestation of His power, then nothing in Nature is too mean for scientific study.[21] The universe and the insect, the macrocosm and microcosm alike, are indications of "divine Reason, running like a Golden Vein, through the whole leaden Mine of Brutal Nature."

Up to this point we have been concerned in the main with the directly felt sanction of science through Puritan values. While this was of great influence, there was another type of relationship which, subtle and difficult of apprehension though it be, was perhaps of paramount significance. It has to do with the preparation of a set of largely implicit assumptions which made for the ready acceptance of the scientific temper characteristic of the seventeenth and subsequent centuries. It is not simply that Protestantism implicitly involved free inquiry, *libre*

19. Walter Pagel, "Religious motives in the medical biology of the seventeenth century," *Bulletin of the Institute of the History of Medicine*, 1935, 3, 214–15.

20. *Usefulness of Experimental Natural Philosophy*, 53; *cf.* Ray, *Wisdom of God*, 132; Wilkins, *Natural Religion*, 236 ff.; Isaac Barrow, *Opuscula*, iv, 88 ff.; Nehemiah Grew, *Cosmologia sacra* (London, 1701), who points out that "God is the original End," and that "we are *bound* to study His works."

21. Ray, *Wisdom of God*, 130 ff. Max Weber quotes Swammerdam as saying: "ich bringe Ihnen hier den Nachweis der Vorsehung Gottes in der Anatomie einer Laus." *Wissenschaft als Beruf*, 19.

examen, or decried monastic asceticism. These are important but not exhaustive.

It has become manifest that in each age there is a system of science which rests upon a set of assumptions, usually implicit and seldom questioned by the scientists of the time.[22] The *basic* assumption in modern science "is a widespread, instinctive conviction in the existence of an *Order of Things,* and, in particular, of an Order of Nature."[23] This belief, this faith, for at least since Hume it must be recognized as such, is simply "impervious to the demand for a consistent rationality." In the systems of scientific thought of Galileo, Newton, and of their successors, the testimony of experiment is the ultimate criterion of truth, but the very notion of experiment is ruled out without the prior assumption that Nature constitutes an an intelligible order, so that when appropriate questions are asked, she will answer, so to speak. Hence this assumption is final and absolute.[24] As Professor Whitehead indicated, this "faith in the possibility of science, generated antecedently to the development of modern scientific theory, is an unconscious derivative from medieval theology." But this conviction, prerequisite of modern science though it be, was not sufficient to induce its development. What was needed was a constant interest in searching for this order in nature in an empirico-rational fashion, that is, an *active* interest in this world and its occurrences plus a specific frame of mind. With Protestantism, religion provided this interest: it actually imposed obligations of intense concentration upon secular activity with an emphasis upon experience and reason as bases for action and belief.

Even the Bible as final and complete authority was subject to the interpretation of the individual upon these bases. The similarity in approach and intellectual attitude of this system to that of the contemporary science is of more than passing interest. It could not but mould an attitude of looking at the world of sensuous phenomena which was highly conducive to the willing acceptance, and indeed, preparation for, the same attitude in science. That the similarity is deep-rooted and not superficial may be gathered from the following comment upon Calvin's theology:[25]

> Die Gedanken werden objektiviert und zu einem objektiven Lehrsystem aufgebaut und abgerundet. Es bekommt geradezu ein na-

22. A. E. Heath, in *Isaac Newton: A Memorial Volume,* ed. by W. J. Greenstreet (London, 1927), 133 ff.; E. A. Burtt, *The Metaphysical Foundations of Modern Physical Science* (London, 1925).

23. A. N. Whitehead, *Science and the Modern World* (New York, 1931), 5 ff.

24. *Cf.* E. A. Burtt in *Isaac Newton: A Memorial Volume,* 139. For the classic exposition of this scientific faith, see Newton's "Rules of Reasoning in Philosophy," in his *Principia* (London, 1729 ed.), II, 160 ff.

25. Hermann Weber, *Die Theologie Calvins* (Berlin, 1930), 23.

turwissenschaftliches Gepräge; es ist klar, leicht fassbar und formulierbar, wie alles, was der äusseren Welt angehört, klarer zu gestalten ist als das, was im Tiefsten sich abspielt.

The conviction in immutable law is as pronounced in the theory of predestination as in scientific investigation: "the immutable law is there and must be acknowledged."[26] The similarity between this conception and the scientific assumption is clearly drawn by Hermann Weber:[27]

> . . . die Lehre von der Prädestination in ihrem tiefsten Kerne getroffen zu sein, wenn mann sie als Faktum im Sinne eines naturwissenschaftlichen Faktums begreift, nur dass das oberste Prinzip, das auch jedem naturwissenschaftlichen Erscheinungskomplex zugrunde liegt, die im tiefsten erlebte gloria dei ist.

The cultural environment was permeated with this attitude toward natural phenomena which was derived from both science and religion and which enhanced the continued prevalence of conceptions characteristic of the new science.

There remains a supremely important part of this study to be completed. It is not sufficient verification of our hypothesis that the cultural attitudes induced by the Protestant ethic were favourable to science. Nor, yet again, that the consciously expressed motivation of many eminent scientists was provided by this ethic. Nor, still further, that the cast of thought which is characteristic of modern science, namely, the combination of empiricism and rationalism and the faith in the validity of one basic postulate, an apprehensible order in Nature, bears any other than fortuitious congruency with the values involved in Protestantism. All this can but provide some evidence of a certain probability of the connection we are arguing. The most significant test of the hypothesis is to be found in the confrontation of the results *deduced* from the hypothesis with relevant empirical data. If the Protestant ethic involved an attitudinal set favourable to science and technology in so many ways, then we should find amongst Protestants a greater propensity for these fields of endeavour than one would expect simply on the basis of their representation in the total population. Moreover, if, as has been frequently suggested,[28] the impression made by this ethic has lasted long

26. *Ibid.*, 31. The significance of the doctrine of God's foreknowledge for the reenforcement of the belief in natural law is remarked by H. T. Buckle, *History of Civilization in England* (New York, 1925), 482.

27. *Op. cit.*, 31.

28. As Troeltsch puts it: "The present-day world does not live by logical consistency, any more than any other; spiritual forces can exercise a dominant influence even where they are avowedly repudiated." *Die Bedeutung des Protestantismus für die Entstehung der modernen Welt* (München, 1911), 22: *cf.* Georgia Harkness, *John Calvin: The Man and His Ethics* (New York, 1931), 7 ff.

29. Memorials of John Ray, 18–19; P. A. W. Henderson, *The Life and Times of*

after much of its theological basis has been largely disavowed, then even in periods subsequent to the seventeenth century, this connection of Protestantism and science should persist to some degree. The following section, then, will be devoted to this further test of the hypothesis.

The Puritan Impetus to Science

IN THE beginnings of the Royal Society there is found a closely wrought nexus between science and society. The Society itself arose from an antecedent interest in science and the subsequent activities of its members provided an appreciable impetus to further scientific advance. The inception of this group is found in the occasional meetings of devotees of science in 1645 and following. Among the leading spirits were John Wilkins, John Wallis, and soon afterwards Robert Boyle and Sir William Petty, upon all of whom religious forces seem to have had a singularly strong influence.

Wilkins, later an Anglican bishop, was raised at the home of his maternal grandfather, John Dod, an outstanding Non-conformist theologian, and "his early education had given him a strong bias toward Puritanical principles."[29] Wilkins' influence as Warden of Wadham College was profound; under it came Ward, Rooke, Wren, Sprat, and Walter Pope (his half-brother), all of whom were original members of the Royal Society.[30] John Wallis, to whose *Arithmetica Infinitorum* Newton was avowedly indebted for many of his leading mathematical conceptions, was a clergyman with strong leanings toward Puritan principles. The piety of Boyle has already been remarked; the only reason he did not take holy orders, as he said, was because of the "absence of an inner call."[31]

Theodore Haak, the German virtuoso who played so prominent a part in the formation of the Royal Society, was a pronounced Calvinist. Denis Papin, who during his prolonged stay in England contributed notably to science and technology, was a French Calvinist compelled to leave his country to avoid religious persecution. Thomas Sydenham, sometimes called "the English Hippocrates," was an ardent Puritan who fought as one of Cromwell's men. Sir William Petty was a latitudi-

John Wilkins (London, 1910), 36. Moreover, after Wilkins took holy orders, he became chaplain to Lord Viscount Say and Seale, a resolute and effective Puritan.

30. Henderson, *op. cit.*, 72–3.

31. *Dictionary of National Biography*, II, 1028. This reason, effective also for Sir Samuel Morland's turning to mathematics rather than to the ministry, is an example of the direct working of the Protestant ethic which, as exposited by Baxter for example, held that only those who felt an "inner call" should enter the clergy, and that others could better serve society by adopting other accredited secular activities. On Morland, see the "Autobiography of Sir Samuel Morland," in J. O. Halliwell-Phillipps' *Letters Illustrative of the Progress of Science in England* (London, 1841), 116 ff.

narian; he had been a follower of Cromwell, and in his writings he evinced clearly the influences of Puritanism. Of Sir Robert Moray, described by Huyghens as the "Soul of the Royal Society," it could be said that "religion was the mainspring of his life, and amidst courts and camps he spent many hours a day in devotion."[32]

It is hardly a fortuitous circumstance that the leading figures of this nuclear group of the Royal Society were divines or eminently religious men, though it is not quite accurate to maintain, as did Dr. Richardson, that the beginnings of the Society occurred in a small group of learned men among whom Puritan *divines* predominated.[33] But it is quite clearly true that the originative spirits of the Society were markedly influenced by Puritan conceptions.

Dean Dorothy Stimson, in a recently published paper, has independently arrived at this same conclusion.[34] She points out that of the ten men who constituted the "invisible college," in 1645, only one, Scarbrough, was clearly non-Puritan. About two of the others there is some uncertainty, though Merret had a Puritan training. The others were all definitely Puritan. Moreover, among the original list of members of the Society of 1663, forty-two of the sixty-eight concerning whom information about their religious orientation is available were clearly Puritan. Considering that the Puritans constituted a relatively small minority in the English population, the fact that they constituted sixty-two per cent of the initial membership of the Society becomes even more striking. Dean Stimson concludes: "that experimental science spread as rapidly as it did in seventeenth-century England seems to me to be in part at least because the moderate Puritans encouraged it."

The Puritan Influence on Scientific Education

NOR was this relationship only evidenced among the members of the Royal Society. The emphasis of the Puritans upon utilitarianism and empiricism was likewise manifested in the type of education which they introduced and fostered. The "formal grammar grind" of the schools was criticized by them as much as the formalism of the Church.

Prominent among the Puritans who so consistently sought to introduce the new realistic, utilitarian, and empirical education into England was Samuel Hartlib. He formed the connecting link between the

32. *Dictionary of National Biography*, xiii, 1299.
33. C. F. Richardson, *English Preachers and Preaching* (New York, 1928), 177.
34. Dorothy Stimson, "Puritanism and the new philosophy in seventeenth-century England," *Bulletin of the Institute of the History of Medicine*, 1935, 3, 321–34.
35. Wilhelm Dilthey, "Pädagogik: Geschichte und Grundlinien des Systems," *Gesammelte Schriften* (Leipzig & Berlin, 1934), 163 ff.

various Protestant educators in England and in Europe who were earnestly seeking to spread the academic study of science. It was to Hartlib that Milton addressed his tractate on education and Sir William Petty dedicated his "Advice . . . for the Advancement of some particular Parts of Learning," namely, science, technology, and handicraft. Moreover, it was Hartlib who was instrumental in broadcasting the educational ideas of Comenius and in bringing him to England.

The Bohemian Reformist, John Amos Comenius, was one of the most influential educators of this period. Basic to the system of education which he promulgated were the norms of utilitarianism and empiricism: values which could only lead to an emphasis upon the study of science and technology, of *Realia*.[35] In his most influential work, *Didactica Magna,* he summarizes his views:[36]

> The task of the pupil will be made easier, if the master, when he teaches him everything, shows him at the same time its practical application in everyday life. This rule must be carefully observed in teaching languages, dialectic, arithmetic, geometry, physics, etc.
>
> . . . the truth and certainty of science depend more on the witness of the senses than on anything else. For things impress themselves directly on the senses, but on the understanding only mediately and through the senses. . . . Science, then, increases in certainty in proportion as it depends on sensuous perception.

Comenius found welcome among Protestant educators in England who subscribed to the same values; individuals such as Hartlib, John Dury, Wilkins, and Haak.[37] At the request of Hartlib, he came to England for the express purpose of making Bacon's Solomon's House a reality. As Comenius himself remarked: "nothing seemed more certain than that the scheme of the great Verulam, of opening in some part of the world a universal college, whose one object should be the advancement of the sciences, would be carried into effect."[38] But this aim was frustrated by the social disorder attendant upon the rebellion in Ireland. However, the Puritan design of advancing science was not entirely without fruit. Cromwell founded the only new English university instituted between the Middle Ages and the nineteenth century, Durham University, "for all the sciences."[39] And in Cambridge, during the height of the Puritan influence there, the study of science was considerably augmented.[40]

36. J. A. Comenius, *The Great Didactic,* translated by M. W. Keatinge (London, 1896), 292, 337; see also 195, 302, 329, 341.
37. Robert F. Young, *Comenius in England* (Oxford, 1932), 5–9.
38. *Opera Didactica Omnia* (Amsterdam, 1657), Book II, preface.
39. F. H. Hayward, *The Unknown Cromwell* (London, 1934), 206–30, 315.
40. James B. Mullinger, *Cambridge Characteristics in the Seventeenth Century* (London, 1867), 180–81 *et passim.*

In the same vein, the Puritan Hezekiah Woodward, a friend of Hartlib, emphasized realism (things, not words) and the teaching of science.[41] In order to initiate the study of the new science on a much more widespread scale than had hitherto obtained, the Puritans instituted a number of Dissenting Academies. These were schools of university standing opened in various parts of the kingdom. One of the earliest of these was Morton's Academy wherein there was pronounced stress laid upon scientific studies. Charles Morton later went to New England, where he was chosen vice-president of Harvard College, in which "he introduced the systems of science that he used in England."[42] At the influential Northampton Academy, another of the Puritan educational centres, mechanics, hydrostatics, physics, anatomy, and astronomy had an important place in the time-table. These studies were pursued largely with the aid of actual experiments and observations.

But the marked emphasis placed by the Puritans upon science and technology may perhaps best be appreciated by a comparison between the Puritan academies and the universities. The latter, even after they had introduced scientific subjects, continued to give an essentially classical education; the truly cultural studies were those which, if not entirely useless, were at least definitely nonutilitarian in purpose. The academies, in contrast, held that a truly liberal education was one which was "in touch with life" and which should therefore include as many utilitarian subjects as possible. As Dr. Parker puts it:[43]

> . . . the difference between the two educational systems is seen not so much in the introduction into the academies of "modern" subjects and methods as in the fact that among the Nonconformists there was a totally different system at work from that found in the universities. The spirit animating the Dissenters was that which had moved Ramus and Comenius in France and Germany and which in England had actuated Bacon and later Hartlib and his circle.

This comparison of the Puritan academies in England and Protestant educational developments on the Continent is well warranted. The Protestant academies in France devoted much more attention to scientific and utilitarian subjects than did the Catholic institutions.[44] When the Catholics took over many of the Protestant academies, the

41. Irene Parker, *Dissenting Academies in England* (Cambridge, 1914), 24.

42. *Ibid.*, 62.

43. *Ibid.*, 133–4.

44. P. D. Bourchenin, *Étude sur les académies protestantes en France au XVIième et au XVIIième siècle* (Paris, 1882), 445 ff.

45. M. Nicholas, "Les académies protestantes de Montauban et de Nimes," *Bulletin de la société de l'histoire du protestantisme française*, 1858, 4, 35–48.

46. D. C. A. Agnew, *Protestant Exiles from France* (Edinburgh, 1866), 210 ff.

47. Émile Boutroux, *Pascal*, trans. by E. M. Creak (Manchester, 1902), 16.

48. *Ibid.*, 17; *cf.* Jacques Chevalier, *Pascal* (New York, 1930), 143; Pascal's

study of science was considerably diminished.[45] Moreover, as we shall see, even in the predominantly Catholic France, much of the scientific work was being done by Protestants. Protestant exiles from France included a large number of important scientists and inventors.[46]

Value-Integration of Puritanism and Science

OF COURSE, the mere fact that an individual is *nominally* a Catholic or a Protestant has no bearing upon his attitudes toward science. It is only as he adopts the tenets and implications of the teachings that his religious affiliation becomes significant. For example, it was only when Pascal became thoroughly converted to the teachings of Jansenius that he perceived the "vanity of science." For Jansenius characteristically maintained that above all we must beware of that vain love of science, which though seemingly innocent, is actually a snare "leading men away from the contemplation of eternal truths to rest in the satisfaction of the finite intelligence."[47] Once Pascal was converted to such beliefs, he resolved "to make an end of all those scientific researches to which he had hitherto applied himself."[48] It is the firm acceptance of the values basic to the two creeds which accounts for the difference in the respective scientific contributions of Catholics and Protestants.

The same association of Protestantism and science was marked in the New World. The correspondents and members of the Royal Society who lived in New England were "all trained in Calvinistic thinking."[49] The founders of Harvard sprang from this Calvinistic culture, not from the literary era of the Renaissance or from the scientific movement of the seventeenth century, and their minds were more easily led into the latter than the former channel of thought.[50] This predilection of the Puritans for science is also noted by Professor Morison, who states: "the Puritan clergy, instead of opposing the acceptance of the Copernican theory, were the chief patrons and promoters of the new astronomy, and of other scientific discoveries, in New England."[51] It is significant that the younger John Winthrop, of Massachusetts, later a member of the Royal Society, came to London in 1641 and probably spent some time with Hartlib, Drury, and Comenius in London. Apparently, he suggested to Comenius that he come to New England and

Pensées, trans. by O. W. Wright (Boston, 1884), 224, No. xxvii. "*Vanity of the Sciences.* The science of external things will not console me for ignorance of ethics in times of affliction; but the science of morals will always console me for ignorance of external sciences."

49. Stimson, *op. cit.,* 332.

50. Porter G. Perrin, "Possible sources of *Technologia* at early Harvard," *New England Quarterly,* 1934, 7, 724.

51. Samuel E. Morison, "Astronomy at colonial Harvard," *New England Quarterly,* 1934, 7, 3–24; also Clifford K. Shipton, "A plea for Puritanism," *The American Historical Review,* 1935, 40, 463–4.

found a scientific college there.[52] Some years later, Increase Mather (President of Harvard College from 1684–1701) did found a "Philosophical Society" at Boston.[53]

The scientific content of Harvard's educational programme derived greatly from the Protestant Peter Ramus.[54] Ramus had formulated an educational curriculum which in contrast to that of the Catholic universities laid great stress on the study of the sciences.[55] His ideas were welcomed in the Protestant universities on the Continent, at Cambridge (which had a greater Puritan and scientific element than Oxford),[56] and later at Harvard, but were firmly denounced in the various Catholic institutions.[57] The Reformation spirit of utilitarianism and "realism" probably accounts largely for the favorable reception of Ramus' views.

Value-Integration of Pietism and Science

DR. PARKER notes that the Puritan academies in England "may be compared with the schools of the Pietists in Germany, which under Francke and his followers prepared the way for the *Realschulen,* for there can be no doubt that just as the Pietists carried on the work of Comenius in Germany, so the Dissenters put into practice the theories of Comenius' English followers, Hartlib, Milton, and Petty."[58] The significance of this comparison is profound for, as has been frequently observed, the values and principles of Puritanism and Pietism are almost identical. Cotton Mather had recognized the close resemblance of these two Protestant movements, saying that "ye American puritanism is so much of a piece with ye Frederician pietism" that they may be considered as virtually identical.[59] Pietism, except for its greater "en-

52. R. F. Young, *Comenius in England,* 7–8.

53. *Ibid.,* 95.

54. Perrin, *op. cit.,* 723–4.

55. Theobald Ziegler, *Geschichte der Pädagogik* (München, 1895), I, 108. Ziegler indicates that while the contemporary French Catholic institutions only devoted one-sixth of the curriculum to science, Ramus dedicated fully one-half to scientific studies.

56. David Masson properly calls Cambridge the alma mater of the Puritans. In listing twenty leading Puritan clergymen in New England, Masson found that seventeen of them were alumni of Cambridge, while only three came from Oxford. See his *Life of Milton* (London, 1875), II, 563; cited by Stimson, *op. cit.,* 332. See also A *History of the University of Oxford,* by Charles E. Mallet (London, 1924), II, 147.

57. Heinrich Schreiber, *Geschichte der Albert-Ludwigs-Universität zu Freiburg* (Freiburg, 1857–68), II, 135. For example, at the Jesuit university of Freiburg, Ramus could only be referred to if he were refuted, and "no copies of his books are to be found in the hands of a student."

58. Parker, *op. cit.,* 135.

59. Kuno Francke, "Cotton Mather and August Hermann Francke," *Harvard Studies and Notes,* 1896, 5, 63. See also the cogent discussion of this point by Max

thusiasm," might almost be termed the Continental counterpart of Puritanism. Hence, if our hypothesis of the association between Puritanism and interest in science and technology is warranted, one would expect to find the same correlation among the Pietists. And such was markedly the case.

The Pietists in Germany and elsewhere entered into a close alliance with the "new education": the study of science and technology, of *Realia.*[60] The two movements had in common the realistic and practical point of view, combined with an intense aversion to the speculation of Aristotelian philosophers. Fundamental to the educational views of the Pietists were the same deep-rooted utilitarian and empirical values which actuated the Puritans.[61] It was on the basis of these values that the Pietist leaders, August Hermann Francke, Comenius, and their followers emphasized the new science.

Francke repeatedly noted the desirability of acquainting students with practical scientific knowledge.[62] Both Francke and his colleague, Christian Thomasius, set themselves in opposition to the strong educational movement developed by Christian Weise, which advocated primarily training in oratory and classics, and sought rather "to introduce the neglected modern disciplines, which served their purposes more adequately; such studies as biology, physics, astronomy, and the like."[63]

Wherever Pietism spread its influence upon the educational system there followed the large-scale introduction of scientific and technical subjects.[64] Thus, Francke and Thomasius built the foundations of the University of Halle, which was the first German university to introduce a thorough training in the sciences.[65] The leading professors, such as Friedrich Hoffman, Ernst Stahl (professor of chemistry and famous

Weber, *Protestant Ethic,* 132–5.

60. Friedrich Paulsen, *German Education: Past and Present,* trans. by T. Lorenz (London, 1908), 104 ff.

61. Alfred Heubaum, *Geschichte des deutschen Bildungswesens seit der Mitte des siebzehnten Jahrhunderts* (Berlin, 1905), 1, 90. "Ziel der Erziehung [among Pietists] ist praktische Verwendbarkeit des Zöglings im Gemeinwohl. Der starke Einfluss des utilitaristischen Moments . . . vermindert die Gefahr der Uebertreibung des religiösen Moments und sichert der Bewegung für die nächste Zukunft ihre Bedeutung."

62. During walks in the field, says Francke, the instructor should "nützliche und erbauliche Geschichten erzählen oder etwas aus der Physik von den Geschöpfen und Werken Gottes vorsagen." ". . . im Naturalienkabinet diente dazu, die Zöglinge in ihren Freistunden durch den Anstaltarzt mit naturwissenschaftlichen Erscheinungen, mit Mineralien, Bergarten, hier und da mit Experimenten bekannt zu machen." Quoted by Heubaum, *op. cit.*, I, 89, 94.

63. *Ibid.*, I, 136.

64. *Ibid.*, I, 176 ff.

65. Koppel S. Pinson, *Pietism as a Factor in the Rise of German Nationalism* (New York, 1934), 18; Heubaum, *op. cit.*, I, 118. "Halle war die erste deutsche Universität von ganz eigenartigem wissenschaftlichen und nationalen Gepräge . . ."

for his influential phlogiston theory), Samuel Stryk, and, of course, Francke, all stood in the closest relations with the Pietistic movement. All of them characteristically sought to develop the teaching of science and to ally science with practical applications.

Not only Halle, but other Pietistic universities manifested the same emphases. Königsberg, having come under the Pietistic influence of the University of Halle through the activities of Francke's disciple, Gehr, early adopted the natural and physical sciences in the modern sense of the seventeenth century.[66] The University of Göttingen, an offshoot of Halle, was famous essentially for the great progress which it effected in the cultivation of the sciences.[67] The Calvinistic university of Heidelberg was likewise prominent for instituting a large measure of scientific study.[68] Finally, the University of Altdorf, which was at that time the most conspicuous for its interest in science, was a Protestant University subject to Pietistic influence.[69] Heubaum summarizes these developments by asserting that the essential progress in the teaching of science and technology occurred in Protestant, and more precisely, in Pietistic universities.[70]

Religious Affiliation of Recruits to Science

THIS association of Pietism and science, which we have been led to anticipate from our hypothesis, did not confine itself to the universities. The same Pietist predilection for science and technology was evidenced in secondary school education. The *Pädagogium* of Halle introduced the subjects of mathematics and natural science; stress being laid, in all cases, on the use of object lessons and on practical applications.[71] Johann Georg Lieb, Johann Bernhard von Rohr, and Johann Peter Ludwig (Chancellor of Halle University), all of whom had come under the direct influence of Francke and Pietism, advocated schools of manufacture, physics, mathematics, and economics, in order

66. Heubaum, *op. cit.*, I, 153.
67. Paulsen, *op. cit.*, 120–1.
68. Heubaum, *op. cit.*, I, 60.
69. S. Günther, "Die mathematischen Studien und Naturwissenschaften an der nürnbergischen Universität Altdorf," *Mitteilungen des Vereins für Geschichte der Stadt Nürnberg*, Heft. III, 9.
70. Heubaum, *op. cit.*, I, 241; see also Paulsen, *op. cit.*, 122; J. D. Michaelis, *Raisonnement über die protestantischen Universitäten in Deutschland* (Frankfurt, 1768), I, section 36.
71. Paulsen, *op. cit.*, 127.
72. Heubaum, *op. cit.*, I, 184.
73. Alfred Heubaum, "Christoph Semlers Realschule und seine Beziehung zu A. H. Francke," *Neue Jahrbücher für Philologie und Pädagogik*, 1893, 2, 65–77; see also Ziegler, *Geschichte der Pädagogik*, I, 197, who observes: ". . . einem inneren Zusammenhang zwischen der auf das Praktische gerichteten Realschule und der auf das Praktische gerichteten Frömmigkeit der Pietisten fehlte es ja auch nicht, nur éine ganz einseitig religiöse und theologische Auffassung des Pietismus kann das

to study how "manufacture might be ever more and more improved and excelled."[72] They hoped that the outcome of these suggestions might be a so-called *Collegium physicum-mechanicum* and *Werkschulen.*

It is a significant fact, and one which lends additional weight to our hypothesis, that the *ökonomisch-mathematische Realschule* was completely a Pietist product. This school, which centered on the study of mathematics, the natural sciences, and economics, and which was avowedly utilitarian and realistic in temper, was planned by Francke.[73] Moreover, it was a Pietist and a former student of Francke, Johann Julius Hecker, who first actually organized a *Realschule.*[74] Semler, Silberschlag, and Hähn, the directors and co-organizers of this first school, were all Pietists and former students of Francke.[75]

All available evidence points in the same direction. Protestants, without exception, form a progressively larger proportion of the student body in those schools which emphasize scientific and technologic training,[76] while Catholics concentrate their interests on classical and theological training. For example, in Prussia, the distribution shown in Table 1 was found.[77]

TABLE 1. Attendance at Secondary Schools Differentiated by Religious Affiliations of the Students, Prussia, 1875–6

Religious Affiliation	*Pro-gymnasium*	*Gymnasium*	*Realschule*	*Oberrealsch*	*Höheren Bürger*	*Total*	*General Population*
Protestants	49.1	69.7	79.8	75.8	80.7	73.1	64.9
Catholics	39.1	20.2	11.4	6.7	14.2	17.3	33.6
Jews	11.2	10.1	8.8	17.5	5.1	9.6	1.3

This greater propensity of Protestants for scientific and technical studies accords with the implications of our hypothesis. That this distribution is typical may be gathered from the fact that other investi-

verkennen: im Geist der praktischen Nützlichkeit und Gemeinnützigkeit ist dieser dem Rationalismus vorangegangen und mit ihm eins gewesen, und aus diesem Geist heraus ist zu Franckes Zeiten in Halle die Realschule entstanden."

74. Paulsen, *op. cit.*, 133.

75. Upon the basis of this and other facts, Ziegler proceeds to trace a close "Kausalzusammenhang" between Pietism and the study of science. See his *Geschichte,* I, 196 ff.

76. The characteristic feature of the *gymnasien* is the classical basis of their curricula. Demarcated from these schools are the *Realschulen,* where the sciences predominate and where modern languages are substituted for the classical tongues. The *Real-gymnasium* is a compromise between these two types, having less classical instruction than the *gymnasium* with more attention paid to science and mathematics. The *Ober-realschulen und höheren Bürgerschulen* are both *Realschulen;* the first with a nine-year course, the second with a six-year course. *Cf.* Paulsen, *German Education,* 46 *et passim.*

77. Alwin Petersilie, "Zur Statistik der höheren Lehranstalten in Preussen," *Zeitschrift des königlich Preussischen Statistischen Bureaus,* 1877, 17, 109.

gators have noted the same tendency in other instances.[78] Furthermore, these distributions do not represent a spurious correlation resulting from differences in rural-urban distribution of the two religions, as may be seen from the pertinent data for the Swiss canton, Basel-Stadt. As is well known, the urban population tends to contribute more in the fields of science and technology than the rural. Yet for 1910 and following—the period to which Edouard Borel's study, with results similar to those just presented for Prussia, refers—Protestants constituted 63.4 per cent of the total population of the canton, but only 57.1 per cent of the population of Basel (the city proper) and 84.7 per cent of the rural population.[79]

Martin Offenbacher's careful study includes an analysis of the association between religious affiliation and the allocation of educational interests in Baden, Bavaria, Württemberg, Prussia, Alsace-Lorraine, and Hungary. The statistical results in these various places are of the same nature: Protestants, proportionately to their representation in the population at large, have a much higher attendance at the various secondary schools, with the difference becoming especially marked in the schools primarily devoted to the sciences and technology. In Baden,[80] for example, taking an average of the figures for the years 1885–95, we have Table 2.

However, it must be noted that although the *Realschulen* curricula

TABLE 2

	Protestants, per cent	*Catholics, per cent*	*Jews, per cent*
Gymnasien	43	46	9.5
Realgymnasien	69	31	9
Oberrealschulen	52	41	7
Realschulen	49	40	11
Höheren Bürgerschulen	51	37	12
Average for the five types of schools	48	42	10
Distribution in the general population, 1895	37	61.5	1.5

78. Edouard Borel, *Religion und Beruf* (Basel, 1930), 93 ff., who remarks the unusually high proportion of Protestants in the technical professions in Basel; Julius Wolf, "Die deutschen Katholiken in Staat und Wirtschaft," *Zeitschrift fur Sozialwissenschaft*, 1913, 4, 199, notes that "die Protestanten ihren 'naturgemässen' Anteil überschreiten gilt für die wissenschaftliche und sonstige intellektuelle Betätigung (mit Ausnahme des geistlichen Berufs) . . ." In 1860, Ad. Frantz had already noted the same fact. See his "Bedeutung der Religionunterschiede für das physische Leben der Bevölkerungen," *Jahrbücher für Nationalökonomie und Statistik*, 1868, 11, 51. *Cf.* also similar results for Berlin in *Statistisches Jahrbuch der Stadt Berlin*, 1897, 22, 468–72. Buckle, *op. cit.*, 482, notes that "Calvinism is favourable to science." *Cf.* also Weber, *Protestant Ethic*, 38, 189; and Troeltsch, *Social Teachings* . . . , II, 894.

79. See "Die Bevölkerung des Kantons Basel-Stadt," *Mitteilungen des Statistischen Amtes des Kantons Basel-Stadt*, 1932, 48–49; and the same publication for the years 1910 and 1921.

are primarily characterized by their stress on the sciences and mathemathics as contrasted with the relatively little attention paid these studies in the *gymnasien,* yet the latter type of school also prepares for scientific and scholarly careers. But, in general, the attendance of Protestants and Catholics at the *gymnasien* represents different interests. The relatively large number of Catholics at the *gymnasien* is due to the fact that these schools prepare for theology as well, while the Protestants generally use the *gymnasien* as a preparation for the other learned professions. Thus, in the three academic years 1891–4, 226, or over 42 per cent of the 533 Catholic graduates of the Baden *gymnasien* subsequently studied theology, while of the 375 Protestant graduates, only 53 (14 per cent) turned to theology, while 86 per cent went into the other learned professions.[81]

Similarly, the Catholic apologist, Hans Rost, though he wishes to establish the thesis that "the Catholic Church has been at all times a warm friend of science," is forced to admit, on the basis of his data, that the Catholics avoid the *Realschulen,* that they show "eine gewisse Gleichgültigkeit und Abneigung gegen diese Anstalten." The reason for this, he goes on to say, is "das die Oberrealschule und das Realgymnasium nicht zum Studium der Theologie berechtigen: denn diese ist häufig die Tribfeder bei den Katholiken zum höheren Studium überhaupt."[82]

Thus, statistical data point to a marked tendency for Protestants, as contrasted with Catholics, to pursue scientific and technical studies. This can also be seen in the statistics for Württemberg, where an average of the years 1872–9 and 1883–98 gives the figures[83] in Table 3.

TABLE 3

	Protestants, per cent	*Catholics, per cent*	*Jews, per cent*
Gymnasien	68.2	28.2	3.4
Lateinschulen	73.2	22.3	3.9
Realschulen	79.7	14.8	4.2
Total population, 1880	69.1	30.0	0.7

80. Martin Offenbacher, *Konfession und soziale Schichtung* (Tübingen, 1900), 16. The slight errors of the original are here unavoidably reproduced.

81. H. Gemss, *Statistik der Gymnasialabiturienten im deutschen Reich* (Berlin, 1895), 14–20.

82. Hans Rost, *Die wirtschaftliche und kulturelle Lage der deutschen Katholiken* (Köln, 1911), 167 ff.

83. Offenbacher, *op. cit.*, 18. These data are corroborated by the study of Ludwig Cron pertaining to Germany for the years 1869–93; *Glaubenbekenntnis und höheres Studium* (Heidelberg, 1900). Ernst Engel also found that in Prussia, Posen, Brandenburg, Pomerania, Saxony, Westphalia, and the Rhine Provinces, there is a higher incidence of Evangelical students in these schools which provide a maximum of natural science and technical subjects. See his "Beiträge zur Geschichte und Statistik des Unterrichts," *Zeitschrift des königlich Preussischen statistischen Bureaus,* 1869, 9, 99–116, 153–212.

Nor do the Protestants evidence these foci of interest only in education. Various studies have found an unduly large representation of Protestants among outstanding scientists.[84] If the foregoing data simply provide slight probabilities that the connection we have traced does in fact obtain, Candolle's well known *Histoire des sciences et des savants* increases these probabilities considerably. Candolle finds that although in Europe, excluding France, there were 107 million Catholics and 68 million Protestants, yet on the list of scientists named foreign associates by the Academy of Paris from 1666–1883, there were only eighteen Catholics as against eighty Protestants.[85] But as Candolle himself suggests, this comparison is not conclusive since it omits French scientists who may have been Catholic. To correct this error, he takes the list of foreign members of the Royal Society of London at two periods when there were more French scientists included than at any other time: 1829 and 1869. In the former year, the total number of Protestant and Catholic scientists (who are foreign members of the Society) is about equal, while in 1869, the number of Protestants actually exceeds that of Catholics. But, outside the kingdom of Great Britain and Ireland, there were in Europe 139½ million Catholics and only 44 million Protestants.[86] In other words, though in the general population there were more than three times as many Catholics as Protestants, there were actually more Protestant than Catholic scientists.

However, there are yet more significant data than these which are based on different populations, where influence of economy, political regime, and other non-religious factors may be suspected to prevail over the actual influence of religion. A comparison of closely allied populations serves largely to eliminate these "extraneous" factors, but the results are the same. Thus, on the list of foreign associates of the Academy of Paris, there is not a single Irish or English Catholic, although their proportion in the population of the three kingdoms exceeded a fifth. Likewise, Catholic Austria is not at all represented, while in general Catholic Germany is similarly lacking in the production of scientists of note relative to Protestant Germany. Finally, in Switzerland, where the two religions are largely differentiated by cantons, or mixed in some of them, and where the Protestants are to the Catholics as three to two there have been fourteen foreign Associates, of whom not one was Catholic. The same differentiation exists for the Swiss and for the English and Irish of the two religions in the lists of the Royal Society of London and the Royal Academy of Berlin.[87]

84. For example, Havelock Ellis' *Study of British Genius*, 66 ff., finds that Protestant Scotland produced twenty-one of the outstanding scientists on his list as against one for Catholic Ireland. Alfred Odin finds that among the littérateurs on his list, the predominant emphasis of Protestants is on scientific and technical matters, rather than on literature, properly so-called. See his *Genèse des grands hommes* (Paris, 1895), I, 477 ff., II, Tables xx–xxi.

With the presentation of these data we close the empirical testing of our hypothesis. In every instance, the association of Protestantism with scientific and technologic interests and achievements is pronounced, even when extra-religious influences are as far as possible eliminated. The association is largely understandable in terms of the norms embodied in both systems. The positive estimation by Protestants of a hardly disguised utilitarianism, of intra-mundane interests, of a thorough-going empiricism, of the right and even the duty of *libre examen,* and of the explicit individual questioning of authority were congenial to the same values found in modern science. And perhaps above all is the significance of the active ascetic drive which necessitated the study of Nature that it might be controlled. Hence, these two fields were well integrated and, in essentials, mutually supporting, not only in seventeenth-century England but in other times and places.

Bibliographical Postscript

MAX WEBER's hypothesis of the role of ascetic Protestantism in the furtherance of modern capitalism has given rise to a substantial library of scholarly and polemical works on the subject. By the mid-thirties, for example, Amintore Fanfani could draw upon several hundred publications in his appraisal of the evidence; *Catholicism, Protestantism and Capitalism* (New York: Sheed & Ward, 1935). Weber did not himself conduct a similar inquiry into the relations between ascetic Protestantism and the development of science but concluded his classic essay by describing one of "the next tasks" as that of searching out "the significance of ascetic rationalism, which has only been touched in the foregoing sketch, . . . [for] the development of philosophical and scientific empiricism, [and for] . . . technical development" (*The Protestant Ethic,* 182–183). First published in 1936, the preceding chapter was conceived as an effort to follow this mandate to extend the line of inquiry which Weber had opened up.

The books and papers cited in this chapter have since been supplemented by others bearing on one or another part of the hypothesis connecting Puritanism, Pietism and science. Numerous works have greatly clarified the varieties and shadings of doctrine and values comprised in Puritanism; among these, I have found the following most useful: John Thomas McNeill, *The History and Character of Calvinism* (New York: Oxford University Press, 1954) which shows Calvinism to have

85. Alphonse de Candolle, *Histoire des sciences et des savants* (Geneva-Basel, 1885), 329.

86. *Ibid.,* 330. *Cf.* J. Făcăoaru, *Soziale Auslese* (Klausenberg, 1933), 138–9. "Die Konfession hat einen grossen Einfluss auf die Entwicklung der Wissenschaft gehabt. Die Protestanten wiesen überall eine grössere Zahl hervorragender Männer auf."

87. Candolle, *op. cit.,* 330 ff.

formed the core of English Puritanism and traces its varied consequences for society and thought; William Haller, *The Rise of Puritanism* (New York: Columbia University Press, 1939) which describes in rich and convincing detail how Puritan propaganda in press and pulpit helped prepare the way for the parliamentary rebellion, the radicalism of the Levellers, numerous sectarian fissions, an incipient bourgeois ethic and experimental science; Charles H. George, "A social interpretation of English Puritanism," *The Journal of Modern History*, 1953, 25, 327–342, which tries to identify the major components and the major types of Puritanism; G. R. Cragg, *From Puritanism to the Age of Reason* (Cambridge University Press, 1950), a "study of changes in religious thought within the Church of England, 1660–1700."

These and similar works have shown anew that Puritanism, like most religio-social creeds, was not of a piece. Practically all the scholars who have made intensive studies of the matter are agreed that most of the numerous sects comprising ascetic Protestantism provided a value-orientation encouraging work in science. (See also the note by Jean Pelseneer, "L'origine Protestante de la science moderne," *Lychnos*, 1946–47, 246–248.) But there the near-unanimity ends. Some have concluded that it was the more radical sectarians among the Puritans who did most to develop an enlarged interest in science; see, for example, George Rosen, "Left-wing Puritanism and science," *Bulletin of the Institute of the History of Medicine*, 1944, 15, 375–380. The biochemist and historian of science, Joseph Needham, comments on the close connections between the Diggers, the civilian wing of the Levellers, and the new and growing interest in experimental science, in his collection of essays, *Time: The Refreshing River* (New York: The Macmillan Company, 1943), 84–103. Others hold that the climate of values most conducive to an interest in science was found among the *moderate* Puritans, as exemplified by Robert Boyle. See James B. Conant, "The advancement of learning during the Puritan Commonwealth," *Proceedings of the Massachusetts Historical Society*, 1942, 66, 3–31); and for a more generally accessible though less detailed discussion, the same author's *On Understanding Science* (New Haven: Yale University Press, 1947), 60–62. R. Hooykaas, the distinguished Dutch historian of science, reports that his biography of Boyle's scientific and religious orientations confirms the principal findings set out in the foregoing chapter: R. Hooykaas, *Robert Boyle: een studie over Natuurwetenschap en Christendom* (Loosduinen: Kleijwegt, 1943), Chapters 3–4 which analyze Boyle's convictions that the study of natural philosophy is a religiously-founded moral obligation (especially as these are developed in Boyle's *The Christian Virtuoso, shewing, that by being addicted to experimental philosophy a man is rather assisted than indisposed to be a good Christian*, 1690), that empiricism and not merely rationality is required to comprehend God's works, and that tolerance, not persecution, is the policy appropriately governing relations with even the most fanatic sects.

The evidence in support of both the competing premises—that the chief locus of interest is to be found among the radical or the moderate Puritans—is still insufficient to justify a firm conclusion. Detailed distinctions among the various Puritan sects of course serve to specify the hypothesis more rigorously but the data in hand do not yet allow one to say, with any confidence, which of these were most disposed to advance the science of the day.

A recent group of studies provides substantial documentation of the ways in which the ethos of one of these Puritan sects—the Quakers—helped crystallize a distinct interest in science. In much the same terms set forth in the preceding chapter of this book, Frederick B. Tolles, *Meeting House and Counting House* (Chapel Hill: University of North Carolina Press, 1948), 205–213, derives the marked interest of Quakers in science from their religious ethos. Less analytically and, at times, even tendentiously, Arthur Raistrick, *Quakers in Science and Industry, being an account of the Quaker contributions to science and industry during the 17th and 18th centuries* (London: The Bannisdale Press, 1950) emphasizes the *fact* of the large proportion of Quaker members of the Royal Society and the *fact* of their extensive work in science. But as Professor Hooykaas properly notes, these unanalyzed facts do not themselves indicate that the distinctive participation of Quakers in scientific activity stemmed from their religious ethic; it might well be that it reflected the widespread tendency of well-to-do Englishmen, who included a disproportionately large number of Quakers, to turn their interest to matters of natural philosophy (R. Hooykaas, in *Archives Internationales d'Histoire des Sciences*, January, 1951). In a compact and instructive paper, however, Brooke Hindle goes on to show that the religious ethic did play this role among the Quakers of one colonial area; *cf.* his "Quaker background and science in colonial Philadelphia," *Isis*, 1955, 46, 243–250; and his excellent monograph, *The Pursuit of Science in Revolutionary America*, 1735–1789 (Chapel Hill: University of North Carolina Press, 1956).

It may be remembered that one of the principal hypotheses of Chapter XVII* held that it was the *unintended and largely unforeseen consequences* of the religious ethic formulated by the great Reformist leaders which progressively developed into a system of values favorable to the pursuit of science (39, this book; *cf.* F. S. Mason, "The scientific revolution and the Protestant Reformation. I. Calvin and Servetus in relation to the new astronomy and the theory of the

* See *Social Theory and Social Structure*.

circulation of the blood. II. Lutheranism in relation to iatrochemistry and German nature philosophy," *Annals of Science*, 1953, 9, 64–87, 154–175).The historical shaping of this ethic was doubtless partly in response to changing social, cultural and economic contexts but partly, also, it was an immanent development of the religious ideas and values themselves (as Wesley, above all other Protestant leaders, clearly perceived). This is only to say again that the rôle of ascetic Protestantism in encouraging the development of science did not remain fixed and unchanging. What was only implicit in the sixteenth and early seventeenth centuries became explicit and visible to many in the later seventeenth and eighteenth centuries. Several recent studies confirm this interpretation.

Based upon a close scrutiny of primary sources and present-day research, Paul H. Kocher's *Science and Religion in Elizabethan England* (San Marino, California: The Huntington Library, 1953) testifies to the long distance scholars have come since the day when they considered only the sources of conflict between science and religion as though conflict were plainly the *only* relation which could, and historically did, subsist between these social institutions. In contrast, this monograph shows that there was ample room for the science of Elizabethan England to develop within the bounds set by the religious doctrine of the time. Nor was this simply a matter of religion *tolerating* science. For the period before 1610, Kocher can find no convincing evidence "for or against" the hypothesis that Puritanism provided a more "fertile soil for natural science than . . . its rival religions in England." (17)* The data for this early period are inadequate to reach a sound conclusion. But, he goes on to say,

> We can see from our vantage point in the twentieth century that Puritan worldliness was ultimately to aid science more than Puritan otherworldliness was to inhibit it, in proportion more perhaps (though this is much less certain) than could Anglican doctrine or practice. But the effects of such impetus were to become visible only gradually as Puritanism developed. The Elizabethan age came too early to afford concrete evidence for distinguishing and weighing against each other the contributions of Puritans and Anglicans to science. (19)

Considered in terms of the immanent dynamic of the religious ethos, however, Kocher's contrast between the "worldliness" and "otherworldliness" of successive generations of Puritans is more seeming than real. For, as Weber was able to show in detail, "worldliness" was his-

* Numbers in parentheses refer to pages in the work cited.

88. Should it be asked why I did not make use of the later and amply-documented book, M. McLachlan's *English Education under the Test Acts* (1931), I

torically generated by the originally "otherworldly" values of Puritanism, which called for active and sustained effort in this world and so subverted the initial value-orientation (this process being an example of what he called the *Paradoxie der Folgen*). Manifest conformity to these values produced latent consequences which were far removed in character from the values which released them.

By the eighteenth century, this process of change had resulted in what has been described by Basil Willey as "the holy alliance between science and religion." (*The Eighteenth Century Background,* New York: Columbia University Press, 1941.) Just as Robert Boyle in the seventeenth century, so Joseph Priestley, the scientist and apostle of Unitarianism, in the eighteenth, symbolized and actualized this alliance.

The later connections between science and religion in England from the late eighteenth to the mid-nineteenth century have been painstakingly examined in the monograph by Charles C. Gillispie, *Genesis and Geology: a study in the relations of scientific thought, natural theology and social opinion in Great Britain, 1790–1850* (Cambridge: Harvard University Press, 1951). Concerned less with the role of religion in the recruitment and motivation of scientists than with the grounds on which the findings of geology were regarded as consistent with religious teachings, Gillispie traces the process through which these tended to become culturally integrated.

When the paper which forms the present chapter was written in 1936, I relied almost entirely on Irene Parker's pioneering study (1914) of the role of the Dissenting Academies in advancing the new scientific education of the eighteenth century.[88] The import of her study is not basically changed but is substantially developed and somewhat modified in the remarkable study by Nicholas Hans, *New Trends in Education in the Eighteenth Century* (London: Routledge & Kegan Paul, 1951). Hans bases part of his study upon a statistical analysis of the social origins, formal education and subsequent careers of some 3,500 individuals who formed the intellectual élite of that century, the basic data having been systematically assembled from the individual biographies in that almost inexhaustible mine of materials for historical sociology, the *Dictionary of National Biography.*[89] Only a few of his numerous pertinent findings will be summarized here. He finds, for example, that the Dissenting Schools and Academies produced about 10 per cent of the élite which, as Hans observes, "was far above their relative strength in the total population of England in the eighteenth century." (20) Nevertheless, he notes, as we have seen to be the case,

could only reply, in the words of another "harmless drudge," "Ignorance, Madam, pure ignorance." It should be added, however, that McLachlan is in fundamental agreement with the major conclusions of Irene Parker.

89. Studies in historical sociology have only begun to quarry the rich ore available in comprehensive collections of biography and other historical evidence. Although

that religious "motives" were not alone in making for the emergence of modern education (and specifically, of scientific education) in this period; with religion were joined "intellectual" and "utilitarian" motives. Thus, while "the Puritans promoted science as an additional support of Christian faith based on revelation, the deists looked upon science as the foundation of any belief in God." (12) The three types of motivation tended to reinforce one another: "The Dissenters, as well as many Puritans within the Church, represented the religious motive for educational reform. The idea of *propagatio fidei per scientia* found many adherents among the Dissenters. The intellectual and utilitarian reasons were put into full motion by secular bodies and teachers before the Dissenting Academies accepted them wholeheartedly." (54)

It is in this last respect that Hans finds it necessary to dissent from the thesis put forward by Irene Parker (which I adopted in my own paper), holding that she attributes almost exclusive influence to the Academies in advancing modern education in the eighteenth century. His corrective modification appears, on the ample evidence, to be thoroughly justified. Furthermore, it serves to clarify a problem which, at least one student of the matter can report, has long been troublesome and unresolved. This is the well-recognized fact that certain extreme forms of Calvinist dissent were for a long time inimical to the advancement of science, rather than conducive to it. As Hans now points out, "although the Calvinist tradition was essentially progressive it easily degenerated into narrow and intolerant dogmatism." (55) The Baptists, for example, were thoroughly "averse to the new learning from conviction and only late in the century joined other Dissenters [particularly the Presbyterians and Independents] in promoting the reform." (55) One wing of nonconformity, in short, adhered literally to certain restrictive tenets of Calvinism and it was this subgroup that manifested the hostility to science which has for so long been found in certain fundamentalist sects of Protestantism. Figuratively, it can be said that "Calvinism contained a seed of modern liberal education but it required a suitable environment to germinate and grow." (57) And, as we have seen, this social and cultural context was progressively provided in England of the time.

statistical analyses of such materials cannot stand in place of detailed qualitative analyses of the historical evidence, they afford a *systematic* basis for new findings and, often, for correction of received assumptions. At least, this has been my own experience in undertaking statistical analyses of some 6,000 biographies (in the D.N.B.) of those who comprised the élite of seventeenth-century England; of the lists of important discoveries and inventions listed in Darmstädter's *Handbuch zur Geschichte der Naturwissenschaften und der Technik*, and of 2,000 articles published in the *Philosophical Transactions* during the last third of the seventeenth century. (*Cf.* Merton, *Science, Technology and Society in Seventeenth-Century England*, 1938, Chapters II–III.) The most extensive use of such statistical analyses is found in P. A. Sorokin, *Social and Cultural Dynamics* (New York: American Book Co., 1937). Of

Supplementing these studies of the changing relations between Puritanism and science in England is the remarkable study by Perry Miller of these relations under the special conditions afforded by New England. (*The New England Mind: The Seventeenth Century*. Reissue. *The New England Mind: From Colony to Province*. Cambridge: Harvard University Press, 1954.) This comprehensive work demonstrates the notable receptivity to science among the theocratic leaders of the colony and the ensuing process of secularization, with its emphasis on utilitarianism. For a short but instructive comparison of the interpretation advanced by Perry Miller and that advanced in the preceding chapter, see Leo Marx, *Isis*, 1956, 47, 80–81.

As we have seen from the data assembled by Alphonse de Candolle —see pages 54–55 of this book—the connections of ascetic Protestantism and interest in science evidently persisted to some extent through the nineteenth century. Candolle's data have lately been examined again, with the same conclusion. See Isidor Thorner, "Ascetic Protestantism and the development of science and technology," *American Journal of Sociology*, 1952, 58, 25–33, esp. at 31–32. Thorner has also analyzed the data presented by P. A. Sorokin as a basis for questioning this hypothesis and finds that the data are actually in accord with it; *ibid.*, 28–30. For Sorokin's critique, see his *Social and Cultural Dynamics*, II, 150–152.

In another, searching review of Candolle's materials, Lilley has indicated their limitations as well as their uses. S. Lilley, "Social aspects of the history of science," *Archives Internationales d'Histoire des Sciences*, 1949, 28, 376–443, esp. 333 ff. He observes that the correlations between Protestantism and science may be spurious since "on the average the commercial and industrial classes [who have a greater interest in science] have tended to be Protestant in persuasion and the peasantry and more feudal types of landowners to be Catholic." We have taken note of this limitation (52) and have accordingly compared the interest in scientific subjects of Protestants and Catholics drawn from the same areas (52, 54). Lilley also criticizes Candolle's work for failing to take account of historical change in these relationships by lumping together, "without distinction, the whole period from 1666

course, the preparation of statistical summaries of this kind have their hazards; routinized compilations unrestrained by knowledge of the historical contexts of the data can lead to unfounded conclusions. For a discussion of some of these hazards, see P. A. Sorokin and R. K. Merton, "The course of Arabian intellectual development: a study in method," *Isis*, 1935, 22, 516–524; Merton, *op. cit.*, 367 ff., 398 ff.; and for a more thorough review of the problems of procedure, Bernard Berelson, *Content Analysis* (New York: The Free Press, 1951). Numerous recent studies of the social origins of business élites in the historical past have utilized materials of this sort: see the studies by William Miller, C. W. Mills, and Suzanne Keller instructively summarized by Bernard Barber, *Social Stratification* (New York: Harcourt, Brace & World, 1957).

to 1868." Presumably, religious affiliations in the latter and more secularized period would represent less by way of doctrinal and value commitments than in the earlier period; purely nominal memberships would tend to become more frequent. This criticism also has force, as we have seen. But as Lilley goes on to observe, further evidence in hand nevertheless confirms the underlying relationship between ascetic Protestantism and science, although this relationship may be masked or accentuated by other interdependent social and economic changes.

That the relationship persists to the present day in the United States is indicated by a recent thorough-going study of the social antecedents of American scientists, from 1880 to 1940. R. H. Knapp and H. B. Goodrich, *Origins of American Scientists* (Chicago: University of Chicago Press, 1952). Their evidence on this point is summarized as follows:

> Our data have shown the marked inferiority of Catholic [academic] institutions in the production of scientists [but not of other professionals; for example, lawyers] and, on the other hand, the fact that some of our most productive smaller institutions are closely connected with Protestant denominations and serve a preponderantly Protestant clientele. Moreover, the data presented by Lehman and Visher on the "starred" scientists [i.e. the scientists listed in *American Men of Science* who are judged to be of outstanding merit], although limited, indicate very clearly that the proportion of Catholics in this group is excessively low—that, indeed, some Protestant denominations are proportionately several hundred times more strongly represented. These statistics, taken together with other evidence, leave little doubt that scientists have been drawn disproportionately from American Protestant stock. (274)

Much the same impression, but without systematic supporting data, has been reported by Catholic scientists. "Father Cooper says he 'would be loath to have to defend the thesis that 5 per cent or even 3 per cent of the leadership in American science and scholarship is Catholic. Yet we Catholics constitute something like 20 per cent of the total population.' " J. M. Cooper, "Catholics and scientific research," *Commonweal*, 1945, 42, 147–149, as quoted by Bernard Barber, *Science and the Social Order*, 136. Barber also cites a similar observation by James A. Reyniers, Director of the Lobund Laboratories of Notre Dame University and by Joseph P. Fitzpatrick, S.J.; *ibid.*, 271.

This review of the more recent literature on the subject rather uniformly confirms the hypothesis of an observable positive relationship between ascetic Protestantism and science. The data provided by any one of these studies is typically far from rigorous. But this is, after all, the condition of most evidence bearing upon historically changing relations between social institutions. Considering not this study or that, but the entire array, based upon materials drawn from varied sources, we would seem to have some reasonable assurance that the empirical relationship, supposed in the foregoing study, does in fact exist.

But, of course, the gross empirical relationship is only the beginning, not the end, of the intellectual problem. As Weber noted, early in his celebrated essay on *The Protestant Ethic,* "a glance at the occupational statistics of any country of mixed religious composition brings to light with remarkable frequency a situation which has several times provoked discussion in the Catholic press and literature, and in Catholic congresses in Germany, namely, the fact that business leaders and owners of capital, as well as the higher grades of skilled labor, and even more the higher technically and commercially trained personnel of modern enterprises, are overwhelmingly Protestant." (35) The fortuity that comparable statistics on the religious composition of scientists are not ready to hand but must be laboriously assembled for the present and partially pieced together for the past does not make the empirical finding any more significant in itself (though it may commend to our respectful attention the arduous labors of those doing the spadework). For, as we have seen in examining the status of empirical generalizations (in Chapter II*), this only sets the problem of analyzing and interpreting the observed uniformity, and it is to this problem that the foregoing essay has addressed itself.

The principal components of the interpretation advanced in this essay presumably do not require repetition. However, a recent critique of the study provides an occasion for reviewing certain empirical and theoretical elements of the interpretation which can, apparently, be lost to sight. In this critique—"Merton's thesis on English science," *American Journal of Economics and Sociology,* 1954, 13, 427-432—James W. Carroll reports what he takes to be several oversights in the formulation. It is suggested that the heterogeneity of the beliefs included in Protestantism generally and in Puritanism specifically has been overlooked or imperfectly recognized. Were the charge true, it would plainly have merit. Yet it should be observed that the hypothesis in question is introduced by a chapter which begins by noting "the diversity of theological doctrines among the Protestant groups of seventeenth-century England" and continues by considering the values, beliefs and interests which are common to the numerous sects deriving from Calvinism (Merton, *Science, Technology and Society in Seventeenth-Century England,* Chapter IV, 415 ff.). And, as may be seen from this bibliographical postscript, historical scholarship has more

* See *Social Theory and Social Structure.*

thoroughly established the similarities, and not only the differences, among the Puritan sects stemming from ascetic Calvinism.

Carroll goes on to say that the evidence for the connection between the norms of Puritanism and of science provides only an empirical similarity between the two (or what is described as a Comtean "correlation of assertions"). But this is to ignore the demonstrated fact that English scientists themselves repeatedly invoked these Puritan values and expressly translated them into practice. (*Cf. ibid.*, Chapter V.)

That the Puritan values were indeed expressed by scientists is in fact implied in Carroll's next suggestion that no basis is provided in the study for discriminating between the "rationalizations" and the "motives" of these scientists. This touches upon a theoretical problem of such general import, and widespread misunderstanding, that it is appropriate to repeat part of what was said about it in the earlier study. "Present-day discussions of 'rationalization' and 'derivations' have been wont to becloud certain fundamental issues. It is true that the 'reasons' adduced to justify one's actions often do not account satisfactorily for this behavior. It is also an acceptable hypothesis that ideologies [alone] seldom *give rise* to action and that both the ideology and the action are rather the product of common sentiments and values upon which they in turn react. But these ideas can not be ignored for two reasons. They provide clues for detecting the basic values which motivate conduct. Such signposts can not be profitably neglected. Of even greater importance is the rôle of ideas in directing action into *particular* channels. *It is the dominating system of ideas which determines the choice between alternative modes of action which are equally compatible with the underlying sentiments.*" (*Ibid.*, 450.)

As for distinguishing between the expression of reasons which are merely accommodative lip-service and those which express basic orientations, the test is here, as elsewhere, to be found in the behavior which accords with these reasons, even when there is little or no prospect of self-interested mundane reward. As the clearest and best-documented case, Robert Boyle can here represent the other Puritans among his scientific colleagues who, in varying degree, expressed their religious sentiments in their private lives as in their lives as scientists. It would seem unlikely that Boyle was "merely rationalizing" in saying "that those who labour to deter men from sedulous Enquiries into Nature do (though I grant, designlessly) take a course which tends to defeat God. . . ." (Robert Boyle, *Some Considerations Touching the Usefulness of Experimental Natural Philosophy*, Oxford, 1664; 2d edition, 27). For this is the same Boyle who had written religious essays by the age of twenty-one; had, despite his distaste for the study of language, expressed his veneration for the Scriptures by learning Hebrew, Greek, Chaldee and Syriac that he might read them in their early versions; had provided a pension for Robert Sanderson to enable him to continue writing books on casuistry; had largely paid for the costs of printing the Indian, Irish and Welsh Bibles and, as if this were not enough, for the Turkish New Testament and the Malayan version of the Gospels and Acts; had become Governor of the Corporation for the Spread of the Gospel in New England and as a director of the East India Company had devoted himself and his resources to the diffusion of Christianity in these areas; had contributed substantially to the fund for printing Burnet's *History of the Reformation;* had published his profession of faith in *The Christian Virtuoso* and, quite finally, had provided in his will for endowment of the "Boyle lectures" for the purpose of defending Christianity against unbelievers. (This is the compact record set forth in A. M. Clerke's biography of Boyle in the *Dictionary of National Biography.*) Although Boyle was foremost in piety among Puritan scientists, he was still only first among equals, as witness Wilkins, Willughby and Ray among many others. So far as any historical record of words and action can permit us to say, it would appear that scientists like Boyle were not simply "rationalizing."

Carroll's final criticism, if intended conscientiously and not frivolously, exhibits a melancholy degree of immunity to commonplace and inconvenient facts of history. He observes that in showing the original membership of the Royal Society to have been preponderantly Protestant, the essay under review does not examine the possibility that the "invisible college," from which the Society stemmed, was part of a widespread Protestant movement of reform and that known Catholics were consequently banned from membership. That *Protestants* comprised the original membership of the Royal Society goes, one would suppose, without saying; in that day and age of the 1660's, in spite of the later political traffic of Charles II with the Catholicism of Louis XIV, Catholics would scarcely have been granted the prerogative of founding an association under the auspices of the Crown. The fact which is of more than passing interest is not, of course, that the Society was preponderantly *Protestant,* but that it was preponderantly *Puritan.* As for the observation that avowed Catholics were banned from academic posts, it evidently needs to be recalled that the Test Act of 1673, though later occasionally nullified in particular instances, excluded Nonconformists and not only Catholics and Jews from the universities. Yet, although this remained in force into the nineteenth century, Nonconformists continued to provide a large fraction of the men of science.

This short review of the most recently accumulated evidence suggests that, however contrary this may have been to the intentions of the Great Reformers, the ascetic Protestant sects developed a distinct predilection for working the field of science. In view of the powerful crosscurrents of other historical forces, which might have deflected this early orientation toward science, it is notable that the association between ascetic Protestantism and science has persisted to the present day. Profound commitments to the values of ascetic Protestantism have presumably become less common, yet the orientation, deprived of its theological meanings, evidently remains. As with any hypothesis, particularly in historical sociology, this one must be regarded as provisional, subject to review as more of the evidence comes in. But as the evidence now stands, the fact is reasonably well established and has definite implications for the broader problem of the connections between science and other social institutions.

The first of these implications is that, in this case at least, the emerging connections between science and religion were indirect and unintended. For, as has been repeatedly said, the reformers were not enthusiastic about science. Luther was at best indifferent; at worst, hostile. In his *Institutes* and his *Commentarie upon Genesis,* Calvin was ambivalent, granting some virtue to the practical intellect but far less than that owing to revealed knowledge. Nevertheless, the religious ethic which stemmed from Calvin promoted a state of mind and a value-orientation which invited the pursuit of natural science.

Second, it appears that once a value-orientation of this kind becomes established, it develops some degree of functional autonomy, so that the predilection for science could remain long after it has cut away from its original theological moorings.

Third, this pattern of orientation, which can even now be detected statistically, may be unwitting and below the threshold of awareness of many of those involved in it.

Fourth and finally, the highly visible interaction of the institutions of science and religion—as in the so-called war between the two in the nineteenth century—may obscure the less visible, indirect and perhaps more significant relationship between the two.

4 *Science and Economy of Seventeenth-Century England*

ROBERT K. MERTON

THE INTERPLAY BETWEEN socio-economic and scientific development is scarcely problematical. To speak of socio-economic influences upon science in general unanalyzed terms, however, barely poses the problem. The sociologist of science is specifically concerned with the *types* of influence involved (facilitative and obstructive), the *extent* to which these types prove effective in different social structures and the *processes* through which they operate. But these questions cannot be answered even tentatively without a clarification of the conceptual tools employed. All too often, the sociologist who repudiates the mythopoeic or heroic interpretation of the history of science lapses into a vulgar materialism which seeks to find simple parallels between social and scientific development. Such misguided efforts invariably result in a seriously biased and untenable discussion.

Formulation of the Problem

WE BEGIN by noting three common but unsound postulates. The first and most illusive is the identification of the personal motivation of scientists with the structural determinants of their research. Second is the belief that socio-economic factors serve to account exhaustively for the entire complex of scientific activity; and third is the imputation of "social needs" where these needs are, in any significant sense, absent.

Reprinted from Social Theory and Social Structure, *by Robert K. Merton (New York: The Free Press of Glencoe, 1957, revised edition), Chapter XIX, pp. 607–627.*

Clark's recent critique[1] of Hessen's essay may be taken to illustrate the confusion which derives from loose conceptualization concerning the relations between the motivation and the structural determinants of scientists' behavior. Clark tends to restrict the role of socio-economic factors in science to that of utilitarian motives of scientists and, correlatively, to identify "the disinterested desire to know, the impulse of the mind to exercise itself methodically and without any practical purpose" with scientific activity unconditioned by socio-economic elements.[2] Thus, to illustrate Newton's disinterestedness (in this sense of the word), Clark cites a frequently-reported anecdote to the effect that a friend to whom "he had lent a copy of Euclid's *Elements* asked Newton of what 'use or benefit in life' the study of the book could be. That was the only occasion on which it is recorded that Newton laughed."[3] Granting the reliability of this tale, its relevance to the issue in question is negligible, except on the assumption that people are invariably *aware* of the social forces which condition their behavior and that their behavior can be understood only in terms of their conscious motivations.

Motives may range from the desire for personal aggrandizement to a wholly "disinterested desire to know" without necessarily impugning the demonstrable fact that the thematics of science in seventeenth century England were in large part determined by the social structure of the time. Newton's own motives do not alter the fact that astronomical observations, of which he made considerable use,[4] were a product of Flamsteed's work in the Greenwich Observatory, which was constructed at the command of Charles II for the benefit of the Royal Navy.[5] Nor do they negate the striking influence upon Newton's work of such practically-oriented scientists as Halley, Hooke, Wren, Huyghens and Boyle. Even in regard to the question of motivation, Clark's thesis is debatable in view of the explicit awareness of many scientists in seventeenth century England concerning the practical implications of their research in pure science. It is neither an idle nor unguarded generaliza-

1. G. N. Clark, *Science and Social Welfare in the Age of Newton* (Oxford, 1937). See B. Hessen, "The social and economic roots of Newton's *Principia*," *Science at the Cross Roads* (London, 1931).

2. *Ibid.*, p. 86; and throughout Ch. 3.

3. *Ibid.*, p. 91. The original, slightly variant, version is in the *Portsmouth Collection.*

4. See the correspondence between Newton and Flamsteed, quoted extensively in L. T. More, *Isaac Newton* (New York, 1934), Ch. 11.

5. It was interest in the improvement of navigation which, according to Flamsteed, the first Astronomer Royal, led directly to the construction of the Greenwich Observatory. (Incidentally, Colbert proposed the Paris Observatory for the same purpose.) A Frenchman, Le Sieur de St. Pierre, visited England and proposed "improved" methods of determining longitude at sea. Flamsteed indicated in an official report that this project was not practicable, since "the lunar tables differed from the heavens." The report being shown to Charles, "he, startled at the assertion of the fixed stars' being false in the catalogue; said with some vehemence, 'he must have

tion that *every English scientist of this time* who was of sufficient distinction to merit mention in general histories of science at one point or another explicitly related at least some of his scientific research to immediate practical problems.[6] But in any case, analysis exclusively in terms of (imputed) motives is seriously misleading and tends to befog the question of the modes of socio-economic influence upon science.[7]

It is important to distinguish the personal attitudes of individual men of science from the social role played by their research. Clearly, some scientists were sufficiently enamored of their subject to pursue it for its own sake, at times with little consideration of its practical bearings. Nor need we assume that *all* individual researches are directly linked to technical tasks. The relation between science and social needs is two-fold: direct, in the sense that some research is advisedly and deliberately pursued for utilitarian purposes and indirect, in so far as certain problems and materials for their solution come to the attention of scientists although they need not be cognizant of the practical exigencies from which they derive.

In this connection, one is led to question Sombart's generalization that seventeenth century technology was almost completely divorced[8] from the contemporary science, that the scientist and inventor had gone their separate ways from the time of Leonardo to the eighteenth century. To be sure, the alliance of the two is not equally secure in all social structures but the assertion of Sombart (and others) that seventeenth century technology was essentially that of the empiric seems exaggarated in view of the many scientists who turned their theoretical knowledge to practical account. Wren, Hooke, Newton, Boyle, Huyghens, Halley, Flamsteed—to mention but an illustrious few—devoted themselves to the prosecution of both theory and practice. What is more important, scientists were uniformly confident of the practical fruits which their continued industry would ensure. It was this conviction, quite apart from the question of its validity, which partly influenced their choice of prob-

them anew observed, examined, and corrected, for the use of his seamen.' " Whereupon it was decided both to erect the Observatory and to appoint Flamsteed the Astronomer Royal. See Francis Baily, *An Account of the Rev'd John Flamsteed, compiled from his own manuscripts* (London, 1935), p. 37. To be sure, Flamsteed's salary was but 100 pounds a year. He was privileged to provide himself with all requisite instruments—at his own expense.

6. Documentation supporting this statement may be found in my *Science, Technology and Society in Seventeenth-Century England* (Bruges, 1938).

7. For a systematic treatment of this problem, see Joseph Needham, "Limiting factors in the advancement of science as observed in the history of embryology," *Yale Journal of Biology and Medicine,* 1935, 8, 1–18.

8. See Werner Sombart, *Der moderne Kapitalismus* (Munich, 1921), I, 466–67. The metaphor is highly appropriate in view of the remark by Oldenburg, the quondam secretary of the Royal Society, that the natural philosophers sought "the Marriage of Nature and Art, [whence] a happy issue may follow for the use and benefit of Humane Life," *Philosophical Transactions,* 1665, 1, 109.

lems. The grain of truth in Sombart's thesis is reduced to the fact that these men of science were concerned not with advancing the development of industrial machinery for factory use—since this had not developed sufficiently to claim their interest—but with innovations which implemented commerce, mining and military technique.[9]

Within this context, Clark's criticism of Hessen narrows down to a repudiation of the thesis that economic factors are *alone* determinant of the development of science. In company with Hessen I hasten to assent to this judgment. The primitive thesis of exclusively economic determination is no more intrinsic to Hessen's analysis, as he himself indicates (*op. cit*, p. 177), than to the work of Marx and Engels.

There remains the third problem—of ascertaining social needs—which can best be handled in specific empirical terms. The widely accepted notion that need precipitates appropriate inventions and canalizes scientific interests demands careful restatement. Specific emergencies have often focused attention upon certain fields, but it is equally true that a multitude of human needs have gone unsatisfied throughout the ages. In the technical sphere, needs far from being exceptional, are so general that they explain little. Each invention *de facto* satisfies a need or is an attempt to achieve such satisfaction. It is also necessary to realize that certain needs may not exist for the society under observation, precisely because of its culture and social structure.[10] It is only when the goal is actually part and parcel of the culture in question, only when it is actually experienced as such by some members of the society, that one may properly speak of a need directing scientific and technological interest in certain channels. Moreover, economic needs may be satisfied not only technologically but also by changes in social organization. But given the routine of fulfilling certain types of needs by technologic invention, a pattern which was becoming established in the seventeenth century; given the prerequisite accumulation of technical and scientific knowledge which provides the basic fund for innovation; given (in this case) an *expanding* capitalistic economy; and it may then be said that necessity is the (foster) mother of invention and the grandparent of scientific advance.

9. Franz Borkenau has perceived this necessary distincton: "Die Naturwissenschaft des 17. Jahrhunderts stand nicht im Dienste der *Industriellen Produktion*, obwohl sie das seit Bacons Zeiten gewünscht hätte." *Der Uebergang vom feudalen zum bürgerlichen Weltbild* (Paris, 1934), 3. (Italics supplied.)

10. For a lucid discussion of needs, see Lancelot Hogben's introduction to the volume edited by him, *Political Arithmetic* (New York, 1938).

11. The difference in costs of land and water transportation is strikingly, though perhaps exaggeratedly, indicated by Petty. "The water carriage of goods around the Globe of the Earth is but about double of the price of Land Carriage from Chester to London of the like goods," *Phil. Trans.* 1684, 14, 666.

Transport and Science

THE burgeoning of capitalistic enterprise in seventeenth century England intensified interest in more adequate means of transport and communication. St. Helena, Jamaica, North America were but the beginnings of England's great colonial expansion. This and the relatively low cost of water-transport[11] led to the marked growth of the merchant marine. More than forty per cent of the English production of coal was carried by water. Similarly, internal trade enhanced the need for improved facilities for land and river transport. Proposals for turnpikes and canals were common through the century.

Foreign trade was assuming world-wide proportions. The best available, though defective, statistics testify to these developments. Imports and exports increased by almost 300 per cent between 1613 and 1700.[12] Wheeler, writing at the very beginning of the century, observed that for approximately sixty years, not four ships of over 120 tons carrying capacity had sailed on the Thames.[13] At Elizabeth's death there were only four merchant ships of 400 tons each in England.[14] The number of ships, particularly those of heavy tonnage, increased rapidly under the Commonwealth, partly in response to the impetus provided by the Dutch War. Ninety-eight ships, with a net tonnage of over 40,000, were built within one decade (1649–59).[15] Adam Anderson notes that the tonnage of English merchant ships in 1688 was double that in 1666,[16] and Sprat claims more than a duplication during the preceding two decades.[17] The official report on the Royal Navy submitted by Samuel Pepys in 1695 comments upon the notable naval expansion during the century. In 1607, the Royal Navy numbered forty ships of 50 tons and upwards; the total tonnage being about 23,600 with 7,800 manning the ships. By 1695, the corresponding figures were over 200 ships, with a tonnage of over 112,400 and with more than 45,000 men.

A substantial element in the heightened tempo of shipbuilding and the increased size of ships was, as Sombart has suggested, military necessity. Though the growth of the merchant marine was considerable, it did not match that of the Royal Navy,[18] as is evidenced by the com-

12. See the actual figures in E. Lipson, *The Economic History of England* (London, 1931), II, 189.

13. John Wheeler, *Treatise of Commerce* (Middleburgh, 1601), 23.

14. Sir William Monson, *Naval Tracts* (London, 1703), 294.

15. The tonnage figures do not include 17 ships for which the data are not available. Adapted from M. Oppenheim, A *History of the Administration of the Royal Navy and of Merchant Shipping* (London, 1896), 330–37.

16. Adam Anderson, *Origin of Commerce* (Dublin, 1790), III, 111.

17. Thomas Sprat, *History of the Royal-Society of London* (London, 1667), 404.

18. Werner Sombart, *Krieg und Kapitalismus* (Munich, 1913), 179 ff.

parative statistics assembled by Sombart. Military exigencies often prompted increased speed in shipbuilding as well as improvements in naval architecture.

> Shipbuilding was furthered by military interests in three ways: more and larger ships were demanded, and above all, they were required within a shorter period. The requirements of the merchant marine could have been satisfied by handicraft methods of shipbuilding for yet another century. But these methods became discountenanced by the growing demands of the war marine; first in the construction of warships themselves, and then, of all ships, as the merchant marine was drawn into the stream of development. . . .[19]

Though Sombart tends to exaggerate the role of military exigencies in fostering more efficient methods of shipbuilding, it is clear that this factor combined with the intensified need for a larger merchant marine to accelerate such developments. In any event, available statistical data indicate a marked expansion in both mercantile and military marine beginning with the late sixteenth century.[20]

These developments were accompanied by increased emphasis upon a number of technical problems. Above all, the increase of commercial voyages to distant points—India, North America, Africa, Russia—stressed anew the need for accurate and expedient means of determining position at sea, of finding latitude and longitude.[21] Scientists were profoundly concerned with possible solutions to these problems.[22] Both mathematics and astronomy were signally advanced through research oriented in this direction.

Napier's invention of logarithms, expanded by Henry Briggs, Adrian Vlacq (in Holland), Edmund Gunter and Henry Gellibrand, was of

19. *Ibid.*, 191.

20. "Nos recherches [based on an examination of port-books] montrent à l'evidence que le commerce et la navigation de l'Angleterre faisaient de grands progrès au déclin du XVIe et pendant la premiere moitié du XVIIe siècle. On n'exagère guère en disant que la navigation anglaise a quadruplé, sinon quintuplé de 1580 jusqu'à 1640." A. O. Johnson, "L'acte de navigation anglais du 9 octobre 1651," *Revue d'histoire moderne*, 1934, 9, 13.

21. Hessen, *op. cit.*, 157–58.

22. In a paper read before the Royal Society by Dr. Bainbridge, it was stated: "Nullum est in tota ferè mathesi problema, quod mathematicorum ingenia magis exercet, nullum, quod astronomiae mágis conducit, quam problema inveniendi meridianorum sive longitudinum differentias." From the minutes of the Royal Society as transcribed in Thomas Birch, *History of the Royal Society of London* (London, 1757), IV, 311. Among the aims of the Society as stated by Oldenburg in the preface to the ninth volume (1674) of the *Philosophical Transactions* are: "spreading of practical mathematiques in all our Trade-towns and ports: making great rivers navigable; aiding the Fishery and Navigation; devising means of fertilizing barren lands, and cultivating waste lands; increasing the Linnen-trade; producing Latton [*sic*] and salt and saltpetre of our own."

aid to astronomer and mariner alike.[23] Adam Anderson possibly reflects the general attitude toward this achievement when he remarks that "logarithms are of great special utility to mariners at sea in calculations relating to their course, distance, latitude, longitude, etc."[24] Sprat, the genial historiographer of the Royal Society, asserted that the advancement of navigation was one of the chief aims of the group.[25] Hooke, the irascible "curator of experiments" of the Society, who was at once an eminent scientist and probably the most prolific inventor of his time, wrote in this same connection:

> First it is earnestly desired that all observations that have been already made of the variation of the magnetical needle in any part of the world, might be communicated, together with all the circumstances remarkable in the making thereof; of the celestiall observations for knowing the true meridian, or by what other means it may be found. . . . But from a considerable collection of such observations, Astronomy might be made available of that admirable effect of the body of the earth toucht by a loadstone, that if it will (as is probable it may) be usefull for the direction of seamen or others for finding the longitude of places, the observations collected, together with good navigation, which they [the Royal Society] engage to doe soe soon as they have a sufficient number of such observations. . . .[26]

A ballad written shortly after the Society began to meet at Gresham College reflects the popular appreciation of this interest, as is manifest in the following excerpt:[27]

> This College will the whole world measure
> Which most impossible conclude,

23. Published in his *Mirifici logarithmorum canonis descriptio* (Edinburg, 1614). It is to be noted that Briggs, who was the first to make Napier's work appreciated and who in 1616 suggested the base 10 for the system of logarithms, wrote several works on navigation. Likewise, that Gellibrand was probably the first Englishman to correct Gilbert's conclusion that magnetic declination is "constant at a given place," by discovering the secular variation of the declination. See his *Discourse Mathematical on the Variation of the Magneticall Needle* (London, 1635).

24. *Op. cit.*, II, 346. Anderson notes likewise Sir Henry Savile's "noble establishment [in 1630] of two professors of mathematics in the University of Oxford; one of which was for geometry, and the other for astronomy. . . . Both which branches of mathematics are well known to be greatly beneficial to navigation and commerce." *Ibid.*, I, 177.

25. Sprat, *op. cit.*, 150.

26. Robert Hooke, *Papers*, British Museum, Sloan MSS. 1039, f. 112. See also Hooke's A *Description of Helioscopes, and, Some Other Instruments* (London, 1676), postscript.

27. *In praise of the choice company of Philosophers and Witts, who meet on Wednesdays, weekly, at Gresham College*, By W[illiam?] G[lanville]. *Cf.* Dorothy Stimson, "Ballad of Gresham College," *Isis*, 1932, 18, 103–17, who suggests that the author was probably Joseph Glanville.

And navigation make a pleasure,
By finding out the longitude:
Every Tarpaulian shall then with ease
Saile any ship to the Antipodes.

Meeting officially as the Royal Society or foregathering at coffee-houses and private quarters, the scientific coterie discussed without end technical problems of immediate concern for the profit of the realm. Hooke's recently published diary discloses the varied pressures exerted upon him by the Society, the King and interested nobles to devote his studies to "things of use."[28] He would frequently repair to Garaways or Jonathans, the coffee-houses in Change Alley, where, with Christopher Wren and others of their company, he would "discourse about Celestiall Motions" over a pot of tea while at nearby tables more mundane speculations engrossed the attention of stock-jobbers and lottery touts. Problems considered at Garaways were often made the object of special inquiry by the Society. In short, the prevailing picture is not that of a group of "economic men" jointly or severally seeking to improve their economic standing, but one of a band of curious students coöperatively delving into the arcana of nature. The demands of economically-derived needs posed new questions and emphasized old, opening up fresh avenues of research and coupling with this a persistent pressure for the solution of these problems. This proved largely effective since the scientist's sense of achievement was not exclusively in terms of scientific criteria. Scientists were not immune from the interest in social acclaim, and discoveries which promised profitable application were heralded far beyond the immediate circle of virtuosi. Scientific achievement carried with it the seldom-undesired privilege of mingling with persons of rank; it was, to some extent, a channel for social mobility. The case of Graunt is well-known. Similarly, Hooke, the son of a humble curate of Freshwater, found himself the friend of noblemen and could boast of frequent chats with the King. The untutored reactions of the laity to the different orders of scientific research might be represented by the contrasting responses of Charles II to the "weighing of ayre," the fundamental work on atmospheric pressure which to his limited mind seemed nothing but childish diversion and idle amusement, and to directly utilitarian researches on finding the longitude at

28. *The Diary of Robert Hooke*, ed. by H. W. Robinson and W. Adams (London, 1935). For example, note the following entries: "At Sir Fr. Chaplains. Lodowick here about Longitude. Affirmed 3000 pound premium and 600 pounds more from the States," p. 160. "To Garaways with Sir Ch: Wren, mett Clark and Seignior, discoursed about watches for pocket and for Longitude. . . . Resolvd to complete [measuring?] degree. New Clepsydra ship, New Theory of sound," 221. "To Sir J. Williamson. He very kindly called me into his chamber. Spoke to me about the . . . Experiment, admonisht me to be diligent for this year to study things of use, to make the Kings Barometer . . .," 337.

sea, with which he was "most graciously pleased." Attitudes such as these served to guide a considerable part of scientific work into fields which might bear immediate fruit.[29]

A Case: Problem of the Longitude

THIS engrossing problem of finding the longitude perhaps illustrates best the way in which practical considerations focused scientific interest upon certain fields. There can be no doubt that the contemporary astronomers were thoroughly impressed with the importance of discovering a satisfactory way of finding the longitude, particularly at sea. Time and time again they evince this predominant interest. Rooke, Wren, Hooke, Huyghens, Henry Bond, Hevelius, William Molineux, Nicolaus Mercator, Leibniz, Newton, Flamsteed, Halley, La Hire, G. D. Cassini, Borelli—practically all of the leading astronomers and virtuosi of the day repeatedly testify to this fact.

The various methods proposed for finding longitude led to the following investigations:

1. Computation of lunar distances from the sun or from a fixed star. First widely used in the first half of the sixteenth century and again in the latter seventeenth century.
2. Observations of the eclipses of the satellites of Jupiter. First proposed by Galileo in 1610; adopted by Rooke, Halley, G. D. Cassini, Flamsteed and others.
3. Observations of the moon's transit of the meridian. Generally current in the seventeenth century.
4. The use of pendulum clocks and other chronometers at sea, aided by Huyghens, Hooke, Halley, Messy, Sully, and others.

Newton clearly outlined these procedures, as well as the scientific problems which they involved, upon the occasion of Ditton's claim of the reward for an accurate method of determining longitude at sea.[30] The profound interest of English scientists in this subject is marked by an article in the first volume of the *Philosophical Transactions*, describing the use of pendulum clocks at sea.[31] As Sprat put it, the Society had taken the problem "into its peculiar care." Hooke attempted to improve the pendulum clock and, as he says, "the success

29. In this connection, see Adam Anderson's remarks on the Royal Society: ". . . its improvements in astronomy and geography are alone sufficient to exalt its reputation, and to demonstrate its great utility even to the mercantile world, without insisting on its many and great improvements in other arts and sciences, some of which have also a relation to commerce, navigation, manufactures, mines, agriculture, &c.," *Origin of Commerce*, II, 609.

30. William Whiston, *Longitude Discovered* (London, 1738), historical preface.

31. Major Holmes, "A Narrative concerning the Success of Pendulum Watches at Sea for the Longitude," *Phil. Trans.*, 1665, 1, 52–58.

of these [trials] made me further think of improving it for finding the Longitude, and . . . quickly led me to the use of Springs instead of Gravity for the making a Body vibrate in any posture. . . ."[32] A notorious controversy then raged about Hooke and Huyghens concerning priority in the successful construction of a watch with spiral balance spring. However the question of priority be settled, the very fact that two such eminent men of science, among others, focused their attention upon this sphere of inquiry is itself significant. These simultaneous inventions are a resultant of two forces: the intrinsically scientific one which provided the theoretical materials employed in solving the problem in hand, and the nonscientific, largely economic, concern which served to direct interest toward the general problem. The limited range of practicable possibilities leads to independent duplicate inventions.

This problem continued to fire scientific research in other directions as well. Thus, Borelli, of the Royal Academy of Sciences at Paris (organized at the suggestion of the perspicacious Colbert), published an offer in both the *Journal des Sçavans* and the *Philosophical Transactions* to explain his method of making large glasses for telescopes or even to send glasses to those persons who were not in a position to make them, so that they might "observe the eclipses of the Satellites of Jupiter

32. Richard Waller, *The Posthumous Works of Robert Hooke* (London, 1705), Introduction. Galileo had apparently described a pendulum clock in 1641; Huyghens' invention in 1656 was independently conceived. Huyghens went on to invent the watch with a spring mechanism. See his description of the invention in the *Phil. Trans.* 1675, 11, 272; reprinted from the *Journal des Sçavans*, Feb. 25, 1675. This led to the notorious dispute between Hooke and Oldenburg, who defended Huyghens' priority in actual construction. It is of some interest, in connection with the question of pecuniary motivation, that Hooke, at the meeting of the Society following that at which Huyghens' communication concerning his "new pocket watch" was read, mentioned "that he had an invention for finding the longitude to a minute of time, or fifteen minutes in the heavens, which he would make out and render practical, if a due compensation were to be had for it." Whereupon Sir James Shaen promised "that he would procure for him a thousand pounds sterling in a sum, or a hundred and fifty pounds per annum. Mr. Hooke declaring that he would choose the latter, the council pressed him to draw up articles accordingly, and to put his invention into act." *Cf.* Birch, *op. cit.*, III, 191. For further details, see Waller, *op. cit.*, Introduction.

33. *Phil. Trans.*, 1676, 11, 691–92.

34. See Leonard Olschki *Galileo und seine Zeit* (Halle, 1927), 274 and 438, and the chapter on "Die Briefe über geographische Ortbestimmung." This method did not enable sufficient precision to be of much practical use. In the paper discussing his discovery of an unusual spot on Jupiter and fixing the period of the planet's rotation, Cassini observes that "a Travellour . . . may make use of it [the rotation] to find the Longitudes of the most remote places of the earth," *Phil. Trans.*, 1672, 7, 4042. In his discussion of the inequality of the time of rotation of the spots in different latitudes, he indicates the importance of this fact for a more precise determination of the longitude, *ibid.*, 1676, 6, 683. The announcement of his discovery of the third and fourth satellites of Saturn begins thus: "The Variety of wonderful Discoveries, which have been made this Century in the Heavens, since the invention of the Telescope, and the great Utility that may possibly be drawn therefrom, for perfecting natural Knowledg, and the Arts necessary to the Commerce and Society of Mankind,

which happen almost every day, and afford so fair a way for establishing the Longitudes over all the Earth." Moreover, "the Longitudes of places at Sea, Capes, Promontories, and divers Islands being once exactly known by this means, would doubtless be of great help and considerable usefulness to Navigation."[33]

It is precisely these episodes, with their acknowledged practical implications, which clearly illustrate the role of utilitarian elements in furthering scientific advance. For it may be said, upon ample documentary grounds, that Giovanni Domenico Cassini's astronomical discoveries were largely a result of utilitarian interests. In almost all of Cassini's papers in the *Transactions* he emphasizes the value of observing the moons of Jupiter for determining longitude, by means of the method first suggested by Galileo.[34] It is perhaps not too much to say that from this interest derived his discovery of the rotation of Jupiter, the double ring of Saturn, and the third, fourth, fifth, sixth, and eighth satellites of Saturn[35] for, as he suggests, astronomical observations of this sort were "incited" because of their practical implications. Lawrence Rooke, who was one of the original company constituting the Royal Society, often noted the "nautical value" of these observations.[36] Flamsteed frequently noticed the usefulness of observing the satellites of

has incited Astronomers more strictly to Examine, if there were not something considerable that had not been hitherto perceived." Translated from the *Journal des Sçavans,* April 22, 1685; reprinted in *Phil. Trans.,* 1696, 16, 79. In the presentation of Cassini's tables for the eclipses of the first satellites of Jupiter, it is remarked that beyond doubt observations of these eclipses best enable the use of portable telescopes for finding the longitude. "And could these satellites be observed at Sea, a Ship at Sea might be enabled to find the Meridian she was in, by help of the tables Monsieur Cassini has given us in this volume [*Recueuil d'observations faites en plusieurs Voiages pour perfectionner l'Astronomie & la Geographie*], discovering with very great exactness the said Eclipses, beyond what we can yet hope to do by the Moon, tho' she seem to afford us the only means Practicable for the Seaman. However before Saylors can make use of the Art of finding the Longitude, it will be requisite that the Coast of the whole Ocean be first laid down truly, for which this Method by the Satellites is most apposite: And it may be discovered, by the time the Charts are compleeted; or else that some Invention of shorter Telescopes manageable on Ship-board may suffice to shew the Eclipses of the Satellites at Sea . . .," *Phil. Trans.,* 1694, 17, 237–38. The latter part of this quotation definitely and lucidly illustrates the way in which scientific and technical research was "called forth" by practical needs. Worthy of note is the fact that Halley was commissioned by the Admiralty "to continue the Meridian as often as conveniently may be from side to side of the Channell, in order to lay down both coasts truly against one another" as well as to observe "the Course of the Tides in the Channell of England. . . ." See his letter of June 11, 1701 to Burchett in *Correspondence and Papers of Edmond Halley,* edited by E. F. MacPike (Oxford, 1932), 117–18.

35. The third (Tethys) and fourth (Dione) satellites were discovered in 1684; the fifth (Rhea) in 1672; the sixth (Titan) and eighth (Japetus) in 1671.

36. See "Mr. Rook's Discourse concerning the Observations of the Eclipses of the Satellites of Jupiter," reprinted in Sprat, *op. cit.,* 183–90. Rooke was Gresham professor of astronomy from 1652 to 1657 and Gresham professor of geometry from 1657 until his death in 1662.

Jupiter, because their eclipses "have been esteemed, and certainly are a much better expedient for the discovery of the Longitude than any yet known."[37]

Newton was likewise deeply interested in the same general problem. Early in his career, he wrote a now famous letter of advice to his friend, Francis Aston, who was planning a trip on the Continent, in which he suggested among other particulars that Aston "inform himself whether pendulum clocks be of any service in finding out the longitude." In a correspondence which we have reason to believe ultimately led Newton to the completion of the *Principia*, both Halley and Hooke urged Newton to continue certain phases of his research because of its utility for navigation.[38]

In 1694, Newton sent his well-known letter to Nathanael Hawes outlining a new course of mathematical reading for the neophyte navigators in Christ's Hospital, in which he criticized the current course, saying in part, that "the finding the difference of Longitude, Amplitude, Azimuts, and variation of the compass is also omitted, tho these things are very useful in long voyages, such as are those to the East Indies, and a Mariner who knows them not is an ignorant."[39] In August, 1699, Newton made public an improved form of his sextant (independently invented by Hadley in 1731), which in conjunction with lunar observations might enable the finding of the longitude at sea. He had already presented the initial outlines of his lunar theory in the first edition of the *Principia*. Furthermore, it was upon Newton's recommendation that the Act of 1714 was passed for a reward to those persons who should devise a successful method for ascertaining longitude at sea.[40] In the course of these activities, Newton was demonstrating his awareness of the utilitarian implications not only of much of his own scientific work, but also that of his contemporaries.

37. *Phil. Trans.*, 1683, 12, 322. Flamsteed elaborated this view more pointedly in other papers on the same subject. See *Phil. Trans.*, 1685, 15, 1215; XVI (1686), p. 199; XIII (1683), p. 405–7. In passing it might be noted that Leibniz invented a portable watch "principally designed for the finding of the longitude." See his paper in *Phil. Trans.*, 675, 10, 285–88.

38. This is the type of evidence which G. N. Clark overlooks entirely when he writes that "the one piece of evidence which can be adduced to show that during his great creative period he [Newton] was actuated by an interest in technology is the letter to Francis Aston . . .," *op. cit.*, 67. See Hooke's letter to Newton (Jan. 6, 1680) in which he writes: ". . . the finding out the proprietys of a curve made by two such principles [one of which was the hypothesis of attraction varying inversely as the square of the distance!] will be of great concerne to mankind because the invention of the longitude by the heavens is a necessary consequence of it." See the letter in W. W. Rouse Ball, *An Essay on Newton's Principa* (London, 1893), 147. Likewise, Halley, in his letter of July 5, 1687, writes: "I hope . . . you will attempt the perfection of the Lunar Theory, which will be of prodigious use in navigation, as well as of profound and subtle speculation." Complete letter is quoted *ibid.*, 174.

39. Newton's letters to Hawes are published in J. Edleston, *Correspondence of Sir Isaac Newton and Professor Cotes* (London, 1850), 279–99. An examination of

Newton's lunar theory was the climactic outcome of scientific concentration on this subject. As Whewell suggests,

> The advancement of astronomy would perhaps have been a sufficient motive for this labour; but there were other reasons which would urge it on with a stronger impulse. A perfect Lunar Theory, if the theory could be perfected, promised to supply a method of finding the Longitude of any place on the earth's surface; and thus the verification of a theory which professed to be complete in its foundations, was identified with an object of immediate practical use to navigators and geographers, and of vast acknowledged value.[41]

Halley, who had decided that the various methods of determining longitude were all defective and had declared that "it would be scarce possible ever to find the Longitude at sea sufficient for sea uses, till such time as the Lunar Theory be fully perfected," constantly prompted Newton to continue his work.[42] Flamsteed, and (from 1691 to 1739) Halley, also endeavored to rectify the lunar tables sufficiently to attain "the great object, of finding the Longitude with the requisite degree of exactness." Observations of the eclipses of the moon were recommended by the Royal Society for the same purpose.[43]

Another field of investigation which received added attention because of its probable utility is the study of the compass and magnetism in general. Thus, Sprat specifically relates such investigations by Wren to current needs when he states that "in order to Navigation he [Wren] has carefully pursu'd many Magnetical Experiments."[44] Wren himself, in his inaugural address as Gresham professor of astronomy, strikes the same keynote. The study of the magnetic variation is to be pursued diligently for it may prove of great value to the navigator, who may thus be enabled to find the longitude, "than which former Industry

the scientific preparation which Newton deemed necessary for a properly trained mariner finds that it includes a smattering of a substantial part of the physical research most prominently prosecuted during this period. In the list Newton mentions the subjects and problems with which not only he was chiefly concerned in the course of his own scientific career, but also his confreres. He indicates further that he was far from unaware of the practical bearings of the greater part of his abtruse discussions in the *Principia;* for example, his theory of the tides, the determination of the trajectory of projectiles, the lunar theory, and his work in hydrostatics and hydrodynamics.

40. Edleston, *op. cit.,* LXXVI. The importance attributed to the solution of this problem may be gauged from the rewards offered by other governments as well. The Dutch had sought to persuade Galileo to apply his talents to its solution; Philip III of Spain also offered a reward and in 1716, the Regent Duke of Orleans established a prize of 100,000 francs for the discovery of a practical method.

41. William Whewell, *History of the Inductive Sciences* (New York, 1858), I, 434.

42. *Correspondence and Papers of Edmond Halley,* 212.

43. *Phil. Trans.,* 1693, 17, 453–54.

44. *History of the Royal-Society,* 315–16.

hath hardly left any Thing more glorious to be aim'd at in Art."[45] La Hire, remarking that nothing is so troublesome on long sea voyages as the variation of the needle, states that "this put me upon finding out some means independent from Observations to discover the variations at Sea."[46] Henry Bond, Hevelius, Molineux, and Mercator were likewise interested in the study of magnetic phenomena with the same general aim in view.[47] Halley, in the famous paper in which he made known his theory of four magnetic poles and of the periodic movement of the magnetic line without declination, emphasized repeatedly the utilitarian desirability of studying the variation of the compass, for this research "is of that great concernment in the Art of Navigation: that the neglect thereof, does little less than render useless one of the noblest Inventions mankind has ever yet attained to." This great utility, he argues, seems a sufficient incitement "to all philosophical and Mathematical heads, to take under serious consideration the several Phenomena. . . ." He presents his new hypothesis in order to stir up the natural philosophers of the age that they might "apply themselves more attentively to this useful speculation."[48] Apparently the currently assiduous work in this field was not sufficient to satisfy his standards. It was for the purpose of enriching this useful speculation that Halley was given the rank of a captain in the navy and the command of the *Paramour Pink* in which he made three voyages. One outcome was Halley's construction of the first isogonic map.

Thus we are led to see that the scientific problems emphasized by the manifest value of a method for finding longitude were manifold. If the scientific study of various possible means of achieving this goal was not invariably dictated by the practical utility of the desired result, it is clear that at least part of the continued diligence exercized in these fields had this aim. In the last analysis it is impossible to determine with exactitude the extent to which practical concern focused scientific attention upon certain problems. What can be conscionably suggested is a certain correspondence between the subjects most intensively investigated by the contemporary men of science and the problems raised or emphasized by economic developments. It is an inference—usually supported by the explicit statements of the scientists themselves—that these economic requirements or, more properly, the technical needs deriving from these requirements, directed research into particular channels. The finding of the longitude was one problem which, engrossing the attention of many scientists, fostered profound developments in astronomy, geography, mathematics, mechanics, and the invention of clocks and watches.

45. Christopher Wren, *Parentalia* (London, 1750), 206.
46. *Phil. Trans.*, 1687, 16, 344–50.
47. See *Phil. Trans.*, *passim*; *e.g.*, 1668, 3, 790; 1670, 5, 2059; 1674, 8, 6065.
48. "A Theory of the Variation of the Magnetical Compass," *Phil. Trans.*, XIII 1683, 13, 208–21. See also his addendum, *ibid.*, 1693, 17, 563–78.

Navigation and Science

ANOTHER navigational problem of the period was determining the time of the tides. As Flamsteed indicated in a note appended to his first tide-table, the error in the almanacs amounted to about two hours; hence a scientific correction was imperative for the Royal Navy and navigators generally.[49] Accordingly, from time to time, he drew up several tide-tables accommodated to ports not only in England but also in France and Holland. This work was the continuation of an interest in providing a theory of the tides emphasized by the Royal Society from its very inception. The first volume of the *Transactions* included several papers presenting observations of the time of the tides in various ports. Boyle, Samuel Colepresse, Joseph Childrey, Halley, Henry Powle, and most notably, John Wallis made contributions to this subject.

Newton took up the task as a further basis for the verification of the general law of attraction and, as Thomson remarks, "his theory of the tides is not less remarkable either for the sagacity involved, or for its importance to navigation." His theory accounted for the most evident aspects of the tides: the differences between the spring and neap tides and the morning and evening tides, the effect of the moon's and sun's declination and parallax, and the tides at particular places, making use of the observations of Halley, Colepresse and others to check his calculated results.[50] Halley, seeking as always to minister to the marriage of theory and practice, was not slow to inform the Lord High Admiral of the "generall use to all shipping" to be derived from these researches.[51] It was not, however, until the work of Euler, Bernoulli, and D'Alembert, and later of Laplace, Lubbock, and Airy, that the theory could be applied with sufficient precision to promise service for practical purposes. Again, one can correlate scientific interests—in this instance, the study of so esoteric a subject as the theory of attraction—with economic exigencies.

Another problem of grave concern for maritime affairs was the depletion of forest preserves to the point that eventually unseasoned wood had to be used in the construction of ships. Timber had become relatively scarce, both because of its use as fuel and its rapid consumption in the naval wars and in the rebuilding of London. The solution to the fuel problem was partially solved by the use of coal for various industries—such as brass and copper casting, brewing, dyeing and ironware, though not for the production of raw iron. The depletion of

49. *Phil. Trans.*, 1683, 13, 10–15; for later tables see *ibid.*, 1684, 14, 458 and 821; 1685, 15, 1226; 1686, 16, 232 and 428.

50. *Principia Mathematica* (London, 1713; second edition), Bk. III, Prop. XXIV, XXXVI and XXXVII.

51. *Correspondence*, 116.

timber so jeopardized shipbuilding that the commissioners of the Royal Navy appealed to the Society for suggestions concerning the "improvement and planting of timber." Evelyn, Goddard, Merrit, Winthrop, Ent, and Willughby contributed their botanical knowledge toward the solution of this problem, their individual papers being incorporated in Evelyn's well-known *Sylva*. Not unrelated to such practical urgencies, then, is the fact that one of the "chief activities" of the Society was the "propagating of trees." Furthermore, says Sprat, the members of the Society "have employ'd much time in examining the Fabrick of ships, the forms of their Sails, the shapes of their Keels, the sorts of Timber, the planting of Firr, the bettering of Pitch, and Tarr, and Tackling."[52] This led not only to the study of silviculture and allied botanical studies, but also to investigations in mechanics, hydrostatics and hydrodynamics. For as Newton noted in his letters to Hawes, the solution of such problems as the determination of the stress of ropes and timber, the power of winds and tides and the resistance of fluids to immersed bodies of varying shapes would be of great utility for the mariner.

Moreover, when one compares the requisites of a man-of-war as enumerated by Sir Walter Raleigh in his *Observations on the Navy* at the beginning of the century with the types of research conducted by the Society, it becomes apparent that all the major problems had become the object of scientific study. Raleigh lists six desirable qualities of a fighting ship: strong build, speed, stout scantling, ability to fight the guns in all weather, ability to lie easily in a gale, and ability to stay well. Contemporary scientists attempted to devise means of satisfying all these requirements. In many instances they were led to solve derivative problems in "pure science" in the prospect of using their knowledge for these purposes. Thus, Goddard, Petty, and Wren investigated methods of shipbuilding with the object of improving existing procedures. Hooke was ordered by the Society to determine the most "stout scantling" by testing the resistance of the "same kinds of wood, of several ages, grown in several places, and cut at different seasons of the year."[53] At times in coöperation with Boyle, Hooke performed numerous experiments to "try the strength of wood," and of twisted and untwisted

52. Sprat, *op. cit.*, 150.

53. Birch, *op. cit.*, I, 460.

54. *Cf.* Hessen, *op. cit.*, 158–59.

55. *The Petty-Southwell Correspondence*, ed. by the Marquis of Lansdowne (London, 1928), 117; Birch, *op. cit.*, I, 87.

56. Sprat, *op. cit.*, 250. See Hooke's letter to Boyle in the latter's *Works*, V, 537.

57. "It being a Question amongst the Problems of Navigation, very well worth resolving, to what Mechanical powers the Sailing (against the wind especially) was reducible; he [Wren] shew'd it to be a Wedge: And he demonstrated how a transient Force upon an oblique Plane, would cause the motion of the Plane against the first Mover. And he made an Instrument, that Mechanically produc'd the same effect, and shew'd the reason of Sayling to all Winds.

cords. These experiments were in progress at the time Hooke arrived at the law which bears his name (*ut tensio sic vis*).

In order to discover way of increasing the speed of ships, it is necessary to study the movement of bodies in a resistant medium, one of the basic tasks of hydrodynamics.[54] Accordingly, Moray, Goddard, Brouncker, Boyle, Wren, and Petty were concerned with this problem.[55] In this instance, the connection between a given technical task and the appropriate "purely scientific" investigation is explicit. Petty, at the time he wrote that "the fitts of the Double-Bottome [ship] do return very fiercely upon mee,"experimented in hydrodynamics to determine the velocity of "swimming bodies." The general connection is established by Sprat in his description of the instruments of the Society:

> [There are] several instruments for finding the velocity of swimming Bodies of Several Figures, and mov'd with divers strength, and for trying what Figures are least apt to be overturn'd, in order to the making of a true theory, of the Forms of Ships, and Boats for all uses.[56]

Christopher Wren, who was for Newton one of "the greatest Geometers of our times," also investigated the laws of hydrodynamics precisely because of their possible utility for improving the sailing qualities of ships.[57] And Newton, after stating his theorem on the manner in which the resistance of a fluid medium depends upon the form of the body moving in it, adds: "which proposition I conceive may be of use in the building of ships."[58]

The Society maintained a continued interest in under-water contrivances, ranging from diving bells to Hooke's proposal of a full-fledged submarine which would move as fast as a wherry on the Thames. A committee on diving considered leaden "diving boxes" and Halley's "diving bell," which were tested in the Thames and, with more convenience to the spectators than the diver, in a tub set up at one of their weekly meetings. Wilkins laid great stress on the feasibility and advantage of submarine navigation which would be of undoubted use in warfare, would obviate the uncertainty of tides and might be used

"The Geometrical Mechanics of Rowing, he shew'd to be a Vectis on a moving or cedent Fulcrum. For this end he made Instruments, to find what the expansion of Body was towards the hindrance of Motion in a Liquid Medium; and what degree of impediment was produc'd by what degree of expansion: with other things that are the necessary Elements for laying down the Geometry of Sailing, Swimming, Rowing, Flying, and the Fabricks of Ships." Sprat, *op. cit.*, 316. Once again we see how the immediate technical aim leads to the study of derivative problems in science.

58. *Principia Mathematica*, Bk. II, Sect. VII, Prop. XXXIV, Scholium. To my knowledge, Newton's remark has not been noticed heretofore in this connection. It reads: "Quam quidem propositionem in construendis Navibus non inutilem futuram esse censeo."

to recover sunken treasures.[59] Hooke linked many of his experiments on respiration with technical problems deriving from such efforts.

Wilkins introduced the "umbrella anchor" to the Society; a device "to stay a ship in a storm." Wren proposed "a convenient way of using artillery on ship-board," and Halley, pointing out that England "must be masters of the Sea, and superior in navall force to any neighbour," described a method of enabling a ship to carry its guns in bad weather.[60] Petty, fondly hoping "to pursue the improvement of shipping upon new principles," built several of his double-bottomed boats with which the Society was well pleased. Unfortunately, his most ambitious effort, the *St. Michael the Archangel,* failed miserably, which led him to conclude that both the fates and the King were opposed to him.

The Society periodically discussed means of preserving ships "from worms," a problem which proved greatly disturbing both to the commissioners of the Royal Navy and to private shipowners. Newton had evidenced interest in this same vexing problem, asking Aston to determine "whether the Dutch have any tricks to keep their ships from being all worm-eaten." No appreciable progress resulted from these discussions, however.

In general, then, it may be said that the seventeenth century men of science, ranging from the indefatigable virtuoso Petty to the nonpareil Newton, definitely focused their attention upon technical tasks made urgent by problems of navigation and upon derivative scientific research. The latter category is difficult to delimit. Although it is true that a congeries of scientific research may be immediately traced to technical requirements, it appears equally evident that some of this research is a logical development of foregoing scientific advance. It is indubitable, however, in the light of what the scientists themselves had to say about the practical implications of their work, that practical problems exercised an appreciable directive influence. Even that "purest" of disciplines, mathematics, was of primary interest to Newton when designed for application to physical problems.[61]

Some attention was likewise paid to inland transportation although to a less extent than to maritime transport, possibly because of the

59. John Wilkins, *Mathematical Magick* (London, 1707; 5th edition), Ch. 5. As early as 1551, Tartaglia had suggested a largely effective means for raising sunken ships to the water's surface. Several patents had been granted for "diving engines" since at least 1631. By the help of one of these devices "and good luck," says Anderson, Sir William Phipps "fished up" nearly 200,000 pounds sterling in pieces of eight from a Spanish fleet which had been sunk off the West Indies. See *Origin of Commerce*, III, 73. Hooke and Halley, as well as several others, responded to this success with new devices for recovering treasures from the deep.

60. Wren, *Parentalia*, 240; *Correspondence . . . of Halley*, 165.

61. E. A. Burtt, *The Metaphysical Foundations of Modern Physical Science*, 210.

62. D[aniel] D[efoe], *Essays upon Several Projects* (London, 1702), 73 ff.

greater economic significance of the latter. The growing interior traffic demanded considerable improvement. Such improvements, said Defoe, are "a great help to Negoce, and promote universal Correspondence without which our Inland Trade could not be managed."[62] Travelling merchants, who might carry as much as a thousand pounds of cloth, extended their trade all over England,[63] and required improved facilities. Because of the "great increase of carts, waggons, &c., by the general increase of our commerce," says Adam Anderson, the King (somewhat optimistically, no doubt) ordered in 1662 that all common highways be enlarged to eight yards. Characteristically, contemporary scientists also sought to overcome technical difficulties. Petty, with his keen interest in economic affairs, devised several chariots guaranteed to "passe rocks, precipices, and crooked ways."[64] Wren endeavored to perfect coaches for "ease, strength and lightness" and, as did Hooke, invented a "way-wiser" to register the distance travelled by a carriage.[65] Wilkins, possibly following Stevin's invention of a half century earlier, described a "sailing Chariot, that may without Horses be driven on the Land by the Wind, as ship are on the Sea."[66] Likewise, the Society delegated Hooke, at his own suggestion, to carry on "the experiment of land-carriage, and of a speedy conveying of intelligence."[67] Such efforts indicate the attempts of scientists to contribute technological props to business enterprise; in these particular instances to facilitate the possible extension of markets, one of the primary requirements of a nascent capitalism.

The Extent of Economic Influence

IN a sense, the foregoing discussion provides materials which only illustrate the connections we have been tracing. We have still to determine the extent to which socio-economic influences were operative. The minutes of the Royal Society as transcribed in Birch's *History of the Royal Society* provide one basis for such a study. A feasible, though in several obvious respects inadequate, procedure consists of a classification and tabulation of the researches discussed at these meet-

63. Daniel Defoe, *Tour of Great Britain* (London, 1727), III, 119–20.

64. *Petty-Southwell Correspondence,* 41, 51 and 125. "And it seems to me [writes Petty] that this carriage can afford to carry fine goods between Chester and London for lesse than 3*d* in the pound." With all due honesty, Petty admits that this "Toole is not exempt from being overthrowne," but adds comfortingly, "but if it should bee overthrowne (even upon a heape of flints) I cannot see how the Rider can have any harme."

65. *Parentalia,* 199, 217 and 240.

66. Wilkins, *op. cit.,* Bk. II, Ch. 2.

67. Birch, *op. cit.,* I, 379 and 385; Hooke, *Diary,* 418. This subject was discussed at some fifteen meetings of the Society within a three-year period.

ings, together with an examination of the context in which the various problems came to light. This should afford some ground for deciding the *approximate* extent to which extrinsic factors operated.

Meetings during the four years 1661, 1662, 1686, and 1687 will be considered. There is no reason to suppose that these did not witness meetings typical of the general period. The classification employed is empirical rather than logically ordered. Items were classified as "directly related" to socio-economic demands when the individual conducting the research explicitly indicated some such connection or when the immediate discussion of the research evidenced a prior appreciation of such a relation. Items classified as "indirectly related" comprise researches which had a clear-cut connection with current practical needs, intimated in the context, but which were not definitely so related by the investigators. Researches which evidenced no relations of this sort were classified as "pure science." Many items have been classified in this category which have (for the present-day observer) a conceivable relation to practical exigencies but which were not so regarded explicitly in the seventeenth century. Thus, investigations in the field of meteorology could readily be related to the practical desirability of forecasting the weather but when these researches were not explicitly related to specific practical problems they were classified as pure science. Likewise, much of the work in anatomy and physiology was undoubtedly of value for medicine and surgery, but the same criteria were employed in the classification of these items. It is likely, therefore, that if any bias is involved in this classification, it is in the direction of overestimating the scope of "pure science."

Each research discussed was "counted" as one "unit." It is obvious that this procedure provides only a gross approximation to the extent of extrinsic influences upon the selection of subjects for scientific study, but when greater precision is impossible one must rest temporarily content with less. The results, as summarized in Table 1, can merely suggest the relative extent of the influences which we have traced in a large number of concrete instances.[68]

From Table 1 it appears that less than half (41.3%) of the investigations conducted during the four years in question are classifiable as "pure science." If we add to this the items which were but indirectly related to practical needs, then about seventy per cent of this research had no explicit practical affiliations. Since these figures are but grossly approximate, the results may be summarized by saying that from forty to seventy per cent occurred in the category of pure science and cor-

68. For a more complete discussion of the procedure used and a detailed classification of the categories, see my *Science, Technology and Society in Seventeenth-Century England*, Ch. 10. Appendix A provides illustrations of the items classified in the various categories.

relatively that from thirty to sixty per cent were influenced by practical requirements.

Again, considering only the research directly related to practical needs, it appears that problems of marine transport attracted the most attention. This is in accord with the impression that the contemporary men of science were well aware of the problems raised by England's insular position—problems both military and commercial in nature—and were eager to rectify them.[69] Of almost equal importance was the influence of military exigencies. Not only were there some fifty years of actual warfare during this century, but also the two great revolutions in English history. Problems of a military nature left their impress upon the culture of the period, including scientific development.

Likewise, mining, which developed so markedly during this period, as we may see from the studies of Nef and other economic historians, had an appreciable influence. In this instance, the greater part of scientific, if one may divorce it from technologic, research was in the fields

TABLE 1. Approximate Extent of Socio-Economic Influences upon the Selection of Scientific Problems by Members of the Royal Society of London, 1661–62 and 1686–87

	Total for the four years					
	Number			*Per cent*		
Pure science	333			41.3		
Science related to socio-economic needs	473			58.7		
Marine transport		129			16.0	
Directly related			69			8.6
Indirectly related			60			7.4
Mining		166			20.6	
Directly related			25			3.1
Indirectly related			141			17.5
Military technology		87			10.8	
Directly related			58			7.2
Indirectly related			29			3.6
Textile industry		26			3.2	
General technology and husbandry		65			8.1	
Total	806			100.0		

of mineralogy and metallurgy with the aim of discovering new utilizable ores and new methods of extracting metals from the ore.

69. See, for example, Edmond Halley's observation: "that the Inhabitants of an Island, or any State that would defend an Island, must be masters of the Sea, and superior in navall force to any neighbor that shall think fitt to attack it, is what I suppose needs no argument to enforce." In his paper read before the Royal Society and reprinted in *Correspondence . . . of Halley,* 164–65.

It is relevant to note that, in the latter years considered in this summary, there was an increasing proportion of investigation in the field of pure science. A conjectural explanation is not far to seek. It is probable that at the outset the members of the Society were anxious to justify their activities (to the Crown and the lay public generally) by obtaining practical results as soon as possible; therefore, the initially marked orientation toward practical problems. Furthermore, many of the problems which were at first advisedly investigated because of their utilitarian importance may later be studied with no awareness of their practical implications. On the basis of the (perhaps biased) criteria adopted in this compilation, some of the later researches would arbitrarily be classified as pure science.

On the grounds afforded by this study it seems justifiable to assert that the range of problems investigated by seventeenth century English scientists was appreciably influenced by the socio-economic structure of the period.

5 *Science in the French Revolution*

CHARLES COULSTON GILLISPIE

SCIENCE PRESENTS ITSELF to history under two aspects: first in its own evolution and secondly, in its accommodation by culture. Which is the more important is a question of perspective, and which the more interesting a matter of taste. The evolution of scientific ideas relates the progress of science to nature, and is the more elegant and precise a subject. But its cultural history relates it to society, and it is that to which we are asked to address ourselves. I think it obvious that science, which is about nature, cannot be determined in its content by the social relations of scientists. At most it may be touched in style, in pace, and—within limits imposed by the logical interdependence of the sciences—in order of development. I should like to lay down, therefore, that I mean the term "French science" in the sense of scientific culture.

The Revolutionary torment in France was the birth agony of our democratic world. It drew science, along with everything of public moment, into its own great double movement. On the one hand, from the fall of the Bastille in 1789 to the fall of Robespierre in 1794, the tide moved irresistibly leftward to culminate in the Jacobin Dictatorship of the First Republic. This was messianic democracy, the most passionate attempt in history to realize moral ideals of virtue, of justice, equality and dignity, in political institutions, and to root them in nature. Once affairs were engrossed by Jacobinism, science was bound to incur enmity. This is not because scientists were reactionary or unpatriotic. On the contrary, they pressed into service on a scale unequalled until the twentieth century. But in its intrinsic combination of assurance and irrelevance, science stood across the cosmic ideals of

Reprinted from Behavioral Science, *vol. 4, no. 1 (January, 1959), pp. 67–101.*

the Republic in a posture nonetheless insulting for being unpremeditated. Without ceasing to be science and becoming moral philosophy, it could not give the Republic the nature it needed, while the Republic to be true to its inspiration in the Enlightenment could ask no less. Thereafter—though on some matters simultaneously—the Revolutionary movement in its second phase was, not so much back to the right, as forward—or inward—toward the integration of authority, the equalization of employments, the rationalization of institutions, and their summons to public service and (especially in the case of science) to education.

Not only was the French scientific community the most brilliant in the world, it was also the most highly institutionalized—a circumstance which permits us to be definite about its history. On the 8th of August 1793, on the rising curve of the Terror, the Jacobin Convention abolished the learned academies as incompatible with a Republic. Counter-revolutionary writers always describe this measure as a simple act of intellectual vandalism. Moreover, there is reason to think that the Academy of Sciences, despite its healthy condition compared to the humanistic academies, was the primary target.

Three distinct themes recur in the campaign of slander. First, these writings express resentment for the new chemistry directed expressly against the person and the influence of Lavoisier, who was pursued by his detractors, not in the humble guise of chemist, nor simply as a financier, but as the arrogant spokesman and evil genius of science. (In France much scholarly labor has been expended to show that Lavoisier was executed as a tax farmer. And I always think it a reassuring—though somewhat puzzling—implication that the wanton execution of a financier should be taken to present a less serious moral problem than the destruction of a scientist. But the distinction is unreal. Lavoisier won odium as he won reputation both for his science and his public work.)

Secondly, a gentler sentiment appears, which might seem inharmonious amid the muttering hostility of the *sans-culottes*. Enthusiasm for natural history was unanimous. The paradox, however, was only apparent. For the Revolution, which suppressed organized physics, provides institutional testimony to the deep instinct of romanticism to seek shelter in the humane metaphors of biology, which proposed organism rather than mechanism as the model of order. The Convention transformed the old *Jardin du Roi* into the modern *Muséum d'histoire naturelle,* and established twelve chairs of biological science. This was a truly munificent provision. It made possible the great age of comparative anatomy and the magnificent tradition of experimental biology in the 19th century.

Finally, running through the attack on the Academy was a political assertion of the sort which rings of injured interests. Science was undemocratic in principle, not a liberating force of enlightenment, but a stubborn bastion of aristocracy, a tyranny of intellectual—and especially mathematical—pretension stifling civic virtue and true productivity, drawing a veil of obscurity between nature and the people.

Humanitarianism

IN OBJECTIFYING, science alienates, and these were passionate outbreaks of the tension which must always exist between man's science and his desire to participate morally and through consciousness in the cosmic process. They were responses loosed by social cataclysm, but conditioned by the terms in which Newtonian science had been presented to the Enlightenment. For the position was anomalous. Science basked and prospered, as everyone knows, in the atmosphere of its own prestige. But what is not appreciated is how it was compromised by the deep commitment of the Enlightenment to the moralization of Nature.

Particularly was this so at the heart of our civilization, in France. For on the one hand, the Cartesian imperative to order and unity, confronted by all Newton's loose ends, was what carried the rational genius of France to the leadership of science. That critical instinct for mathematical clarity and theoretical elegance saved French science, humanitarian though it was, from submersion in the Baconianism which vulgarized the English tradition, and it was certainly this which was responsible for the poverty of British achievement in the abstract reaches after Newton. But if French science had its head in the celestial mechanics of Laplace, its footing was in the popular utilitarianism of the *Encyclopedia,* and this proved treacherous ground. For when that great undertaking is examined critically for its philosophy of nature, it appears less as an instance of Newtonianism, than as an attempt to draw its teeth and humanize it. In its theory of matter, in its enthusiasm for natural history, and in its half-Baconian, half-democratic sentimentality about humility and true knowledge in technology, that philosophy of nature presaged the pattern of events which engulfed the scientific community in the Revolution.

In the *Encyclopedia,* for example, chemistry is still subjective, an operational mode of communion with nature. It proposes to replenish the Newtonian destitution of matter by addressing itself to the sympathies in nature, the active principles at work in the world, not to the mass of bodies but to their organization. Chemistry is intimacy with nature, the poor man's manual metaphysics, whereby that artisan, in whose skills true wisdom lies, manipulates reality, not in the humiliating abstractions of mathematics, but with his own hands. "Chemistry *unites* the two tongues—the popular and the scientific." This was the chemistry taught in public courses, where pharmacists learnt their art. In such eyes not only does Lavoisier insult and betray these hopes, but worse, the new nomenclature comes as a deliberate injury, making a mystery of the craftsman's livelihood, reducing him to dependence on the scientist.

This archaic chemistry of a world alive was congruent with Diderot's implicit conception of the entire *Encyclopedia*, as an immense substitution of organismic concepts and technology for Newtonianism at the center of science. For Diderot—and after him the whole romantic tradition—rejected the claim of mathematics to be the language of science. Geometry falsifies by depriving bodies of the qualities in which alone they exist for a sympathetic science. The mathematical spirit is worse than inhumane. It is arrogant, "prideful," the science of the infinite. But Diderot was no Pascal to agonize over infinity. He dismisses it, with perfect nonchalance, as uninteresting. Since knowledge must have bounds, let them be at the limits of the useful. And Diderot restores the mind to the coziness of a finite cosmos by wrapping science tight around humanity. In place of the image of matter in motion through a Euclidean void, Diderot—the advance guard for Goethe and the romantics—substitutes the model of the universe as a cosmic polyp: time, its life unfolding; space, its habitation; gradience, its structure, which embodies the twin ideas of universal sensibility and evolution. Thus will biology supersede physics in a world we can fit. And Diderot uses a second metaphor: the swarm of bees. For the solidarity of the universe is social. On a cosmic scale, it is that community which the social insects know, and society becomes continuous with nature, the home of virtue, in an unbroken order—and reading this we see where Marx and all the social naturalists come from.

In such a situation communication is direct, experiential. It does not lie through mathematics. It lies, instead, through craftsmanship, and so—as in Bacon—truth opens to the common touch and right method dispenses the ordinary man from the need for genius. To the dignification of craftsmanship by science corresponds, therefore, a reciprocal democratization of science, one which entrained an inevitable cheapening of expectations. For this is the final consequence of that 18th-century humanitarianism which would retrieve from the cold abstractions of classical physics a science warming to man. It assimilates the whole of science to its applications, makes it only the rationalization of technology, and seeks in practice to obey Diderot's injunction to keep man at the center—not only man but everyman. It proposes that dream of a citizen's science which the Revolution turned into actual measures.

All these ideas would never have broken out of the realm where such hopes and resentments always circle—except for the Revolution, when they provided the pattern in which statesmen responded to a political campaign against science founded in real interests. It is obvious that the men who then came to power, their ears attuned to nature by Diderot and Rousseau, did not understand the conditions of scientific culture. The murmurs of 1789 were a chorus by 1793: Intellect is the enemy of liberty; erudition is unsuited to a Republic; Robespierre rejects Condorcet's proposal to base education on science, as tending toward an intellectual aristocracy, and prefers a Spartan education in civic virtue. And it is obvious that these misunderstandings arose from a peculiarly damaging moral enthusiasm for nature. For any glimpse of science itself could come only with the shock of betrayal of humane values, which—so the scientizing moralists had taught—derive through science from Nature herself.

In this context, the Revolution gave the artisans of Paris their opportunity for a decisive offensive to reverse the aphorism by which science governs the arts. A revolt of technology was one among the many rebellions swelling into Revolution. Specifically, the trouble went back to the Academy's responsibility for refereeing the monopolies and subsidies by which the State encouraged invention. This invidious role earned the Academy deep Gallic hatred among the artisans whose work it judged. A quotation from their journal will give the temper —that of Carlyle's revolutionary Paris. After dismissing theoretical science as sophistical, it writes:

> The Arts are more sure and their beneficence more certain. How blameworthy, then, are these abusive and tyrannical organizations which violate the holiest of liberties, the liberty of thought, of inventive GENIUS; and which subject those truly privileged by NATURE —the artisans—to these obstructive regulations—to these harsh, impatient censors, whose ignorance and inquisitorial jealousy seeks first of all to humiliate and waylay real TALENT.
>
> How cruel and vexatious, are these exaggerated pretentions of academic bodies. How revolting, how destructive of industry, is this tyrannical control which WEALTH vests in these usurious vampires, these despotic drones, always ready to devour the honey gathered by the bees, and who take advantage of their position to engross even the hives.
>
> All of which they accomplish by reducing artisans to degrading and ruinous occupations, fatiguing and repelling their zeal and courage by all manner of disgusts and frustrations—
>
> In order to appropriate their inventions and ideas for themselves —to protect the pride of these privileged scientists, or to protect their *interests* in existing enterprises.

These were not people whom it was safe to have offended, and the fall of the Academies set in train liquidation of the entire structure of French science. Rebellion was followed by purge. Direction of the metric project, for example, was assumed by a provisional committee composed of former academicians. Prieur de la Cote-d'Or, a member of the governing Committee of Public Safety, sat for the regime and wielded its fearful authority. Like Carnot he was a military engineer who had indulged vague scientific ambitions with no success. There he sat, as a colleague of Laplace, Lavoisier, and Lagrange, and he may not have felt at ease. On 23 December, they—together with Coulomb and Delambre—were removed by a degree of the Committee of Public Safety. The ground was that the state must assign missions only to men "worthy of confidence by their Republican virtues and hatred of Kings." And from the acts of the regime, it is indeed possible to construct the Jacobin philosophy of science. It was clearly a forerunner of the Marxians', as was Jacobinism of Marxism. Theirs would be a science which, as to its technological aspect, would be a docile servant, and as to its conceptual, a simple extension of consciousness to nature, the seat of virtue, attainable by any instructed citizen through good will and moral insight. "Let anyone be a savant who wishes"—in Cassini's summary of the Revolution (he had been ousted as Director of the Observatory by his research assistants)—"O Happy Liberty."

Throughout the period the spokesmen for science conducted its defense with neither skill nor dignity. One searches in vain for a single appreciation of science as a simple intellectual good. Instead, the Academy gave the case away and addressed vulgar Baconian platitudes about science and the trades to statesmen who were being told by the artisans themselves that science was throttling creative industry, and who had been taught by the *Encyclopedia* that it is the practical man who knows, not the theorist. Lavoisier and the others spared no efforts to draw off the menaces of the popular societies of technology: they even joined and sat endlessly on platforms listening to reports on silkworms and waterwheels, participating in proceedings which were a travesty of the Academy's own tradition. As the Academy's days ran out, and these bodies moved in for the kill, Lavoisier desperately wrote speeches to put in the mouths of sympathetic deputies—and then ran about the streets trying to find these few champions to send them to the Convention for some crucial vote. At the very moment of the Academy's dissolution, Lavoisier warned that if France abused the devotion of her scientists, she would lose their services. Nothing of the sort happened, of course. Reflecting on the behaviour of the scientific community under the Revolutionary and Napoleonic regimes, one is reminded of the Schumpeter thesis about the bourgeoisie needing a master. As if to refute Lavoisier—indeed, even as he moved ever closer to the guillotine—scientists were mobilized. They served brilliantly and performed prodigies in the famous effort of war production celebrated in every textbook. France did not lose their services. Only science did.

And yet, it is difficult to see how this mattered to science, deeply though it mattered to scientists. For a feature of the impersonality of science is that it does emerge from all the vicissitudes it inspires. Even the great tragedy of Lavoisier was personal, after all. It is fair to say that he had already made his contribution to science. Before ever he fell victim to the Revolution, he had already begun to fall victim to that subtler fatality which turns the creative scientist into a statesman of science, into a public figure.

Rationalism

MOREOVER, the liquidation of the science of the old regime is only one aspect of its story in the Revolution. The other, the rationalist tradition of scientific interpretation, though temporarily overwhelmed, had too its origin in the Enlightenment and came into its own after the fall of Robespierre. This was the tradition, after Locke, of the associationist—or the positive—psychology, according to which man is what he makes of his experience. It would base society, not on virtue, but on talent. It was the tradition, too, in which scientists (as distinct from *philosophes*) themselves participated—D'Alembert, Lavoisier, Condorcet. And after Thermidor the authorities hastened to make amends. Upon the *tabula rasa* left by the Jacobins they erected a new set of scientific institutions: *écoles centrales*, the *École normale*, the *École polytechnique*, the medical faculties of Paris, Strasbourg, and Montpellier, the *Conservatoire des Arts et Métiers*. Only the *Muséum d'histoire naturelle* emerged flourishing from the Terror. Other schools were revived: the *École des mines*, the *École des ponts et chaussées*, and the *Collège de France*. Finally, at the summit was created the *Institut de France*. Thus, France was endowed at one stroke with her scientific institutions, and the first generation who taught and studied in them assured the restoration of her scientific leadership and its enlargement through the early 19th century.

It was a remarkable effort, animated by a consistent philosophy, which was nothing less than to unify the sciences through a common conception of the nature of scientific explanation, and in so doing to link them, both institutionally and philosophically, to realization of the idea of progress. So for a time science was conceived as a function of its educational mission. The *École normale* assembled the first scientific faculty, to be taken over by *Polytechnique*, as distinguished a faculty man for man as has ever existed. For the first time, students were offered systematic technical instruction, directed toward engineering, to be sure, but on a high theoretical plane, and under the foremost men: Lagrange, Laplace, Monge, Prony, Berthollet, Guyton, Chaptal. The students, able and eager, chosen by competition, were immensely exhilarated by the sense of being conducted at once to the very forefront of scientific conquest, and at being told that the future of the Republic, which is to say mankind, depended on how they acquitted themselves in so exposed a situation.

But *Polytechnique* had an equal influence on its teachers. If one were to read only the research memoirs of the Institute in its first ten years, up to 1805, and compare them to those of the old regime, one would conclude that French science had gone down with its Academy. Quite erroneously—for the explanation is that scientists were communicating, not primarily with their colleagues, but with their students. *Polytechnique* made scientists into professors—again for the first time. It brought Laplace back to Paris from the refuge he had taken in 1793. It brought Lagrange back to mathematics from the preoccupation which had enveloped him since completing *Mécanique analytique* ten years before. Monge drew his descriptive geometry together only for his course. So, too, did Laplace come to write the *Système du Monde* and the *Essai sur les probabilités*. Cuvier's *Leçons d'anatomie comparée* were given at the *Collège de France*. Lamarck first presented the idea of evolution as the framework for his lectures at the *Muséum*. In short, the systematic treatise displaced the research memoir for a time.

It was the necessity to reorganize science for teaching which produced this general movement of rationalization. In the philosophy of science, this movement takes its place between the Enlightenment and positivism, introducing a displacement towards action, a great enrichment of detail, and a certain access of rigor. It involved all the sciences in a common preoccupation with method, with classification, with nomenclature. The author of a treatise, at once investigator and professor, would address himself to his entire science, which he would expound according to whatever principles resolved it into a rational body of knowledge. He would present them, not just as an authority, but argumentatively, as an advocate. His claim to originality lay, less in this or that discovery, than in having discerned the principles. So Cuvier founds the method of comparative anatomy in the principle of subordination of parts. So Berthollet does not just state a concept of mass action—he makes it (unfortunately, as it happens) the principle which will rationalize chemistry according to affinities in the circumstances of reaction. I cannot, of course, undertake a rehearsal of all the scientific literature. But it is astonishing that nearly all the titles are variations on a single theme—for example, Prony's "*Philosophical* Mechanics—or *Rational Analysis* of the Different *Parts* of the Science of Equilibrium and Movement." "Philosophical" and "Analytical" are the key words, for underlying them all was an implicit conception of scientific explanation as a kind of cosmic education.

In this light science itself is positive knowledge, of course. It is its *function* in the world which is educational (or philosophical—the words are 18th-century synonyms) and its mode of procedure which is analytical. First science seeks to discern the elements of a complex subject. These once found, it ranges and classifies them according to the logic of nature underneath the welter of phenomena. Next, it establishes a systematic nomenclature designed to fix the things in the name, cement the memory to nature, and fasten the idea to its object. So the human understanding will be led toward a rational command over every department of nature by following its inherent order. Scientific explanation, then, consists in resolving a subject into its elements in the objective world, in order to reassemble its images in the mind according to the principles of the associationist psychology. The inspiration was algebra. But the model was botany.

In fact, of course, this confidence in the universality of analytic method rested on a semantic illusion. What is an algebraic process in Lagrange is simple taxonomy among the naturalists. But the difficulty was concealed by a very worthy commitment to the unity of science: to that kind of tolerance and mutual respect which rescues communication from specialization by means of the comparative approach. So Monge gave credit to Linnaeus for his idea of grouping surfaces into natural families. So Bichat brings anatomy to the instruction of physiology by founding histology in the classification of tissues. So Berthollet looks to statics for the idea of chemical masses in equilibrium. It was Lagrange who had suggested to Lavoisier that the future of chemistry lay in turning it into a material algebra, and this truly was Lavoisier's inspiration.

The force and range of the work were remarkable, then. I do not think it would be an exaggeration to call this French essay in rationalization the last thrust of the Enlightenment, by which the Enlightenment returned whence it had originated, and repaid the debt it had incurred to scientific culture a century before. It imparted a unity to scientific effort which it was not to know again. Not only so, but it may be that this is the last phase in the history of science which belongs to humanism. For as a matter of principle, no one of these savants limited his vision by his science. "It is certainly," wrote Cabanis (who founded moral philosophy in physiology), "a magnificent and beautiful conception, to consider all the arts and all the sciences as forming a community, an indivisible whole, limbs from the same great trunk, united by a common origin, and still more by the fruit they are destined to bear: the progress and happiness of mankind."

6 *American Indifference to Basic Science during the Nineteenth Century*

RICHARD HARRISON SHRYOCK

THE GENERAL INDIFFERENCE to basic research displayed in the United States during the greater part of the nineteenth century is a significant phenomenon in the history of modern science. In contrast, the American record in applied science—in technology, inventions, surgical practice—was a distinguished one. This technological achievement was important. It exercised a world influence, and its merit should not be underestimated by assumptions about the superiority of "pure science" per se. It is an even more serious mistake, however, to assume the superiority of applied science; and this is what Americans tended to do—almost unconsciously—during the past century. This error was recognized by American scientists themselves when, from time to time, they urged that more heed should be given to fundamental investigations.[1]

If the lag in basic research in the United States could be ascribed simply to a lack of facilities in a new country, any further analysis of the phenomenon would be of little interest. No one expects a relatively poor and isolated people to contribute much to the arts and sciences. This was the situation of various colonial populations during recent times; but the Americans had risen above this level even before the end of the eighteenth century. As early as 1760, for example, Philadelphia had become one of the chief centers of the Enlightenment in the English-speaking world.[2]

Reprinted from Archives Internationales d'Histoire des Sciences, *no.* 28 (1948–1949), *pp.* 3–18.

1. This appeal was still being made at the end of the century. See, e.g., William Pepper, *Higher Medical Education*, Lippincott, Philadelphia, 1894, pp. 1 ff.

2. Carl and Jessica Bridenbaugh, *Rebels and Gentlemen: Philadelphia in the Age of Franklin*, Reynal and Hitchcock, New York, 1942, pp. 361 ff.

A century later, by 1860, the United States had attained a population of over thirty millions and wealth was accumulating from the exploitation of vast natural resources. Earlier European enthusiasm for the American experiment in "democracy" was now tempered by realistic criticisms of the crudity or naïveté of American society. But it is misleading to think of the United States, at midcentury, as a "new," primitive, or unsophisticated country in any ordinary sense of those terms. It possessed the basic conditions and facilities usually considered essential to cultural activity. Large cities, wealth, learned institutions, intellectual freedom, and an optimistic faith in "progress"—all these were available. Close cultural contacts had long been maintained with both Britain and France, while, at the same time, nationalistic feeling demanded national originality.[3] In certain fields, notably in literature, the results were all that might have been anticipated. The generation which produced Emerson, Whitman, and Poe was capable of intellectual activity of a high order.

Yet, in science, the story was a far different one. None of the American cities had become research centers in the sense that were London, Paris and Berlin, and few individuals were devoted primarily to research careers. The exceptions, such as Joseph Henry in physics and Joseph Leidy in biology, only proved the rule. Here, then, was a country well equipped to pursue basic investigations; but which actually neglected them, and excelled only in their applications. An analysis of this paradox may throw some light upon the influence which society exerts upon science.

It has been suggested that Americans found it so easy to borrow science from abroad that they felt no need to provide it for themselves. Tocqueville, in 1835, advanced this as one of the factors explaining the failure of Americans to cultivate what he termed "theoretical" science. "I am convinced," he wrote, "that if the Americans had been alone in the world . . . they would not have been slow to discover that progress cannot long be made in the application of the sciences without cultivating the theory of them. . . ." This conjecture, in the nature of the case, cannot be proved or disproved; but Tocqueville himself went on to note the danger that in a too "close adherence to mere applications, principles would be lost sight of, and when the principles were wholly forgotten, the methods derived from them would be ill pursued." He even cited isolated Chinese civilization, "absorbed in productive industry," as an actual instance of this outcome.[4] One

3. The demand for national originality was stimulated by the Revolution, and later by British criticisms, and led to some exaggerated claims concerning national achievements. See Merle Curti, *The Growth of American Thought,* Harper and Brothers, New York, 1943, pp. 245 ff.

4. Alexis de Tocqueville, *Democracy in America,* translated by Henry Reeve, D. Appleton, New York, II, 1904, p. 518.

might have feared a similar result, if America had been isolated. In any case, the mere availability of European theoretical science cannot, in itself, explain American indifference to its cultivation. Over and over again, in Europe, the availability of more advanced, basic science in one country served as a stimulus—rather than as a deterrent—to its cultivation in another. There must have been other factors in American civilization which explained the difference in the American reaction.

A second thesis, commonly held, is that Americans were necessarily preoccupied with "subduing Nature" or "conquering a continent."[5] They were so busy laying the material bases of civilization that they had little time or interest left for cultural matters. Unfortunately, "subduing Nature" is a vague phrase. If it implies that most men were actually engaged in clearing a wilderness or building homes, the thesis is absurd except for the times and places of original settlement. A variation on this theme is the assertion that since Americans were the descendants of immigrants and pioneers, they all inherited the attitudes of immigrants and pioneers—with a resulting neglect of cultural activities. To put it more definitely: it may be that the primitive conditions of early settlement left some imprint upon later states of mind. Having once turned, necessarily, to the mother lands for science and art, the people of the United States may have formed a habit of dependence which persisted into years of wealth and maturity. It is difficult to generalize about this, since some Americans always showed an exaggerated awe for things European, while others displayed an equally exaggerated and even boastful independence. The inherited awe did lead at times to slavish imitation in the arts and sciences, but this did not prevent real achievement in certain aspects of culture. It is difficult to see why science should have been the peculiar victim of a persisting colonial state of mind.

If the phrase "conquering a continent" implies only that most Americans were busy exploiting natural resources, one wonders why they *were* so preoccupied. The mere existence of these resources did not necessarily lead to such activities among European settlers, as is evident from the very different history of minority groups of French peasants in Quebec and of German peasants in Pennsylvania. Various qualities which the dominating British peoples brought with them—puritanism, capitalistic acquisitiveness, relatively free attitudes and institutions—all interacted with the environment to produce social democracy and the desire of every man to "get ahead" in a material way.

5. See, e.g., Henry B. Parkes, *The American Experience*, Alfred A. Knopf, New York, 1947, pp. 9 ff.

6. Tocqueville, *op. cit.*, pp. 524 ff.

7. See Perry Miller, *The New England Mind*, Macmillan, New York, 1939.

From this situation sprang the apparent materialism, the vulgarization, and the "commercial mindedness" which European travelers noted and which many a native critic deplored.

Tocqueville, probably the ablest foreign observer, believed that the combination of democracy and of economic opportunities inevitably led to a neglect of theoretical science and to a cultivation of technology. The masses, favored by equalitarian opportunities, sought in science only the immediate means to exploiting natural wealth in the interest of their own comfort. On the other hand, there was no class in a democratic society which possessed either the tradition or the leisure to cultivate the more contemplative, theoretical aspects of science. "In aristocratic ages," declared the French observer, "science is more particularly called upon to furnish gratification to the mind; in democracies to the body." He believed, nevertheless, that basic science would emerge in the United States and in other democracies, if the "constituted authorities" would deliberately encourage those individuals who were inclined to it. Technology, meanwhile, would take care of itself.[6]

Tocqueville's analysis of the American situation remains one of the most valid which has been provided. Subsequent experience suggests, however, that it is incomplete in some respects and calls for refinement in others. His view that intellectual interests in general, as well as science in particular, were inhibited by American conditions, cannot be entirely reconciled with the record. Theology had flourished in colonial New England, reaching a high point in Jonathan Edwards.[7] Even basic science had flourished in the Philadelphia of 1750, reaching its high point in Franklin.[8] And soon after Tocqueville wrote, there ensued the literary "flowering" of the Emerson-Poe-Whitman era. Tocqueville offers no explanation of the peculiar neglect of basic science in the nineteenth century. Some refinement of certain of his categories, such as those of "classes," "the multitude," and "utility," may assist in explaining this neglect in further discussion below.

A third thesis relates inertia in the natural sciences to the clerical control of colleges and universities. Such control continued longer in American institutions than it did in outstanding Continental universities. And it was only after about 1875, when the direct influence of the Protestant churches upon universities was declining, that there was a marked increase in the latter's scientific activities. But this sequence does not prove that the churches had discouraged research up to that

8. See I. Bernard Cohen, "How Practical Was Benjamin Franklin's Science?," *Pennsylvania Magazine of History and Biography* (Philadelphia), vol. 69, Oct. 1945, p. 284.

time. Science had flourished in a Protestant environment during the sixteenth and seventeenth centuries in Europe,[9] and this tradition was maintained to some degree in American colleges during the ensuing period.[10] After about 1780, Protestant clergy in the United States did display some fear of science because of the spread of deism and unitarianism. The subsequent religious "revivals" produced an orthodox atmosphere in the colleges which was hostile to "free thinkers." French medicine of 1840 was occasionally condemned as materialistic, and the opposition to "Darwinism" after 1860 is well known.

But there is little evidence that the churches directly opposed scientific work in the better schools of this period. Professional opportunities were doubtless best for scientists who were sympathetic with, or at least not hostile to, the prevailing forms of Christianity; but there were many such men who could have pursued basic research without fear of ecclesiastical opposition. The view that science revealed the glory of God persisted, and resulted in some encouragement of science teaching in even the most orthodox circles. Moreover, independent institutes and academies in the larger cities functioned in an entirely secular environment. Even if some clergymen distrusted science, it must be remembered that there was no state church in the United States and that there was therefore no common ecclesiastical policy. In the presence of various denominations and their rival institutions, there was room for liberalism in certain schools if this was denied in others.

One concludes that religious enthusiasms inhibited science only in a rather negative and indirect manner. The quasi-religious influence of German idealism enabled some science professors to hide their lack of originality behind a sonorous *Naturphilosophie*. But not until about 1890 did an extreme form of philosophical idealism, "Christian Science," become a popular cult which openly opposed scientific interests. Meanwhile, the churches indirectly inhibited the support of science by focusing public interest on religious issues and by attracting the more able men into clerical careers. At Yale College in 1850, for example, it was found that the most promising graduates who entered the professions chose theology or law in preference to medicine.[11] Philanthropy, as well as personnel, flowed into ecclesiastical channels. As late as 1891, it was openly stated that American theological schools possessed a total endowment of $18,000,000; whereas medical schools could boast of only $500,000.[12] The endowment of departments or in-

9. Jean Pelseneer, "L'Origine Protestante de la Science Moderne," *Lychnos* (Uppsala), 1946–1947, pp. 246–248.

10. Theodore Hornberger, *Scientific Thought in the American Colleges, 1638–1800*, University of Texas Press, Austin, 1945, pp. 80 ff.

11. *Transactions*, American Medical Association, vol. 4, 1851, p. 409.

stitutes of the physical sciences at this time was also negligible. Such contrasts, however, were the result rather than the cause of the greater prestige of religious interests.

One step removed from religious influence was the operation of a moral attitude which, after 1800, was most marked among the Anglo-Saxon peoples. This was the opposition to the dissection of the human body. How far this expressed traditional moral feeling, and to what extent it reflected a growing Victorian prudishness, is difficult to say. The inhibiting influence of this attitude was limited to the medical field; but therein it was of real significance, because it prevented autopsies at the very time when studies in pathologic anatomy were essential to the whole advance of basic medical science. Popular feeling also discouraged the training of physicians for clinical research, because medical students were denied access to hospital wards. This denial apparently resulted from equalitarian feeling for the rights of the poor, who must not be embarrassed as the mere subjects of student observations. Somewhat akin to the opposition to dissections was the anti-vivisection movement of the later nineteenth century, which handicapped the biological sciences and was also primarily an Anglo-Saxon phenomenon. This dangerous type of sentimentality is still much alive in the United States today.[13]

Quasi-moral or prudish attitudes indirectly handicapped certain other types of research in the United States, such as that on venereal disease.[14] But since influences of this sort related only to special aspects of medical science, they cannot account for the neglect of basic science in general.

A clue to the explanation of this larger phenomenon, however, is finally provided in the contrast noted at the start; that is, in the difference between American attitudes toward pure and applied science. This implies that utility was highly appreciated, but that there was little interest in abstract studies offering no promise of immediate usefulness. In this contrast one has a formula which clarifies the entire situation. It explains why national patent law was so planned as to encourage originality among inventors, and why successful men of this type —from Robert Fulton to Thomas A. Edison—were so highly esteemed. It also explains the aid granted to geologic surveys and to agricultural studies, by federal and state governments which would do almost noth-

12. G. M. Gould, "The Duty of the Community to Medical Science," *Bulletin,* The American Academy of Medicine, vol. I, Aug. 1893, p. 331.

13. Richard H. Shryock, *American Medical Research: Past and Present,* Commonwealth Fund, New York, 1947, pp. 20, 61, 68.

14. Shryock, "Freedom and Interference in Medicine," *Annals,* The American Academy of Political and Social Science (Philadelphia), vol. 200, 1938, p. 44.

ing to assist research in astronomy, mathematical physics, or physiology. A clear test of governmental attitudes was afforded by the persistent refusal of Congress to support national scientific organizations, such as the National Institution in Washington (1840–1847); and it was only through the gift of an Englishman that the Smithsonian Institution was finally established there.[15]

Private philanthropy was no more inclined to aid basic science than was government. When hospitals were subsidized, this was to provide for the sick poor; when universities were endowed, this was for the education of youth. Such research as was done in these institutions was incidental to their main functions. Professors in the arts colleges—including scientists—were mildly esteemed as teachers, but viewed otherwise as rather useless and "impractical" persons. Neither the salary nor the prestige accorded "M. le professeur" or the "Herr Doctor Professor" in Europe were granted to his American colleagues. Willard Gibbs, the greatest mathematical physicist in the United States, taught for years at Yale without receiving any salary whatever. Since "practical" success was apt to be measured in terms of money, this lack of adequate income did not help the professor's social standing. It is no wonder that scientists and other academic persons played no such roles in government or public affairs, as did their European colleagues.

Gibbs, for personal reasons, was content in this situation. He had his own income and cared little about public recognition. But not many scientists could subsidize their own research; and even those who could do this, were discouraged by public attitudes from devoting themselves to "useless" studies. Dr. Samuel Jackson, of the University of Pennsylvania, pointed out in 1840 that even a wealthy physician must cultivate a fashionable practice, and thereby neglect research, if he wished to secure either public or professional recognition.[16] And the best known physical scientists were those who, like Benjamin Silliman of Yale, gave a large part of their time to teaching, geologic surveys, and popular lecturing.[17]

The blighting effect of utilitarianism was evident in the diversion of men who showed promise in basic research into applied activities. Silliman may have illustrated this, though one cannot be certain in individual cases. European professors were puzzled by American students who, after beginning well abroad, lapsed into mediocrity upon returning

15. Madge E. Picard, "Government and Science in the United States: Historical Backgrounds," *Journal of the History of Medicine* (New York), vol. 1, April 1946, p. 254.

16. *Address to the Medical Graduates of the University of Pennsylvania,* T. K. Collins, Philadelphia, 1840, p. 15.

17. John F. Fulton and E. H. Thomson, *Benjamin Silliman,* Henry Schuman, New York, 1947, *passim.*

18. Edgar Goldschmid, "Contributions des Etats-Unis à l'Anatomie pathologique au début du xixe siècle," *Archives internationales d'Histoire des Sciences,* 1948, no. 3, pp. 479 ff.

home. And one recalls cases in which Americans, inspired by European science, actually began to make basic contributions; but never went on to a fulfillment of the potentialities so revealed. Thus Samuel Gross did original work in pathologic anatomy and published in 1839 the first real text on that subject in any language; but he became so involved in teaching and surgical practice that the time he could give to autopsies became quite inadequate.[18] Hence his ultimate contribution to the field was less than that of his great European contemporaries.

The same withering influence was to be observed in the failure to recognize such basic work as Americans did accomplish, despite all handicaps. Such exceptional men as Gibbs, or the physiologist William Beaumont, were more appreciated in Europe than in their own country. Joseph Henry, who pursued basic work in electro-magnetics which paralleled that of Faraday, was not as well known in America as was Samuel Morse; although the latter simply applied Henry's discoveries to the making of a commercially successful telegraph.[19]

Was the extreme emphasis upon utility an old or new note in American science? Among 18th century scientists, especially in the American Philosophical Society, the conscious direction of research toward human welfare had been demanded in the traditional Baconian manner. William Smith, provost of the University of Pennsylvania, declared in 1790 that: "The man who will . . . point out a new and profitable article of agriculture and commerce, will deserve more from his fellow citizens and from heaven, than all the Latin and Greek scholars, or all the teachers of technical learning, that ever existed, in any age or country."[20] And the motivation of many of the colonial planters and merchants, who pursued science as an avocation, was plainly a utilitarian one. Against this background, American attitudes during the 19th century seemed to represent a persistence of those of the earlier period.

A closer analysis, however, reveals that in the later age subtle changes occurred in the nature of science and also in the composition of society. In the first place, and despite formal statements to the contrary, some of the best work of the older leaders had been of a basic sort, or had been—in any case—motivated by pure intellectual curiosity. This was true, for example, of the ablest Philadelphians—of Bartram, Rittenhouse, Benjamin Rush and Franklin.[21] Since science had then been relatively non-technical and could be cultivated by self-trained men, there was

19. Bernard Jaffe, *Men of Science in America,* Simon and Schuster, New York, 1944, pp. 193 ff.

20. Whitfield J. Bell, "The Scientific Environment of Philadelphia, 1775–1790," *Proceedings,* American Philosophical Society (Philadelphia), vol. 92, no. 1, 1948, pp. 10, 11.

21. Rush attempted to solve what he believed were the basic problems in pathology, though his approach now seems speculative; see Shryock, "Benjamin Rush from the Perspective of the Twentieth Century," *Transactions,* College of Physicians of Philadelphia, 4 series, vol. 14, no. 3, Dec. 1946, pp. 114 ff. On Franklin's basic research, see Cohen, *op. cit.*

nothing to prevent gifted persons from pursuing it in their spare time. If their interests were abstract or "impractical," that was their own concern as amateurs and who was to say them nay? Either basic or applied studies could be made, under these circumstances, as their authors desired.

Between 1800 and 1850, however, science rapidly became relatively complex and technical. This trend required more formal training, more specializing, and more professionalization.[22] And when men made their living as scientists, they could not pursue their own curiosity with the same freedom as could amateurs. If the surrounding society would support pure research—as in the German universities—well and good. But if society disdained this, as in the United States, it was apt to be neglected. Thus changes in the nature of science itself accounted in part for the American situation under discussion.

In the second place, American society underwent a transformation, between 1780 and 1830, from a relatively aristocratic to a relatively democratic one. Prior to the latter year, there had been some "patrician" support of science as well as of art—a persistence of the European type of patronage.[23] Although the merchants and planters of 1750 or 1800 possessed a middle-class background, they aspired to gentility and represented in a sense the American aristocracy. Even those individuals in this class who pursued science themselves were really patrons, since they supported the work from income derived from other sources.

This situation changed when, after about 1830, equalitarianism ("Jacksonian democracy") permeated American society. There was a "decline of aristocracy" in science as well as in politics. Wealthy gentlemen who devoted themselves to pure research became rare, and individuals who did so—for example, the psychologist Dr. James Rush—were apt to be viewed as eccentric.

Tocqueville recognized the connection between equalitarianism and the neglect of basic science during the 1830's. By that time, American apologists were proclaiming that applied science was the peculiar glory of a democracy, since only in such a society would science be put to work for the benefit of the masses.[24] But it is doubtful if the emphasis upon utility was to be ascribed directly to "the multitude" or to the "working classes" to which Tocqueville referred. There is every evidence that business men became a more dominant class between 1830 and 1900 than they had been before that time. Equalitarianism influenced them, not by weakening their power, but by making this more economic

22. J. D. Bernal, *The Social Function of Science*, Macmillan, New York, 1939, pp. 28, 29, notes professionalization, and adds that it was accompanied by the withdrawal of scientists from "the direction of State and of industry." Perhaps this was because government and business were also becoming more complex.

23. Curti, *op. cit.*, pp. 213 ff.

than social in nature. Technology made possible new and greater fortunes, at the same time that equalitarianism undermined certain social ideals which had tempered the older quest for wealth.

In other words, such categories as "middle class" or even "business men" are too inclusive for the present purpose. The textile manufacturers of 1850 or the railroad magnates of 1875 apparently sought wealth as a means to power, and accumulated more than could be used for personal comfort or even for social display. Their neglect of the genteel tradition and its cultural implications probably resulted from a number of circumstances. They were *nouveau riches,* in a day when equalitarianism had robbed gentility of much of its older prestige. Moreover, they were business men at a time when business—like science—was becoming more complex and technical. The very scale and tempo of industrial expansion, especially after 1850, was such as to absorb all a man's time and interest. We do not yet have adequate studies of the thought of this class—of their folk lore, ideals, rationalizations, and so on. But they seem to have been more ruthless than were the older merchants or planters in demanding efficiency and profits.[25]

Far from challenging this sense of values, the masses made it their own. Could not even the common man "get ahead" in the American environment by emulating the millionaires? "Thus the ideals of our business leaders became the ideals of the great majority of the people, though only a few were themselves endowed with the talent for leadership."[26] Under these circumstances, "radical" movements like Marxian socialism made small appeal; hence the masses received no such indoctrination in the values of science as might have come from that source. They either accepted unconsciously the business man's evaluations, or remained indifferent to the whole matter.

Industrial leaders, on the other hand, were in a position to support basic science if they had so desired. Without waiting for mass enlightenment or possible government aid, they could have provided philanthropic endowments or direct subsidies for corporation research. But of all intellectual or cultural activities, theoretical science was the least likely to appeal to them. The most ruthless magnate might enjoy literature of some sort and appreciate the ornamental quality of the fine arts, but why should he encourage the "idle curiosity" of research men? Such curiosity, to him, was neither interesting nor ornamental. There was nothing in his own experience to suggest that out of basic science would come ultimately—applications and profits. And al-

24. *Ibid.*, pp. 333 ff.
25. Thomas E. Cochran and William Miller, *The Age of Enterprise: A Social History of Industrial America,* Macmillan, New York, 1942, pp. 67 ff.
26. *Ibid.*, p. 153.

though this probability already had been demonstrated in the past, the self-made industrialist was not usually familiar with so "impractical" a subject as the history of science.

Even if they had known the ultimate value of theoretical science for technology, it is doubtful if "big business men" would have supported the former as long as the connection seemed remote. Living in a highly competitive, "get-rich-quick" environment, they were interested in immediate rather than in ultimate results. There was an analogy here with the common tendency of most Americans, farmers as well as manufacturers, to exploit the resources of the moment without regard to the long-run consequences.

It is true that merchants and manufacturers were sometimes defended against those who condemned their materialism, and that this defense included the claim that their activities encouraged science. The view here advanced by conservative Americans was, strangely enough, similar to that emphasized by many later Marxists; that is, that the greater part of science had in the past grown out of response to economic needs.[27] But even in such analysis, it was clear that Americans conceived of science primarily in terms of its applications, and were hardly conscious of theoretical science as such.[28]

The proof of the rôle of business men in the American neglect of basic research, prior to about 1900, is to be observed in their support of this same process after that time. The rather sudden emergence of basic science in the United States can be largely ascribed to the support of business leaders—support which was provided only after science had reached a point where its implications for technology became more apparent. There was, between 1900 and 1940, a gradual increase in popular appreciation of the relations between basic and applied research, and a consequent growth of government aid for both procedures. But this was preceded by more immediate aid extended to basic science by private wealth. Indeed, both popular interest and government subsidies may be interpreted as a response to the earlier research achievements of the universities, foundations, and corporations which were financed so largely in this recent era by business leaders.[29]

In conclusion, we may revert for a moment—for comparative purposes—to one other phenomenon for which Tocqueville's analysis failed to account. This was the successful cultivation of American literature during the nineteenth century, in contrast to the relative neglect of basic science. The French observer was correct in believing that equalitarianism would influence literature, but was misled in the view that the quality

27. *Cf.* Bernal, *op. cit.*, pp. 20 ff.; and J. Pelseneer, "Les Influences dans l'Histoire des Sciences," *Archives Internationales d'Histoire des Sciences*, no. 3 (1948), p. 352 (Nouvelle Série d'*Archeion*, tome XXVII).

28. See, e.g., "The American Merchant," *The Knickerbocker or New York Monthly Magazine*, vol. 14 (1839), pp. 10 ff., 118.

of this art would decline under the leveling impact of democracy. There were, of course, critics who lamented the influence of commercialism and of vulgar taste upon the arts; and there was certainly much that was banal in the popular literary output. But what happened in the long run was that superior literature—as it has been judged abroad, as well as in the United States—was diverted and modified, rather than inhibited by mass tastes and the dominance of business ideals.

As soon as literary forms of high quality could be adapted to the tastes of a relatively large public, there was an actual advantage to authors in the widespread literacy of a democracy. Earlier than in any other country, the United States produced a literature written by the middle classes and aimed at the masses. Outstanding authors, from Bryant to Mark Twain, consciously cultivated forms which were—to use Poe's phrase—"more appreciable by the mass of mankind."[30] There was no analogy to this situation in basic science, since the latter was not susceptible to democratic adaptations. "Popular science" was not, and cannot be, basic science; whereas literature may be simultaneously popular and of high quality.

In summing up this analysis, one observes that the very middle class which did much for science during the early modern centuries, neglected in the United States to aid fundamental studies throughout most of the 19th century. It is not surprising, in view of all the circumstances noted, that "pure" science (motivated by pure curiosity) made little headway. But it is rather striking that so little basic research grew out of extensive applied activities. It seems clear, therefore, that economic interests and technology will not of themselves lead automatically into basic investigations. Other variables influence the outcome.

The industry of the 19th century (including American) needed basic research as well as technology. Indeed, this need was probably greater than that of any preceding economic order. Yet the industrialists were more responsible than any other class, at least in the United States, for all the indifference which has been described. This apparent paradox cannot be fully explained by reference to the laissez faire principles upheld in Britain and America during this period. These principles were opposed to government planning and even to government aid for science, but would not in themselves have discouraged private support.[31] In the United States, moreover, the laissez faire philosophy was hardly established until after mid-century, and it will be noted that it was never sufficiently strong to prevent the considerable government assistance to applied science which has been mentioned.

29. Shryock, *American Medical Research,* pp. 53 ff., 90 ff.

30. Clarence Gohdes, *American Literature in Nineteenth-Century England,* Columbia University Press, New York, 1944, pp. 11, 12.

31. *Cf.* Bernal, *op. cit.,* pp. 27, 28.

A more complete explanation of American neglect may be found in the analysis of simultaneous changes in the nature of science and in the composition of society, each of which interacted on the other in a complex manner. A more technical and specialized science, on the one hand, and a more equalitarian and industrialized society on the other, each played a part in producing the end results.

It seems clear that industrial society failed—to a large extent in Britain, but even more completely in the United States—to support basic science from any motive whatsoever. Where more or less aristocratic institutions survived, however, there was maintained, or evolved, a tradition of "pure" science which resulted in much basic research. (Whether, in certain countries, more such research also grew out of technology than was the case in the United States, is a question which cannot be considered here.)

Aristocratic support was never very strong in the United States and largely disappeared there after 1830. It was more persistent in England, but flourished most in the German and other Continental universities. The latter system, with its support of "pure" science, doubtless suffered from a lack of planning and coordination, and perhaps also involved some studies which led nowhere in terms of ultimate social values. But it had the merit of assuring complete freedom to individual scientists; and this permitted the greatest among them to open up hitherto unknown and therefore unpredictable areas which subsequently proved of the greatest utility.

In a word, aristocratic support enabled basic science to advance to a technical point at which its value for industry became apparent even to the industrialists. At present, in the United States, leading industrialists and the scientists associated with them are fully aware of the significance of fundamental studies. More than this, some of these leaders are convinced that those studies which are motivated by pure curiosity often prove of the greatest utility—even if measured in crass financial terms.[32] But the chief emphasis upon a "pure" science has developed since about 1890 in the research institutes and in the universities. It is probably true that in both government and industrial science, the major American emphasis continues to relate to applied studies, or to such types of basic investigation as seems to promise most for utility in the near future.

32. For an illustration of this, see the remarks of Irving Langmuir in the *Dedication* volume, Lilly Research Laboratories, Eli Lilly and Co., Indianapolis, 1934, pp. 7 ff.

7 *Soviet Scientists and the Great Break*

DAVID JORAVSKY

MIDWAY THROUGH THE FURY of his first Five Year Plan Stalin singled out 1929 as "the year of the great break [*perelom*]," the year of shattering transformation on all fronts of socialist construction. He had in mind the beginning of "the decisive offensive of socialism on the capitalist elements of town and country," and of course he did not mean to suggest that the offensive would be completed in 1929. The shattering and transforming, he made clear, had only begun. Academic historians, who like to speak of the period as a watershed or turning point, ought to concede that Stalin's more violent image is more appropriate for the crisis of forced industrialization and collectivization, though they are probably right in shunnnig his effort to dramatize the great break by marking it with the number of a single year. On "the scientific front of the cultural revolution" the great break took about two and a half years, from the end of 1929 to the middle of 1932—which is short enough, considering the magnitude of the changes accomplished, to require no further dramatization. In this brief period "the scientific change-over [*smena*]" from "bourgeois" to "red" specialists, and the accompanying search for a suitable philosophy or ideology of science, reached a crisis, a breaking point, by which past trends were selected, some for destruction, others for increasing dominance over a generation of Soviet scientists and philosophers of science. In the present essay the break in "the scientific change-over" is examined; the cor-

Reprinted from The Russian Intelligentsia, *edited by Richard Pipes (New York: Columbia University Press, 1961), pp. 122–140.*
This essay is a version of a chapter from the author's book, Soviet Marxism and Natural Science, 1917–1932, *published by Columbia University Press in the United States, and by Routledge and Kegan Paul in England in 1961. The chapter as it will appear in that book is extensively documented.*

responding break in the Soviet Marxist philosophy of science is left for treatment elsewhere.

From its start the Bolshevik regime had been convinced that Russia's scientific and technical personnel were inadequate for "the construction of socialism." They were far too few, and most of them were basically hostile or at least skeptical toward Bolshevism. But their work was essential for the mere existence of the Soviet state, while its improvement was dependent on them in a painfully paradoxical manner: the great army of "red" specialists that was supposed to emerge from the peasant and proletarian supporters of the regime could receive its training only from these "bourgeois" specialists. The result of this paradox, clearly perceived by the Bolsheviks, was a cautious mixture of policies in the search for the scientific changeover, and a correspondingly slow rate of progress, throughout the period of the New Economic Policy (NEP). Forceful measures were not used against the old specialists, who were assured by word and deed that as long as they worked conscientiously at their individual trades and abstained from political action, they were free to have whatever ideology they liked. Kalinin, indeed, in a speech to a convention of doctors at the end of 1925 (*Izvestiia*, No. 287), implied that they could have whatever political ideology they liked, even anti-Communist, as long as they did not act on it. He and other high Soviet offcials appealed to the older specialists on a level where Bolshevik ideals coincided with those of Russian scientists: heavy financial support for scientific research and training, the dissemination of scientific knowledge to the widest possible audience, and the practical application of scientific discoveries to the modernization and strengthening of the native land. Invidious comparisons with the record of the pre-Revolutionary regime were made on all three scores, and some older scientists were brought round to praise the Bolsheviks. For example, the biochemist A. N. Bakh, who had been a member of the People's Will when Lenin was an eight-year-old, and, more significantly, a Socialist Revolutionary at the time of Lenin's Revolution in 1917, became the president of VARNITSO, the Association of Scientific and Technical Workers for Support to Socialist Construction; and the physicist O. D. Khvol'son, who had been past fifty in 1909, when Lenin attacked him as a fideist, lent his considerable reputation to support of the Union of Scientific Workers, which, like all Soviet trade unions, taught its members to identify the improvement of their lot with the enhancement of the Soviet regime. In order to broaden its appeal, the Union abstained almost completely from the dissemination of Marxism as a philosophy of natural science, but even so, at the end of 1927, the last year of NEP, it could claim barely six percent of all scientific workers as its members. Moreover, the Communist Academy's societies of materialist scientists, which *had* campaigned for the acceptance of Marxism as a general philosophy, had a combined membership by the most generous estimate of only a few hundred. Obviously, peaceful persuasion was a slow method of transforming mature scientists into dedicated supporters of the regime. Yet precisely such scientists, the government was convinced, were essential to the success of the Plan.

The progress of the new "red" specialists was hardly more encouraging, as the Plan began in 1928. The law required that the sons and daughters of manual workers and peasants be favored in admissions to higher schools, and there was an impressive network of *rabfaki* or "workmen's faculties" to give them the academic prerequisites for higher education. But this system had been in effective operation only six years, not enough time for even the first contingent of predominantly proletarian students to have received graduate degrees. They were reaching the universities as undergraduates, but in 1928 the children of non-proletarian elements were still the majority of first-year graduate students. To the Bolshevik authorities this seemed the main reason that the majority of graduate students in the natural sciences either were uninterested in Bolshevik ideology or shared their "bourgeois" professors' skepticism. Beginning in 1927 all graduate students were required to pass an examination in Marxism, but when an examiner asked a future mathematician for an appraisal of dialectical materialism, his question was answered with another: "Why should I bother with such nonsense?"[1] Nor was this "contradiction between the ever-growing role of the scientific worker in socialist construction and his ideological and socio-political backwardness" the only cause of deepening Bolshevik anxiety. The elementary problem of numbers threatened to get out of hand; the rate of production of new technicians and scientists seemed to be falling hopelessly behind the staggering increase in the Plan's demand for them.

"Military measures" designed to achieve "maximum results in the shortest time" seemed the only way out, as Andrei Vyshinsky, then an important official in higher education, wrote in 1928. The Party's Central Committee ordered a detachment of one thousand Communists to be enrolled in higher schools in the fall of 1928 with scant regard for academic prerequisites. A spirited public drive was set in motion to "renew" the Academy of Sciences, that is, to swell its ranks with many new pro-Soviet members nominated by institutions outside the hitherto autonomous Academy. In the universities professors of ten or more years' standing were ordered to undergo "re-election": in public meetings their fitness both as specialists and as "social men" (*obshchestvenniki*) was to be examined by colleagues, students, and representatives

1. See M. N. Pokrovskii, "O podgotovke nauchnoi smeny," *Kommunisticheskaia revoliutsiia,* 1929, No. 13, pp. 62–64.

of the Party. Of course, membership in the Union of Scientific Workers or in VARNITSO counted heavily in a candidate's favor, especially since the Union, early in 1929, formally established the Marxist-Leninist *Weltanschauung* as a requirement for membership. In short, efforts to push forward new "red" specialists, and pressure on mature specialists to give up "neutralism" or "the so-called simply legal relationship to the Soviet regime" were considerably intensified in 1928 and the first half of 1929.[2]

Still, as the Soviet regime approached its supreme test, the drive for the "solid" collectivization of agriculture, it showed growing dissatisfaction with the progress of the scientific change-over. Professors were reportedly sneering at the dispatch of "the thousand" as an effort "to prove the theorem that any illiterate can become a university student." And it appeared that such professors usually had little to fear from the re-elections. In most institutions there had as yet been none; in others, all candidates, regardless of their "social physiognomy," were being "re-elected" by a formal ritual of meetings and eulogies; and when there were genuine "re-elections," the Communists on the spot (students for the most part) tended toward one of two extremes, equally denounced as deviations in the central press. Either they attacked the professorial candidates as if they were *lishentsy* (a Soviet neologism for such people as priests and former gendarmes, who were deprived of civil rights), or, more often, the Communists succumbed to the mysteries of the academic guild and agreed to use professional competence as the sole basis for judging the candidates.

The renovation of the Academy of Sciences also moved forward haltingly. In January, 1929, A. M. Deborin and V. M. Friche, the chief Soviet Marxists in the fields of philosophy and art criticism, though nominated to membership by many Communist institutions, were voted down by the Academy, while the mathematician N. N. Luzin, an intuitionist in the theory of mathematics, was elected qua philosopher. To be sure, the Academy quickly reconsidered this affront to the Bolsheviks, and elected the two Marxists at an extraordinary session in February. At the public celebration that followed, the Secretary of the Academy, the 67-year-old orientalist S. F. Ol'denburg, reassured the Bolsheviks: "We feel still more our close connection with public opinion; we feel that there is no 'we' and 'you,' but only 'we.'" Apparently not all shared this feeling. A meeting of the Academy's graduate students (*praktikanty*) seems to have resolved that compulsory training in Marxism should *not* be made a part of their program. And in the fall of 1929 a special investigating commission descended on the Academy

2. See the articles by K. V. Ostrovitianov, I. K. Luppol, and A. Ia Vyshinskii, in *Nauchnyi rabotnik*, 1929, No. 7–8; *Antireligioznik*, 1929, No. 6; and *Nauchnyi rabotnik*, 1930, No. 1.

and dismissed at least one hundred and twenty-eight people, some of whom appear to have been prosecuted subsequently in secret before administrative tribunals.[3] For Soviet scientists the great break had begun in earnest.

Without access to the archives one cannot know much about secret arrests and punishments, but the public record does reveal some things. In the first place it shows pretty clearly that mass terror, which had not been used against scholars, made its appearance after the purge of the Academy of Sciences in November, 1929. From the spring of 1928, to be sure, there had been intensely publicized trials of specialists accused of wrecking, that is, of activity "bringing economic and political harm to the Soviet state with the purpose of sapping its power and preparing for an anti-Soviet intervention." Such loud admonitory shouts at the "bourgeois" specialist did not cease at the end of 1929, but they were joined by less definite warnings, by obscure, unexplained acts of terror. For example, the only published allegations of malfeasance in the Academy of Sciences concerned some historical documents found in the Academy library; they were supposedly "of political importance," and the Librarian, the historian S. F. Platonov, was "relieved of all administrative posts" for failing to turn them over to the proper agency and for allowing unauthorized people to see them. Platonov's further fate was not publicized, nor was any light shed on the misdeeds or the further punishment of those dismissed with him. A similar obscurity, an equally ominous omission of specifics, characterized most of the references in 1930 and 1931 to "methods of terrorizing the accomplices of counter-revolution," methods that were being used, if we can believe the journal of the Communist Academy (1931, No. 1, p. 83), against "that whole upper-echelon bourgeois intelligentsia, which, though not caught *flagrante delicto*, fosters wrecking activity by its sympathy or by its neutrality."

There was no longer any question whether antipathy to Communism was permissible. The great outcry was against "apoliticism" and "neutralism." VARNITSO took scientific workers out on the streets of Moscow to demonstrate against these sins, and a 38-year-old Bolshevik mathematician warned against the most refined kind of "apoliticism": verbal endorsements of Marxism-Leninism unaccompanied by deeds to prove sincerity. "Never," he exclaimed, "has the class struggle in science been carried on with such bitterness as just now. Never has the demand for *our* science, a science that really serves socialist construction, been as great as today. Whoever now is not with us, whoever is still neutral, is against us."

3. See the reports in *Izvestiia*, 13 January, 10 September and 16 November 1929. See also the reports in *Nauchnyi rabotnik*, 1929, Nos. 1, 2, 3, 4, and 1930, No. 1.

Nor were lesser Bolsheviks the only ones to erase the former distinction between political loyalty and ideological solidarity. Lunacharsky, the former Commissar of Education who, early in 1928, had given scientists one of the last assurances of their right to reject Marxism, in 1930 spoke to them with a new toughness, saying nothing of rights but only of obligations. Perhaps Karl Radek's exhortation, "On One Side of the Barricades or the Other," which appeared in the journal VARNITSO in August, 1930, was the most revealing. "*The broad mass of specialists*," he wrote, underscoring the extensiveness of such feelings, "stunned by shootings and arrests, dash off in various directions, and frightened by the hostile atmosphere that events have created about them, do not know where to submit, but try in the meantime to hide their heads under their wings, in expectation of better times."

Radek warned such "Philistines" that "the mistrustful attitude of the working class cannot be assuaged by correct declarations of loyalty or by silence." Still, declarations of loyalty abounded, culminating at the end of 1930 in a birthday greeting from the Unions of Scientific and Educational Workers to the Star Chamber itself: the Unions thanked the OGPU, on the occasion of its thirteenth anniversary, for purging their ranks of those not worthy of the honorable title of Soviet scientific worker.

One also gets a vivid sense of the hostile atmosphere surrounding the mature specialist in 1930–1931 from the plays and novels of the time. Gorky's irritation with the Russian intelligentsia was given new expression in a play about a counter-Revolutionary engineer, whose villainy is inspired by arrogant pride in his special knowledge and by contempt for the working man's ignorance. "The workers have seized state power," he explains to his wife, "but they can't manage. . . . In general, the dictatorship of workers, and socialism, are fantasies, illusions, which we the intelligentsia involuntarily support by our work. . . . Machinists, housepainters, weavers, they're not capable of state power; it must be taken over by scientists [*uchenye*], by engineers."

When his troubled wife asks whether he is not two-faced, he exclaims: "Am I two-faced? Yes! Any other way is impossible! . . . The role of the defeated, the prisoner's role, is not my role!" This was one of Gorky's exceptional individuals (gone wrong, to be sure), but a similar sense of outraged pride was presumed to be hidden behind the mask of complaisance worn by the ordinary "Philistine" (*obyvatel'skii*) specialist, the type who had no stomach for plots against the Soviet state but might, if the proletarian dictatorship relaxed the threat over him. Anyone educated before 1917 was a "bourgeois" specialist unless he proved himself otherwise. He belonged to that "long-winded, weak-kneed intelligentsia [as Gorky described the type in *Izvestiia* (12–13 December 1929)] . . . which met the October Revolution with passive sabotage or with active, armed resistance, and which in part continues to struggle 'in word and deed' against the Soviet regime even to the present day, wrecking consciously and unconsciously."

Was the suggestion in Gorky's final phrase—that a specialist might *unknowingly* commit the capital crime of wrecking—simply an extravagant flight of rhetoric? Probably not, for the central problem of a very popular play written in 1930 by Afinogenov was precisely a scientist's unwitting wrecking. The play, significantly entitled *Fear,* dealt with an eminent professor of human physiology (was Pavlov the author's original inspiration?) who was not one of the conscious enemies of the Soviet state, but played into their hands by working out a theory that fear was the essential motive in the behavior of Soviet people. His theory was "exposed" in a public meeting by an old Bolshevik woman, whose rich political experience compensated for her lack of formal education. When she concluded her indictment of the professor with a cry to the audience on stage to be vigilant and merciless toward the class enemy, the real audience in the theater responded with loud applause (*Pravda,* 7 January 1932). The response lends verisimilitude to the dramatist's picture of "public opinion" (*obshchestvennost'*) working with the OGPU to make the professor a Soviet patriot or else break him. Certainly, the *vydvizhentsy* in the audience (a Soviet coinage for proletarians "pushed up" into scientific or other responsible work) must have felt a tightness in their throats at the symbolism of the play's ending. The professor, converted, promises to give a public criticism of his wrecking theory and to hand over the keys of all the offices to the *vydvizhentsy* in his institute.

One naturally wonders whether the great break of 1929–1931 was actually the triumphal completion of the scientific change-over, as this play suggests. Were Russia's mature scientists actually forced to choose between full-throated Bolshevism and self-destruction by "wrecking"? Did the actual management of scientific institutes and university departments pass in this brief period to the "pushed up" new generation of "red" specialists? Certainly there is nonfictional evidence that suggests an affirmative answer to both questions.

"The thousand" of the academic year 1928–1929 (that is, the detachment of Communists pushed into universities and institutes with little regard for academic prerequisites) were joined by two thousand more in 1929–1930, and by still more in the next two academic years. Even without such special detachments, the staggering over-all expansion of higher education between 1928 and 1932 (the student body trebled and the teaching staff doubled) suggests that the older specialists were being "dissolved in a sea of new forces," as a report of the State Planning Commission put it in 1930. "A young man who studies our science," said the Bolshevik mathematician who helped write the report, "has every chance of becoming a professor at twenty-five."

At the same time, drastic measures were taken to reduce the sense of need for highly trained professors. In November 1929 the Party's Central Committee ordered "continuous productice practice" for students in higher technical education, with the result that abstract, theoretical subjects, the stronghold of the old specialists, were pushed into the background. Some institutes even abandoned courses in theoretical physics and chemistry altogether, brushing aside as reactionary—or worse—the professors who protested that technicians rather than engineers would be the result. New methods of teaching and grading in "brigades" were designed to get round the need for individual *expertise* in students as well as teachers. Even in research it seemed that the masses might break down the *tsekhovshchina* or guild-like seclusion of the old specialists. When T. D. Lysenko, a virtually unknown 31-year-old seedman from the Caucasus, failed to impress a scientific convention in 1929 with a report of his experiments on plant physiology, he got some dirt farmers to try them out, and the Commissariat of Agriculture was so impressed that in 1931 it ordered collective farms to experiment with Lysenko's allegedly new methods "on a mass scale. . . . Only in this way," the Commissar of Agriculture declared, "will the enterprise be set up in a really scientific way, in a really revolutionary way." Such things seem to confirm the playwright's vision of the great break as a time when the old specialists surrendered the keys of their institutes to the *vydvizhentsy*, to those pushed up from bench and plow.

Moreover, one can find real analogies to the fictional indictment of a scientist for unwitting wrecking, in apparent confirmation of the notion that the older scientists were actually forced to choose between full-throated Bolshevism and self-destruction by "wrecking." The calm, objective tone of a physician's pamphlet on the control of venereal disease, for example, aroused intense anger in the *Kazan Medical Journal* (1931, No. 4–5). The pamphlet did not sufficiently extol Soviet accomplishments in this field or sufficiently berate bourgeois failures.

> In our time [the reviewer lectured] the time of the socialist offensive, when all hostile class forces are resisting desperately, the pen is obliged to shoot just as accurately and truly as the revolver. Paper, the printed word, speech—all are weapons that must guard our life, our construction, our philosophy from all sides. . . . We say loud and clear: it is not only useless to write such "scientific" works [as the pamphlet under review], but also harmful and criminal.

It may be that the "criminal" physician survived this rhetorical fusillade, but more than verbal guns seem to have been used against some statisticians in the State Planning Commission. Apparently they did not satisfy the Party's insistence that the goals of the Plan must be the scientific prediction of mathematicians no less than the passionate desire of "shock-brigaders" (*udarniki*). To be sure, such statisticians like the physician writing on public health or the fictional professor in *Fear,* were on the dangerous frontier between the natural and the social sciences, but one finds an attempted crusade against wrecking in the mathematical theory of statistics, too. B. S. Iastremskii, a 53-year-old insurance specialist who had vainly criticized the established authorities in statistical theory before the Revolution, won a following at the time of the great break. He helped "unmask the wreckers" in the Planning Commission, and then teamed up with two young Bolshevik mathematicians to produce an allegedly revolutionary *Theory of Mathematical Statistics* in 1930. If other mathematicians were obliged to support this book in a milieu where the distinction between pistols and pens was lost, then it might seem reasonable to suppose that "the proletariat on the front of the cultural revolution," to quote a young Bolshevik mathematician in 1930, was indeed "storming heaven itself," forcing the mature scientist "to place not only himself as citizen [*obshchestvennik*] but also his science in the service of socialist construction, to reconstruct it."

Yet there is evidence that requires major amendments to these simple conclusions. Mathematicians apparently were *not* forced to accept the new theory of statistics; at the end of 1930 a supporter of the theory complained that O. Iu. Shmidt, the most important Bolshevik "on the scientific front" and a mathematician himself, was indifferent to the new theory, and within a few years it appears to have died altogether. Lysenko's program for boosting yields did not turn into an attack on geneticists until 1936. The pistol-waving review cited above was not at all typical of the journals on natural science; throughout the great break they continued to print specialized articles, and showed the influence of the times only by occasional editorial declarations of Soviet loyalty and by considerable transformations of their editorial staffs. Looking through such journals, one begins to understand why Radek reported that "the broad mass of specialists" had responded to the clamorous demands for positive proofs of patriotism in a "Philistine" (*obyvatel'skii*) way, with little more than correct declarations of loyalty, or silence. In fact this conclusion is suggested by the Bolsheviks' exasperated repetition of the warning that such a flaccid response was not enough, that "Philistine" specialists could not escape history, or, to use Radek's industrial paraphrase of Lincoln, that they would be "cast aside by the flywheel of history." Perhaps the "Philistine" specialist had a firmer grip on the dizzily spinning Russian flywheel—than Radek himself? To ask this in sneering malice is to take sides, and not with the liberal opponents of regimentation, but with the type whose only cause was self-preservation, who would not sacrifice himself in any cause.

Direct evidence of the extent and geological force of this "swamp" (*boloto*)—if the metaphor may be shifted to the standard Bolshevik pejorative for the passive, adaptive, self-centered type—is provided by the scientific conventions of 1930. They were the highpoints of a drive to capture the scientific societies, whose virtually complete autonomy had aroused only headshakes and grumblings before 1930. Now, the Bolsheviks proclaimed, this autonomy was to be destroyed. But for all such loud talk the conventions of 1930 were much the same as those of previous years: a great mass of special papers were read, most of them trivial or repetitive, as in scholarly conventions the world around. The great break at each convention was a keynote speech, with a corresponding resolution, on the role of science in the construction of socialism, and, when the convention broke up into its many sections, perhaps two or three papers on the dialectical materialist reconstruction of science.

The pettiness both of the victories claimed and of the rebuffs lamented is the most striking characteristic of the Bolshevik reports of these conventions. The Congress of Physiologists, for example, was pictured as making history in its handling of the Pavlov *affaire*. He stayed away from it, as he consistently did from all but foreign conventions, to demonstrate his disapproval of the Soviet regime. Previous congresses of physiologists had elected him honorary chairman *in absentia*, but the Congress of 1930 passed him by, and elected to its "honorary presidium"—the entire Political Bureau of the Communist Party. As against this victory, a defeat: the famous Professor A. F. Samoilov, whose report on electrical methods in physiological research was the most memorable event at the convention, dismissed dialectical materialism "with genial irony." In Baku, to take another example, a Congress of Pathologists adopted the proper resolutions and elected to its presidium the 36-year-old Dr. S. G. Levit, one of the leaders of the drive for a dialectical-materialist reconstruction of science. But then a foreign professor who told the convention that science should be free both of religion and of Marxism was duly applauded at the end of his speech. Perhaps, the Bolshevik reporter noted hopefully, the audience did not understand him, for he spoke in German.[4] In Odessa, where the physicists held a pleasant August meeting, there seem to have been no such contretemps. The 50-year-old Academician A. F. Ioffe, a universally respected physicist and a genuinely enthusiastic supporter of the Soviet cause (though not of dialectical-materialist reconstructions of physics) gave a keynote speech appealing for planned research to aid

4. See B. M. Zavadovskii, "Itogi IV vsesoiuznogo s'ezda fiziologov," and S. Vail', "II vsesoiuznyi s'ezd patologov," *Estestvoznanie i marksizm*, 1930, No. 2–3, pp. 142–165.

industrial expansion. Clearly, Ioffe and Levit belonged to a tiny band of prophets in a heathen land, where principled and outspoken opponents were even rarer, and certainly much hardier. Reading through the reports of the conventions, one senses a vast flaccidity silently, perhaps indifferently, absorbing a few brave Bolshevik speeches.

There was one illuminating exception to this rule, the rebellion of the Congress of Mathematicians in June 1930 and the related trouble of the Moscow Mathematical Society in December. The turmoil in statistics had nothing to do with these events, and other substantive issues relating to mathematics were only slightly involved. The "Moscow school" was famous for its otherworldly absorption in pure theory, and D. F. Egorov, the 61-year-old chief of the "school," would not criticize this tradition or declare his interest in serving the Five Year Plan. As this fact suggests, the main source of trouble seems to have been a general stiff-necked nonconformity in Egorov and an equally stiff-necked liberalism in his colleagues. He scandalized the Bolsheviks by refusing to join the Union of Scientific Workers, while remaining an elder of the Orthodox Church, and the Moscow Mathematical Society not only kept him on as president but listed émigrés in its membership. Already in the "re-elections" of 1929, Bolshevik graduate students at the University of Moscow singled him out for attack, and it seems that he was removed from control of the Mathematical Institute. Still, he was a leading figure at the All-Union Congress of Mathematicians in June, 1930, and it may well be that his example was a contributing cause of the rebellion in the Congress. It refused to send greetings to the Sixteenth Congress of the Communist Party, which was then in session. A complete revolution was not attempted; the mathematicians did resolve to aid socialist construction, though cautioning that theoretical work should not be neglected in the interests of immediate practicality. But 1930, the year of savage class warfare in the villages and "shock brigades" in the towns, was hardly a time even for a limited rebellion, which was continued, moreover, by the refusal of the Moscow Mathematical Society to expel Egorov.

The climax was reached at the end of 1930, when Egorov told a meeting of the Society that "nothing else but the binding of a uniform *Weltanschauung* on scientists is genuine wrecking." It is significant that the Bolsheviks present could not agree on the appropriate reaction. The one who took the floor tried to smooth over the clash with talk of a misunderstanding, for which he was subsequently accused of "rotten liberalism" and "Maecenasism." Bolsheviks with the proper "irreconcilability" (*neprimirimost'*) took action after the meeting was over. Egorov was arrested. But his colleagues in the Society, including a member of the Communist Youth, silently defied the terror by holding a regular business meeting. (They were expected to condemn the arrested man and engage in "self-criticism" for resisting Bolshevization so long.) Thereupon five mathematicians, styling themselves an "Initiating Group for the Reorganization of the Mathematical Society," published a denunciation of the Society's belief that " 'one can be an Egorov by conviction yet work honorably with the Soviet regime.' " They could hardly have given a better characterization of the government's policy toward "bourgeois" scientists during NEP, but they lashed this belief as "Philistinism [*obyvatel'shchina*], hiding in its corner from the class struggle, and decorating this corner with scientific aestheticism instead of the canary of the rank-and-file Philistine." The other mathematicians, however, would not yield to revolutionary appeals any more than to terror. The wretched affair had reached a climax without issue. The Society was not reorganized but simply ceased functioning for more than a year, in the course of which Egorov died. The place and cause of his death are not in the public record, nor are the steps leading to the revival of the Society in 1932. One supposes that the locked opposition of the intransigent Bolshevizers and unyielding liberals gave way to some such complex adjustment of principle and reality as had already made the functioning of the other scientific societies possible.[5]

Until 1932 it was still not clear that an adjustment would be made even with the complaisant majority of the old specialists. The Bolshevizers kept up the struggle for something more than complaisance, and looked beyond the scientific societies to the places where scientific work was done. A young Bolshevik biologist told a meeting of the Communist Academy in January 1931:

> However strange it may be, in the fourteenth year of the Revolution, though we have at our disposal a colossal apparatus of scientific establishments, museums, laboratories, observatories, etc., in essence we do not possess them at all. It would seem to me . . . that the Association of Natural Science [of the Communist Academy] should set itself the organizational and ideological task of entering, of penetrating these institutes, these observatories and laboratories, through the cells of atheists that exist there, the sections of VARNITSO that exist there, all the circles of political or other character that exist there, so that we will have at each institute some cells on which we can rely in our work.

As if in response to this suggestion, the Party's Central Committee decreed on March 15, 1931, that the Communist Academy should establish its "methodological control" over the most important *vedomstvennye* (ordinary state) scientific research establishments.

5. The most detailed account of the Egorov affair is in the declaration of the Initiating Group for the Reorganization of the Mathematical Society. See *Nauchnyi rabotnik*, 1930, No. 11–12. Cf. also the report in *VARNITSO*, 1930, No. 11–12,

They were to submit their research plans for approval by the Academy and admit the representatives of the Academy to the drawing up of future plans.

The old distinction between the special network of Marxist-Leninist establishments and that of *vedomstvennye* (ordinary state) scientific establishments was thus to be erased; the Communist Academy, which had previously been the directing center only for the former, was now apparently to become the center for all. Clearly, there was a conflict here with the competence of the Academy that Peter the Great had founded, the Academy of Sciences of the USSR; and one wonders what dreams of supreme power may have come to the young Bolsheviks in the Communist Academy during 1931, as they began their work of establishing "methodological control." Their victory over the societies must have had a paradoxically sobering influence, for it had been too easy. Under the aegis of the same decree, they launched a new drive against the scientific societies. VARNITSO and the Union of Scientific Workers called a public meeting, where, we are told (in the journal VARNITSO, 1931, No. 3, p. 59) various speakers "completely exposed the protective colors in which the societies have redecorated themselves, using Marxist terminology for this and also some change of leadership." Forcibly converted, Russia's mature scientists were coming under the entirely logical suspicion of hypocrisy, and Soviet Marxists seemed about to follow the depressing example of Spanish Christians: forcefully ridding themselves of Moors and Jews, they felt a natural compulsion to known the hidden thoughts of Moriscos and Marranos.

In 1932, however, the frenzy of Stalin's first Five Year Plan was at last spent. Everywhere Lenin's famous slogan in an earlier lull was revived: "Better less, but better." Criticisms and cautions that had formerly been denounced as deviations now came from the Central Committee itself. In June A. I. Stetskii, the head of the Central Committee's Propaganda and Agitation Department, called off the ardent young Bolsheviks who had been establishing the Party's "methodological control" even in engineering and ichthyology. In *Pravda* (4 June 1932) he voiced the familiar complaint of "a purely verbal, formal, declaratory" endorsement of Marxism, but he did not conclude, as he had a year before (in the journal of the Communist Academy), that the pressure on the old specialists must be increased, the struggle intensified. He concluded that arduous, prolonged work at the specific material of the sciences was necessary, so that the non-Party specialist might be convinced of Marxism in terms of his own specialty. And foreshadowing the fate of the Communist Academy itself, Stetskii

and E. Kol'man, "Vreditel'stvo v nauke," *Bol'shevik,* 1931, No. 2. See also *Uspekhi matematicheskikh nauk, Novaia seriia,* 1946, Vol. I, No. 1 (11), p. 236.

called on Communist scientists to leave their special societies at the Academy and dissolve into the broad societies of non-Party scientists. Three months later the Central Executive Committee of the Soviet Union called for an end to "shock brigade" methods of effecting the cultural revolution in higher education. The intervention of student and Party organizations in the management of the higher schools was to stop, in part—we may suppose—because the political reliability of the rectors had been tested in the fire of the past few years. The over-emphasis on "continuous productive practice" was to end, for this practice had been producing technicians rather than engineers. Of course, the professors who had been called reactionary for predicting as much were not explicitly vindicated, very likely because the Soviet economy needed technicians anyhow, and because professors needed to learn respect for the Party's decrees. The "brigade" methods of teaching and grading were to be dropped, and the individual responsibility of students to individual teachers was firmly re-established. Professional competence was to be the only basis for filling vacancies and giving promotions in faculties, and graduate students were to be appointed only by the faculties concerned, and only from graduates of higher schools. Evidently the day of the *vydvizhentsy*, the "pushed up" ones, had passed.

In 1933, indeed, the tables seemed to be turned altogether: Communists in fields of higher learning were subjected to "re-elections" at the hands of professors. More precisely, the Party organizations in institutions of higher learning examined their members at public meetings, to which non-Party specialists were urgently invited in order to help weed out unworthy Communists. The actual results were hardly revolutionary. Many of the professors at these meetings were discreetly silent, and only a few Party members suffered expulsion (e.g., a laboratory assistant for having his son circumcised with religious rites, a graduate student for giving flippant answers to questions on Leninist theory). Most, including the leading Bolshevizers of higher learning, were merely given a forum in which to tell the story of their lives as model Communists. (I. I. Prezent, for example, who had abandoned the Deborinites and the Mendelian biologists when they lost favor, and was now becoming Lysenko's chief advocate, was the star of the meeting that expelled the religious laboratory assistant and the flippant graduate student.[6]) But, however trifling the actual purge of the new

6. See *Front nauki i tekhniki*, 1933, No. 12, pp. 115–122. For convenient evidence of Prezent's Deborinite and Mendelian past, see M. L. Levin's speech in O. Iu. Shmidt (ed.), *Zadachi marksistov v oblasti estestvoznaniia* (Moscow, 1929), pp. 88–89.

7. This has been the main stress in Soviet comments on the great break in the cultural revolution. See, e.g., E. Iaroslavskii (Yaroslavsky), *O roli intelligentsii v SSSR* (Moscow, 1939), pp. 25 *et seq.*

"red specialists" may have been, its staging gave ceremonial recognition to the continuing importance of the old non-Party specialists.

Of course, it would be quite wrong to imagine that the great break had passed without trace. The older generation of scientists, still the possessor of essential knowledge, was still in charge of many university faculties and scientific research institutes; and most were still felt to be ideologically alien, though all but a very few indomitable spirits now refrained from the expression of any but the correct ideology. What was new was a fundamental transformation in the intellectual autonomy of these old specialists. In principle, they had lost it altogether; to use a favorite expression of the day, they had "disarmed themselves" (*razoruzhilis'*) before the Party's Central Committee. In practice they still enjoyed an almost unimpaired autonomy in their subject matter, and an immeasurable autonomy in ideology—immeasurable because of the mask of silence and possible hypocrisy that covered it. How long this incongruity of principle and practice would continue depended on the Central Committee's assessment of changing necessities and possibilities. Aside from the "disarming" of the old specialists (at least in principle), the Committee had gained an enormous number of new scientists in training, most of them from social classes that it hoped would produce great specialists who would be genuinely Bolshevik in ideology.[7]

The reader may feel inclined to conclude that the Bolsheviks had become fatally embroiled with the law of enforced belief, as we may call the truism that hypocrisy and unthinking zealotry spring up where heresy and freethinking are cut down. Ultimately, it would seem, one of the two would have to die, either creative thought or thought control. But we cannot rest with this generality, for the silence of the Central Committee on the substantive issues of natural science was one of the most striking characteristics of the great break. They had razed the walls of academic autonomy, but one senses a moat of irrelevance still lying between their ideology and the natural sciences, bridgeable, perhaps, but hardly by the primitive zealots who assumed control of the Bolshevik ideology during the great break. Moreover, on the scientists' side, enforced belief might become genuine in time; the Church that Egorov refused to quit had been forced on his ancestors by an earlier state.[8] General considerations tell us little; to assess the meaning of the great break for Soviet scientists, we must examine the

8. In Russia, as in the rest of Europe, Christianity was a minority faith until it was established by royal edict. Vladimir, the Primary Chronicle tells us, ordered the chief heathen idol flogged through the streets and thrown in the Dnieper. "And then through the whole town Vladimir sent these words: 'If anyone does not come to the river tomorrow morning [for baptism], be he rich or poor, or beggar or slave, he will be my enemy.' "

further history of relations between Soviet scientists and Soviet ideologists.

ALMOST thirty years have passed since the great break. To what extent has the conversion of Soviet natural scientists become genuine? And how has this conversion—in part, enforced appearance, in part, inward reality—affected the Soviet Marxist philosophy of science and the natural sciences themselves? A reliable answer to these questions depends on research that has only been started. What follows, accordingly, is a tentative sketch, a series of hypotheses, to be modified or destroyed by further research.

The general philosophy that triumphed during the great break was a congeries of rather vague formulas that enjoyed little refinement or change as long as Stalin monopolized the right to "develop Marxism further." But scientists were expected to prove the genuineness of their conversion to this philosophy by helping the Party's philosophers apply it to the special sciences. In fact there seems to have been more conflict than co-operation between scientists and philosophers, a conflict that has fluctuated with the alternating waves of Communist confidence and mistrust in the population at large, scientists included. The mistrust that characterized the great break burst out again (probably with greater violence) during the great purges and the treason trials of 1936–1939, and again at the culmination of Stalinism in the period 1947–1953. At such times crusading ideologists have been ascendant, angrily charging scientists with subservience to bourgeois theories and with resistance to the Bolshevization of science. In the alternating periods of comparative Communist trust in scientists as in the population at large, scientists have been ascendant, reproaching the ideologists for discrediting Marxism by illiterate attacks on proven theories. And in such periods of relative trust, most of the ideologists have hastened to forget or forswear illiterate attacks, echoing the scientists' insistence on technical competence in the philosophical analysis of the natural sciences.

This pattern can be most clearly and completely established in the case of physics. Ideological attacks on relativity and quantum mechanics—and on the eminent physicists who stood up for their science—flared up during the great break, during the purges, and again in the late 'forties and early 'fifties. In the intervals such attacks nearly disappeared, and eminent physicists set the tone of philosophical discussion on their subject. Indeed, the pattern can be illustrated by tracing the double and triple tergiversations of individual ideologists (e.g., A. A. Maksimov or V. E. L'vov) as one period has given way to another. The only men who were consistently hostile to the new physics, regardless of the Party's fluctuating attitudes toward scientists, were not ideologists but a tiny group of old physicists, now deceased.

To the extent that other natural sciences have caught the ideologist's eye (a small extent for most fields during most of Soviet history), much the same pattern can be established. Genetics and its allied disciplines are the great exception. To be sure, there have been zigzags in the ideologists' evaluation of this science too, and they can be correlated with the major fluctuations in the Party's willingness to trust scientists. But this correlation is of secondary importance in genetics. Cutting through such zigzags and fluctuations, the long-run curve of Lysenko's "Michurinism" rose steadily until 1953 and has declined only slightly since.

The great break enabled Lysenko to emerge from obscurity as the Commissar of Agriculture's "revolutionizer" of agricultural science, but the end of the great break did not hurl him back into obscurity as it did many other "revolutionizers." Within the agricultural field he continued to build his reputation and his following until 1936, when a new spasm of mistrust seized the Communist Party, and Lysenko began his full-scale assault on genetics. (At the same time he appropriated the reputation of the recently deceased Michurin, who, while he lived, seems to have been cold to Lysenko's advances.) By 1940 Lysenko was on the verge of complete triumph, but in the preceding year Stalin had signaled a *détente* by expressing confidence in the loyalty of Soviet specialists, and an uneasy truce ensued between Lysenko and the geneticists. In 1948, after Zhdanov had called Bolshevik ideologists once more to battle against bourgeois culture, Lysenko rushed the geneticists with a weapon that no other Bolshevizer of natural science has ever wielded: the explicit approval of the Party's Central Committee, in the face of which nearly all public opposition to "Michurinism" collapsed. The thaw following Stalin's death revived public opposition to Lysenko, but the thaw has not caused "Michurinism" to melt away. In the present period of comparative calm for other natural sciences, conflict continues between the geneticists, who have recruited some allies from other fields of academic natural science, and the "Michurinists," who are deeply entrenched in the agricultural schools and experiment stations, and seem still to enjoy the favor of practical politicians as well as ideologists.

It is the present writer's hypothesis that the Soviet Union's chronic agricultural crisis has given Lysenko's crusade its lasting seriousness, in contrast to the intermittent vaporings of would-be Bolshevizers in other fields of natural science. Bolshevik efforts to increase agricultural productivity have been partly a struggle with technical backwardness, and partly a struggle with traditional peasant notions of proper social organization. Lysenko, by enlisting seedmen and stockbreeders in the holy cause of scientific socialist agriculture, has offered help on both levels of the chronic agricultural crisis. He has come forward as the mobilizer of agricultural science against poor varieties of plants and animals, the recruiter of peasant scientists en masse for the improvement of socialist agriculture. He has recruited them by elevating their traditional lore to the status of a comprehensive theory of life, and by exploiting their rough contempt for the subtleties of the academic scientist.

The most serious objection to this hypothesis is that the practical successes of "Michurinism" could not have won agriculturists and Party chiefs to support Lysenko, since these practical successes have been incorrectly explained, vastly overrated, or altogether illusory. Though serious, this objection is not unanswerable. As long as Lysenko's opponents have been unable to consign all his claims to the third category, he has had ground for rebuttal. Moreover, as the point was reached where nearly every seedman or stockbreeder, when offering an improved variety, did so in the name of "Michurinism" and in defiance of the geneticists, it became impossible for the *latter* to claim practical successes for themselves.

If this hypothesis concerning "Michurinism" proves true, then the turmoil in Soviet biology can be little affected by trends in the philosophy of science. It can die down as (or if) Soviet agriculture approaches Khrushchev's goal: to "overtake and surpass" American levels of productivity. For the rest, answers to the questions raised on page 126 above must be even more conjectural. It may be that the cyclical interaction of natural science and Soviet Marxism is doomed to continue. But there are other possibilities. If the present rapprochement of Soviet ideologists and natural scientists is the result of mutual trust that will prove permanent, then the atmosphere that nourishes primitive Bolshevizers will be steadily dissipated, and Soviet Marxism will begin to face its supreme test as a philosophy of natural science. Losing its past function as a crude faith for bellicose ideologists and a scourge for suspect scientists, it will have to speak ever more rationally, subtly, meaningfully about natural science, or develop into a philosophy that can.

8 *How Free Is Soviet Science?*
Technology under Totalitarianism

LEOPOLD LABEDZ

NOW THAT THE INITIAL REACTION to the Sputniks has worn off, it may be worthwhile taking a glance at Soviet scientific and technological achievement in the longer perspective.

Most of the discussion in the West has so far centered on the achievements of Soviet science and education, rather than on the prospects of space exploration. Fear of the military threat implicit in these achievements is balanced by hope that the increased prestige of science in Soviet society will make for greater rationality and perhaps even undermine the totalitarian structure of the dictatorship. Figures concerning the number of scientists and engineers turned out yearly by Soviet schools and universities have suddenly become a subject of general interest and concern, resulting in all sorts of wild comparisons. While previously Soviet technological progress was ignored, the Sputniks have by now generated fear, so that even some of the more ridiculous claims of Soviet propaganda have gained currency.

The public was evidently caught unprepared for the event; and yet there was really no reason for surprise. It had been known for years that, alongside heavy industry, Soviet policy has consistently given priority to technical education, that science gets preferential treatment as long as it does not clash with Communist ideology, and that Soviet scientists are showered with privileges, high incomes, and considerable prestige. It was also known that Soviet scientists had achieved considerable success in the realm of military technology (notably jet aircraft), that in the field of thermonuclear reactions they had already caught up with America in 1953, and that their attitude toward basic

Reprinted from Commentary, *25, no. 6 (June, 1958), pp. 472–480. Copyright American Jewish Committee.*

research differed from that of Mr. Charles Wilson. Even the earth satellite itself was clearly foreshadowed in the Soviet press, which last year gave particular prominence to the centenary of the "grandfather of the Sputnik," K. E. Tsiolkovsky. On the eve of the launching, the review *Oktyabr* (September 1957) wrote, "There is intensive preparation going on for launching artificial earth satellites."

Western attitudes throughout this period were characterized by incredulity as to the potentialities of Soviet science and technology. This was not simply the result of ignorance, nor was it primarily determined by the traditional stereotypes of "Russian backwardness." The operative factor was rather a belief that totalitarianism was "killing" science. In this respect Lysenko did more harm to Western ballistics than to Soviet biology, for he encouraged skepticism in the West as to the position of science in a totalitarian state. The Lysenko affair from the start led to certain misconceptions concerning the "death of science in Soviet Russia." Details were represented as the whole picture. Oversimplified dichotomies concerning freedom and scientific efficiency, vague assertions about the incompatibility of totalitarianism and fundamental research, provided an excuse for complacency. It is not only in the Communist countries that ideological fervor has a distorting effect on public opinion.

Whatever the political system, science is not easily extinguishable in modern industrial society. The "Aryanization" of physics cost Hitler Germany a great deal, but did not prevent her from achieving remarkable technological results. Stalin's great purge in the 30's certainly affected the Soviet scientific elite. (It also decimated the military elite, but that did not prevent later successes by the Soviet army.) But although Soviet scientists were suffering from political persecution, and research was harassed, the consequences did not affect all branches of science uniformly, and the continuity of research and teaching was only broken sporadically.

Two factors seem to have been overlooked in the general indignation about the persecution of Russian scientists and the practice of deciding scientific controversies by decree. One is that the regime was often ready to "rehabilitate" imprisoned scientists, when the need for their services was realized. Some of the most eminent names among Soviet scientists come to mind: Professor Ramzin, the chief defendant in the celebrated "wreckers" trial of the so-called Industrial party in 1930, was later released from prison, became the head of the Institute of Thermodynamics, and eventually (in the early 40's) received a Stalin Prize for his work. A. N. Tupolev, the aircraft designer, though imprisoned in the 30's, was later reinstated and contributed greatly to the development of Soviet air power; he designed the TU 104 and TU 114 jet airliners, and in 1956, as an Academician and Stalin Prize winner, accompanied Khrushchev on his visit to Britain. L. D. Landau, the child prodigy of Soviet physics, was also persecuted during the Great Purge in 1936–38, but later organized a remarkable school of theoretical physics around him. Such examples, which can be multiplied, tended to be disregarded in the justifiable indignation over the fate of Vavilov and other biologists.

The second point is more paradoxical and is connected with the unintended consequences of political actions. It has often been maintained that the purge greatly harmed the Soviet military performance in the last war. It is arguable, however, that its effects were not entirely one-sided. A good deal of dead wood may well have been incidentally cut out during that operation. It is indeed more than probable that had the veteran Marshal Budenny, for example, been included among the victims of the purge, the early Soviet operations in the Ukraine in 1941 would have been conducted more effectively. In general, the more experienced senior officers, who were replaced in the purge by younger commanders, might have found it more difficult to adjust themselves to the conditions of modern warfare. It is difficult to say whether the loss in experience was counterbalanced by the gain in adaptability, but the balance sheet is certainly incomplete without taking into account the unintended assets as well as the known liabilities. The same applies to science, where one can also detect some paradoxical consequences of the purges, although in this case they were, of course, more detrimental, since science depends to a greater degree on continuity and past experience. Even so, the commonly accepted picture is incomplete. The mechanism of the purge is more generally apprehended than its various social consequences, including those affecting science.

An example may serve as illustration. In Czarist Russia the study of both mathematics and statistics had been on a very high level, and during the 20's Soviet statisticians continued the tradition of their predecessors. The Moscow Institute of Current Research became a flourishing center of statistical investigations. At the beginning of the planning era, in 1929–30, the Institute was abolished and its head, Professor N. D. Kondratyev, and most of his colleagues were purged because they refused to juggle with statistics. Ever since, Soviet statisticians have been in trouble. Theirs was a delicate subject which, on the practical plane, involved dealing with data concerning the rate of investment and other touchy subjects. The demographic effects of forcible industrialization and political purges were another painful topic. In 1937 the population census was suppressed, the directors of the Census Bureau were executed as "enemies of the people," and many statisticians were arrested. In 1949–50 there came another wave of persecution of statisticians, who were told that they did not conform to the "class character of statistics" and were not sufficiently "party-conscious." Here was what seemed like a perfect example of a process on which a Western author at the time made the following comment:

> Since science is to some extent an interconnected web of activities and theories, the several parts stimulating or retarding one another, political control of even a few areas of science may diffuse its harmful effects to other areas. . . . Scientific fanatics and quacks—men like Lysenko in genetics—take over when political authority demands what competent scientists cannot conscientiously give it—particular substantive theories or results on order. Where scientific authority is endangered or destroyed, competent men fear to take a position on scientific theory itself, for the demands of political authority are changeful and make almost any position insubstantial. Even further, in such a situation, competent men avoid a scientific career altogether. In all societies, men seek a *relatively* "safe" occupational career. A "flight from science," perhaps only to the more applied branches of scientific technology, as was the case in Nazi Germany, may be one of the unintended consequences of the extension of political control of science in Russia.[1]

What really happened in Russia did not quite conform to this pattern. No doubt immense harm resulted from the rigidity of the system, doctrinaire dogmatism, the terror, isolation from the outside world, and the rest of the totalitarian set-up. But there was certainly no "flight from science." A scientific career was, and still is, immensely attractive to Soviet youth, and the risks involved were not greater than in other occupations. According to the Soviet Statistical Yearbook, the number of scientific workers increased from 98,000 to 224,000 between 1940 and 1955, and to 240,000 by 1957 (*Vestnik Akademii Nauk*, No. 10, 1957). Nor was the element of relative "safety" negatively correlated with "pure science," and positively with applied science. The situation was less simple. It is true that the terror produced a flight from danger zones, but the risk was unevenly distributed in different fields at different times, and the party line *on science* was often dictated by a strange mixture of ideological and pragmatic considerations. Its unintended consequences were not always one-sided. Thus, while it suppressed statistics, it did not prevent mathematicians from pursuing that abstract and innocent game, the theory of probability. To be sure, in 1949 the statisticians were condemned for supposing that their discipline could be based on that doctrine, and statistics was defined as a "social science based on Marxist-Leninist theory." But the mathematicians were left alone. The consequences can only be conjectured, but it is probable that when statistical investigations

1. Bernard Barber, *Science and the Social Order* (Free Press), 1952, p. 82.

became more risky, potential students turned to mathematics. A sort of artificial natural selection must have been operative: what was lost in statistics was gained in mathematics. The Russians already had a fine tradition in the "probabilist" branch of mathematics, and this tradition was further developed during the Soviet period. A. N. Kolmogorov became a world authority on the subject, and the school of Soviet mathematical probabilists flourished, while the statisticians were prevented from using probability techniques in industry, to the obvious detriment of Soviet technology. But when in 1955 the Soviet leaders finally realized the potentialities of "quality control" techniques, Bulganin made an earnest appeal for their development in Soviet industry. Since during the quarter century when statisticians were being persecuted, mathematics had provided a refuge, the necessary cadres existed and could be trained to do the job without undue delay.

THIS case suggests the possibility that the flight from risks may sometimes have taken a direction other than that from pure to applied science. The case of cybernetics likewise shows that the interconnection of scientific activities provides means for the partial absorption of political shocks. When cybernetics was first developed in the West, Soviet publications, particularly *Voprosy Filosofii*, ridiculed it as a utopian and anti-humanist attempt on the part of the decadent imperialists to replace men by machines. But when it dawned on the Soviet leaders that mechanical brains could provide immense advantages, and automation became a Soviet watchword, Norbert Wiener ceased to be a bogey-man and cybernetics was "rehabilitated." The neighboring fields of more abstract disciplines provided the men in this case too, and the first Soviet electronic counting machine (called BESM) was quickly built for the Soviet Academy of Sciences. *Voprosy Filosofii* now claimed that the proper development of cybernetics was only possible on the basis of Marxism-Leninism; popular pamphlets were produced to explain the significance of cybernetics and electronics under such slogans as "Let's look into the future"; and *Vestnik Akademii Nauk* claimed that a translating machine built in the Soviet Union was superior to any in the West!

All this is not to say that physicists are now free to pursue their studies without interference. Stern warnings still abound in the Soviet press that "in circumstances of acute ideological warfare, there can be no question of any kind of neutrality or non-participation in politics among Soviet scientists" (*Vestnik Akademii Nauk*, No. 10, 1957). And the organ of Agitprop gives examples of such participation; they consist of attending various seminars where the history of the party, and dialectical and historical materialism, are taught in a form especially adapted for scientists. The topics dealt with include: "The

Achievements of Modern Cosmogony and Dialectical Materialism," "The Formation of a Materialist World Outlook While Teaching Physics," "The Inculcation of a Materialist World Outlook in Lectures on Natural Science," and so on (*V Pomoshch Politicheskomu Samoobrazovaniyu*, No. 3, 1958). But the same organ, which serves the cause of ideological instruction, indicates that the attendance at such *Politgramota* (i.e. political education) circles has been slack, and that "many of the engineers and technical workers have only a poor knowledge of the works of the classics of Marxism-Leninism." Attendance at such "ideological services," which waste the time and energy of technical and scientific cadres, has no doubt fallen off somewhat since Stalin's death. It is interesting to note that *Kommunist* (No. 1, 1958) pointed out that the number of hours devoted to the "social sciences" (i.e. Marxism-Leninism) at the Moscow Institute of Physics and Technology is lower than that devoted to "social sciences" (i.e. the humanities) at the Massachusetts Institute of Technology. According to *Kommunist*, the figures are 496 and 960 hours, i.e. 23.3 per cent and 45.1 per cent of the curriculum, respectively. Even if the figures should be wrong, the emphasis is significant. (A Western comparison by Alexander G. Korol in *Soviet Education for Science and Technology* gives the opposite impression.)

It is common knowledge that since Stalin's death, Soviet scientists have been given more leeway in planning research regardless of State plans. The secretary of the Academy of Sciences, Professor A. V. Topchiev, stresses that after 1953 "a new stage in planning research" led to better results (*Vestnik Akademii Nauk*, No. 11, 1956). The allocation of research funds continues to be generous. According to *Pravda* (February 6, 1956) it was 13.6 billion roubles for that year, or 2.4 per cent of the estimated total budget. The Minister of Higher Education, V. Eliutin, has stated in *Pravda* that there are 767 university-level schools in the USSR (this includes institutions providing evening and correspondence courses) and that 260,000 specialists graduated from them in 1956–57, of whom 71,000 were engineers.[2] Among the higher educational establishments 179 are devoted to engineering and technology, while 138 are polytechnical institutes. All-round technical education is widespread, starting with the so-called "polytechnization" of teaching in the schools, and with 4,000 technical schools (*technikums*) giving technical training to graduates from secondary schools (*Vestnik Vysshey Shkoly*, No. 11, 1957). In general, Soviet scientists do not think that basic research is "something you are doing when you do not know what you are doing," to quote Mr. Charles

2. For an analysis of the problem of scientific and technological manpower, see: *Professional and Scientific Personnel in the USSR*, by Nicholas de Witt (Harvard University, 1956).

Wilson's celebrated general (motors) definition. On the contrary, they insist that scientific research should not be restricted by the drive for immediate practical results. This is not always the view of the government, but it is interesting to note the words of the president of the Soviet Academy of Science, Professor A. N. Nesmeyanov, on the subject: "Science has all its roots in practice. Its fruits belong to practice. But the linking of science with practice must not be short-sighted. Often the results of abstract science, precisely because they are all-embracing, are more practical than the results of applied science" (*Vestnik Akademii Nauk*, No. 2, 1957).

The quality of Soviet natural science can, of course, only be assessed by specialists in the various disciplines, but it seems to vary from one field to another. It is in line with routine Soviet procedure that there is an exoteric and an esoteric view of it, one designated for mass propaganda, and the other for the initiated. The same Professor Nesmeyanov, who on the occasion of the 40th anniversary of the revolution uniformly praised the achievements of the various branches of Soviet science in *Pravda* and *Komsomolskaya Pravda*, gave a different evaluation in the above-mentioned *Vestnik*, a periodical with a restricted scientific audience. There he indicated that Soviet science is strong in physics, theoretical mechanics, and mathematics; weaker in astronomy and chemistry; and weak or deficient in biology, geology, and certain technical disciplines. This is a more severe assessment than the one given by *Fortune* (No. 2, 1957). As to the "general sciences," by which the Russians mean philosophy, economics, history, literature, law, and languages, the Professor considers them below standard, and dismisses them in one sentence as "very patchy."

FOR propaganda purposes, Soviet publications are much less candid and eagerly seize upon any Western statement praising their efforts or achievements, while dismissing more critical assessments. Thus *Kommunist* noted with satisfaction the laudatory side of the estimate of Soviet science in the *Fortune* article, while dismissing all its other comments as malicious, false, and demagogic. Similarly, I. Kuzminov in *International Affairs* (No. 3, 1958) quoted with approval Dr. Libby's statement in an interview given to the *U.S. News and World Report* (August 1957) in which he "acknowledged the superiority of the Soviet Union in training scientific and technical personnel." But nothing is said about the problems now besetting Soviet higher education which shows some symptoms of over-expansion. In guarded words *Planovoe Khozyaistvo* (September 1957) intimated that a level has been reached which "permits at the present time the fulfillment of requirements for specialists . . . in all branches of the national economy and culture." As a result the number of specialists trained in the higher educational establishments shows a tendency to decline; this is most probably a deliberate policy, independent of the lower output of university candidates now graduating from secondary schools, who are fewer because they come from age-groups affected by the war deficiency in the birth rate. There also seem to be surpluses of specialists in some fields, while shortages still remain in others.[3]

Because of the over-expansion of the Soviet educational system (for uneven development is not confined to the capitalist economy), a new phenomenon has appeared in the Soviet economy: unemployment among the young. This will probably result (for other, more weighty, reasons too) in a certain shift in Soviet science, with greater emphasis being given to quality rather than quantity. Competition for admission to Soviet higher educational establishments is becoming fiercer, and entrance requirements are being raised, a phenomenon not confined to the Soviet Union. The result will be a further differentiation between various levels of technological education. A Western critic, Mr. A. G. Korol, has suggested that "when ultimately the Soviet people gain a decisive voice in the allocation of their intellectual and physical resources . . . their schools will produce better but fewer engineers." It would seem that this aim can be achieved without the Soviet people having a say in the matter, simply through the need to re-allocate manpower resources and to adjust the educational system to the march of economic development.

An important consideration in analyzing Soviet science has to do with the nature of scientific work in general. Like other human activities, science has become a field of specialization, i.e. increased division of labor, on the basis of large-scale organization bureaucratically controlled, and is planned to a far greater degree than was the case previously. The administration of research and planned discovery, with laboratory assistants carrying out work previously done by the "stars" —all these characteristics of contemporary scientific work fit the general bureaucratic character of the Soviet system. The Soviet Academy of Sciences has become a gigantic organization planning, co-ordinating, and controlling scientific activities.[4] But contrary to a widespread notion, the problem of financing research does not differ fundamentally in the Soviet Union from the situation in other, less bureaucratic societies. It is only in the fields to which the government assigns first priority that finance ceases to be a problem. The multiplication of staffs in response to real or supposed organizational needs is indeed more accentuated in the USSR than in the Western countries. An-

3. W. M. Matthews, "Youth Unemployment in the Soviet Union," unpublished manuscript, Oxford University.

4. See Alexander Vucinich, *The Soviet Academy of Sciences* (Stanford University Press, 1956). The Academy now has 105 scientific research institutes in which more

other penalty of the bureaucratic organization of science—pseudo-research—is also not a Soviet monopoly, but it is magnified by totalitarian interference with scientific authority.

This last factor is the real crux of the matter in the prevalent Western criticisms of the Soviet attitude to Science. Crucial it is, but perhaps not quite in the manner often supposed. Freedom of research may be an important value, but is not the only factor yielding results. The moral or political issue of the value of the individual is not to be confused with the factual problem of the indispensability of the individual scientist for the eventual outcome of research. Some scientists are irreplaceable, others are not. But in the long run all large-scale organizations disregard the former assumption. In terms of individual freedom, the lonely inventor, such as Lopatkin in Dudintsev's protest novel, *Not by Bread Alone*, gets all the sympathy; in terms of technological efficiency, the issue of his dialogue with the bureaucrat, Drozdov, is much less clear-cut. It may well be that there is some point in Drozdov's argument that improvements in pipe production can be more efficiently attained through the routine work of bureaucratically controlled organization than through the eccentric activities of an individual genius. In any case, since this Soviet novel shows that bureaucratic routine also includes stealing ideas, there is little room for argument so far as the efficiency of the system is concerned. That does not detract from the value of the individual or social protest embodied in Lopatkin, but the author may be thought to have confused the issue by presenting his hero's problem as ultimately dependent on the ideologically acceptable question of greater or lesser technological efficiency. The point was not lost on such Soviet critics as Surkov, who triumphantly counterposed the Sputnik to Lopatkin with his home-produced gas pipes.

All this is not to say that totalitarian interference with science is a negligible factor and that on balance regimented science is more effective than free science. As Stepan Dedijer has rightly observed, there are two specific factors which have to be taken into account in assessing Soviet scientific successes: "First, it is a very large country; and secondly, it started with a scientific tradition."[5] In this respect Sir Eric Ashby, who has made sober and timely pronouncements about the potentialities of Soviet science,[6] may well have underestimated the degree of harm done by what he himself called "the medieval technique of appeal to authority and indictment of heresy." Without the totalitarian strait-jacket, Russian science might well have developed

than 15,000 persons are employed (*Vestnik Akademii Nauk*, No. 10, 1957).

5. *Bulletin of Atomic Scientists*, September 1957.

6. Eric Ashby, *Scientist in Russia* (London, 1947); see also B. H. Liddell Hart (Ed.), *The Soviet Army* (London, 1956, pp. 452–460).

faster. Yet the relationship of Soviet science to Communist ideology is a topic which has confused many analysts. The official case is, of course, that "the Sputniks are not only a demonstration of the tremendous possibilities inherent in the Socialist means of production. They are also a reaffirmation of the correctness of the Marxist-Leninist materialist philosophy."[7] As against such assertions, a Western critic, after a careful survey of various scientific disciplines in the Soviet Union, comes to the equally orthodox conclusion that "the close link between science and dialectical materialism exercises, if anything, a detrimental effect. . . . The great practical achievements in technology, therefore, have come to pass not *because* of the alliance between Soviet physical science and dialectical materialism but *in spite* of the entanglement of science with ideology."[8]

One can no doubt go on counterposing Soviet technological achievements to Soviet interference with scientific theory, but after a while this exercise becomes fruitless. It is evident that such interference has had a detrimental effect on Soviet achievements in many theoretical fields. The wrecking of genetics; the persecution of biologists like Vavilov, Zhrebak, and Orbeli for their "Morganist-Weissmanist" deviations; the Pavlovian stranglehold on psychology, with the resulting ban on psychoanalysis; the purge of Sergeyev and other medical research workers; and the long delay in recognizing the achievements of modern theoretical physics arising from the "idealistic" work of Einstein or Heisenberg—clearly show that Leninist ideology has had a stifling influence on the development of scientific theory in the Soviet Union. Professor Wetter is right in saying that the achievements have come in spite, and not because of, ideological entanglement. Yet, while the power of logic is not the same as the logic of power, neither is perfectly consistent in the real social context. The scientists are not absolutely dependent on the logic of the dialectical *Weltanschauung* in their work, nor are the *apparatchiki* ready to endanger their power by sticking to the logic of ideology, when they realize that to do so is inadvisable on pragmatic grounds. A given scientific theory is then permitted, and by a "creative development" Leninism is reinterpreted so that it becomes possible to claim that the two are not incompatible, but on the contrary that the newly accepted theory could be only developed on the basis of "the most progressive scientific ideology."

The scientists, on their part, are far from being ideal humanists of the Leonardo variety. Only a few are philosophically inclined. Most of them, like their counterparts in the West, know more and more about less and less. Indeed, they know even less about general ideas outside their province than their Western colleagues. If left in peace,

7. Prof. L. Ilyichov, in the journal *International Affairs* (Moscow, March 1958, p. 11).

very few would busy themselves with questions of the philosophy of science, an occupation pursued everywhere by philosophers rather than by scientists. As matters stand, top-ranking scientists can go about their affairs unconcerned about the compatibility of their notions with Marxism, provided that its current interpretation does not interfere with their work. Some may accept its general ideas, not bothering about their consistency, while others may believe that dialectical materialism has no application to natural science, but that its scientific character has been demonstrated in the field of social analysis. In general, the "new Soviet intelligentsia" stands in vivid contrast to its predecessor, the pre-revolutionary Russian intelligentsia. It is, on the whole, narrow in outlook, unconcerned with general ideas, and displays all the usual traits connected with technical specialization and bureaucratic dependence. In the Soviet system, the old function of the intellectual has disappeared and his place has been taken by mental technicians. It is not the scientists or the engineers who display residual features of historical continuity, but the *literati* and the students, whose impetus, however, is spent when they are drawn into the administrative routine of daily life, with all its personal worries. Under such conditions, the meaning of ideology for a technical expert may remain ambiguous if its content does not clash too obviously with his interests. Even in its Stalinist or Khrushchevian form it may serve the function attributed by Max Weber to all religions and ideologies: to "make sense" of the world as a total experience.

In this respect "Marxist" ideology, in its Sovietized form, is peculiarly suitable for narrow specialists: it provides them with a *Weltanschauung*, a substitute for religion, in which science is worshipped, and which also satisfies the same need for a mind squeezed by specialization that "primitive magic" did for less sophisticated brains. Its value for a generally critical outlook can be exaggerated. Max Weber has argued that the "magical image of the world" prevented the emergence of science in China, while Protestantism was particularly consonant with scientific activity because of its stress on rationalism and empiricism. Others have narrowed down the beneficial influence of Protestantism on the emergence of science to one of its variants, Calvinism. Whatever the historical value of these propositions, they have little relevance to the totalitarian experience. "The magical image of the world" may have harmed, but did not preclude, scientific activity in Russia. With the industrialization of China, natural science will play a greater role there, though the Maoist outlook is hardly less "magical" than the Confucian. Yet both empiricism and rationalism (when they do not clash) perform a particular function in "scientific

8. Prof. G. Wetter, in the periodical *Soviet Survey* (London, March 1958, p. 59).

ideology": rationalism is reduced to the role of non-substantive or formal rationality; empiricism—to a peculiar type of pragmatic adjustment within dogmatic doctrinal limits. Ever since Engels, "dialectical pragmatists" have been fond of repeating that "the proof of the pudding is in the eating," but they usually behave as if the taste of the pudding is fixed once and for all. Yet the thesis that science is incompatible with totalitarianism is false as a factual proposition. If Calvinist theology can be considered to have been beneficial to science at a certain time, there is no reason why science cannot coexist (uneasily) with Communist doctrinairism. It has done so in the past, and the post-Stalinist reorientation has made some of the points of conflict less acute.

Another familiar thesis relevant to the problem of the Soviet political structure is that of the growing dependence of politicians on experts. Unmodified, it fits a contemporary totalitarian state even less than Max Weber's doctrines on the subject fitted Imperial Germany. The growing complexity of life and knowledge in modern society makes expert advice indispensable, but this does not lead to *political* dependence on experts. Generalities about industrialization, higher educational levels, and the growing aspirations of managerial strata, have been utilized since Stalin's death to render respectable hopeful forecasts about a democratic evolution of the regime, of which the collective leadership was to be a preliminary step.[9] Since then, the managers have been hit by decentralization measures, and there is no more talk about collective leadership. As to the experts, Khrushchev may well hold with a 19th-century German petty despot, the Elector of Hanover, that, like professors and prostitutes, they can be found on every street corner. Their behavior clearly does not cause him much concern. They are competent, often justifiably proud of their achievements, intensely patriotic, and frequently display the same primitive arrogance so characteristic of Khrushchev himself, despite the fact that unlike him, they are not the products of *Rabfaks* (Workers' Colleges). Characteristic in this respect is a speech made by the leading Soviet atomic expert, Academician I. V. Kurchatov, whose lectures once caused such a stir at Britain's atomic power plant in Harwell. Speaking at the March 1958 session of the Supreme Soviet which also witnessed Khrushchev's formal assumption of supreme power, and answering President Eisenhower's message to Bulganin which stressed that the United States had not exploited its temporary monopoly of atomic arms, he asserted that "when the Soviet Union began to accumulate atomic weapons, there were so few in the United States that they could not play any serious role in a war. And hydrogen arms were first built in the USSR and not the USA" (Pravda, April 1, 1958).

9. See Isaac Deutscher, *Russia after Stalin* (1953).

If it was unrealistic to suppose that state control of science would prevent the Soviet scientists from achieving spectacular results, it is equally unrealistic to expect that contacts with the West will appreciably affect their attitude towards the Soviet system. They have, of course, a stake in its rationalization and in the avoidance of excesses similar to those practiced in the past. But it is Western scientists who now pointedly remark that the difference in pay between a scientist and an unskilled worker is so much higher in the Soviet Union! If Semenov can study chemistry without bothering about the currently orthodox party line on the "theory of resonance," if Bogolubov, Alexandrov, and Fok can occupy themselves with mathematics, without relating it to dialectical materialism, if Kapitza, Blokhinstev, or Landau are able to pursue their researches in nuclear physics unconcerned with the philosophical implications of the indeterminacy principle—in short if the causes of scientific disaffection have been somewhat alleviated —there is no reason to be seen why Soviet scientists should not provide their government with efficient service.

The threat to ideology from the point of view of the regime does not come from the philosophical implications of the natural sciences, but from the humanities and the social sciences. This explains the contrast, often noticed by Western observers, between the levels of the natural and the social sciences in the Soviet Union. As a rule, the closer the subject is to sensitive ideological points, the smaller its chance of unfettered development. In this respect physics appears to be situated fairly far from the magnetic pole, while genetics and psychology are much closer, and sociology is almost on top of it. The hard core of the dogma is sensitive to independent research in these latter fields. While in medieval times Catholic theology was especially touchy about the findings of natural science which seemed to contradict its view of the universe, Communist ideology is particularly vulnerable to critical findings about social structure and development. Since it pretends to be *the* science of society, Soviet Marxism cannot be basically modified without undermining the position of its guardians.

9 *Can Prediction Become a Science?*

S. LILLEY

I SUPPOSE THE MOST PROMINENT psychological feature of the world of 1946 is the very prevalent feeling of uncertainty about the future. Not that uncertainty about the future is a new feature in human life —far from it. But the uncertainties of today are in some ways of a quite new type. For the last 250 generations or so, a man could never be sure if his country was going to be involved in a war; but today he has the added uncertainty that he can hardly visualise what that war will be like—atomic, bacteriological or what? For 250 years or so, the average working-man could not be sure that he would have a job next year; but today he has also serious doubts as to whether the particular job in which he is skilled will exists at all in ten years' time.

Doubts and uncertainties of this type are a quite new feature of social evolution. Before about 1700 they hardly existed, and only in the last fifty years at the most have they become really pressing. And the reason is not far to seek. It lies in the increasing speed of technological advance.

Until the great industrial changes of the eighteenth century, technological progress was so comparatively slow that human thought and human institutions could be left to adapt themselves without conscious effort. They lagged a little behind the ideal form required to fit the advancing technology, but the lag was seldom enough to lead to serious dislocation. It was possible to wait till an invention had become successful or a technique had spread through a large part of an industry, and then begin to look for *ad hoc* methods of adapting the social or industrial structure or the psychological outlook of the people

Reprinted from Discovery, *November, 1946, pp. 336–340.*

concerned to the new requirements. Only once in a few centuries did such passive tactics lead to any serious crisis.

Today the situation is very different. If we wait for an invention to be made, or a technique to become generally economic, then, far too often for comfort, it will prove to be too late to make any smooth adjustments of either institutions or modes of thought. I am far from believing that the atomic bomb is the major cause of present international tension. But in so far as it is a contributory cause, the reason is that the world left consideration of the problem of how to reorganise itself against the threat of atomic destruction till *after* the atomic bomb became a fact. What a difference it would have made in 1946, and even in 1939, if some twenty years ago men had begun to work out the problems of organising a world in which such technological mass-destruction weapons had been developed—always assuming what it is the purpose of this article to consider: the possibility of such discussion in advance.

That is one of the major problems before mankind at present: to predict inventions and technological developments *before* the fact—or to live in perpetual misery of uncertainty. Can we learn to predict technological changes, and the problems they will create, with sufficient accuracy to permit us to adjust in anticipation our social forms and mental outlooks? It is at least arguable that when social evolution presents mankind with a wholly new problem of this sort, the very same changes that produce it also provide the means for its solution—assuming that men are willing to recognise and develop those means. And indeed, *a priori* there is every reason to believe that a science concerned with the prediction of the technological future is now a possibility. For after all, the studies of economic and technological historians have demonstrated that the technological developments of the past have almost always had well-defined causes which were observable to the seeing eye many years before the development actually took place.

There are causes of a technical nature; in the growth of fundamental scientific knowledge, the appearance of suitably accurate machine tools or appropriate materials, the advances of auxiliary techniques, and so on. In the past the necessity for so many correlated advances before an invention could become generally practicable has led to a time lag, often of the order of fifty years, between the first efforts of a lone inventor and the successful social application. That time lag gives one possible method of prediction—by considering undeveloped inventions and answering such questions as "Are materials of the strength and lightness that this idea obviously requires likely to be developed or not?" But that particular method of prediction is likely to be much less useful in a world in which revolutionary inventions like radar or the atomic bomb can be developed into established techniques in something like five years. So far as technical causes are concerned, the prophet is likely to be driven back a further stage to the thorough examination of the results of fundamental research, to see which of them point to the production of inventions likely to be encouraged by the second group of causes.

Social Needs and Demands

THAT second group of eauses is, of course, the social group. Inventions do not mature into socially used techniques unless there is a need for them within the social framework of the times. More than that, they do not mature unless there is an effective social demand for them. And an effective social demand is often very different from a real need. There is no need for an atomic bomb, but the arrangement (or derangement) of human institutions in the present-day world is such that there is a very high degree of effective social demand for it. Similarly the great need for revolutionary inventions (perhaps of the moving-platform type now well-established among predictors) to solve London's internal transport problem remains unsatisfied, because in present circumstances there is no way in which the need can be transformed into an effective social demand. And lastly an invention matures only if the power behind this effective social demand is great enough to overcome the power behind conservative influences, such as the desire to protect capital from obsolescence when a better potential technique appears.

Attempts at Prediction

A PRIORI, then, it seems quite reasonable to suggest that a scientific examination of the interplay of these various forces—scientific and technical possibility, effective social demand and resistances from vested interests and other sources—would lead to a body of theory on the basis of which future technological developments could be predicted with the necessary accuracy. There have, of course, been many attempts at such prediction. Gilfillan, in an article in *Technological Trends* (1937),[1] has made an analysis of the degree of success attained in some of them. Table 1 shows a statistical analysis of three typical cases.

Judging by the avoidance of blunders, that record is a very good one. On the other hand the proportion of definite positive successes is not very high—certainly not high enough to be very useful in making social and political decisions. The figures in the last column, obtained by adding the probables and half the doubtfuls to the definite successes,

1. *Technological Trends and National Policy, including the Social Implications of New Inventions.* (1937, U.S. Government Printing Office, Washington.)

look encouraging at first glance. But this first glance is deceptive. The figures represent the proportion of probable successes at some unspecified time in the future. If technological prediction is to be a useful tool in social construction, the predictions must have dates attached. So far few have dared to predict with that precision and their success has not been high.

But it has to be remembered that these, and the countless other essays in prediction, are merely the sporadic, unco-ordinated attempts of more or less isolated individuals. There has been as yet no important attempt to undertake in detail the fundamental analysis that would be required to found a science of prediction. The various attempts, though often made by men expert in the scientific method, have not yet been carried out on an essentially scientific basis. The best that can be said is that they are the products of acute intuition, modified by a rational, but far from systematic, survey of the available evidence. Again, there has been nothing in the nature of a "school of prediction"—no bringing together of a suitably balanced team of scientists, technologists, historians and sociologists to give those several years of work that are essential to the foundation of any science. Nor has there been any notable development of that other and looser form of co-operation that is frequently so fruitful, in which several workers in a field publish results, criticise and comment on the work of others, and by their mutual interaction

TABLE 1

Date of Prediction	*Proved Right by 1937* (A) %	*Probably Destined to Be Right* (B) %	*Proved Wrong by 1937* (C) %	*Probably Destined to Be Wrong* (D) %	*Still Doubtful in 1937* (E) %	*Estimated Eventual Success* (= A + B + ½E) %
1906	46	24	21	—	9	74
1915	28	48	—	—	24	88
1920	38	29	8	3	22	78

advance the science they are interested in. The results of Table 1 have been obtained by isolated individuals, using a method which is no more than rationally controlled intuition. The question is: could the teamwork of many, devoted to a really scientific analysis of the causation involved in technological progress, give the sort of result that is required? Could it make possible predictions of 95 or even 99% accuracy, on which it would be possible to take social action?

H. G. Wells was one of those who maintained that it could. He made an inspiring appeal for a science of prediction in a Friday evening lecture to the Royal Institution in 1902. That paper was regarded so highly at the time that not only was it published in *Nature* and the

Proceedings of the Royal Institution, but it was also selected for printing in the *Annual Report of the Smithsonian Institution* as one of the important general papers on scientific matters for the year. "I must confess," said Wells,

> that I believe quite firmly that an inductive knowledge of a great number of things in the future is becoming a human possibility. So far nothing has been attempted, so far no first-class mind has ever focused itself upon these issues. But suppose the laws of social and political development, for example, were given as many brains, were given as much attention, criticism and discussion as we have given to the laws of chemical composition during the last fifty years—what might we not expect?

Wells himself made many predictions, with about average success. His greatest contribution, in fact, lies not in his own predictions, but in his faith that a science of prediction was possible and in his appeals to have it made an actuality. It will be in keeping, therefore, with his spirit, if we use his own works to illustrate some of the problems that a science of prediction must face. I am not concerned here with his futuristic novels—though it is worth noting that in one of these, *The World Set Free* (1914), he predicted the atomic bomb, and did so on the basis of the scientific material contained in Soddy's *Interpretation of Radium*—rather I am concerned with his more sober attempts to forecast the future, such as *Anticipations* (1902). The first chapter of that book is largely devoted to the effects that motor cars, buses and lorries would have on the structure of society. It contains its blunders, and certainly his suggested future of the railways is wrong; yet in the main the prediction is excellent. This exemplifies one of the major lessons that can be drawn from many prophetic essays—that a very high standard can be attained, even with present imperfect methods, in predicting the effects that will arise when an invention, already technically achieved but only in limited practical use, develops into a widespread technique. On the other hand, Wells fails abysmally with another invention in transport that was just around the corner in 1902—the aeroplane. That it is practicable, he agrees; but he continues: "I do not think it at all probable that aeronautics will ever come into play as a serious modification of transport and communication . . ." his chief argument being man's "considerable disposition towards being made sick and giddy by unusual motions, and however he soars he must come to earth to live."

His predictions about war are particularly interesting. His forecasts of individual technical weapons are almost all wrong. His flying machines are based on balloons having contractile gas containers, which can be used in the same way as a fish's swimming bladder to raise or lower the ship. These will be ranged in lines fore and aft of the aircraft, and by contracting them at one end and expanding them at the other the ship will be tilted. A suitable disposition of planes will then produce from this tilt a swooping motion; and that will be the chief mode of progress. His airships will not be able to carry much armament; ramming will be their chief weapon, and in terms of ramming instead of machine-gun fire, he visualises something very like the battles of fighters that we are now accustomed to. He thinks the submarine will not be an effective weapon. He foresees tanks, but does not think them of high value; he conceives them to be very slow and unwieldy. He puts tremendous emphasis on highly manœuvrable bicycle troops.

Yet out of all this he obtained in 1902 a picture of future war that describes 1939–45 much better than 1914–18. The point is that, though his forecasts of weapons fail from attempts to describe them too minutely, nevertheless the general nature of those weapons is correct. His aircraft are technically absurd, but suppose he had merely said, "There will be aircraft," as anybody in 1902 was justified in saying. That would have been sufficient. Given the aircraft, he accurately predicts aerial warfare, high-explosive and incendiary bombing, co-operation between land and air forces, and the tremendous advantages that will accrue to the side that obtains mastery of the air. Though his tanks are inadequate and his cycle troops never in fact achieved decisive importance, those two concepts, along with some others, enabled him to predict the highly mobile nature of modern war. His general picture of the increasing use of technical weapons allowed him to foresee something of total warfare, and even to catch a glimpse of the role of scientists, technicians, engineers and agriculturists in the war that was then thirty-seven years ahead.

All this points to one elementary lesson about the methods of technological prediction. We may predict almost every individual invention wrongly; yet if we are genuinely following out the general technological trends of the time, and if we do so over a sufficiently wide field, the result may be an excellent prediction of the total effect. That this is no accident is shown by the fact that the prediction could have been done just as well without reference to particular inventions at all. Wells would have reached his forecast of the general nature of war, if his argument had been something like this:

> Technological trends of the present will clearly soon provide means of flight. They will greatly increase speed of manoeuvre on land. They will provide transport that will enable armies to be greatly increased in size. They will put more emphasis on the productive forces behind the fighting line. . . . Therefore the general nature of a future war will be . . .

(and I suspect that in his own mind the argument did run on those lines, the detailed inventions being an afterthought designed to tickle the palates of certain types of readers).

General versus Specific Predictions

THE moral for forecasters is: Do not predict individual inventions in detail—that is usually a waste of time. Concentrate on two things; first, the extrapolation of present trends (for example, the further development of already existing petrol-driven land transport in the case under consideration); second, predictions of the form "it will become possible to fly." The chance of success for any one of them might be small; but the probability of *at least one* succeeding was very high.

Later in the same book Wells provides another excellent illustration of this point. In 1902 it would have been folly to predict which (if either) of Britain or U.S.A. would take the lead in flight or television or straight line mass-production, or any one particular line of development. Wells chose a much better method, when he wrote:

> The United States of America are rapidly taking, or have already taken, the ascendancy in the iron and steel and electrical industries out of the hands of the British; they are developing a far ampler and more thorough system of higher scientific education than the British, and the spirit of efficiency percolating from their more efficient businesses is probably higher in their public services. These things render the transfer of the present mercantile and naval ascendancy of Great Britain to the United States during the next two or three decades a very probable thing.

Note that here we have a correct prediction with *date attached*. Britain today might be in a less unpleasant situation if that prediction had been heeded (which brings me to remark parenthetically that one of the great advantages to be derived from prediction is that, within certain limits, we can use the results of prediction to falsify those of its features which may be undesirable).

Gilfillan, in the article already referred to, discusses a very similar aspect of prediction—what he calls the *principle of equivalent invention.* The predictor need not be concerned with how a technical problem is to be solved, but only with whether it will be solved or not. He instances the problem of flying in fog (he was writing in 1937) and points out that there were about twenty-five likely means of solving it. These included fog dispersal, several radio suggestions which virtually cover radar, and to pick two instances which, so far as I know have not been developed, trailing a television transmitter below the clouds and the use of infra-red light to view the ground. Of the twenty-five possible methods, not all were likely to be blind alleys or eccentric products of brainstorms, or even unpracticable for unforeseen technical reasons. Hence it was possible to conclude almost with certainty that the problem would soon be solved. I suppose that we can now say that it has largely been solved, and I note that about ten of the suggested methods which Gilfillan lists are now in use. The point is that we are not concerned with the prediction of inventions, only with their effects. And the fact that there are almost always many possible inventions that could lead to the same desired effect enormously increases the chance of successful prediction. The more widely varied the methods are, the less chance of some common unseen factor rendering them useless in practice, and so the greater chance of accurate prediction.

I said earlier that there has as yet been no important attempt to use the joint efforts of many experts for prediction. There is, in fact, one partial exception to that remark—the work embodied in *Technological Trends* (1937), the U.S. Government report in which Gilfillan's article appeared. This was one of the results of the New Deal attempts to overcome the effects of the economic crisis of the 1930's. It was intended to help New Deal planning by predicting the trends of technology for the next few years and discovering what their social effects would be.

I have to confess that it was no more than moderately successful. Many of the discussions of quantitative trends with established techniques (as opposed to qualitative leaps to new techniques) were quite near the mark. On the other hand, there was surprising failure to predict, or even to notice the possibility of, some of the more radical inventions which were just around the corner. The section on air transport, for example, makes no mention of jet propulsion or of helicopters, although both were then under active experimentation and current results were good enough to require at least very careful consideration.

The remarks on the use of radio for safety could not be stretched to cover the implications of radar (but note how Gilfillan in another article took radar in his stride). In almost every respect the article on flight is too conservative—"Technical developments of the planes themselves are slowing down" is a phrase only too typical of the general tone. In the same way, the article on power fails to mention atomic energy, even in a section that briefly discusses such possibilities as tidal power, wind power, power from thunderstorms (*sic*), direct use of solar energy and photo-chemical possibilities. If thunderstorms were worth mentioning (only to dismiss them, of course), then surely the then familiar knowledge of nuclear energy would have made it worth while at least to consider whether methods of releasing and controlling it were feasible. The gas turbine is also omitted from the list of prime movers, though it was by that date in limited use and showing much promise.

Technological Trends would at first sight seem to lessen our hopes that a science of prediction is possible. But I believe that fuller consideration of this report will show good reasons why it was not more successful. To begin with it was only superficially a product of teamwork. There was some committee work, but in most cases the articles appear to have been written by separate experts working individually. Again, there is little sign in it of any attempt to establish general laws of causation and then apply them to prediction. The earlier articles do consider causation in a general, but not very deep, sort of way. But the predictions in particular fields seem to have been made independently of that general analysis, and to have used no more subtle methods than what I have called "rationally controlled intuition."

The Results of Bias

WHAT is, perhaps more important is that the opinions expressed in *Technological Trends* are only too clearly influenced by the circumstances in which the book was written. The 1930's was a period of slowing down in technological advance—at least in radical technological advance, though the rate of minor improvement was still going up. Major technological changes were less frequent than in, say, the period 1880–1910. The slowing down was due to a reduction in incentive, arising in turn from the world economic depression and (as we now know as a result of war-time revelations) from the conservative influence of monopolies seeking to protect capital and the machinations of international cartels. A few decades earlier, whenever science uncovered a new possibility, there was a rush to exploit it in practical form; in the 1930's it was more often left alone, at least by corporations in command of adequate resources. *Technological Trends* was clearly written on the assumption that this situation would continue—which we now know to have been a false assumption. Now it might be suggested that the authors of that work could have seen the causes of the current retardation, deduced that it was abnormal, and corrected their predictions accordingly; at least, that they could have stated an alternative set of predictions allowing for a revival of the rate of progress. But that would have implied at least a measure of criticism of the existing social structure of the U.S.A. With a very few notable exceptions, the authors of *Technological Trends* were not the type of men to take such a course. Indeed it was unlikely that an established government would pick men who would produce a critical report. Though liberal and progressive in form and method, the New Deal policy was, after all, essentially conservative. Its business was to solve the most pressing economic problems facing the United States without making or suggesting fundamental social or economic changes.

That conception is reflected in *Technological Trends*. To take but one example, the problem of technological unemployment receives a large proportion of the space. One gets the impression that it is regarded by most of the authors, not as an evil of social organisation, but as a more or less inevitable result of technical advance. To think otherwise would be to make those unwanted criticisms—but to accept it as inevitable is to hope (though perhaps not very consciously) that there will not be too much of that advance, and so to risk biassing one's predictions.

Of course, the main reason why *Technological Trends* failed in its predictions was that it failed to foresee that the depression of the 'thirties would be transformed into the war of the early 'forties, and that the war would greatly decrease the influence of the retarding forces. Again one might expect the ideal predictor to have recognised the signs that war within a few years was highly probable and to have varied his predictions accordingly or give alternatives. But again the same factor comes in: to have done so would have been to criticise the existing régime.

These reasons for the failure of *Technological Trends* (if I am right in my analysis) point to one of the greatest difficulties to be faced if a real science of prediction is to be created. Some means must be found of freeing the investigators from those influences, both internal and external, which prevent them from making fearless criticisms of that which exists and so tend the bias their estimates. The condition is easily stated; the method of achieving it much more difficult.

I have ranged the evidence strongly against the possibility of prediction being successful enough to be useful. And I have then tried to explain it away—with what success, it is for the reader to judge. Yet I must confess that I remain on the side of H. G. Wells. I believe that successful prediction is possible. If I am right, then the examples of the past, from Wells himself to *Technological Trends* will in their various ways help to show the conditions on which successful prediction depends. It depends on more scientific methods of causal analysis than have hitherto been used, and on a concentration of far more serious attention on the subject. It depends on finding means of freeing the investigators from the effects of undue social conservatism. It depends on finding solutions to a great many other problems, of which I shall mention only one more.

Gilfillan claims that, with notable exceptions like Wells's *Anticipations*, attempts at prediction of military matters have been singularly unsuccessful. And he gives as one of the reasons that "the latest developments and projects are held so closely secret, that it is impossible for an outside author to outguess the silent experts." At a time like the present, when military secrecy is tending to spread into many fields of science and invention and to restrict free publication of matter that is very far from being directly military, those difficulties might well make prediction impossible in any technological field. The moral does not need emphasising.

In conclusion, let me stress again that I am not here concerned with these technological-sociological investigations and the hoped-for science of prediction as academic exercises—though that would not prevent some whose outlook was mainly academic from giving great help. I am concerned with it as a way of improving the conditions of life, specifically as an aid to planning the course of social evolution in such a way as to avoid those now too familiar conflicts between technological instruments on the one hand and human thought and social institutions on the other. I am thinking of it as one element in the process, which has been becoming more important in the last few decades, by which mankind is coming more and more to control its own social evolution, instead of leaving it to the interplay of blind and uncomprehended social and economic forces.

Technological prediction should be an important element in helping to plan the future, to control the future of human society. At the same time the fact of that planning should help to make the prediction easier.

10 *Public Opinion about Science and Scientists*

STEPHEN B. WITHEY

FOR THE PEOPLES of the United States the impact of science has been so widespread and continuous that science has become woven, like a basic thread, into the material of existence and it is difficult to tear it away from the rest of society in order to evaluate and assess it.

Science covers the advent of antibiotics, polio vaccine, and the notion of adaptation energy. Science includes electronics and its results in altered patterns of communication and entertainment, automation, and shifting patterns of job demand. Fusion and fission have changed the nature of the international threat and also formed a new basis for international cooperation. Child rearing, education, and human relations are also affected by science and are in the same state of flux that describes much of the present scene.

Definition of Science

THUS, when we look at public opinion about science, the initial question might well be: What is this thing about which we have an opinion? The public is by no means clear, nor is anyone else. In the recent study conducted by the Survey Research Center,[1] reference was

Stephen B. Withey, "Public Opinion about Science and Scientists," Public Opinion Quarterly, *vol. 23 (1959), pp. 382–388. Reprinted by special permission from a copyrighted issue of the* Quarterly.

This is a revised version of a paper read at the May 1959 meetings of the American Association for Public Opinion Research. The talk was part of a symposium on "The Nature and Implications of Current Attitudes toward Science and Scientists."

1. A two-part study done in early 1957 and mid-1958 by the Survey Research Center of the University of Michigan for the National Association of Science Writers (NASW), in a contract with New York University, and supported by the Rockefeller Foundation.

made to both basic and applied work, in any area of inquiry about nature, from the stars and astrophysics to the mind and psychology. In fact, science was defined for respondents in this way:

> It includes everything scientists discover about nature—it could be the discoveries about the stars, or atoms, about the human body or the mind—any basic discovery about how things work and why. But science also includes the way in which this information is used for practical purposes—it might be a new way of curing a disease, or the invention of a new auto engine, or making a new fertilizer.

Very few indeed had any cognitive content over so broad a spectrum. Probably not more than 12 per cent of the adult population really understands what is meant by the scientific approach. For about two-thirds science is simply thorough and intensive study, which is, in a way, an adequate label. But the sensitive reader of interviews is aware, in the responses of most people with this point of view, of a lack of insight and understanding. It seems as though these people see a stage-set façade, a house front without a house. What happens backstage is only surmised. Things and people emerge center stage, but where they come from is a matter of imagination. A full quarter freely admitted that they did not know what was meant by studying something scientifically. Only about 1 in 10 talked at all about controlled experimentation, scientific method, measurement, systematic variation, theory, or similar notions.

About half the adults said that science could study anything! One quarter were not sure of limitations, and another quarter felt science had real limitations, usually in areas of human experience or behavior. When respondents were *directly* queried, only about 1 in 2 would go along with the notion that we can ever really understand human behavior through scientific study.

Contact with Science

APART from the problem of vagueness of definition and haziness on detail, one might ask: How much contact does the public have with science? One can give only a partial answer. There are no clear figures on the number who run into science in their occupations, nor are there data on hobbies and avocations, or night courses in subjects like electronics. But something is known about the public's contact with science in formal education and through the mass media.

About half the adult public (47 per cent) have had some course in science at the high school level or above. Only about 1 in 9 have had science courses at the college level. It would be an unusual individual, in these days, who acquired any feeling for theory or research in science without being exposed to at least one college-level course in science. The number who have this sort of sophistication is quite small. It is, however, on the increase, having tripled during the last four decades. About 1 in 5 among the current twenty-to twenty-five-year-olds could be so classified.

On the other hand, contact with science through newspapers, magazines, radio, and television is, as one might expect, much broader. When asked to recall some item of science news, *2 out of 3* could recall at least one specific science or medical news item that they had seen or heard through one of these media. Medical items far outweighed other items of scientific work, with heart disease and cancer being the topics most frequently remembered. Further inquiry clearly showed that newspapers are the major carriers. TV and magazines run about an equal second. It should be stressed, however, that this recall measure of information was pre-Sputnik.

It is also clear that mention of science in the mass media is infinitesimal compared to the volume of "raw science." In 1955, science news took less than 5 per cent of the space in a sampling of twenty-six dailies. There has been only a very slight increase in the years 1939 to 1951. Since Sputnik it appears, from a NASW study of 240 managing editors,[2] that science coverage has materially increased. Two-fifths of these editors reported double the coverage; another two-fifths reported increases around 50 per cent. Even so, the space devoted to science is below that devoted to international or domestic affairs, sports, women's news, and most other news or feature categories.

Level of Information

THE public's level of information on scientific developments, in general, is an unknown, but there are some gross measures. Four informational items on polio, fluoridation, radioactivity, and space satellites were asked in the months before satellite launchings. One in eleven could answer *none* of them. Only one in six could answer all four questions on these relatively well-publicized topics. Much more was known about the two that dealt with medical or health topics than the two more closely related to physics.

With the first satellite launchings, however, these startling probes into space became known to almost everybody. Nevertheless, only a minor increase occurred in the number who had some understanding of the scientific purposes behind rocketry and space ventures. The number changed from about 1 in 5 to about 1 in 4. Yet the number who had any detailed information—about 1 in 10—showed no increase.

2. Study reported by Hillier Krieghbaum, Department of Journalism, Communication Arts Group, New York University.

These people presumably knew such details from information about the International Geophysical Year.

Pre- and post-Sputnick studies showed no increase in the over-all number who reported reading scientific works or news of scientific events. The only exception was an increase in the proportion of college women who reported reading science items.

A certain amount of engrossment in scientific matters, as well as a background that makes such attention rewarding, is required to maintain a moderately high level of information on scientific developments. A high level of information, as grossly measured, i.e. knowledge in the four subject areas, was related to *multiple* use of the media. This finding implies that exposure to specific developments is, for the well-informed, often a matter of repetition and elaboration. Also, a high level of information was significantly related to a background of some education in science, in which passing high school science courses was a minimum.

Most of the relationships for such measures as education, income, sex, and occupation were predictable: the better-educated seem to have more information on science and more interest in it; men show some small superiority over women in information and interest; the better-educated tend to hold the better occupations; younger people (say under 45) seem to have more information and interest than their elders; metropolitan suburbanites are in general better informed than their city neighbors or more distant suburbanites; men are more interested than women in nonmedical science, but women are more interested than men in medical science.

Interest in Science

STUDIES have, on several occasions, tried to track down lines of interest in news and feature stories. Some of these seem to reinforce and substantiate one another. In a study of adults in the greater metropolitan area of Detroit,[3] fourteen headlines were presented and reader interest was measured on a quantified scale. These samples of typical headlines excluded specific professional areas of interest, personalities, and known topics of special interest, such as sports. On a factor analysis four factors were drawn out, the first of which one might label "disaster interest"—catastrophes, accidents, dangers, etc. The second factor is hard to label but includes items of neutral emotional interest and minor personal relevance. Threatening items appeared to have a low factor loading in this case. The headline "Detroiters Spend Weekends Simi-

3. S. B. Withey and J. M. McLeod, "Detroiters and Their Newspapers," Survey Research Center of the University of Michigan, August 1957 (mimeographed), but this analysis is not part of that report.

larly" had the highest loading on this factor. A third factor was "science." Although no medical items fell into this cluster, it clearly formed a center of interest for a sector of the reading public. The fourth factor happened to be "area saliency." Headlines that had a definite location (e.g. Switzerland, South Africa, Cincinnati) had the highest loadings, while those without a specific location (e.g. "Radiation Dangers Explored") had loadings near zero. Thus it can be said that scientific topics, not including medical subjects, and separate from interest in tragedy or disaster, or intellectual interest in affairs and events of the day, comprise an area of motivated interest for certain persons. It would be surprising if this interest spread across the whole domain of scientific inquiry; it already tends to exclude medicine, but it does, by the study definition, refer to atomic science, physics, and space developments, since these were the subjects covered in the headlines that were included in this factor. Perhaps behavioral science would make up another factor loading if such items had been included.

In the 1957 science study, headlines dealing with some aspect of scientific inquiry were presented to test reader interest. Half the sample were given flat, unimaginative headlines—the type that might be written by a social scientist. The other half were given headlines on the same topics written by journalists with an eye to securing the interest and attention of readers.

The first interesting finding was that headlines that aroused interest on the part of more than 50 per cent of respondents when presented in drab form aroused little more interest when dressed up in journalese. No item that did not deal with human beings attracted more than about 1 reader out of 5, except for one journalist-written headline: "New Chemical Theory Doubles Mileage of Gasoline"—a twist that makes molecules directly relevant to human experience. No headline that carried an implied threat became more interesting when the writing style was changed. Whatever interest such a headline attracted in one style it attracted in the alternate form within 4 or 5 percentage points.

Three kinds of science treatment seemed to increase reader interest: *actuality,* or reporting an accomplished fact rather than a hypothesis or future possibility; *specificity* rather than generality or abstractness; and *relevance* of the scientific event to human behavior or welfare.

In the same study an attempt was made to probe the motivations behind attention to science and scientific events. This is not simple, because the motivations undoubtedly vary from topic to topic and event to event. Almost half the adult population responded that they wanted to keep up with events—science in this context is one with any other headline or significant development of the day. The next most frequent rationale was, however, less general. About *1 in 4* said that they paid attention to science because "science may determine whether my family and I, and the world itself, will survive." Around 1 in 7 saw science as helpful or interesting. Only a tiny minority saw it as exciting.

The Net Impact of Science

THE great majority of persons (83 per cent) felt that the world is better off because of science. About 1 in 10 offered some qualification that led them at least to question whether the world is not worse off because of science. The contributions of science to the armaments of war were seen as the major area of bad effects of science. Most of the developments in atomic science were seen as falling in this area of threat. Also, most of the new advances in space were seen much more in the area of an international race with Russia, and the total context of the "cold war," than as advances in science per se. The positive contributions of science were seen as improved health, a higher standard of living, and the industrial and technological improvements that underlie these.

The responsibility for the bad effects of science was not readily laid on the shoulders of any special group. Only 12 per cent blamed scientists themselves. Instead, vague but sinister, powerful, and selfish groups with special interests were referred to.

The launchings of satellites have not created any change in the number who see the world as worse or better off because of science, nor in the attribution of responsibility for these effects. There has been no really significant shift, since Sputnik, in variety of attitudes toward science that will be reported, except for an increase of about a quarter in the number saying that a growth in science means that a few people could control our lives. Most of this change has occurred, however, among those who have little knowledge of science and little interest in it.

TABLE 1. Reactions to Statements about Science (*in per cent*)

	Agree	*Disagree*
Science is making our lives healthier, easier, and more comfortable	92	4
One of the best things about science is that it is the main reason for our rapid progress	87	6
One trouble with science is that it makes our way of life change too fast	47	46
Science will solve our social problems, like crime and mental illness	44	49
The growth of science means that a few people could control our lives	40	52
One of the bad effects of science is that it breaks down people's ideas of right and wrong	25	64

Pre- and post-Sputnik measures show no significant changes in the number who feel that science will solve our social problems, will make life better, will break down morality, or that scientists want to make life better for people. There was an increase in the number of "poor attenders" to science who felt that scientists pry into what is not their business, and that science changes our way of life too fast, but this change was small and occurred in a minority category. The current divisions of opinion on these issues are presented in Table 1. The items were presented as statements with which respondents could only agree or disagree.

There has been no change at all in the public's attitudes and stereotypes about scientists themselves since Sputnik. Public reactions are given in Table 2.

TABLE 2. Reactions to Statements about Scientists (*in per cent*)

	Agree	*Disagree*
Most scientists want to work on things that will make life better for the average person	88	7
Scientists work harder than the average person	68	25
Scientists are apt to be odd and peculiar people	40	52
Scientists are not likely to be very religious people	32	53
Most scientists are mainly interested in knowledge for its own sake; they don't care much about its practical value	26	65
Scientists always seem to be prying into things they really ought to stay out of	25	66

In summary, it might be said that for the public the caduceus of medicine sits proudly at the top of the totem pole of science. Under it comes the composite face of the contribution of science to human living and welfare; this figure sits astride the rather grotesque face of science's part in the military. Below are other minor figures noticed by some but overlooked by others.

If the inquisitive observer watches the worshippers and the more casual passersby, he notices respect and appreciation, but little real curiosity and interest, and he can overhear a certain amount of distrust and apprehension expressed in subdued conversations. The scientists are seen as well-meaning, brilliant, hard-working people. Their less appealing characteristics—off-beat and peculiar—are elicited only when the observer probes. On the surface the natives are quiet, supportive, and appreciative but there is some questioning, some alert watching, and considerable mistrust. The public will wait and see; they have no reason to do anything else, and many have no other place to turn.

11 *The Origins of U.S. Scientists*

H. B. GOODRICH, R. H. KNAPP, AND GEORGE A. W. BOEHM

THE MAKING OF A SCIENTIST has always been a more or less mysterious affair. The origins of the famous scientists of the past show no particular pattern: some were well educated and some poorly, some trained in science and some untrained, some guided toward science from childhood and some impelled into it fortuitously at a relatively late age. Today science is so complex and formidable a discipline that it might seem there is no room for happenstance or deviation in the development of a scientist; it would almost appear that a candidate must be specially prepared for this esoteric calling from birth, like the Spartans for soldiering. Asked where U.S. scientists come from, the average person would probably say that they flow mainly from the major centers of American intellectual activity and are prepared predominantly in our great universities and special scientific schools.

The facts say otherwise, as a recently completed study makes clear. The survey, a statistical analysis that took some five years, produced several significant surprises. This article will summarize some of our principal findings and conclusions; a detailed report of the survey will shortly be published in book form.

In 1946 a committee of the Wesleyan University science faculty was appointed to study the undergraduate training of U.S. scientists. Supported by the university trustees and subsequently by the Carnegie Foundation, the project rapidly grew in scope. By the time it was completed, it had become a broad survey in cultural anthropology—an examination of the undergraduate ecology of the nation's scientific manpower.

Reprinted with permission from Scientific American, *vol. 185, no. 1 (July, 1951).*

The first step was to define "scientist." For want of a better measure we chose as the subjects of the study all persons who had received doctorates in the natural sciences and were listed in *American Men of Science*. (We had decided, for various reasons, to exclude social scientists.)

The next step was a statistical survey aimed mainly at finding out just what to investigate. Tabulations from the third (1921) and seventh (1944) editions of *American Men of Science* showed that in each decade from 1880 to 1930, the last year for which complete listings were available, the number of scientists roughly doubled. In certain fields the increase was more spectacular than in others: physics and biology closely followed the general rate of increase; geology, mathematics and astronomy suffered a relative decline in numbers; the course of psychology was somewhat erratic; chemistry, from the turn of the century, rose at an accelerating pace, outstripping the growth rate of any other field. In general, the scientific fields that offered the brightest hope of employment and good pay, especially through the opening of industrial applications, attracted the most people.

The preliminary survey also showed that individual undergraduate institutions varied greatly in their output of future scientists. The output of a given institution of course fluctuated from decade to decade with changes in teaching and administrative staffs, but it became clear that some colleges consistently produced a larger proportion of scientists than others, at least since the First World War. Moreover, these institutions were highly productive not just in one field of science but in various fields.

Accordingly we decided to study the productiveness of the undergraduate colleges in more detail and to determine why some turned out relatively many scientists, others very few. As the index of a college's performance in this respect we used the number of graduates per thousand who subsequently earned a Ph.D. in science. In coeducational schools we considered only male graduates, since relatively few women obtain doctorates in science. We also restricted the study to men who graduated from college between 1924 and 1934, in order to obtain a peacetime picture without the dislocations in education caused by the two world wars. As a check on the validity of our use of the listings in *American Men of Science* we computed a test index based on the list of doctorates in the natural sciences compiled by the National Research Council. This index had a high correlation with the one obtained from *American Men of Science*, confirming the validity of the latter.

What does the index show? The first surprise is that small liberal arts colleges are far and away the most productive sources of future scientists among U.S. institutions. Of the 50 leading institutions in this respect (*i.e.*, those that turn out the largest proportion of graduates

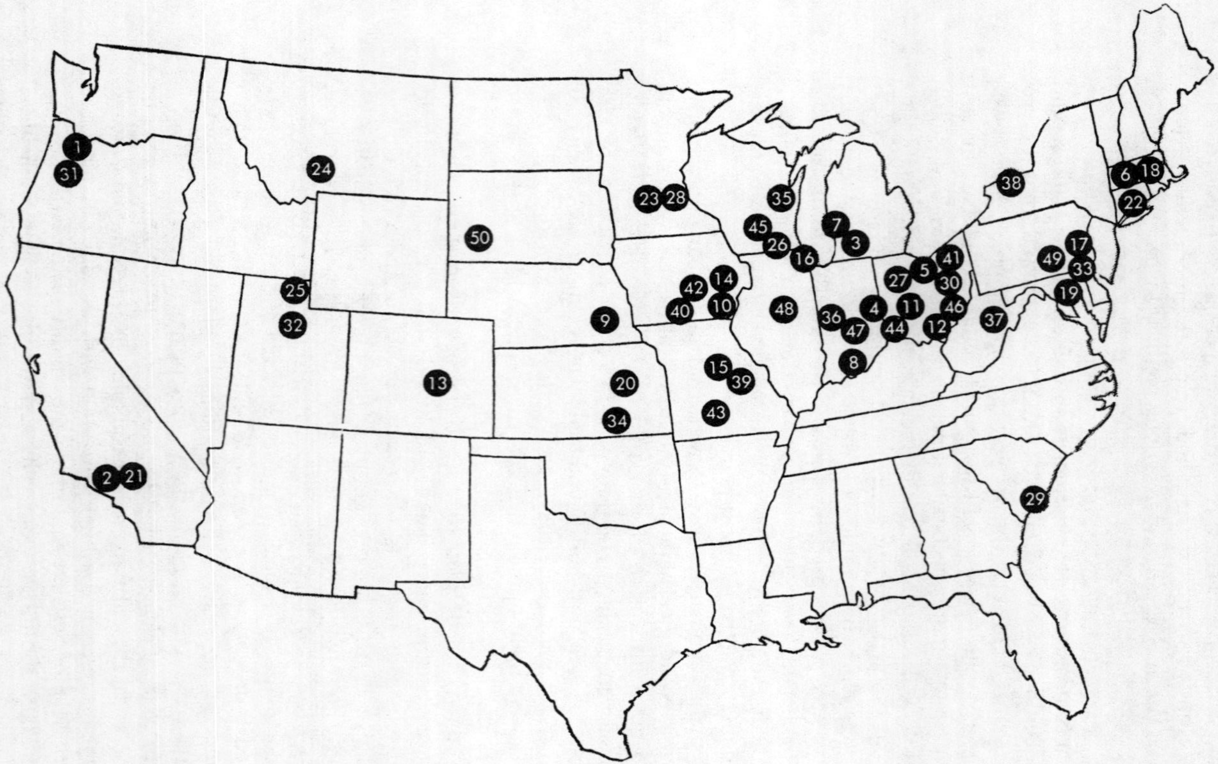

FIGURE 1

Fifty best producers of natural scientists are listed in Table 1. In the third column from the left is the number of graduates per thousand who went on to take a doctor's degree in a natural science. Each institution is located by a number on the map.

TABLE 1

1. Reed	Ore.	131.8	26. Beloit	Wis.	32.9
2. California Institute of Technology	Calif.	70.1	27. Bluffton	Ohio	31.8
3. Kalamazoo	Mich.	66.3	28. Carleton	Minn.	31.6
4. Earlham	Ind.	57.5	29. Charleston	S.C.	31.6
5. Oberlin	Ohio	55.8	30. Wooster	Ohio	31.4
6. Massachusetts State	Mass.	55.6	31. Willamette University	Ore.	31.2
7. Hope	Mich.	51.1	32. Brigham Young University	Utah	30.4
8. DePauw University	Ind.	47.6	33. Swarthmore	Pa.	30.2
9. Nebraska Wesleyan University	Neb.	47.4	34. Southwestern	Kan.	30.1
10. Iowa Wesleyan	Iowa	45.5	35. Lawrence	Wis.	29.9
11. Antioch	Ohio	45.1	36. Wabash	Ind.	29.9
12. Marietta	Ohio	45.1	37. West Virginia Wesleyan	W. Va.	29.8
13. Colorado	Colo.	43.9	38. Rochester, University of	N.Y.	28.2
14. Cornell	Iowa	41.2	39. Westminster	Mo.	28.0
15. Central	Mo.	39.9	40. Simpson	Iowa	27.6
16. Chicago, University of	Ill.	39.9	41. Hiram	Ohio	27.4
17. Haverford	Pa.	39.4	42. Grinnell	Iowa	27.3
18. Clark University	Mass.	39.0	43. Drury	Mo.	26.5
19. Johns Hopkins University	Md.	37.3	44. Miami University	Ohio	26.4
20. Emporia	Kan.	36.5	45. Wisconsin, University of	Wis.	26.2
21. Pomona	Calif.	36.0	46. Muskingum	Ohio	25.7
22. Wesleyan University	Conn.	34.3	47. Butler University	Ind.	25.4
23. St. Olaf	Minn.	34.2	48. Eureka	Ill.	25.0
24. Montana State	Mont.	33.9	49. Lebanon Valley	Pa.	24.7
25. Utah State Agricultural	Utah	33.4	50. South Dakota School of Mines	S.D.	24.6

who become scientists), 39 are small liberal arts colleges (see Table 1). Only three large universities appear on this list of leaders, and only two technological institutions; the others among the 50 are three state agricultural schools and three small universities that lean toward technology.

For some of the smaller institutions on the list the number of graduates and scientists is too small to make the indexes statistically reliable. But the striking accomplishment of the 39 liberal arts colleges as a group is beyond dispute, as rigorous statistical methods demonstrate.

The second striking fact, which may surprise some, is that the institutions which lead in the production of scientists are mainly concentrated in the Middle West (see Figure 1 and Table 1). That this region is particularly productive of scientists is confirmed by a study of all the 500 institutions for which indexes were computed. In that ranking the Middle West and the Pacific coast lead the nation, with the Middle Atlantic States and New England next and the South last.

The significance of this situation is underlined by the fact that in the production of graduates entering some other professions, such as the law, the ranking is quite different. According to a survey made before World War II, the U.S. Northeast is the region most productive of future lawyers; of the 35 undergraduate institutions that led in this respect nearly two-thirds were in New England, New York and Pennsylvania.

Our next step was to compare groups of institutions, classified according to type, in their output of future scientists. Here the top position was taken by the state-supported agricultural colleges, which as a group had an average index of 19.8 scientists per 1,000 male graduates. The liberal arts colleges came next; the average index of a group made up of 153 privately endowed, non-Catholic colleges graduating from 30 to 200 students a year was 17.8. A group made up of 50 eminent universities stood only in third place, with an average index of 13.8. The leading engineering schools as a group (excluding California Institute of Technology, which occupies a class by itself) produced only 6.4 scientists per 1,000 graduates. The lowest-ranking group was that composed of all the Catholic institutions in the U.S.; their average index was 2.8.

In the case of the agricultural colleges, which achieve top rating here, we must take into account that almost every student in these colleges majors in some kind of scientific work, whereas in the other types of schools on the average only one student in three is a science major. Taking this factor into consideration, it again appears that in proportion to the number of undergraduates studying sciences, the liberal arts colleges are the most productive of scientists.

The low ranking of the technical schools in this hierarchy can be explained by their vocational emphasis; their training is mainly for engineers, not scientists. An engineer receiving a bachelor's or master's degree in a technical school is ready to take a job and does not usually go on to get a Ph.D. On the other hand, a physicist, chemist or mathematician may be severely handicapped in his profession unless he continues his education through the doctoral level.

Probing more deeply by looking into the differences between individual colleges and universities, we found some factors which seemed significant. Besides the factor of geographic location, already mentioned, we discovered that the intellectual quality of the student body in a college and the cost of attending the institution were related to the college's production of future scientists. Colleges that had a high average student intellect, as measured by the American Council on Education psychological tests, tended to show a high production of scientists. As for cost of attendance, the relatively inexpensive and the relatively expensive schools were less productive than those of moderate cost. We believe that the failure of high-cost institutions to achieve distinction in the production of scientists is attributable to the fact that the relatively wealthy students who attend them do not, as a class, turn to science. Evidence which we have assembled indicates that scientists are rarely drawn from homes of wealth. The economic prospects of the scientific profession offer too little inducement to wealthy youngsters; they prefer to maintain their economic standing by going into law, medicine, business management or other work with a greater financial reward than science.

These were the major findings of our statistical study. Next we got down to cases. Knapp took a year's leave of absence to visit and study at first hand 22 selected liberal arts colleges, some prominent, others obscure. One of the most rewarding investigations was that of Reed College.

This small college in Oregon with a total enrollment of only about 600 has been far and away more productive of future scientists than any other institution in the U.S. Since its founding in 1911, Reed has had a brilliant record of achievement, though from early days it has labored under financial handicaps. Between 1925 and 1940 it produced 12 Rhodes Scholars. During the 1924–1934 period that we studied, 44 per cent of Reed's students majored in a physical or biological science. The college's claim to distinction is not confined to the natural sciences; it is probably as well known for its graduates who have done outstanding work in the social sciences. Though salaries have been relatively low, many top-notch men have come to Reed to teach and have stayed there, disdaining more lucrative positions. Among the students, most of whom commute from nearby Portland, the campus hero is the scholar. The curriculum is organized to foster maximum individuality of instruction, and teachers and pupils alike carry on a tradition of disputatiousness which in many another institution might be a sign of disorganization and dissatisfaction.

Yet anyone who is tempted to draw generalizations from Reed, as far as productivity of scientists is concerned, should consider Iowa Wesleyan, which stands at the opposite end of the spectrum in almost every way. Iowa Wesleyan, like Reed, has an enviable record in production of scientists. But unlike Reed, it has had little else to recommend it. During the depression it was on the verge of closing its doors, and its regional accreditation was withdrawn for several years. Its student body was of undistinguished quality. Its faculty, which had almost no voice in the administration, was perennially disgruntled and appallingly underpaid; the turnover was so rapid that most of the teachers might almost have been taken for transient guests. The region from which Iowa Wesleyan draws its student body is ground down by an endemic economic depression. But in this setting two men stand out like knights in shining armor. One of them, a competent physics professor and successful inventor, designed equipment for Admiral Byrd's first Antarctic expedition. The other, a chemist, invented a successful process for making patent leather. Consequently to a great many Iowa Wesleyan students a scientific career seems to hold out great opportunities, and they try to emulate their local heroes.

In addition to investigating individual institutions, we examined the records of approximately 200 professors and attempted to determine what factors, personal and pedagogical, influence students to take up careers in science. This was done by direct investigation and by questionnaires sent to former pupils.

Statistical examination of the ratings assigned by students and by the investigator indicated first that a successful teacher of science usually is not especially distinguished for his mastery of superficial pedagogic skills. Rather, the successful teachers are marked by three cardinal traits: masterfulness, warmth and professional dignity. It would appear that the success of such teachers rests mainly upon their capacity to assume a father role to their students, in the best sense, and to inspire them to an emulation of the teacher's achievements.

In the light of our studies, what environments are most conducive to the production of American men of science? Our evidence points to the fact that the most productive type of institution is a small liberal arts college, especially at a certain stage of its evolution. The typical U.S. liberal arts college was originally founded by a Protestant sect to train clergy and teachers. It drew most of its student body from the surrounding area and the economic middle and lower-middle class. In the second stage it becomes secularized but continues to draw its students mainly from the same population as before. Eventually such a college may develop into a heavily endowed institution of high reputation, attracting a wealthier class of students. But our statistical and case studies show that those liberal arts colleges that are in the second stage of this evolution are most productive of scientists. Among U.S. colleges those in the East and South are generally older than those in the West. Thus many of the Eastern and Southern schools have passed through the highly productive second stage, while the Western colleges are now in the midst of it. Then, too, the frontier traditions of the West, based on intimate association with the natural universe, seem conducive to the development of scientific interest. One might say that as frontier regions enter the first stages of intellectual development, they turn with particular enthusiasm to the pursuit of science, even though their largely agrarian way of life offers few local prospects of professional employment.

Though some of our conclusions may be tentative and others clearly speculative, our survey has established certain facts that are pertinent to the present manpower emergency. Certainly the clear demonstration of the contributions of smaller liberal arts colleges to the scientific profession should be of considerable interest to those formulating our national policies.

12 *The Younger American Scholar:*

ROBERT H. KNAPP AND JOSEPH J. GREENBAUM

Introduction

TWO PURPOSES have motivated the execution of this study; first the desire to compile a roster of younger American scholars and scientists who, while not yet attaining mature levels of intellectual accomplishment, show promise of such achievement in the future; and second the wish to determine how those patterns and characteristics of undergraduate institutions may be associated with varying success in the production of such individuals.

Considering that since the last war, the federal government and several national and professional organizations have been compiling and maintaining rosters of various intellectual specialists, especially those in the fields of science, it may be asked: "What is the need for a new roster?" The point, however, is that all these rosters are primarily lists of those who have already established themselves, to some extent, in their respective intellectual disciplines. Such rosters usually do not include the names of younger people who are just starting to do professional, intellectual work. The main purpose of listings already in existence is not to spot individuals who show signs of future intellectual distinction, but rather to identify those who can successfully carry out particular intellectual tasks, here and now. The maintaining of this type of roster is, of course, essential. But equally important is the early identification of those individuals who show promise of future intellectual achievements of note. For it is clear that if we are to maintain

Reprinted from The Younger American Scholar: His Collegiate Origins, *pp. 1–101, by Robert H. Knapp and Joseph J. Greenbaum by permission of The University of Chicago Press.*

and increase the intellectual strength of the country, we must have some way of recognizing those with high research and scholarship potential, so that we may support and encourage their growth and help to actualize their intellectual potential. It is for this reason that we have undertaken the job of compiling another national roster.

From such a roster, not only will we be able to spot those with intellectual promise, but we should also be able to discover which American undergraduate institutions are most fruitful in the present-day production of these individuals. Once having compiled our roster, we need only identify the collegiate origins of those listed in it in order to assess the achievement of particular institutions and to obtain an intimation at least of the demographic and sociopsychological factors which favor superior productivity of scholars.

The identification of these variables will enable us to draw conclusions which should have important implications for the general picture of higher education in this country. Thus we will be able to obtain information concerning the distribution of educational subsidies, especially those granted by private and governmental agencies. In this connection, it will be interesting to find out, for example, whether graduates of institutions producing a large number of distinguished individuals, as measured by Ph.D. awards and university fellowships, receive a proportional number of awards from private foundations; or whether there are geographic and other determinants more strongly influencing the distribution of privately controlled subsidies. In a like manner, the pattern of government subsidization of higher education can be evaluated. Such evaluations as these will point out whether there are any serious major weaknesses and discrepancies in the present graduate subsidization programs and perhaps suggest new sources of intellectual productivity that should be more highly cultivated.

Similarly, this study will throw light on such important educational problems as: whether private or public institutions are differentially productive of scholars: the relation of cost of attendance to intellective achievements; recent general trends in regard to the study of the sciences, the humanities, and the social sciences; the contribution of women to scholarship and their fields of greatest strength; the comparative achievements of various types of undergraduate institutions, e.g. liberal arts colleges, universities, teachers' colleges, in the production of scholars and scientists. Besides highlighting these and similar problems of higher education, the design of this study will enable us to gain fresh insights into the geographical patterning of ideological subcultures in America. Do the different geographical regions of the nation favor scholarship and study in different disciplines? If so, which? Are certain sections of the country predominant in the production of scholars? If so, which? By obtaining answers to such questions as these, we should be able better to discern those cultural factors that are determining these regional differences and thus gain a clearer picture of the cultural differentiation of the nation. Similarly, knowledge about the relative importance and frequency of choice of study of the humanities, the sciences, or social sciences, should reveal the intellectual tendencies of different regions of the nation during the past decade. In short, our data will not only yield us insights into current trends in higher education, per se, but should also show how higher education is woven into the general cultural fabric of the nation.

Before concluding this brief introductory chapter, we would like to consider two general problems that beset and limit our study. First and most important, there is the problem of deciding whom to include in the roster. Since we are trying to identify those individuals who show promise of future intellectual achievements, we cannot turn to any source that lists those of present-day professional accomplishments, e.g., *The American Men of Science*. Rather, we must find sources that will suggest names of those with future scholastic promise. This is not an easy task. One method that suggests itself is to consult with the various undergraduate and graduate faculties and ask them to choose from among their recent graduates those who have been most promising. While such an evaluative procedure is open to many difficulties of reliability of sampling and judgmental validity, probably the most serious difficulty from our viewpoint is the prohibitive cost in time and money that this procedure would entail. Because of these practical considerations, we were forced to rely solely on easily obtained, non-judgmental, quantitative data.

The second problem to be noted is that no matter what source is employed, there is no way of evaluating future validity today. Clearly this is so, for the sources from which we draw our roster names are only indicators, or "best bets," that the individuals involved will in the future turn out to be scholars and scientists of merit. Thus we must wait some ten years or more in order to evaluate the final wisdom of our choices of source-data for compiling the roster. This does not mean, however, that the validity of our findings in regard to current trends in the production of scholars cannot be evaluated. For in this case, some congruence theory of truth can be applied to the data; such that if our various sources all yield the same production patterns when compared with the several independent variables of this study, we can assume with high confidence the validity of our results. As we shall see, such congruence is amply demonstrated in our results.

The Compilation of the Roster

IDEALLY, the best way to have accomplished the general and specific aims of this investigation would have involved compiling a roster of every recent college graduate who showed signs of promise in any of the many diversified fields of academic endeavor. Obviously, such an approach was neither practical nor feasible, because of the limitations of time and finances. Neither was it methodologically necessary. It is clear that valid conclusions about current trends in collegiate production of scholars and scientists can be arrived at by adequately sampling the population of recent graduates who have shown in their post-baccalaureate work indications of future intellectual achievements. However, because of the multi-purposed nature of the present investigation, we were obliged to compile the roster so that it would not only be representative of the population of younger American scholars and scientists, but so that it would also include as many such individuals as time, money, and other practical considerations would allow. The roster as it now stands is not a definitive inclusive list of all individuals with promise of scholarly achievement, but, as shall be seen, it does approach this aim; and certainly it is more than just a statistically sufficient, representative sample of such persons.

The following operational principles guided our decisions concerning which individuals to include in the roster.

1. Since this investigation is concerned with a contemporary picture of higher education, it was decided to include only those persons who have in recent years won some post-baccalaureate distinction. The calendar year 1946 was chosen as the base limiting point for our study. It was felt that any distinctions won previous to this time would not reflect the present state of collegiate productivity of scholarship, but rather the confusion of the war years, or the state of higher education of a decade (or more) ago.

2. No winners of post-doctorate distinctions were included in the roster. It was believed that in such cases it would not be possible to evaluate whether the collegiate or graduate institution was more influential in encouraging and shaping the individuals' intellectual interests and achievements. Further, since one of the main aims of the roster is to spot those individuals of promise rather than present mature accomplishments, the including of post-doctorate awardees in the roster would take us far beyond the scope of the investigation.

3. It was decided, because of the practical limitations of time, research personnel, and finances, to concentrate our efforts primarily on those fields of intellectual endeavor of a scholarly nature. Thus no attempt was made to include in our roster individuals who had won distinctions in the fields of medicine, law, or theology, if these distinctions were solely confined to these professions and not earned in competition with the general body of graduate students.

4. Such practical considerations, as the ready availability of the data, their completeness, etc., played an important part in deciding which

individuals to include in the roster. While it would have been very valuable to have graduate faculty ratings, scholarship board evaluations, and the like, on the individuals in our roster, such qualitative assessments were not easily obtainable; and thus it was decided to include only those individuals whose distinctions could be determined from records available to the public.

5. Through necessity, having precluded the use of qualitative judgments in compiling the roster, it was decided that the best public indications of future scholarly or scientific promise were the winning of some fellowship, scholarship, or prize of a graduate level in open competition, or the attainment of a Ph.D. or its equivalent (e.g., the Sc.D.).

6. In order to maximize the discriminatory value of the roster, only those fellowship awardees were included whose fellowships were in the amount of $400.00 or greater (or the equivalent in tuition, traveling expenses, etc.) for any single academic year.

7. Finally, in accordance with principle 4 (i.e., the ready availability of the data), it was necessary to exclude all awards that called for service, as central listings of such awards were seldom available. Thus teaching and research assistantships were not included in the roster.

In the terms of the foregoing principles for compiling the roster, four distinct sources were employed. They were (*a*) university fellowship and scholarship awardees; (*b*) recipients of governmental fellowships; (*c*) holders of private foundation fellowships; and (*d*) recipients of Ph.D.'s. The methods and rationale for determining which universities, government agencies, and private foundations to use as sources in compiling the roster is given below.

There are over 130 institutions in the nation qualified to grant a Ph.D. If our task were simply to obtain a representative sample of the doctorate graduates and fellowship holders from these universities, the best sampling procedure to follow would be to arrange a multi-category stratification of the university population (e.g., according to size, geographical location, type of administrative control, and the like), and then draw a random and proportional sample from each of these strata and substrata. But since the purpose in compiling this roster is to approach comprehensiveness as well as representativeness of the population, proportional stratified sampling was not adequate to our needs. Fortunately, however, a study of the Ph.D. degrees granted over the past six years shows that a small number of institutions consistently account for the majority of doctorates that are granted each year. In 1950, of the 130 universities that awarded the Ph.D., 25 institutions accounted for nearly 75 per cent of all the Ph.D.'s that were conferred. Furthermore, these same 25 institutions similarly account for nearly three-fourths of all the Ph.D.'s granted since 1946. We felt, therefore, that if we drew only from these 25 institutions, we would have both a fairly representative group of universities as well as a more comprehensive sampling of the parent population. It seems safe to assume also, that since these 25 institutions account for the majority of the Ph.D.'s that are annually granted, these same universities should account for at least 80 per cent of the recent university fellowships and scholarship holders.

Table 1 shows a rank ordering, in terms of Ph.D. degrees conferred, of the 25 largest Ph.D. degree-granting institutions for 1950. From this group of institutions, data on Ph.D. awards and university fellowships were obtained.

It is clear from the table that this sample is not completely representative of the population of universities in the nation. For example, the geographical distribution of graduate institutions is somewhat distorted. The South is represented only by the universities of North Carolina and Texas. Other variables, such as size of institution, religious affiliation and the like, are not representatively sampled. Thus it is likely that such systematic biases as the fact that graduates from certain colleges may tend to take their advanced work at small universities; that geographical proximity in selecting graduate schools may operate in certain sections of the country; that selective drawing power may characterize graduate schools with religious affiliations, and other similiar biases have not been controlled in our sample. But, in spite of these difficulties, it does not seem likely that our sample of the Ph.D. and university fellowship holders is seriously biased, for it not only includes approximately 75 per cent of the population, but also universities from nearly all the major geographical sections of the country, as well as both private and public institutions, and those with and without religious affiliations.

TABLE 1

1. Harvard University
2. Columbia University
3. University of Wisconsin
4. University of Chicago
5. University of California (Berkeley)
6. Ohio State University
7. University of Illinois
8. Cornell University
9. University of Michigan
10. New York University
11. Yale University
12. Stanford University
13. University of Minnesota
14. State University of Iowa
15. Purdue University
16. Massachusetts Institute of Technology
17. University of Pennsylvania
18. Northwestern University
19. University of Southern California
20. Iowa State College
21. University of North Carolina
22. University of Texas
23. Johns Hopkins University
24. Catholic University of America
25. Princeton University

There are several hundred private foundations in the nation that in some form support graduate study through fellowship programs. These foundations range from such nationally known organizations as the Social Science Research Council to small local organizations that only incidentally and occasionally offer fellowships for graduate study. Clearly, it would be a futile task to attempt to trace down all recipients of awards from such a large number of private foundations. Furthermore, since the majority of these foundations award fellowships on a restricted basis and not in open scholastic competition, to include such organizations as sources in compiling our roster would not necessarily give us a listing of promising scholars and scientists. Therefore, in deciding upon which foundations to use as sources of private fellowships, we limited ourselves to those organizations that select their candidates solely on the basis of scholarship and promise, and in open scholastic competition. Following informal advice from the Social Science Research Council, we sampled the nine largest private foundations. Table 2 lists the foundations selected for our study. Certainly 9 out of 200 foundations is not a large sample, but if we redefine our population so as to include only those foundations that have national unrestricted fellowship programs, our sample is probably well representative and quite inclusive.

There are three major fellowship-granting agencies of the federal government: the U.S. Public Health Service, the Atomic Energy Commission, and the U.S. State Department under the Fulbright Program. Since all three of these agencies grant awards based solely on scholarly merit, and since central listings were available for all the agencies, it was decided to include the entire population of government fellowship holders in the roster.

TABLE 2

1. American Scandinavian Foundation
2. American Association of University Women
3. American Council of Learned Societies
4. American Philosophical Society
5. John Hay Whitney Foundation
6. National Research Council
7. Rhodes Scholarship Trust
8. Social Science Research Council
9. Woodrow Wilson Fellowship Foundation

The Calculation and Distribution of Indices

BY FOLLOWING the principles and drawing from the sources indicated above, we were able to compile a roster of over 25,000 individuals who showed some signs of intellectual promise. For each of these individuals, we obtained the following information: (1) name and sex, (2) undergraduate institution of study, (3) baccalaureate year, (4) graduate institution of study, (5) major field of study, (6) nature of scholarship awarded, and (7) Ph.D. and year, if granted.

However, 67.7 per cent of the 25,000 individuals received their baccalaureates before 1946. And since, as has already been pointed out, the paramount purpose of this study is to describe the collegiate origins of promising younger scholars and scientists, it was decided to use only those individuals in our statistical analysis who received their baccalaureate in or after 1946. In the light of the interests motivating this study, we further restricted our analysis to only those awardees who graduated from colleges in this country. The statistical analyses that are to follow are thus based on some 7,000 individuals, who come from a large variety of different colleges and universities spread over the nation. In fact, every state in the union is represented by some undergraduate institution, and, in all, there are 562 colleges which have graduates since 1946 appearing in our roster. This is a large proportion of all accredited undergraduate institutions in the country. The 1948 edition of *American Colleges and Universities* lists over 800 such institutions. In terms of this reference, our roster thus includes individuals graduating from about two-thirds of the accredited American colleges and universities.

The first step in the analysis was to order the rosterees in terms of their undergraduate institutions of instruction. (In cases where any individual attended more than one college, he was assigned to the college from which he received his bachelor's degree.) After recording this information, we next computed estimates of the total number of graduates, both male and female, for each institution for the six-year period from 1946 to 1951. This estimate was arrived at by multiplying the sum of the graduating classes for the years 1948 and 1950 (as given in *Earned Degrees Conferred by Higher Educational Institutions* of the U.S. Office of Education) by a factor of 2.5. The reader will wonder why a factor of 3.0 was not employed in estimating the total graduates for the six years. Available records show that almost all institutions increased progressively in size from 1946 to 1951. Since our information on the classes of 1948 and 1950 came from the third and fifth years of the six-year series, we would have falsely overestimated the total number of graduates were the factor of 3.0 employed. By checking on several institutions whose records were available, we concluded that the factor of 2.5 would yield a more accurate estimate in most cases. In 22 cases we were unable to obtain complete information on the size of either the 1948 or 1950 class, and because of this fact these 22 institutions were dropped from our analysis. In all, then, we were able to make estimates of the size of the graduating class for some 540 schools for the period specified.

Once having estimated the size of the graduating classes, we were ready to compute one of the basic comparative statistics used throughout the study. This statistic is an *index*, relating the kind, source, or total number of awardees for each institution to the total size of the graduating class for that institution from 1946–51. If, for example, we were interested in obtaining for an institution its production rate of Ph.D. graduates, we would take the total number of Ph.D. awardees for that institution, divide this number by the total number of graduates of the institution for the six years, and the resulting quotient would then express the number of Ph.D. awardees per 1,000 graduates for that institution. Thus by computing such indices, we were able to take into account the differences in the size of various undergraduate institutions and to arrive at a single statistic which allowed us to make meaningful comparisons among the institutions.

For each institution we computed several varieties of such indices of distinction. First, an *All Male* index, that is, the total number of male awardees to the total size of the six-year male graduating class; and, similarly, an *All Female* index was computed for those institutions having women graduates. Indices were also calculated by sources of awards for male graduates. Thus, male Ph.D., male Private Fellow, male Government Fellow, and male University Fellow indices were computed. These indices were arrived at in a manner similar to that described above. That is, the total number of awardees for each of the several sources was divided by the institution's total graduates over six years. In a like manner, indices for male graduates according to field of study were calculated, viz., a male science index, a male humanities index, and a male social science index. In all, then, for each coed institution in our sample, we computed nine indices, and for all male institutions, a total of eight indices. While for all women's colleges, we calculated only one—an All Female index. It will be seen, however, in a later chapter that composite indices on women according to their field of study and source of award were also computed. The reason individual indices by field of study and source of award were not calculated for women graduates was because of the small number of women awardees, such that a breakdown by field or source could only be highly unreliable for any particular college. On the other hand, a composite index combining institutions, because of the greater number of individuals involved, more closely approached levels of adequate reliability. In fact, though individual male indices were computed by field and source for the various undergraduate institutions, we used composite male indices in our analyses, again because of their greater reliability.

Let us now look at the frequency distributions of undergraduate institutions on each of these indices (Table 3).

TABLE 3

	Science		*Social Science*		*Humanities*	
1.	Cal. Tech.	38.2	Swarthmore	19.9	Haverford	23.7
2.	Swarthmore	34.6	Reed	17.7	Kenyon	20.2
3.	Chicago	27.7	Chicago	14.6	Oberlin	12.2
4.	Reed	26.5	Univ. of the South	13.4	Queens	11.9
5.	M.I.T.	19.4	Carleton	12.8	Julliard	11.5
6.	Augustana	15.8	Oberlin	12.2	Princeton	11.2
7.	Oberlin	15.4	Antioch	11.8	Carleton	9.7
8.	Johns Hopkins	14.6	Haverford	11.1	Grinnell	9.1
9.	Antioch	13.8	Princeton	10.3	Knox	8.9
10.	Cooper Union	13.5	Yale	9.5	Reed	8.8
11.	Carleton	12.8	Pomona	9.3	Yale	8.4
12.	Purdue	12.5	Harvard	9.2	Harvard	8.0
13.	Cornell	12.0	Wesleyan	8.8	Trinity	7.5
14.	Brooklyn	11.5	Allegheny	7.8	Swarthmore	6.6
15.	Wooster	11.1	Grinnell	7.3	Chicago	6.3
16.	Princeton	10.9	Williams	7.1	Columbia	6.3
17.	Berea	10.6	Queens	7.1	Antioch	5.9
18.	DePauw	10.4	Monmouth	6.9	St. Olaf	5.8
19.	Harvard	10.1	Wabash	6.9	Hamilton	5.5
20.	Beloit	9.8	Amherst	6.1	Amherst	5.5

Discussion

THE results presented in the preceding sections of this report offer a number of occasions for comment, discussion, and speculation, especially, as we shall see, since several of our findings are quite at variance with results obtained in similar studies for an earlier period of time. But perhaps the first matter which should be commented upon is the concentration of scholarly creativity among a surprisingly small group of institutions. As we have noted in the text of earlier chapters, but 50 or so institutions out of the more than 800 granting baccalaureate degrees attain a rate of male production in excess of 10 per thousand, and among this group the range is from 10 to approximately 60. On the other hand, all the remaining institutions are crowded within a narrow range lying from 0 to 10 per thousand, with very heavy concentration at the lower value.

Thus it may be pointed out that the production of promising young scholars, in general, rests on a particularly narrow base within the American system of higher education, with some three score institutions, at the most, showing significant and impressive rates of production, while among the remainder the dedicated younger scholar is a rare exception among their graduates. Of course, it may be that this pattern of achievement is to be expected. In a certain sense, our indices of scholarly production are measures of creativity. Among individuals we suspect that this attribute of mind and character is so distributed that a very few possess creativity in a high degree while the majority possess it scarcely at all. Here, we see that this pattern is manifest among institutions as we believe it to exist among individuals. On the other hand, there can be little doubt that among the graduates of our unproductive institutions, there exist individuals of high talents who, exposed to a different sort of community influence, and favored with more stimulating educational opportunities, might well contribute significantly to American scholarship. While there are probably certain merits in specialization, so marked a degree of specialization in the production of younger scholars as we have found seems to us to approach a monopoly and to leave undeveloped and unproductive large segments of the American system of higher education. We strongly suspect that under proper circumstances, effective recruitment of younger scholars could be accomplished from many institutions now virtually barren of productivity.

A second matter, particularly worthy of discussion, is the pattern obtaining in the production of scientists. As with all types of scholars, we have noted that those receiving awards in science tend to come in highly disproportionate numbers from institutions of high costs. In fact, for every sample, the highest fifth for cost of attendance is at least twice as productive proportionally as the remaining institutions. Indeed, the factor of cost attendance proved consistently the most significant variable in differentiating productive from unproductive institutions. It will, therefore, come as a surprise to the reader to know that this is quite at variance with results obtained in earlier studies of the undergraduate origins of American scientists, and constitutes a result so divergent from earlier findings as to suggest that some major change has occurred in the American educational system within the past decade, at least respecting the undergraduate recruitment of scientists. In order that this divergence of findings may be clearly comprehended, it is necessary for us to review the results of our earlier studies.

Five years ago, under the chairmanship of Dr. Hubert Goodrich, a committee was established at Wesleyan to undertake the examination of the undergraduate origins of American scientists. Employing the third and seventh editions of *American Men of Science*, a listing was made of all individuals with doctorate degrees together with their undergraduate origins. A series of productivity indices, comparable to those employed in this study computed for undergraduate institutions, was then computed.[1] The baccalaureate classes covered by this index were for the years 1924–34. Briefly, our findings showed that the most

1. See R. H. Knapp and H. B. Goodrich, *The Origins of American Scientists* (Chicago: University of Chicago Press, 1952).

productive institutions were those of modest cost, middle and far western location, and drawing their students in good part from semi-rural regions. These findings prompted us to formulate a "grass roots" hypothesis of the origin of American scientists, a proposition consistently supported by our data. Noteworthy among our findings was the mediocre or inferior achievement of institutions of high cost, and the general inferiority of universities as compared with smaller liberal arts colleges. Our present results indicate that institutions of high cost are now unusually productive; that the geographic gradient earlier observed is now less sharply defined; and that universities and liberal arts colleges cannot now be differentiated in productiveness. Of these discrepancies, that involving cost of attendance is most challenging.

It seems to us that four considerations may be advanced to account for the reversal of findings here noted. First, it might be argued that the methods of sampling employed in the assembling of our roster introduced an advantage to eastern institutions of high cost and a disadvantage to middle western and far western institutions of lower costs. Thus our failure to include assistantships and possibly our sampling of only the largest 25 graduate institutions might be invoked in support of this position. Indeed, had we included assistantships, at the cost of greatly extending the labor and time required for our study, we might expect middle western institutions and possibly larger state universities of modest cost to benefit more than eastern institutions of high cost. However, it is important to note here that all of our types of awards including doctorates showed the marked superiority of institutions of high cost. Moreover, the differential between the top fifth and the remaining institutions with respect to the production of scientists is so marked that we cannot believe that our present findings are merely an artifact of sampling procedures.

The second explanation which might be invoked and to which we are strongly inclined to subscribe is the proposition that eastern institutions of high cost have in the past two decades undergone something of a revolution in character, both with respect to intellectual climate and with respect to student clientele. Three factors appear to us to have operated to bring about this change. The first is the existence of the G.I. bill, which has enabled, in the period covered by this study, students of superior abilities and motivations to attend these high-cost institutions. Second, apart from the veterans' program, eastern institutions of high cost now generally support scholarship programs at a level unknown or unenvisioned during the period of our earlier study. Third, the high general prosperity enjoyed by the nation in recent years has, we believe, made possible the attendance of individuals, previously excluded for economic reasons, at these institutions of high cost. Twenty years ago, the student bodies of these institutions were probably more distinguished for their ability to defray cost of attendance than for their serious intellectual dedication. In the most recent years, however, these three factors may have operated to open these institutions to individuals of superior abilities and scholarly motivation. With an abundance of student quality in attendance, the superior educational opportunities offered in this type of institution have been well employed, with the result that such institutions have achieved an exceptional record in the production of younger American scholars, including scientists.

The third explanation which might be invoked to explain the superior achievements of our institutions of higher cost lies in the altered condition of science. We think it likely that the dramatic achievements of science in the late war and the increased prestige and financial return which the profession has lately enjoyed, have made it an acceptable calling for economic and social classes which previously rarely selected this vocation. Thus in our earlier study we were able to demonstrate that scientists tended in the main to be recruited from humble circumstances, and that among institutions of high cost and wealthier student clientele, few students elected to concentrate in science, being strongly disposed to careers in law, medicine, and managerial professions. Possibly, the altered prestige and rewards of science have made the calling more acceptable to these groups, with the result that scientists are being recruited more frequently from these wealthier classes.

Finally, it may be that the intellectual appeals of science have somewhat altered. Twenty years ago, much of the scientific training and orientation in America was pragmatic in spirit and applied in content. Most scientists, we believe, would defend the proposition that science has taken a marked turn for the theoretical and intellectual in the past two decades, a process of maturation greatly stimulated by recent achievements in physics. Thus, the profession of science may have come to offer greater appeal to students of purely intellectual motivations with the consequent decline in the recruitment of students and pragmatic orientation, and an increase in their recruitment from among the ranks of those dedicated to pure scholarship.

In this connection, we should make some mention again of our earlier study. Though our "grass roots" principle seems most successfully established, we were able to note a small group of liberal arts colleges whose achievements in science were highly distinctive while their pattern could not well be encompassed with our "grass roots" scheme. Among these institutions, including Reed College, Oberlin, Antioch, and Swarthmore, we felt a different principle was operating. Here, it seemed to us, the superiority of the institution in the production of scientists was not attributable to its special dedication to science as among our "grass roots" institutions, but rather to their singular hospitality to intellectual values in general. We termed them institutions of "general intellectuality," not a happy phrase, but one intended to convey the fact that their scholarly achievements were spread across the entire spectrum of learning instead of being confined to science, and that the climate of values sustained by the institutions elevated the scholar and the intellectual to the position of "culture hero." It is most interesting to note that while our "grass roots" institutions fail in our present study to achieve the eminence they earlier enjoyed, Reed, Swarthmore, Oberlin, and Antioch stand at the very top of our list; Swarthmore and Reed occupying respectively the first and second positions. Thus, it may be that the changed pattern in the production of scientists may partly be described as a decline in the effectiveness of our "grass roots" institutions and the emergence in a number of eastern institutions of high cost of the spirit of "general intellectuality." This shift is probably largely due to the changed ethos of these eastern institutions induced in large part by their altered student clientele.

Oddly, the production of social scientists offers comparatively little occasion for discussion or comment. The general findings of our study seem to apply throughout to the social sciences, but in virtually no respect may this particular field of scholarship be singled out for distinctive comment. Thus, while the humanities and natural sciences show differential patterns for region or cost categories, the social sciences occupy a sort of colorless middle ground. This much, however, may be said: While the natural sciences draw most broadly from the American educational system, the social sciences have a far less extensive base; less extensive in that the number of institutions making significant contributions in this area is much smaller than in science. Even so, however, the base is less narrow than that of the humanities, and we may again point out the middle station of the social sciences.

Of the three fields of study, considered in our analysis, the humanities are conspicuous for the small number of institutions which make effective contribution to this area. Here we may note that among our universities, the highest fifth with respect to cost of attendance is approximately five times as productive of awardees in humanities as the remaining institutions. Moreover, with respect to geographic location, institutions of high production are sharply concentrated in the northeastern sections of the country and especially in New England. Thus the recruitment of scholars for this field is established upon a singularly narrow base within the American system of higher education. It is very doubtful, indeed, whether this concentration is to be desired, since it means that scholarship in this area is being advanced in great proportion by a small and geographically concentrated segment of our nation. It seems to us that this concentration is so marked as to threaten the intellectual health and vigor of these areas of learning.

One factor which prompted our initial interest in this study was its relevance to certain aspects of regionalism in the American nation. In our earlier study of scientists, we were impressed with certain irresolvable geographic differences in the production of scientists, differences so marked and unrelated to any obvious educational factors that we were impelled to invoke the cultural values of several geographic regions in explanation. But since we knew only the patterns obtaining with respect to science and had no knowledge of the origins of those pursuing social sciences or humanities, much of our discussion had to be speculative. It has been, therefore, with the greatest of interest that we have noted certain proportional differences in scholarly production from region to region. Thus the high proportion from the middle western regions electing the pursuit of science and the low proportion in humanities was quite consonant with our view that the regional culture of the middle west supported most strongly the pragmatic orientations of science. Conversely, we had every reason to believe that the eastern sections of the country would show relatively less distinction in science and a stronger inclination to pursue humanities. In this respect we were again confirmed by our present findings. However, certain other trends were quite unanticipated. The South, for example, which proved in our earlier study of scientists singularly unproductive, we had thought might achieve distinction in the humanities or social studies. This proves not to be the case. Indeed, while the southern states are generally unproductive in all areas, it is in science that relatively they achieve their greatest distinction. Similarly, we had expected that Catholic institutions would be marked by relatively large contributions to the field of humanities. In this speculation, however, we were again mistaken. Catholic institutions, though exceptionally unproductive in all areas of scholarship, achieve their best record in the sciences. It cannot be finally decided here whether the discrepancies between our present findings and these earlier hypotheses may be accounted for by historical changes in the past two decades, or whether our original estimates were incorrect, though we are inclined to subscribe to the latter view. Finally, we should note the discrepancy between the earlier high standing of the Mountain and Pacific regions, and their distinctly inferior status in our study. This constitutes a major reversal of pattern which is somewhat surprising, though the relatively high achievements of these far western regions like the middle western regions, in science is in accordance with expectation.

At the conclusion of this chapter, we should like to review certain limitations and anticipate possible criticisms of this study. At the outset, we made note of the inherent difficulties in assessing the contributions of undergraduate institutions to the ranks of American scholars in a period so recent as that covered by the present study. Of necessity, we had to rely upon sources which indicated, in the main, promise rather than achievement. In so doing, every effort was made to select sources which would be as free from arbitrary bias as possible. Indeed, the fact that each of our four types of awards shows essentially the same pattern in relation to our independent variables provides a sort of self-confronting validation of our procedure. Still, it must be frankly admitted that were this period to be examined a decade hence, after present graduates have established themselves securely in their scholarly profession, more confident and complete results could be obtained.

One criticism of the design of our study which was advanced on a number of occasions, and which we feel has considerable merit, is directed toward our failure, as we have previously noted, to include those individuals who have received assistantships at universities since 1946. The omission of this class of awards from our roster was motivated by several considerations. First, the number of such individuals would have at least doubled the size of our roster and greatly compounded the time and cost required for the study. This is especially so, in view of the fact that central files of such awards are not kept in many institutions, and such listings could only be obtained from the several departments within each institution. Again, there is the question as to whether assistantships in general represent as distinguished an award as fellowships which involve no service requirement. We are inclined, admitting many exceptions, to think not. However, the failure to include assistantships has probably worked somewhat to the disadvantage of institutions in the southern and western sections of the country and of institutions of lower cost. We profoundly doubt, however, whether the inclusion of such assistantships would materially alter the main results of our study and may point with some confidence to the pattern obtaining among Ph.D. awards where solid achievement is measured rather than the mere indications of promise.

A second criticism which might be directed against our study rests in our failure to sample all universities granting doctorate degrees. As we have noted, however, the 25 universities employed produced approximately 75 per cent of the doctorates awarded during the period of our study. The remaining 80-odd institutions granting doctorates, therefore, account for but a quarter of the total, and the costs in obtaining these data did not seem warranted. With respect to university fellowships, it is safe to say that our sampling includes 80 per cent of those granted by American universities. Again, we are confident that the inclusion of the remaining 20 per cent would not alter the basic results we have obtained, while the cost of such a procedure would have been very considerable.

A third criticism might be directed against our neglect to take into account the quality of individuals produced by different institutions.

This omission, however, could not be readily prevented for lack of any reliable method of assessing the individual recipient and his merit. Such a program of assessment might, to be sure, be undertaken, but it would necessarily involve an extensive program of investigation quite beyond the scope of the present study. It might be thought that some rough assessment of merit might be obtained by noting individuals with more than one distinction. Actually this was considered, but was rejected for two reasons. First, less than one in ten of the individuals dealt with in this study had received more than a single distinction of note. Second, individuals with more than one distinction seem to us more commonly those individuals of greater maturity rather than greater merit; that is, they have had time to receive two rather than just one award. These considerations, therefore, have led us to base our analysis upon the number of individuals rather than the number of awards, and to regard all awards as of equal weight. It seems clear to us that any adjustment which might have been made, short of a very extensive program of individual assessment, would not alter our general conclusions in the slightest.

Finally, we may note some criticisms, entirely legitimate, which may be advanced against our statistical procedures. We have pointed out earlier that the nature of our distributions of indices was such that conventional modes of statistical analysis could be employed only with great caution. However, we have employed composite indices as measures of central tendency and have employed chi square on occasion as a test of significance. Though both these procedures are open to criticism, the results obtained in our study appear to us so definitive in most cases, that these objections, though technically just, have little relevance to our conclusions. Again, our failure to undertake an elaborate statistical analysis of the interactions of our several independent variables seems to us excusable in view of the very cumbersome nature of our data and the fact that such interaction analyses if accomplished could scarcely expand our present knowledge greatly.

13 *Undergraduate Origins of American Scientists*

JOHN L. HOLLAND

THE ROLE of undergraduate institutions in the production of scientists and scholars is ambiguous, despite two comprehensive studies by Knapp *et al.* These studies suggest that certain colleges and universities are more productive of scientists and scholars than others —a kind of "institutional productivity" hypothesis. The results of the present study, however, argue for an opposed hypothesis, or a "student quality and motivation" hypothesis; namely, differential institutional productivity appears to be a function of the concentration of bright students at certain institutions and of differences in student motivation for scientific and scholarly achievement. Specifically, bright students congregate in institutions with high indices of scientific productivity; their explanations for their choice of college seem consistent with a need for scientific achievement, and they have fathers whose choice of occupation appears conducive to intellectual achievement. In contrast, bright students attend institutions with low indices of productivity in fewer numbers; they give explanations for their college choice which are suggestive of a more external or practical orientation to education, and they have fathers whose occupations suggest a fostering of persuasive or leadership skills and achievement.

The first study of this problem, by Knapp and Goodrich,[1] implies that certain colleges and universities are more productive than others of men who are later listed in *American Men of Science*. The 50 colleges in this group, for the period 1924–34, are characterized by their

---

Reprinted by permission from Science, *vol. 126, no. 3271 (Sept. 6, 1957), pp. 433–437.*

1. R. H. Knapp and H. B. Goodrich, *Origins of American Scientists* (Univ. of Chicago Press, Chicago, Ill., 1952).

small size, liberal arts orientation, and geographic location—middle and Far West. In a second investigation, Knapp and Greenbaum studied the undergraduate origins of promising young scholars.[2] For this study, the criterion of productivity is the number of graduate fellowships and scholarships and of Ph.D. degrees awarded per institution in proportion to the total number of their graduates for the period 1946–51. Taken together, these analyses imply that the scientist who has achieved distinction and his younger counterpart, the promising scholar in science as well as in other fields, originate the most frequently in two select sets of 50 colleges and universities and that the special qualities of these institutions facilitate achievement in science or scholarship in general.

The present study reexamines the "institutional productivity" hypothesis in the light of some new evidence obtained in the course of the National Merit Scholarship program. This scholastic talent search yielded several national samples of scholastically superior high-school seniors, with sufficient background data to explore three fundamental questions concerning institutional productivity. (i) Are differential rates of college productivity a function of differential attendance rates by scholastically superior students? (ii) Are differential rates of college productivity a function of student socioeconomic status? (iii) Are differential rates of productivity a function of paternal vocational motivation and the implied attitudes and values concerning science, scholarship, and achievement?

Basic Data

THE high-school senior samples studied in the following sections include four similar groups obtained in the National Merit Scholarship program for the years 1955 and 1956: (i) The winners in the 1955 program, 556 National Merit Scholars; (ii) a 10-percent sample of the Certificate of Merit students, or 430 near winners for 1955; (iii) a 73-percent sample, or 3000, of the 4226 Certificate of Merit winners who replied to a questionnaire concerning present college attendance and scholarships (this sample overlaps sample ii); and (iv) a 10-percent sample of the 7500 finalists for 1957, or 750 Merit Scholars and Certificate of Merit students. Groups i, ii, and iii represent the survivors of a nation-wide testing program. All students ranked at least in the top 5 percent of their graduating class. Their average

2. R. H. Knapp and J. J. Greenbaum, *The Younger American Scholar: His Collegiate Origins* (Univ. of Chicago Press, Chicago, Ill., 1953).

3. J. L. Holland and R. C. Stalnaker, "A descriptive study of talented high-school seniors: National Merit Scholars," unpublished paper; J. L. Holland and J. M. Stalnaker, "The effects of an honorary scholastic award upon subsequent training, scholarship aid, and personal status," *J. Higher Educ.* in press.

verbal and mathematical aptitude scores place them in about the top 2 percent of the graduating high-school population. Group iv represents a sample of the survivors of the 1956 program, in which 162,000 high school seniors participated. The average scholastic ability score for this sample places them in about the top 5 percent of the high-school graduating population. Within these samples, more than 71 percent of each male sample have chosen a scientific career. The concentration in science for females is about 35 percent. More explicit descriptions of these populations are provided in several other reports.[3]

All four samples are biased, since school participation in this program varies from state to state (22 to 80 percent) and since urban and larger schools participate more frequently than rural and smaller schools. No exact estimates of bias exist, but census estimates suggest that about 88 percent of the 1955 senior population did participate.[4] The sampling for the 1956 program is indeterminate with respect to scholastic ability, since students could be nominated by their principal or could enter the program by nominating themselves without regard to their class rank.

Results

ARE *differential rates of college productivity a function of differential attendance rates by scholastically superior students?* A first test of this hypothesis was made by calculating the "expected" percentages of student attendance for both the Knapp and Goodrich and the Knapp and Greenbaum "high productivity" institutions. These estimates were obtained by totaling the college population for each of the two criterion lists of 50 colleges and determining what percentages they represent of the total undergraduate population for the United States.[5] This simple method yielded "expected" percentages for groups of "high" and "low" productive institutions. These percentages were then compared with the "observed or actual" percentages of attendance for the Merit Scholar and Certificate of Merit samples of the 1955 program (groups i and iii). This analysis is contained in Table 1.

The data in Table 1 reveal that these talented groups attend "high" productive schools in frequencies which are 3 to 15 times the "expected" frequencies. For example, 7.3 percent of the Certificate of Merit males would be "expected" to attend the 50 Knapp and Greenbaum "high" productive colleges, and 92.7 percent would be "ex-

4. Dittoed table from the U.S. Department of Health, Education, and Welfare, Office of Education, Research and Statistical Services Branch, Washington, D.C., 1957.

5. *Opening (Fall) Enrollment in Higher Educational Institutions* 1955, U.S. Dept. Health, Educ. and Welfare, Office of Educ., Circ. No. 460 (Washington, D.C., 1955).

pected" to attend all other "low" productive colleges. Their actual attendance, however, at "high" and "low" productive colleges is 42.8 and 57.2 percent, respectively. Their actual attendance, then, at "high" productive colleges (42.8 percent) is about 6 times their expected attendance rate (7.3 percent). Similarly, their attendance at "low" productive colleges is less than their expected attendance. These results are consistent and statistically significant for both criteria. It is interesting to note that these trends hold even when the Knapp-Goodrich criterion is applied to female samples, although that criterion was devised to assess the institutional productivity of male scientists only.

It appears significant, too, that the productivity ratios for "high" productive schools approximate the ratios of students which they attract beyond their expected quotas. For the Knapp and Goodrich criterion, the median productivity ratio of "high" to "low" productive schools is about 4 to 1. High-ability students are attracted to these colleges in ratios of about 3 and 4 to 1. For the Knapp and Greenbaum criterion, the median productivity ratio of "high" to "low" productive institutions is about 4 to 1 for males and about 11 to 1 for females. The corresponding ratios of "observed attendance" of high-ability students to "expected attendance" are about 6 and 12 to 1 for males and females, respectively.

The present results are in agreement with the finding of Knapp and Goodrich, who report a correlation of 0.39 between scholastic ability and institutional indices of productivity for a sample of 50 col-

TABLE 1. Percentage of Expected and Actual Attendance of High-Aptitude Students at "High" and "Low" Productive Institutions. Each 2-by-2 Table Is Significant beyond the 0.001 Level of Significance

	MALES				FEMALES			
Criterion	Cert. of Merit College Attendance		Merit Scholars College Attendance		Cert. of Merit College Attendance		Merit Scholars College Attendance	
	Expected	*Observed*	*Expected*	*Observed*	*Expected*	*Observed*	*Expected*	*Observed*
Knapp and Greenbaum								
High productivity	7.3	42.8	7.3	43.3	1.6	16.2	1.6	24.7
Low productivity	92.7	57.2	92.7	56.7	98.4	83.8	98.4	75.3
Total	100.0	100.0	100.0	100.0	100.0	100.0	100.0	100.0
Knapp and Goodrich								
High productivity	3.9	13.4	3.9	16.2	3.9	14.9	3.9	17.5
Low productivity	96.1	86.6	96.1	83.8	96.1	85.1	96.1	82.5
Total	100.0	100.0	100.0	100.0	100.0	100.0	100.0	100.0

leges. Their limited data did not permit, however, a more complete exploration of this relationship or an estimate of the concentration of bright students in institutions with high indices.

It is possible that student knowledge of the colleges' productivity indices may have a marked effect on their choice of college, so that talented students shop for colleges with high indices. Accordingly, a sample of 7500 finalists (group iv) in the 1956 National Merit Scholarship program was studied in order to explore this possible bias.

The data for this analysis were obtained from scholarship application blanks in response to the query, "Why have you selected the college named on page 1 [the student's choice]?" The students who had selected one of the 50 institutions that ranked highest in productivity in the Knapp and Goodrich study were contrasted with an equal, random sample of students who had selected other colleges and universities. The blanks were reviewed for indications that the students had any knowledge of the productivity studies that are reported here or of similar information, the implication being that such information might have influenced their choice. This search was negative.

To judge from their verbal reports, students select colleges largely from considerations of convenience, cost, familial affiliations, academic reputation, and so on. No student mentioned the productivity articles, but one of the 1628 students sampled appeared to have some secondhand knowledge of this material. It appears, therefore, that the differential attendance of high-ability students at the 50 selected institutions was not influenced by publication of the Knapp studies; an explanation of the differential rate must be sought elsewhere.

Are differential rates of institutional productivity a function of differential socioeconomic status among student populations? Since earlier work by Knapp and Goodrich suggests that scientists may originate more frequently in "lower," rather than in "higher," socioeconomic groups and that conceivably, then, institutions with high indices of productivity may attract larger proportions of students of lower socioeconomic status, a test of this hypothesis appeared desirable.

For this test, the socioeconomic status of groups i, ii, and iv was determined by means of the Minnesota Scale for Paternal Occupations. For males, the distributions of fathers' occupations were found to be not significantly different for "high" and "low" productive institutions. For females, the differences within groups i and ii are not significant. The distributions of status for group iv—a 10-percent sample of 1956 finalists—reveal statistically significant differences. Students selecting "high" productive schools come from higher socioeconomic status than those selecting "low" productive schools. The efficiency of this discrimination is moderate. However, since five out of six tests are clearly negative, there is little evidence that differences in institutional productivity are due to different levels of socioeconomic status among student bodies.

Are differential rates of institutional productivity a function of parental vocational motivation and their implied attitudes and values concerning science, scholarship, and achievement? Restated, students attending "high" and "low" productive colleges may represent different, socially derived familial vocational motivations which might account for the differential productivity of scientists between these groups, and, by implication, "high" and "low" productive colleges may attract different kinds of people, or personalities.

Groups i, ii, and iv were employed to test this hypothesis. The fathers' occupations for these samples were classified in one of the following six vocational interest groups: physical activity (manual and skilled trades), science, welfare or social service, clerical or business detail, persuasive or verbal activity (oral language occupations, including selling, law and supervisory occupations), and esthetic or artistic occupations.[6]

Distributions were then formed for institutions with high and low indices of productivity for each sample. The Knapp-Goodrich criterion was employed in this analysis, since it appears to be the more valid of the two criteria—valid in the sense that it represents an ultimate, rather than an immediate, criterion. The results of this classification for males are shown in Table 2. The results for females are clearly negative for all samples and are not reported.

Table 2 displays two meaningful trends in all three samples. Students attending or selecting "high" productive colleges tend to have fathers who are engaged in occupations characterized by their physical-activity, scientific, or social-service aspects. In contrast, students in "low" productive colleges have fathers who are engaged in occupations which are characterized by their persuasive, sales, supervisory, and leadership orientation. The differences for the "persuasive" category are statistically significant at the 0.01 level for groups i and iv and at the 0.06 level for group ii. The remaining categories are not statistically significant; however, the grouping of the physical-activity, science, and welfare occupations and of the clerical, persuasive, and esthetic categories for an over-all test is significant beyond the 0.05 level for all three samples. It should be emphasized, however, that the total difference between "high" and "low" productive samples is largely due to differences found for the "persuasive" category.

6. Specific criteria for classification were based on six keys from the *Holland Vocational Preference Inventory* and an unpublished inventory, which are in general agreement with classification schemes suggested by E. K. Strong, Jr. [*Vocational Interests of Men and Women* (Stanford Univ. Press, Stanford, Calif., 1943)], L. L. Thurstone ["A multiple factor study of vocational interests," *Personnel J.* 10, 198, (1931)], P. E. Vernon ["Classifying high-grade occupational interests," *J. Abnorm.*

TABLE 2. Percentages of Fathers' Occupations Falling in Various Fields of Work for Students Selecting "High" and "Low" Productive Institutions (Males)

Field of Work	Group i Merit Scholars		Group ii Cert. of Merit		Group iv Finalists	
	(No. = 65) H.P.	*(No. = 337) L.P.*	*(No. = 47) H.P.*	*(No. = 261) L.P.*	*(No. = 76) H.P.*	*(No. = 449) L.P.*
Physical activity	24.6	18.7	19.0	15.5	21.1	17.8
Science	16.9	12.2	21.4	15.5	15.8	12.7
Welfare	12.3	10.9	16.7	9.1	14.5	7.1‡
Clerical	12.3	10.7	4.8	8.3	6.6	7.8
Persuasive	21.5	40.1*	28.6	44.2†	27.6	46.1*
Esthetic	3.1	2.9	4.8	0.0	0.0	1.1
Unclassified	9.2	4.5	4.7	7.4	14.4	7.3
Total percentage	99.9	100.0	100.0	100.0	100.0	99.9

* Difference between percentages significant at 0.01 level.

† Difference between percentages significant at 0.06 level.

‡ Difference between fathers' occupations for students selecting institutions with "high" and "low" indices of productivity is significant at 0.05 level of significance.

The implications of these findings are several. They suggest that the "high" and "low" productive colleges attract somewhat divergent student populations among male, high-aptitude high-school seniors. On the one hand, "high" productive colleges attract students whose fathers work with their hands (machines or tools), with scientific ideas or apparatus, or with people in a social-service sense (especially teaching). Such a background appears conducive to achievement in science or to an emphasis on intellectual attainment. In contrast, "low" productive schools attract students whose fathers' work is characterized more by its oral, persuasive, or leadership activities (in particular, supervisory and ownership positions in business, law, and government). Correspondingly, these backgrounds appear to be less fertile for the development of young scientists.

These parental vocational differences also suggest differential values and attitudes toward achievement in American society. Earlier psychometric analyses indicate that interest in science is associated with theoretical values, with a relative deemphasis on political and economic values; business interests are positively correlated with political and economic values, with a deemphasis on theoretical and scientific values. This evidence has been summarized recently by Roe.[7]

(Soc.) Psychol. 44, 85 (1949)], and others. This method required some subjective judgment, since the reference lists are not exhaustive. Distributions for a sample of 556 ratings are not significantly different for two raters, and the differences for a single category average 2.6 percent.

7. A. Roe, *The Psychology of Occupations* (Wiley, New York. 1956).

The verbal and social quality of the "persuasive" occupations suggests another area of disparity in the backgrounds of students selecting "high" and "low" productive colleges. Studies by Stern *et al.*,[8] Roe, and others consistently reveal that scientists as a group are less social than many other occupational groups, especially men in sales and leadership occupations. It seems reasonable to assume, then, that the children of scientists and workers would acquire somewhat similar social skills and interests and experience which, in some measure, would impel them toward achievement in relatively non-social pursuits. Conversely, the children of fathers who have oral and leadership skills may have their verbal and social skills fostered so that their achievement is directed toward verbal or language fields. Such a formulation implies that good business and government leaders may originate the most frequently in institutions which produce relatively few scientists, and vice versa.[9]

This loose theorizing suggested the need for a more elaborate analysis of the college-choice evidence reviewed previously for information about productivity only; that is, student explanations of college choice for groups selecting "high" and "low" productive colleges should contain differences which are consonant with the foregoing formulations. The test of this formulation follows.

Verbal Explanations of College Choice

FOR this analysis, the verbal explanations of college choice for all 1956 finalists (group iv) selecting "high" productive colleges were compared with those of an equal random sample of students selecting "low" productive institutions. Male student statements were classified by two graduate student judges having no knowledge of either this report or of similar reports. Female statements were classified by a single judge. The discrepancies in the two judges' coding of verbal reports range from 0.4 to 6.6 percentage points.

For males, the statistically significant differences in Table 3 reveal that a college of "academic standing, small size, research reputation, or liberal arts orientation" is desired more frequently by students selecting institutions with high indices of productivity. Students choosing institutions with low indices are less concerned with these qualities and are more concerned with attending a "good college, one which is close to home, or in a desirable location."

The differences for females form a similar pattern. Women selecting "high" productive colleges prefer a college of "small size" and a "liberal

8. G. G. Stern, M. I. Stein, B. S. Bloom, *Methods in Personality Assessment* (Free Press, New York, 1956).

9. In a personal communication R. H. Knapp reports limited confirmation of this hypothesis. With scholastic ability partialed out, Knapp obtains a correlation of

arts orientation." Among women selecting a college with a low index, these qualities are less valued, and attending a "good college, or one of prestige" is held more valuable.

In all, the significant differences in Table 3 appear consistent with the formulation outlined earlier. To interpret, the selection of institutions with high indices appears to be a function of needs for a small social group or organization and a broad, intellectually oriented educational experience. In contrast, the selection of institutions with low indices appears to reflect confidence in living in a large, social group and a more external—perhaps more practical—orientation, including considerations of "prestige" and a "good college."

TABLE 3. Summary of Verbal Reports of College Choice by High-School Seniors Selecting "High" and "Low" Productive Institutions

Reported Reasons for Choice of Institution	Males (No. = 1088)		*Significant Difference*	Females (No. = 540)		*Significant Difference*
	H.P.	*L.P.*	*P*	*H.P.*	*L.P.*	*P*
Good college	47.9	55.1	< 0.05	35.6	45.6	< 0.05
Close to home	14.3	18.4	< 0.07	12.9	10.0	
Low cost	3.9	5.5		2.9	5.9	
Academic standing	20.6	15.6	< 0.05	33.3	27.4	
Small size	27.9	8.5	< 0.001	38.5	15.2	< 0.001
Recommended by friends and others	15.1	15.8		12.6	11.5	
Good faculty	10.3	8.6		4.1	6.7	
Prestige of college	6.3	5.1		0.0	5.2	< 0.001
Desirable location	9.9	13.6	< 0.06	9.6	10.7	
Religious affiliation	4.2	6.3		8.9	11.1	
Physical facilities	6.8	8.1		0.7	1.9	
Research reputation	2.0	0.2	< 0.01	0.4	0.4	
Liberal arts orientation	17.1	11.2	< 0.01	22.2	10.0	< 0.001
Miscellaneous	25.2	18.6		15.9	29.6	

This evidence also tends to negate the assumed effects of the small, liberal arts college strongly implied by the Knapp studies. The student concentration and motivational hypothesis appears more probable; that is, small, liberal arts colleges may attract a disproportionate number of students who are consciously seeking a small, liberal arts college and are relatively homogeneous and potentially high achievers with respect to scientific and scholarly eminence. And while these criterion colleges

—0.40 between the production of scientists and the production of lawyers among a selected group of institutions by employing the Knapp-Goodrich index for scientists and a similar index developed for determining the production of lawyers.

may have beneficial and nurturing qualities, the student talent and motivation they receive by virtue of their reputation and orientation may outweigh the college influence itself. In short, differential student populations among colleges appear as a more probable explanation of differences in productivity than the special qualities of individual institutions.

Discussion

THE validity of this study rests largely on its sampling of high-school seniors in the United States and on several psychological assumptions concerning the relationship of scientific eminence or scholarship to scholastic aptitude and achievement at the high-school senior level. For 1955, the high-school senior samples appear to be approximately representative of high-aptitude seniors, since an estimated 88 per cent of the eligible seniors did participate. Census estimates reveal that the non-participating 12 per cent is concentrated largely in 13 southern states, a bias which suggests the effect of a more representative sample.

It is likely that the concentration of high-aptitude students in "high" productive colleges revealed by this study would be unaffected by a complete sampling, since these institutions occur very infrequently in the South and with great frequency in the Far West and Middle West. This possibility is reinforced by a previous report which reveals a marked tendency for students to attend colleges close to home or in their own geographic region.[10]

With respect to psychological assumptions, it is assumed that scientists and scholars are drawn generally from students who are characterized by their high academic achievement in secondary schools and by their high scholastic aptitude and verbal and mathematical aptitudes. This assumption is supported by extensive studies reviewed by Wolfle.[11]

This study assumes further that early academic achievement is predictive of later scientific eminence (achievement) and scholarship in other areas. For this assumption there is also extensive support. Studies by Cox,[12] Terman,[13] and others reveal that later achievement has many forerunners of which academic achievement is only one.

The present study militates strongly against the "institutional hypothesis." In opposition to the latter formulation, the present evidence suggests that differential productivity is probably a function of differential student populations which may be characterized by (i) their

10. National Merit Scholarship Corporation, *First Annual Report* (Evanston, Ill., 1956).

11. D. Wolfle, *America's Resources of Specialized Talent* (Harper, New York, 1954).

divergent proportions of high-aptitude students and (ii) their differences in motivation (scientific and intellectual activities as contrasted with business and oral activities) as well as by a number of implied differences in attitudes and values. The second hypothesis appears to parallel the work of Stern, Stein, and Bloom, who report marked variability in percentages of personality types or patterns within five divergent institutions. Briefly, high productivity appears to be an expected result of working with an intellectually well endowed student body which tends coincidentally to have a modal orientation toward scientific and scholarly goals. It appears more probable that college productivity indices should be attributed to student characteristics rather than to institutional influences.

This formulation might be modified further by the time dimension. It appears possible that some institutions in their early history may have stimulated in a substantial manner students who later achieved the Ph.D. degree and eminence in science and in other areas and that through such activity these institutions acquired reputations which differentially attracted desired student groups and repelled undesirable ones. As an end-result, investigators are faced with the problem of disentangling the influence of an institution from the quality and character of its student body, both of which are probably changing. In addition, there may be a snowballing effect of college influence, which may be out of proportion to its real effects. This sequence might occur in the following manner. (i) A college influences students to achieve in areas of science and scholarship; (ii) a few talented students respond by attaining fellowships and Ph.D. degrees, and perhaps one becomes a recognized man of science; (iii) prospective students are impressed, possibly by these achievements but more probably by such evaluations as "tough school" and "high standards"; (iv) brighter students with high intellectual goals enroll in great numbers; (v) the odds for the success of college influence are now greater; (vi) there is more achievement; (vii) more bright students enroll, with science or other appropriate scholarship orientation, and students with low academic motivation or motivation for goals which represent the minority of the student body find more suitable institutions; (viii) and so on. This oversimplified analysis suggests that teacher influence may have marked effects on scholar productivity in the early years of the history of a college, but that this influence may decrease to the point where the level and character of the student population become of much greater importance as the institution develops in time.

12. M. Cox, *Genetic Studies of Genius, II: The Early Mental Traits of Three Hundred Geniuses* (Stanford Univ. Press, Stanford, Calif., 1926).

13. L. M. Terman and M. H. Oden, *The Gifted Child Grows Up* (Stanford Univ. Press, Stanford, Calif., 1947).

Summary

AN analysis of the college attendance or college choice for four high-aptitude, high-school senior samples suggests that the differential institutional productivity of scientists and scholars is a function of the differential college attendance, paternal vocational motivations, and their implied correlates among high-aptitude students. This formulation appears to be more probable for males than for females. The institutional productivity hypothesis proposed in previous studies is not supported by the present evidence.[14]

14. I wish to acknowledge the fundamental work of Robert H. Knapp, Hubert B. Goodrich, and Joseph J. Greenbaum, who provided the basic formulation for this problem, in which I merely substituted some new values and samples. I am indebted also to Donald L. Thistlethwaite for a critical review of this paper. This study was partially supported by research grants from the National Science Foundation and the Old Dominion Foundation.

The Social Image of the Scientist and His Self-Conceptions 3

THE six selections in Part 3 of the Reader are concerned with the self-images of the scientist, the problems of identification and strain in his work role, and the conceptions that others have of the scientist's role and personality. These selections might, perhaps, be more accurately described as parts of a "social psychology" of science, for they show the interaction of role and personality rather than being concerned with role-structures alone. The interdisciplinary nature of the type of inquiry being carried on in these studies is revealed, it would seem, in the fact that the several authors come from such different disciplines as medical psychiatry, anthropology, social psychology, sociology, and political science.

In the first selection, Dr. Kubie, a practicing psychoanalyst, raises a series of problems about the scientist's self-images, his personality tensions, and his role conflicts. Dr. Kubie's discussion is based on his own wide experience in psychoanalytic practice and is admittedly not necessarily representative in the sense that all scientists must undergo the tensions and conflicts he describes. How representative these tensions and conflicts are could only be answered by empirical research that has not yet been done. When such research is undertaken, Dr. Kubie's discussion will be a useful source for hypotheses. In the meantime, Dr. Kubie is to be read as describing some problems that are recurrent for some proportion of scientists. His essay is a valuable beginning.

The two studies by Mead and Métraux and by Beardslee and O'Dowd are descriptions of the images of the scientist that are held by American high-school and college students. The Mead and Métraux study takes its data from a nation-wide sample of essays written by high-school students who were allowed to produce the images they have of the scientist's role and personality in response to semi-structured questions. The Beardslee and O'Dowd data are limited to four sets of college students in the northeastern United States. The central finding of the Mead and Métraux study is that while students like the products of science, they are not strongly attracted to becoming either scientists or the wives of scientists. Mead and Métraux suggest a number of ways in which the images about the rewards of being a scientist might be changed so that more people might want to become scientists and so that everyone would understand them better and thus facilitate their work. Beardslee and O'Dowd also show that the image of the scientist has some negative shades for college students. Since it is likely that there are negative shades for all occupational roles, as well as positive ones, it is a special virtue of the Beardslee and O'Dowd study that it makes comparisons between scientists and other occupational types. A fuller understanding of how people see and feel about the scientist's role lies in this direction of comparative analysis, to refine our understanding of the favorable and unfavorable aspects seen in the scientist's and other roles.

Another fascinating source of data about the images of the scientist in contemporary America is science fiction, which is read by many scientists as well as nonscientists. In his use of the content analysis method to discover the image of the scientist in science fiction, Hirsch has shown that here too there are negative shades, but that the businessman fares much worse than the scientist. The scientist and his influence on society is seen, on the whole, in an optimistic light.

In Hall's study of the images that politicians in America have of scientists, there is an important departure from the approach of the previous researches. Hall focuses on a selected group that is of great and direct importance to the scientific enterprise, rather than on the larger and more undifferentiated groups that make up high-school or college students or readers of science fiction. It would be of great value to have the counterpart of Hall's research, a study of how American scientists see politicians, for this would indicate much about the scientists themselves. The ways in which scientists participate in the political process, as they increasingly do, may be much *affected by the images they have of politicians and of themselves in relation to politicians. Do they, for example, perceive themselves as necessarily "more objective" than politicians even in political matters?*

Finally, in the paper by Becker and Carper, which usefully compares graduate students in physiology, philosophy, and mechanical engineering, some of the common and different elements of identification with a professional occupational role are analyzed. It is this kind of comparative study that can highlight the specific images that scientists and would-be scientists have of themselves. The different views of university positions which physiologists and engineers have, for example, are brought out in this study.

14 *Some Unsolved Problems of the Scientific Career*

LAWRENCE S. KUBIE

1. *Problems Arising Out of the Universal Neurotic Process*

IT IS MY THESIS that the life of a young scientist challenges our educational system from top to bottom with a series of unsolved problems which await investigation. Among these are certain subtle problems, arising out of unrecognized neurotic forces, which are basically important both in the choice and in the pursuit of scientific research as a career. This will constitute Part I of this paper. Part II will consist largely of a discussion of socio-economic dilemmas which also influence the scientist's emotional and intellectual career. It will not be the purpose of this paper to argue that every young scientist should be psychoanalyzed; and the reader is asked to keep this disclaimer in mind. Nevertheless, in any multidisciplinary investigation of these complex interrelated problems, I believe that the psychoanalytic study of a random sampling of scientists, both young and old, would be one of the essential instruments.

In this connection it will be argued that Science in the abstract and Scientists as human beings pay a high price for the fact that during the preparation of young people for a life of scientific research their emotional problems are generally overlooked. This discussion will not attempt to outline a full remedy for this neglect, for it would be premature to make such an attempt before more is known about the problem. Here again, however, in exploring possible remedies, a psychoanalytic study of an adequate sample of young scientists would provide

---

Reprinted from American Scientist, *vol.* 42 (1954), *pp.* 104–112.

information which would help towards an ultimate solution. I hope that this paper will also contribute to a more general recognition of the fact that many young scientists require special help in their struggles for emotional maturation.

My own clinical experiences with this group suggest that the emotional problems which arise early in the careers of young scientists are more taxing than are those which occur in other careers. Yet without instruments with which to measure and compare these imponderables, this cannot be proved. Nor can it be claimed that the problems to be detailed below are peculiar either to science in general or to any special field of science. In fact, since the stresses which arise in different careers have never been systematically compared, it cannot even be determined whether or to what extent emotional problems vary from one career to another, either in degree or kind. Therefore, the reader may well ask why this paper is published before such investigations have been carried out. Its justification lies in the fact that such studies will themselves require a large investment of time, money, and trained personnel, none of which will be made available until responsible educators become convinced that such studies are essential enough to justify careful planning, a co-ordinated multidisciplinary approach, and generous financial support. Before such studies are made, all that we can do is to indicate fragmentary observations, which suggest that it would be enlightening to make socio-economic studies of the lives of young and old scientists, plus psychoanalytic studies of a statistically adequate random sample of them.

These investigations would throw light on such problems as: (*a*) the special stresses, both economic and psychological, which occur in the life of the young scientist; (*b*) the great variety of conscious and unconscious forces whose interplay determines a young man's choice of scientific research as a career; (*c*) the interplay of conscious and unconscious forces in his subsequent emotional and scientific maturation; (*d*) how the special stresses which develop later in life react upon the earlier emotional forces which originally turned him towards science; (*e*) how unconscious stresses influence the young investigator's general approach to scientific research and scientific controversy; (*f*) how the unconscious symbolic significance of particular scientific problems and theories can distort the logic and the judgment even of men of exceptional ability. This article will attempt only to illustrate the wide variety of problems which are relevant to these general headings.

As a personal note, I should add that my observations, both on myself and on colleagues in various fields of science, have been made at random over a period of nearly thirty years. They began in the twenties, when I was working in one of the laboratories of the Rockefeller Institute. It became known that I had had some previous training in psychiatry.

Presently I found that if I were to have any time for my own work I had literally to lock my door for a few hours each day. Otherwise, almost every afternoon, young colleagues and sometimes older ones would drift in to talk, not about scientific issues but about their personal problems. At that time, my psychiatric training and experience were limited, and I knew nothing at all about psychoanalysis. Yet these random and unsolicited revelations made it clear not only that, as one would expect, a scientist's ability to endure the prolonged frustration and uncertainties of scientific research depend on neurotic components in his personality (both masked and overt), but also that there are significant relationships between masked neurotic components in the personality of the apparently normal scientist, and such things as (*a*) the field of work which he chooses; (*b*) the problems within that field which he chooses to investigate; (*c*) the clarity with which he habitually uses his native capacity for logical thinking; (*d*) the ways in which he attacks scientific problems; (*e*) the scientific causes which he espouses; (*f*) the controversies in which he becomes entangled and how he fights; and (*g*) the joy or sorrow which is derived from the work itself and also from his ultimate success or failure. Thus over the intervening years I have seen men of imagination and erudition whose scientific lives were nonetheless baffled and unproductive, and also men with lesser gifts who seemed to function freely, creatively, and productively; scientists who were happy in spite of failure, and others who became depressed in spite of acknowledged and recognized success. Although such facts were new to me twenty-five years ago, they had long been an accepted part of human wisdom. This makes it strange that their deeper sources in human nature and their special importance to scientific workers have never been systematically explored. I cannot attempt such an exploration here, but it may be possible to make articulate the challenge to all scientists which lurks in these ancient and unexplored caverns of the human spirit.

1. THE EMOTIONAL EQUIPMENT WHICH THE YOUNG SCIENTIST BRINGS TO HIS CAREER

The young scientist often reaches maturity after a lopsided early development. In this development he resembles many other intellectuals. A typical history is that an intellectually gifted child develops neurotic tendencies which hamper his early aggressive and psychosexual development. If at this point he is intellectually stimulated by one or another of the emotionally significant adults of his life, he is likely to turn away from athletics and the social life which he finds difficult to more bookish activities, thus postponing indefinitely any facing of earlier challenges. If success rewards his consolatory scholarly efforts during adolescence, he may in later years tend to cultivate intellectual activity exclusively. In this way absorption in the intellectual life will frequently be paralleled by an increasing withdrawal from athletic and social and psychosexual activities. As a result, by the time adult life is reached his only triumphs and gratifications will have been won in the intellectual field, his range of skills will have become restricted, and the life of the mind will be almost the only outlet available. Because of the extra drain of the laboratory on the student's time, the young man who sets out to become a scientist spends his adolescence putting every emotional egg in the intellectual basket to a greater extent than is true for most other young intellectuals. By such steps as these, the sense of security and the self-esteem of the young intellectual come to stand on one leg; so that when research is begun he invests in it a lifetime of pent-up cravings. After such a development, it is inevitable that scientific research will be supercharged with many irrelevant and unfulfilled emotional needs; so that the lifework of the young scientist tends to express both the conscious levels of his intellectual aspirations and his unfulfilled instinctual needs and unconscious conflicts.

Even the most brilliant scientific successes cannot solve unconscious personal problems, nor gratify unrecognized instinctual pressures. Whenever anyone works under the whiplash of unsolved unconscious conflicts, whether he is painting a picture, writing a play, pursuing a scientific discovery, or making a million dollars, the individual is prone to work with desperation. If there is failure, he blames unhappiness on his failure. But, to his amazement and dismay, he discovers that depression may follow success no less than failure. Basically, this is because success also leaves his deeper problems unsolved. If we always bear in mind that the pursuit of unconscious and often unattainable needs plays a determining role in the intellectual career, the familiar phenomena of depression attending success would not perplex us. We should wonder rather at the shortsightedness of a process of scientific education in which self-knowledge is the forgotten man, and in which emotional maturation is left to chance.

2. THE CHOICE OF A CAREER

The aspects of this vexing problem which are peculiar to a career in science require certain general considerations. I suppose that it is not inaccurate to say that, of the many unsolved problems of human life, two which are of major importance are how to enable successive generations to learn from the mistakes of their predecessors without repeating them, and how to make it possible for young people to anticipate the future realistically. Not literature nor the arts or formal education has solved these two problems, which are interdependent in every aspect of life. Both are relevant to the choice of a career. When a youth decides to become a doctor, a lawyer, a businessman, or an artist, the decision is not made on the basis of a realistic foreknowledge of what one of these careers would be like as compared with another, nor out of a deep introspective knowledge of himself and of how he would fit into the lifework he has chosen. Even if his own father is a lawyer or a doctor, he will have had an opportunity to observe only the outer aspects of that life, the dramatization of its activities; he cannot have felt its joys and sorrows directly. What he will have experienced vicariously through identification with his parent will depend less upon what that life was really like than upon how it affected him; and upon the subtle balance of conscious and unconscious, hostile and loving components in his identification with the parental figure. Nor do adults know how to communicate the truth about their own adult lives to their children. Consequently, the adolescent's and even the college student's anticipation of the quality of life in any future career is dominated by fantasies. To a remarkable degree this is true even of more familiar and humdrum careers. The quality of adult living belongs to the remote and mysterious future; it is something the flavor of which the child cannot anticipate. Until this obstacle to communication between the generations is overcome, successive generations will continue in the future, as in the past, to make their choice in the darkness of fantasy and confusion. The child of a wealthy broker who was "on the street" took these words literally, as children do, and looked for his father in every pushcart peddler who passed. Although the visual misconception of the child was corrected as he matured, an emotional hangover remained which had an important influence in determining the choice of his subsequent lifework. Usually the less familiar the career which a young man chooses, the greater will be the importance of fantasies, both conscious and unconscious, among the forces which determine the initial choice of a career, and also the subsequent adjustment, happiness, and effectiveness in the one selected.

One natural conclusion to be drawn from these considerations would seem to constitute an argument for the wider use of aptitude testing in the choice of careers. Actual experience, however, and a hard-headed and realistic skepticism make one cautious about expecting too much help from these devices. The most extensive trials of the value of aptitude testing were the so-called "Stanines," which the USAAF developed during the war for the screening of air cadets and their allocation to training as pilots, bombardiers, and navigators. In terms of its relevance to this problem, I would summarize the results of this experience as follows:

a. The tests of aptitudes were remarkably accurate as far as they went.

b. It was possible to sort out those with the automatic speed and motor skills and/or the mechanical precision needed for various tasks.

c. Men were placed accurately on a point scale as to their relevant psychometric and neuromuscular capacities.

d. In this way, the tests selected accurately a small group at one extreme, most of whom would succeed in training, and another small group at the opposite pole, most of whom would fail. (There were exceptions to the results even at both extremes.)

e. As was to be expected, however, the vast majority of the men tested fell into the central zone of the normal curve of distribution, while only a relatively small percentage of the tested population was placed at the two extremes.

f. With rare exceptions, the individuals who fell into the extremes knew their own aptitudes and ineptitudes before going through any tests. From their experiences at play, in sports, in school, and on various jobs, they knew already that they were specially adept or specially maladroit with respect to certain types of activity. Indeed the representatives of the two extreme ends of the scale were usually able to describe their strong and weak points almost as precisely as these could be measured.

g. Consequently, the tests are of greatest use when jobs are scarce in times of peace, or else in times of war when a lad may want desperately to be accepted by a special branch of the service for some particular position, and may therefore exaggerate his native skill or hide his native ineptitude. But at other times when there is no special incentive to deception (beyond the usual need for self-deception), the men at the end zones need no tests.

h. The next important lesson of the entire experiment with the "Stanines" was that for the majority, who fall in the great middle zone of the normal curve of distribution, although their minor variations in aptitudes can be measured with considerable precision by various "human engineering" devices, these variations do not determine either success or failure, happiness or unhappiness in a career. By exclusion, therefore, we may conclude from the results of the "Stanines," that for most of us (that is, for the Average Man) a subtle balance of conscious and unconscious forces determines how effectively we use our native aptitudes, whether intellectual, emotional, sensory, neuromuscular, or any combination of these aptitudes. For most of us it is not the minor quantitative differences in the machine itself, but the influence of these conscious and unconscious emotional forces on our use of the human machine which determines our effectiveness. For me this was the

1. As a sharp contrast to the engineering approach of the aptitude testers, I would cite the studies that Anne Roe has been making since 1946. She has used various projective techniques, certain aptitude and psychometric devices, personal documents, life histories, and personal interviews to study the personalities of various kinds of

ultimate lesson from the experience of the Air Force with the "Stanines," and I believe, furthermore, that this result is what might be expected in any similar effort to predict success and failure in civilian careers by the use of precise aptitude tests alone.

This is the stumbling block against which the aptitude testers always stub their toes, and until they learn how to evaluate with equal precision the influence of unconscious emotional forces, they will continue to mislead young people into thinking that scores on aptitude scales will determine successes and failures, happiness and unhappiness in their lifework.[1]

Although these unconscious, irrational, and symbolic forces are subtle and difficult to describe, they determine how most of us use our equipment, and the fate of our lives under conditions of success as well as failure. We shall attempt to discuss a few of these forces in relation to scientific problems and also to the life of science. To youngsters, the dream of a life of scientific research is charged with complicated and usually unnoticed symbolic connotations, which alter steadily during growth from youth to manhood. Therefore, what science "means" consciously to any mature scientist has as many unconscious layers as the stages of his interest in science. No valid generalizations can be made about this condition until many scientists have been studied analytically and the data collated.

Some of those who show scientific interest and capacity in their youth subsequently lose these qualities completely, whereas others pursue them throughout life. In only a few instances has it been possible to study the evolving symbolic connotations of a scientist's interest in scientific matters. In these few cases the "scientific" interests of early childhood frequently turn out to have been in part a window-dressing for quite different concerns. I cannot overemphasize the importance of keeping the fact in mind that human behavior is like a centipede, standing on many legs. Nothing that we do has a single determinant, whether conscious, preconscious, or unconscious. In singling out certain neglected unconscious symbolic determinants it may often sound as though I were overlooking all of the others. This is only because I want to emphasize the importance of the unconscious forces, precisely because they have been neglected so consistently, and because, as a direct consequence of this neglect, they tend to be destructive.

One of these unconscious forces is the child's fearful and guilt-laden curiosity about the human body, both its tabooed external aspects and its mysterious inner workings. A familiar example of this force may be

scientists, scholars, and artists. Her results are necessarily still fragmentary, but already they offer many suggestive leads, and her bibliographies are essential guides to the scanty literature in this field. The results also indicate how enormous is the amount of work that remains to be done.

noted in the physician whose interest in medicine has some of its roots in the child's buried envy of the doctor who could gratify the forbidden bodily curiosities and enter the sickroom from which the child was excluded. How universal this drive would be, and how it would vary with the age of the child and the quality of his relationships to others cannot be decided by guessing. Nor can I document this forcefully without presenting a mass of clinical data for which there is neither time nor space. Somewhat scattered data, gathered during the occasional opportunities to make analytical studies of various kinds of scientists, have shown that even widely varying forms of scientific interest can serve as an acceptable cover for some of the forbidden concerns of childhood. Furthermore, this tendency to utilize various facets of the outer world as a symbolic projection of inner conflicts does not cease when the child becomes adult, but may continue throughout life. This fact is of more than academic interest, since the scientific activities of the adult can be distorted by the same unconscious childhood conflicts out of which his original interest in science may have arisen. Indeed this must result whenever adult activities continue to represent earlier conflicts, and projections of unconscious personal conflicts can often be recognized even through their adult scientific disguises in the reasoning and experiments of outstanding scientists.

As an example of the role of unconscious residues of childhood's battles, I would cite the gynecologist whose ancient and infantile curiosities were not to be satisfied by the justified activities of his profession, and who was plagued by an insatiable compulsion to visit burlesque shows. One could hardly ask for a better experimental demonstration of the fact that unconscious needs cannot be gratified by conscious fulfillment. A comparable example is found in the X-ray man whose choice of a career was determined predominantly by his unconscious curiosity about the internal structure of his mother's body. In all innocence both men dedicated their lives to the service of childhood cravings which were buried in guilt and fear. It should be our goal to learn how to guide gifted young men so that they will not build their entire lives on such psychological quicksands.

3. NEUROTIC DISTORTIONS OF SCIENTIFIC RESEARCH: GENERAL CONSIDERATIONS

The first step in any program of scientific research is to observe natural phenomena while taking care not to alter these phenomena by the very process of observing them. In spite of the most meticulous care, however, the ever-present unconscious forces of the observer color in some degree the glasses through which he makes even simple observations. Therefore, it is out of such tinted observations that he develops his scientific theories. Initially, these are hypotheses about possible relationships between the observed data. Hypotheses are always more vulnerable to distortion by unconscious processes than are the primary observations themselves. Therefore the next step for the research worker is to test his theories, together with their inevitable distortions, in experiments which either isolate and quantify the original data, or that test the consequences of the derived theories. Without our realizing that this is occurring, the process of investigation tends in this way to balance the distortions introduced by unconscious bias. Once he has set up his initial experiments, however, the scientist again becomes an observer. Now, however, he no longer observes facts in nature, but rather in a milieu which he has created artificially by means of his experiment.

Each successive step in these scientific processes calls forth a greater investment of conscious and unconscious feeling; yet if the experimenter is to be objective about the outcome of his experiments he must somehow manage to climb out of his own psychic skin so as to be able to criticize his own handiwork. This is as essential to objective scientific work as it is to artistic creativeness, but it is never easy, because it is impossible for an investigator to prevent the intrusion of his unconscious biases into such sequences of experiment and observation.

Furthermore, even these steps constitute merely the foundation for another round of observation, theory, and experiment. From experimentally derived observations, come a second order of theories, in which unconscious biases have even greater weight; and these theories in turn must be subjected to new experimental tests, which require still further sequences of observation and of theory. Thus the structure of science adds layer on layer, each burdened by more subtle and complex unconscious emotional investments, demanding of the scientist an ever greater clarity about the role of his own unconscious processes in his conscious theories and experiments, and each requiring an ever more rigorous correction for the influence of unconscious preconceptions.

For none of this self-critique in depth does our educational process prepare us. Yet much of it was implicit in Claude Bernard's *An Introduction to the Study of Experimental Medicine*[2] when he wrote:

> The metaphysician, the scholastic, and the experimenter all work with an *a priori* idea. The difference is that the scholastic imposes his ideas as an absolute truth which he has found, and from which he then deduces consequences, by logic alone. The more modest experimenter, on the other hand, states an idea as a question, as an

2. Bernard, Claude. *An Introduction to the Study of Experimental Medicine.* Translated by H. C. Greene. New York: The Macmillan Co., 1927.

> interpretative and more or less probable anticipation of nature, from which he logically deduces consequences which, moment by moment, he confronts with reality by means of experiment.[3]

Again in another connection, Claude Bernard pointed out that the scientist and the philosopher are subject to the same internal human laws, prey to the same emotions, prejudices, and biases, and that these operate equally in the philosopher and the scientist. The difference is that for the scientist the fact that a theory seems true to him, that it feels true, or even that it is logically or mathematically possible does not make it true. For the scientist, the theory is not true until he has taken it to the laboratory, "leaving his theories in the cloakroom," and subjected it to the ultimate test of the experimental method.[4]

Other observers of the world of science have referred to this fact. Every scientist can read with profit and delight Charles Richet's spirited and witty *Natural History of a Savant*,[5] and Gregg's sage volume of lectures on *The Furtherance of Medical Research*,[6] both of which touch on these questions. More recently R. C. Tolman[7] referred challengingly to "the criteria for selecting diligent and competent scientists, the effects of personal bias on results, the relation between subjective origins and objective outcomes of scientific experiments."[8]

This is a portrait of the ideal scientist, ideally in action. It implies that the subtle interplay of reason and emotion, and of conscious and unconscious forces, are as important in the lives and activities of scientists as of anyone else. If this is true then nothing could be more important to science than that scientists should know themselves in the neo-Socratic or Freudian sense, that is, in terms of the interplay between their own conscious and unconscious processes. Yet, as we have already stated,[9] in the education of the scientist, as of everyone else, self-knowledge in depth is the forgotten man of our entire educational system.[10]

Since the father of modern physiology, a great immunologist, a senior statesman among medical educators, and a great atomic physicist all have recognized the confusing influence of subtle psychological processes in scientific work, then surely it is time for the problem to be made the central focus of a major investigation, in which psychoanalytic techniques will be one of the essential tools.

3. *Ibid.*, pp. 27–28.

4. Osler, Sir William. *The Evolution of Modern Medicine*. New Haven: Yale University Press, 1923.

5. Richet, Charles. *Natural History of a Savant*. Translated by Sir Oliver Lodge. London & Toronto: J. M. Dent & Sons, Ltd., 1927.

6. Gregg, Alan. *The Furtherance of Medical Research*. New Haven: Yale University Press, 1941.

7. Tolman, Richard C. "Physical Science and Philosophy," *The Scientific Monthly*, LVII (August, 1943), 166–74.

The Distortion of the Creative Drive by Neurotic Forces

It is rarely recognized that research makes demands upon the young investigator which may exploit his neurotic vulnerabilities. For instance, a drive for "originality" may cloak a difficulty in mastering existing facts and techniques, or it may serve to disguise an unconscious hostility to all existing authority. How often is this drive for originality naïvely mistaken by teacher and student for creative scientific imagination? How often, therefore, is the young investigator encouraged to penetrate into new territory before he has mastered the terrain from which the expedition must start? It is no answer to these questions to say that the same misinterpretations occur among young artists, writers, and musicians. Fallacious values and goals are destructive whenever they occur and in many different fields of work. Nor does it lessen the significance of any of the examples which follow to dismiss them as psychopathological. Such pathology is only an exaggeration of what occurs in more subtle and disguised forms in everyone. The wider and more easily recognized deviations of pathology illuminate the "normal" for us, and sensitize us to slighter anomalies which we otherwise would overlook.

For instance, unresolved neurotic anxieties may impel one overanxious young investigator to choose a problem that will take a lifetime or, alternatively, may drive another into easy, get-rich-quick tasks, which yield a yearly paper, a yearly acclaim, the yearly promotion. The former tendency to postpone the day of reckoning indefinitely occurs in the young scientist who deals with his anxieties by pretending that they do not exist. The latter is found in the man who finds it impossible to endure suspense and uncertainty for more than a few months. Neurotic anxiety can take either form; and young scientists frequently walk a tightrope between these two alternatives, that is, between the annual piecework type of productivity and the long-drawn-out tasks which postpone indefinitely any ultimate testing of theories against experimental data and observations of nature.

Then there is the battle with phobic indecision over which task to undertake, or how to undertake it: an indecision which may arise not out of an inadequate mastery of specific facts and techniques, but from a

8. In the same paper (p. 4) Dr. Tolman makes a further series of comments that are directly relevant to our problems. "He (the scientist) selects this program . . . not to obtain results . . . but to satisfy his own subjective needs. . . . The origin of problems is a subjective one. . . . On the basis of many such nightly reflections, that which has objective validity is finally abstracted out from the welter of subjective experience in which scientists as well as other human beings are immersed. . . ."

9. Kubie, Lawrence S. "The Problem of Maturity in Psychiatric Research," *Journal of Medical Education,* XXVIII (October, 1953).

10. Lombard, George F. F. "Self-Awareness and Scientific Method," *Science,* CXII (September 15, 1950), 289–93.

general neurotic tendency to obsessional doubting. I have seen this symptom work identical destruction in the careers of a young playwright who could not decide which of two equally good plots to use, and of a young chemist who could not decide which of two equally promising leads to follow.

There was also the scientist who had proved his case, but who was so driven by his anxieties that he had to bolster an already proven theorem by falsifying some quite unnecessary additional statistical data. This was a compulsive act, comparable to a kleptomania by a wealthy man, or to the action of a successful and famous writer who suffered from a compulsion to insert a few words from someone else into everything he wrote.

Again, there is the scientist who is always pursuing a new scientific father. This occurs more frequently than is realized. One outstandingly able young scientist ran through five careers, abandoning each one after a brilliant start just as he reached the point of launching his own independent work. When he could no longer postpone accepting a professorship, he broke down and disappeared from the world of science.

But the most ubiquitous tragedy of all is the anxiety-driven scientist who lives on a treadmill—the man who has tasted what it means to gain temporary easement from his anxieties by doing a fine piece of scientific work, but who thereafter is driven not by a quest for further truth but by an insatiable need to repeat the same achievement in an effort to assuage anxieties whose origins were unconscious. This investigator uses scientific research precisely as the man with a handwashing compulsion uses soap and water, or as an addict uses drugs.

I cannot leave this phase of the problem without referring to one highly technical and complex issue. In psychiatry we recognize certain rough parallelisms between types of illness and types of personality. These can have comparable influence in research. During the exploratory phase, while crude data are being gathered, an investigator ought to be free from rigidity. He should be ready to abandon preconceived objectives and anticipated goals, so that any hints that come from unexpected findings can be pursued. He must be psychologically free to follow uncharted courses. Therefore, premature systematization of the data must be avoided. This requires that type of free and imaginative flexibility which is sometimes attributed to the so-called "hysterical" personality. Later, a more rigid process is required, one which has some of the features of the obsessional neurosis, or even some of the tendency of a paranoid patient to organize his delusions into logical systems. Scientific research thus seems to require that, as the work progresses, the investigator should be free to operate now with one type of personality and now with another. It would be profitable to compare analytically the personalities of those scientists who can change in this way and of those who cannot, especially in relation to their scientific productivity. This would seem to be a problem of basic importance for the optimal use of scientific personnel.

Dr. Anne Roe has given me permission to quote a letter of September 13, 1952, in which she summarizes some of her unique studies in this field:

> Any brief summary of these data is necessarily inadequate, and the generalizations require qualifications; but certain differences among these groups of scientists show up, both on the test material and in the life patterns. These are most striking in interpersonal relations, in the handling of anxiety and aggression, in the patterns on intelligence tests, and in the use of imagery.
>
> The typical physicist and biologist grew up with a minimum of group social activity, entered into heterosexual activities rather late and is now not much interested in any social activities. The psychologists and anthropologists for the most part were early conscious of their own and the family's social status, began dating early and enthusiastically, and are still enormously involved with other persons, one way and another. Both physicists and biologists show an unusual independence of parental ties, without guilt; and present attitudes toward the father are characteristically respectful, but lacking in closeness. Attitudes towards the mother are variable. Many of the psychologists and anthropologists, on the other hand, went through periods of great family dissension, and are still angry with or disparaging of their parents. I am sure it is significant also that in the families of these groups, the mother was most often the dominant character. This was rare in the other groups.
>
> The biologists, as a group, rely strongly and effectively on rational control. This appears in their lives, in their general unaggressiveness and in their unproductiveness and intense concern with form on the Rorschach. The physicists have a good deal of free anxiety, shown in their behavior and on the Rorschach, particularly in the large amounts of *K* and *k*. This is better controlled among the theoretical than among the experimental physicists. The *difference* between biologists and physicists is like the difference between compulsive obsessives and anxiety hysterics (I do not imply that all biologists are obsessives and all physicists hysterics). I am quite sure that there are relationships, of a nature still obscure to me, between the preoccupation with space, the type of symbolization that physics uses (which has spatial concomitants) and the choice of physics as a profession; but I suspect that in so far as space symbolizes distance from other persons (as Schilder says) it is more comforting than anxiety-arousing for these men, and I do not believe that their

disinterest in persons is always compensatory. The psychologists and anthropologists are enormously productive on the Rorschach, quite unconcerned with rational controls for the most part, and intensely preoccupied with persons. Their handling of anxiety is quite varied; but as a group they are much the most freely aggressive, and this often has strong oral elements.

The level of intelligence of my group is extremely high, but there are interesting differences in patterning: the theoretical physicists surpass the others on both verbal and spatial tests; experimental physicists tend to be low on verbal and high on spatial; anthropologists are high on verbal and low on non-verbal; psychologists are high on both; biologists show all combinations, but generally the geneticists and biochemists are relatively higher on non-verbal and others reverse this.

Differences in use of imagery during thinking are also fairly sharp. It is not easy to get a good report on this, and I am not happy about my data; yet they are remarkably consistent. In their conscious thinking, biologists are chiefly visualizers; among physicists the experimentalists rely most often on visual imagery, and the theorists on symbolization (usually mathematical and closely allied to verbal symbolization) or imageless thought. Psychologists and anthropologists rely predominantly on auditory verbal thinking. It occurred to me that it was possible that whatever process was most relied upon during the day would be the one to show up in hypnagogic revery. For those who are strongly visualizers or verbalizers, hypnagogic imagery is usually but not always in the same mode; for those whose dominant mode of conscious thought is symbolic or imageless it may be visual or auditory or symbolic, but usually with other twists to it. It would be interesting to find out if there are similar differences in the dream process.

I should add that the one thing which characterizes every one of my groups of eminent scientists is the high degree of ego-involvement in the vocation, both now and earlier. That this is the major factor in their vocational success seems highly probable; but without a comparison group of less successful scientists I can't be certain about this. The ways in which this came about and the situations that made it possible are, of course, extremely varied. In some instances I can demonstrate quite direct relations between professional activities and specific emotional problems. In others I cannot; and I am not convinced that a genuinely neurotic problem is always

11. Roe, Anne. (*a*) "Psychological Tests of Research Scientists," *Journal of Consulting Psychologists*, XV (1951), 492–95. (*b*) "A Study of Imagery in Research Scientists," *Journal of Personality*, XIX (1951), 459–70. (*c*) "A Psychological Study of Eminent Biologists," *Psychol. Monograph* No. 331 (1951). (*d*) "Group Ror-

involved. But I am convinced that it is the matter of personal involvement that is significant for problems of vocational choice and success.

I have included this long excerpt in spite of the fact that the researches of Dr. Roe in this area are still incomplete and inconclusive —as she herself points out—because even her tentative conclusions are unique, exciting and suggestive, and also because her work gives us an indication of how great an investment of time, effort, personnel, and money an adequate study of this problem would entail.[11]

The Influence of Unconscious Symbolic Processes on the Production of Logical Thought and Logical Error in Scientific Research

It is obvious that conscious emotions which are close to the surface can influence a man's scientific work, especially perhaps those anxious ambitions and pettier jealousies which are bred by certain special economic and professional insecurities which will be discussed in Part II, and which may induce a young scientist to push for quick and showy results. Of far greater importance, however, is the subtler influence of streams of unconscious feelings which may be represented symbolically yet compulsively in the scientific activity of an individual, just as they are represented in neurotic symptoms and in dreams, or in all artistic and literary creativity. At this point, therefore, I must explain what the concept of symbolic representation means in this connection.

The symbols by which we think are multivalent tools, always representing many things simultaneously, some conscious, some preconscious, and some unconscious. In logical thinking, the conscious and preconscious symbolic processes represent external reality without disguises; what we call "logic," therefore, is in essence a coding of relationships which are inherent among such internal and external data as are accessible to our direct perceptual processes. One might almost say that although logic resides in the mind, its roots are in the relations among external facts themselves. It is a neglected consequence of this principle, that it is literally impossible to be "illogical" about accessible data except when one has an unconscious axe to grind. Failures in logic are a measure of man's capacity to deceive himself with unconscious premeditation, by misperceiving observational data and by misusing conceptual data for his own unconscious purposes. Many years ago, William Alanson White warned that when anyone says that two and

schachs of University Faculties," *Journal of Consulting Psychologists,* XVII (1952), 18–22. (*e*) "Analysis of Group Rorschachs of Psychologists and Anthropologists," *Journal of Projective Techniques,* XVI (1951), 212–24. (*f*) *The Making of a Scientist.* Dodd, Mead & Company, New York, 1953.

two are five, he does so because he has to; and that the way to meet this problem is not to teach him to say by rote that two and two are four, but to discover with him why he needs to believe otherwise.

It is an inevitable consequence of these facts that in spite of any degree of intellectual brilliance, individuals whose psychological development has been distorted by unsolved unconscious conflicts will have significant limitations in their capacity to build concepts out of the accessible data of external reality. This, indeed, is the greatest psychological hazard of the young intellectual—the fact that unconscious emotional forces persist in him in the form of unconscious needs and unconscious conflicts over these needs. In some, these forces will be expressed in obvious neurotic symptoms. In others, they cause subtle distortions of patterns of living. Sometimes they are expressed in distortions of artistic or intellectual (in this instance, scientific) activities. Naturally there are varied combinations of these three alternatives; but it is an impressive paradox that among individuals in whom unconscious problems are expressed in obvious neurotic symptoms their scientific work frequently escapes the distortions which occur in other scientists whose unconscious processes have no outlet through overt neurotic symptoms. This is not always the case; but it is frequently true that the masked influence of unconscious psychological forces can warp the thinking of a brilliant investigator even when he shows no overt neurotic quirks.

Let me give a few brief examples of the operation of unconscious conflicts on scientific work and scientific careers.

I have known scientists of great ability whose work nevertheless always tended to be vague and ambiguous. Some of these men unconsciously designed their laborious experiments so as to prove nothing. For unconscious reasons they could not allow themselves to find out the answers to their own scientific questions. Such an unconscious conflict over seeing and/or knowing with a preponderant unconscious need *not* to see and *not* to know, arises in early years. In adult life, it accounts for some tragic failures among scientists of brilliant capabilities. This conflict can also produce nihilistic critics who, however brilliant, may also be essentially destructive. For them it is as though seeing and knowing were transgressions which were endlessly tempting but always forbidden in the end. It is conceivable that adequate psychoanalytic therapy early in their training might have saved at least some of these gifted yet wasted and unhappy lives.

Experiments under hypnosis have demonstrated that unconscious processes can take over the intellectual equipment of a scientist and

12. Jones, Ernest. "The Problem of Paul Morphy—A Contribution to the Psychoanalysis of Chess," *International Journal of Psychoanalysis*, XII (January, 1931),

misuse that equipment for their own unrecognized purposes. Under post-hypnotic suggestion, for instance, highly skilled and experienced mathematicians have been led to attempt to prove theorems which they knew to be absurd or to solve mathematical problems which were known to be insoluble. This is the same type of process by which unconscious conflicts and purposes can lead a neurosurgeon to misapply his technical skills, or by which the subtle reasoning of a chemist or physicist, or the ingenuity of a clinical psychologist in devising or interpreting psychological tests, can be misapplied. Actually this is no more mysterious than is the way in which unconscious processes regularly exploit the need for food or the conventional impulse towards cleanliness. One will find in the literature of psychoanalysis many studies of the effects of unconscious processes in disturbances of various normal activities such as eating, washing, dressing, painting, writing, sports, play, sleep, sex, and excretion; but to my knowledge there are no similar studies of the power of unconscious processes to disturb the equally symbolic methods of scientific research. Yet the surreptitious influence of these forces on scientific activities may determine the success or failure of an entire life.[12]

That ancient tragedy of human nature, the success which brings no joy with it, occurs at least as frequently in the life of the scientific investigator as in art and business. A life of fruitful scientific exploration may end in a feeling of total defeat, precisely because in spite of scientific success the unconscious goals of the search have eluded the searcher. Sometimes at the end of a career this need to reach some still undefined goal has led a successful scientist to turn to a pseudo-scientific investigation of the supernatural. More often it leads to depression and a total arrest of all scientific productivity. Sometimes success breeds panic directly, as was observed in the case of a graduate student in physics, a man of outstanding ability, when the head of the department came up behind him in the laboratory one day, and said, "You handed in the best ——— examination I have ever received." Thereupon the student laid down his apparatus and left the laboratory in a panic, which prevented his returning for several weeks.

When they operate below the level of conscious awareness and therefore are not subject to conscious control and direction, the early patterns of familial loves and hates, of submissions and rebellions, may exercise a profound influence on the later work of a scientific investigator; even to the extent of determining his choice of science as a career, his field of work in science, the problems he chooses, the causes he espouses, and the very experiments which he undertakes. While this

1–23. Reprinted in *Essays in Applied Psychoanalysis* Vol. I, by E. Jones. London: Hogarth Press and the Institute of Psychoanalysis, 1951, pp. 165–96.

fact has long been acknowledged, it has never been appreciated in sufficient detail.

There is, for instance, the force of unconscious imitation, to which all of us are liable, imitation even of those very traits against which we may have rebelled most vigorously in childhood. Manifestations of this, both gross and subtle, occur all around us and in every aspect of life. There is the child of the alcoholic who hated the parent's alcoholism yet becomes an alcoholic. There is the child of a parent with a tyrannical temper, firmly resolved never to raise his voice in anger against his children, yet who hears his father's voice issue from his own mouth, as he yells at his three-year-old son in an automatic imitation of the voice and manner which he had always hated in his own father. A famous professor of biochemistry in one of our leading medical schools was the son of a fundamentalist minister against whose narrow ranting he rebelled. Yet the son spent an entire afternoon ranting, as his father was wont to do, but this time it was against the gentle religiosity of a hapless salesman of scientific apparatus who visited his laboratory. This same professor used every biochemical controversy as a pulpit from which to expound sarcastic diatribes against his colleagues, quite like the paternal sermons which had offended him during his childhood.

Also in rebellion against a fundamentalist background, a famous professor of psychology showed a missionary zeal in defense of a mechanized concept of human behavior, so narrowly partisan, indeed so "fundamentalist," that in essence the concept destroyed the value of his whole theoretical approach, which could otherwise have been of considerable scientific significance. Again, there was an eminent physiologist whose desiccated approach to certain problems bore the destructive imprint of an early conflict over whether or not to join the priesthood.

Such thought-provoking reactions are not rare. Many more examples could be cited, but they would merely serve to illustrate again the fact that the human beings who do research work are the subtle and complex instruments of their unconscious and conscious processes; and that the very content of a scientist's investigations as well as his vulnerability to the emotional stresses of research will reflect in varying manner the influence of those psychological forces which are unconscious residues from the unresolved neurotic problems of his early childhood. This fact indicates than an essential element is left out of the training of Man the Scientist, namely, an opportunity to free himself from bondage to the unconscious residues of his own childhood.

Research is a strange and challenging occupation for any young man to contemplate. We still know far too little about the unconscious components of the forces which lead a man or woman to go into research, or about the influence of the unconscious elements in determining the success or failure of his efforts. All of these problems, with their general as well as their special human significance, should be explored. How to do this is a matter for special consideration, since it presents many difficulties. Perhaps the first step would be to subject to psychoanalytic exploration, and in selected instances to psychoanalytic therapy a random sampling of: (1) promising young men who hope to make scientific research their life work; (2) men who have already devoted many years to research, including (*a*) men who in spite of high native endowments have been unproductive, (*b*) others who have been creative but who have ended up nonetheless in frustration and despair, and (*c*) finally those who have succeeded and who have enjoyed fully the fruits of their achievements.

Just as psychiatry has had to study elations in order to understand depressions, so in such a study it would be important to keep in mind that it is just as important to study successes as failures. In science as in other fields, success or failure cannot be accounted for by differences in intellectual capacity alone. Consequently, an analytical study of those who succeed and of those who fail and of the many gradations between success and failure would be of value not only to science but also to those foundations and universities that wish to use men wisely to advance the frontiers of human knowledge.

To uncover in this way some of the unconscious factors which determine the choice of a career, and to explore the subtler forces which determine whether or not that career will be externally productive and internally fulfilling, would be a major contribution to human wisdom. To the best of my knowledge no such study has ever been made of any occupational or professional group. A start must be made somewhere, and in view of the paramount importance of science in today's world, it might be appropriate to start with scientists. Such an enterprise would merit the support of scientific foundations. The sums which are spent on research are so huge that it would seem to be common sense, business sense, and scientific sense to study the men who expend these investments.

It is probable that among the scientists who read this article, some may have developed articulate insights into unconscious forces which have helped to determine the choice and direction of their careers. If they will share their insight with me, I will be deeply indebted to them. They can do this by writing to me, either anonymously or preferably over their own signatures. I will regard all such data as professionally confidential communications, with the assurance to such readers that the material will not be used at any time without complete disguise, and not without specific permission and approval of the text itself by each informant.

II. *Problems Arising from Socio-Economic Forces*

1. THE STRUGGLE FOR MATURITY

In a recent article the problems of maturity in psychiatric research have been dissected in detail and at length.[13] Much of that discussion is applicable to other fields as well; for it is not in psychiatry alone that men are needed who are mature as human beings, as specialized technicians, and as seasoned tacticians in the practical world. These needs are important for higher education in every scientific discipline, medical and non-medical alike, and indeed throughout the academic world. Let me give an example from one of these other fields.

One of the world's outstanding economists has taught in the academic world, has organized and led his own university department, has done pure research in fundamental economic theory, and has held government posts of great practical and diplomatic importance during times of war and peace, and during economic crisis. At first hand, therefore, he knows the field of economics as a teacher, as a pure scientist, and as an applied scientist. Indeed, his position in economics has closely paralleled that of the scientific medical clinician, in that both are called upon to bridge the gap between pure and applied science by operating effectively in both fields. From his years of varied, successful, and useful experience this economist now views with concern the processes of higher education in his own field. He finds among his purely academic colleagues an almost total lack of something which is equivalent to what in medicine is called clinical maturity. They seem to him neither to derive their problems from the world's needs nor to test their hypotheses against such realities. He points out that this condition is in part due to the fact that all higher education now takes so long that, among those who succeed in attaining the higher ranks, there will automatically be a large proportion of individuals who harbor a secret inclination to retreat from life, and who are relatively deficient in aggressive, outgoing, reality-oriented impulses.

Furthermore, if the young scholar is to remain in academic work he has to teach, and our economist agrees with me that teaching is a dangerous sport for the young. Teaching makes it easy to appear scholarly and to sound profound, and gradually to believe that one really has these qualities. The young teacher usually has too great an edge on his young audience for his own good; students have no way of gauging the adequacy of the experience which lies behind the teacher's words. Anyone who lives in an atmosphere where nobody can answer back, soon begins to feel omniscient, with the result that effec-

13. Kubie, *op. cit.*

tive self-criticism is almost as rare among young teachers as it is among dictators and generals. The fall is painful when those hard realities which have no respect for the pedagogue's fantasies of omniscience are encountered.

When I apply these reflections to my own experience, I think of the contrast between what goes on within me when I am working with a patient in my office, and what goes on within me when I am teaching young medical students. The latter is heady wine. Delusions of psychiatric omniscience could be easily acquired were it not for the fact that in my office every day, hour after hour, I face the toughest and keenest skeptics in the world: namely, a patient in his stubborn unconscious struggles. Such experiences keep the clinician humble; they are also the ultimate test of his theoretical and practical wisdom. What the patient is to the psychiatrist, the challenge of politics and the market place is to the economist. This is the clinic that tests his theories, and the impotence of many economic and political scientists is due to the fact that they are so rarely forced to accept the challenge of testing and implementing their theories. Such economists are comparable to a hypothetical professor of medicine who has never served an internship or treated a patient. Finally, in both fields it takes years to reach full maturity of judgment. Only the slow passing of the years can provide an opportunity to observe the ultimate fates of our pet theories, once they have been put out to work.

These considerations have still another implication for higher education. What happens to the economist who never takes his economics into the market place is precisely what happens to the psychiatrist who never confronts the tough opposition of a sick and needful patient. Without this challenge it is possible for him to feel satisfied with formulations which are superficial, partial, and theoretical. His theories seem to him to represent the truth, because he has never had to subject them to any test more critical than the deceptive test of their internal logical consistency. In any field the theoretician need only be logical; yet internal logic, unless it is buttressed by strict mathematical formulation, is no better proof in science than in a system of paranoid ideas, for such ideas can be logically consistent one with another providing one grants the initial premise. This is true throughout the world of pure science. By contrast, the therapist and all applied scientists must always make their knowledge work, which is a necessary, humbling, and maturing challenge.

This is a long detour, but I have allowed myself to explore it because it is important to realize that comparable problems exist throughout all aspects of higher education. As education requires a longer period, the danger of its becoming more and more remote from reality increases steadily, because: (*a*) the student has so limited an opportunity to confront himself with external reality during the process of education; (*b*) the longer the educational process, the more it tends to select men who secretly want to escape external reality; (*c*) finally, it encourages his vanity by giving him premature opportunities to teach theories which have never been tested. These are the three basic threats to maturity throughout our system of higher education. The appointment of brilliant young theoreticians to professorial chairs merely serves to increase these dangers to the maturity of science in general, at the same time that it stunts the development of the individuals themselves.

2. SPECIAL ASPECTS OF THE ECONOMIC STRESSES IN THE LIFE OF THE YOUNG SCIENTIST

Among the special stresses which beset the life of a scientist, there are external as well as internal forces; and no consideration of the problem would be complete if a discussion of these factors were omitted.

I have frequent opportunities to talk to graduate students who are planning a life of scientific research. On such occasions it distresses me to discover how rarely any of them are facing the future realistically, even with respect to such elementary and basic facts as ultimate financial security and independence. I do not remember one who had included in planning for his career a budget of reasonable living costs for a prospective family, compared with existing academic and research salaries. Nor has it lessened my distress to discover that no one of the scientific mentors of these prospective scientists has felt it to be his duty to confront the student with the basic and inescapable facts.

What then are some of the economic realities which the young scientist fails to anticipate? First of all, he rarely seems to realize that a day in the laboratory is the same for a rich scientist or a poor one, while the price of poverty will be paid by his family at home. He does not accept the full import of the fact that his wife and his youngsters are the ones who will have to spend 24 hours a day in quarters so crowded that they will lack space for peaceful family living and the dignity of privacy. Furthermore, because he is young and takes health and longevity for granted, he rarely includes in his calculations the fact that if he has no personal capital he will have to allocate throughout his life a significant share of his small salary for an over-all insurance program; since insurance will be the only way in which his meager earnings can give his family any protection against illness, ensure the future education of his children, or provide a modest independence for his own old age. Nor does he stop to consider the time lag between inflationary trends and even a minor adjustment of university or laboratory salaries. For instance, in one major university a top professorial salary of $12,000 was established 35 years ago. At that time the entire

$12,000 was usable income, but today over one-quarter of it is required for state and federal income taxes, and 48 per cent of the purchasing power of the remainder has been lost through inflation. Consequently, in that university the salary of the full professor now yields the purchasing power which $4,000 had at the time when the university established its salary scale. Only recently has this university been able to offer salary increases; and even the largest of these nets the recipient only about $1,000 in increased purchasing power, leaving him $7,000 behind his colleague of 35 years ago.

About these problems Alan Gregg recently wrote as follows:

> If it be conceded that it is fair to permit a university professor to marry and have two children, to one of whom he may be allowed to give a professional education, then it is clear that with retirement fixed at sixty-five, the professor, in order to contribute to his children's education till one of them is twenty-five years old, must have his children born before he is forty-one. Since few men are professors before that age and fewer will be in the future—the salary in the ranks under that of professor should be such as a family of four can live on. How many are? The prevailing policy is a stupid one, the result can only be a virtually sterile academic society, a professoriate over-concerned with economic security, and therefore secretly rebellious or timidly resigned, or the academic career open only to those who have inherited money or married it. Men with energy and common sense, but no fortunes of their own will refrain from entering or advising entrance into so timid and defenseless a company. This is particularly serious in our medical schools, because the practice of medicine is so ready an alternative to teaching and research. Five years after graduation the ablest of my contemporaries were making $10,000 a year in practice and have continued at that level or above it. The equally able men who went into teaching or research were at $3000 after six years and a few have worked up to $9500 twenty years later. It may be replied that low salaries eliminate the incompetent. They do not. I may as well say straight out that the incompetents stick on. Low salaries may cause the money-minded and the hard-pressed to go elsewhere. The worst and the best stay on. One cannot pass over the incompetents as negligible. They clog the roads. But the plight of the best is truly distressing. Indeed, I could offer no excuses for what I know, but cannot reveal, regarding the struggles of some of the most eminent scientists in this rich country to raise and educate their none too numerous families. If you detect indignation in these remarks, please remember that the best investigators are men of humility and modesty who do not know how to protect themselves in serving our

highly competitive society. That seems to me society's loss, and its shame.[14]

Dr. Gregg wrote specifically of medical scientists, but his comments are equally valid for others. He challenges us to ask ourselves whether it is wise or even honorable to educate young men to high scientific ideals for a life of science, unless we also give them fair and repeated warning that they may never be able to earn a dignified independence for their own old age, or an education for their children which will be comparable to their own. A failure to warn the prospective student-scientist about the practical problems which lie ahead is like training soldiers for war with no emphasis on the fact that many will be wounded and killed. Yet our system of education for scientists engages in a silent conspiracy to deceive the student by never confronting him frankly with the basic facts about the economics of his future career.

This conspiracy of silence shrouds in obscurity many other important facts about a life devoted to science. There are statistics which show how many financial failures it takes to make one stock market success: where are the statistics to show how many scientific failures it takes to make one Pasteur? It would seem to be the duty of scientists and educators to gather such vital statistics on the life-struggles of a few generations of scientists and would-be scientists, and to make sure that every graduate student of the sciences will be exposed repeatedly to the implications such data may have for his own future.

Otherwise the young scientist has no way of giving substance to his anticipatory dreams of what his life will be. In the absence of explicit and forceful indoctrination in these matters, it is not strange that we so rarely find a student who has asked himself such disquieting questions, as "What security can I anticipate if I reach the top of my field? What can my family count on if I turn out to be one of the good men who nears the top but never quite attains it? What if I am one of those who simply straggle along in the middle of the crowd? What if I am relatively a failure?" Instead, most young men view their prospects solely by identifying with their most successful chiefs, never stopping to consider how many must fail for each one who reaches this goal. And what student stops to consider the consequences of the crucial fact that not even his job tenure will be secure unless he reaches the upper levels of academic rank?

Certain of these facts concerning the developing of the personality of the young scientist and the strain of his economic situation have critical effects upon his marriage. There is first of all the initial poverty and the crowded living quarters, at least in our larger centers, where space is at a premium; the pressure of anxiety about the future

14. Gregg, *op. cit.*, pp. 106–107.

which increases with the passing years as these insecurities slowly come to be appreciated and the consequent tendency to overwork, with the nights burdened by tasks which cannot be completed in the hypothetical working day. There is the inadequate social life; the cramped and hampered sexual life, which may at the same time be a neurotic residue from those early unconscious conflicts which have contributed to shaping the young scientist's life as a whole; the increasing monastic absorption of the man, and the wife's early fading and gradual loss of vitality and of confidence in herself as a woman. These are some of the effects that may occur. Many examples could be given from the lives of prominent scientists and their wives and children which would demonstrate the high price which is paid by these civilian expendables through their sacrifices to that carnivorous god, the scientific career. If such examples were to be described in clinical detail, it would be easy for anyone who knows many scientists to recognize well-known individuals. Many would silently recognize themselves. Therefore a decent respect for the right to privacy precludes detailed descriptions.

3. SPECIAL CAREERISTIC STRESSES

In the young scientist's approach to his career, unrealistic and wishful thinking is not confined to financial matters. He tends to dismiss other painful prospects with a bland assumption that through the intervention of some special magic his life will escape the aches and pains that others have experienced. This was the type of self-deception that precipitated many soldiers into emotional illness. Thus, it was found during the war that the man who went into battle squarely facing the fact that he might be mutilated or killed was less likely to break down under the stress of combat than was the soldier who went into battle with a serene but unrealistic fantasy of his personal invulnerability. Among young scientists we find a similar self-deceiving fantasy: that is, that a life of science may be tough for everyone else, but that it will not be for him. Such an attitude prepares the ground for the high incidence of "nervous breakdowns" among scientists in their middle years.

How, then, can the young scientist be prepared effectively for the stresses which face him, if like most young people he shuts his mind to the implications of most of the warnings which do reach him? Thus he usually hears that even after months and years of hard work and uncertainty the scientist must accept disappointment. But merely hearing this will not sustain his self-confidence through long years of obscurity and disappointment, especially if these years are not crowned with success and recognition. Nor is it easy to help the young scientist to attain emotional maturity, since preparation for his life-work necessitates his remaining for many years in the immature role of the student, economically dependent either on a fellowship or on his family's largesse. This dependence delays his emotional maturation, while at the same time putting him under pressure to engage in get-rich-quick researches on problems which will not take too long to finish. Again the student will often have heard of the pressure that will be brought to bear on him to seek rapid advancement by means of publications; but this will not prepare him for the intensity of the actual conflict between such pressure and his own ideal of mature, slow-paced, and thoughtful work. Furthermore, as the number of young scientists increases, the rivalry intensifies so that it becomes always more difficult to resist career pressures. To deal with them realistically requires that the young man should yield just enough to get ahead, neither defeating himself by blind opposition, nor yet selling out to opportunism. How can his training help him to acquire an inner assurance, the humor, and the equanimity which will make it possible to walk this tight-rope as he attempts to achieve a practical compromise? This is no small order, even for mature and well-established scientists. Surely there is something more than the present *nothing*, which might be tried in an effort to prepare young men to deal with such dominant problems as these.

Paradoxically, a quite opposite difficulty is linked to this same conflict. This hazard, which is psychological rather than practical, arises out of the dangers which are inherent in the ambitious dreams of young men whenever a situation exists which encourages them to dream too long. Because of the length of scientific training and of actual research work, the young scientist may dwell for years in secret contemplation of his own unspoken hopes of making great scientific discoveries. As time goes on his silence begins to frighten him; and in the effort to master his fear, he may build up a secret feeling that his very silence is august, and that once he is ready to reveal his theories, they will shake the world. Thus a secret megalomania can hide among the ambitions of the young research worker, secret fantasies which have their roots in the universal dreams of omnipotence and omniscience of early childhood. Out of these dreams have grown the age-old myths with which childhood fends off its sense of helplessness in an adult world: Hercules, Samson, Paul Bunyan, Jack the Giant Killer, David and Goliath, and the modern Superman. With similar dreams of solving the great riddles of the universe, the immature scientist may hold at bay his secret fears of failure. Thus, the long, silent, waiting years are often infused with unhealthy fantasies which exist side by side

15. There are many examples of scientists who missed making basic discoveries by a hair because of accidental external circumstances: for example, the story of Minkowski's place in the discovery of insulin by Banting and of Freud's role in the application of cocaine in local anesthesia, which was completed by Koller. The opposite is also true: "Eight simultaneous discoveries of the cellular basis of plant and animal life. At least three independent demonstrations of artificial immunity following inocu-

with healthy ambitions. With so many practical anxieties dogging him (especially where teaching ability as such is of little help towards his advancement), it is not strange that young research scientists dream unattainable dreams, live unrealistic lives, overwork desperately, and develop a monastic absorption which strains every human tie. It is in this setting of complex psychological tensions that tantalizing near-misses and heartbreaking setbacks occur.

Another way in which the young scientist tries to ease his anxiety is by making a silent bargain with himself that he will try himself out in the field of research for a limited time. Yet with the passing years, the feeling grows that there is no alternative mode of life, that there is no turning back, and that gradually and insidiously all of his life's eggs have been committed to this one basket. One day he awakens, therefore, to find that the opportunity to change his occupation has passed; the chips are down for good, and his life is now permanently committed to a career which is almost devoid of security of any kind, whether financial or scientific. This moment occurs in the life of almost every young scientist. It is a turning point of acute danger to the mental health of many.

One further careeristic hazard is probably the most threatening of all. For every successful piece of research which yields positive results and for which the scientist therefore receives recognition, advancement, and an increased measure of security, there are, and always must be, hundreds of negative experiments, experiments which merely prove that something is *not* so. These negative experiments clear the road for the steady advance of science, but at the same time they clear the road for the more glamorous successes of other scientists, who may have used no greater intelligence, skill or devotion, and perhaps even less. Indeed, which scientist must content himself with the oblivion of the negative result, and which one will win acclaim for positive results may be largely a matter of accidental timing. Essential new techniques or new facts derived from some other scientific discipline may have been one year away, or ten, or fifty. As in prospecting for gold, a scientist may dig with skill, courage, energy, and intelligence just a few feet away from a rich vein—but always unsuccessfully. Consequently in scientific research the rewards for industry, perseverance, imagination, and intelligence are highly uncertain. Success or failure, whether in specific investigations or in an entire career, may be almost accidental, with chance a major factor in determining not *what* is discovered, but when and by whom.[15]

lation with attenuated cultures of anthrax bacillus. Five officially recorded demonstrations of the clinical value of cowpox vaccinations. Five independent discoveries of the phenomenon of heartblock. Three simultaneous demonstrations of vaso-constrictor nerves. Five independent introductions of ether as a surgical anesthetic." (W. H. Manwaring, *Science*, p. 361, April, 1940; referring to B. J. Stern, *Social Factors in Medical Progress*, quoted by Gregg in *op. cit.*, p. 88.)

Evidently the rewards of a career in science are slow and also uncertain; bad luck can frustrate a lifetime of sacrifice and ability. Every successful scientific career is an unmarked gravestone over the lives of hundreds of equally able and devoted, but obscure and less fortunate, anonymous investigators. Science too has its "expendables"; but these do not earn security or tenure, or veteran's pensions—not even the honors which accrue to the expendable soldiers of war.

This is the ultimate gamble which the scientist takes, when he stakes his all on professional achievement and recognition, sacrificing to his scientific career recreation, family, and sometimes even his instinctual needs, as well as the practical security of money. Yet young students are not warned that their future success may be determined by forces which are outside of their own creative capacity or their willingness to work hard. This situation tends to create a star system in science, not unlike that of the stage; and it is not surprising that the mental illnesses which occur in the two types of careers should have so many points in common.

Furthermore, with the growing complexity of science, the psychological difficulties of the scientist have increased in special ways. In another connection, we have noted that the lengthening of the training period holds young adults in the relatively juvenile and dependent role of student for added years. However, we did not state the fact that this protracted quasi-adolescence also stirs tides of conscious and unconscious rebellion, thus creating internal conflicts in the young scientist's mind as well as external conflicts directed against the "system" and its growing pressures. This lengthening period of subordination, insecurity, and rebellion creates a wholly new group of intense moral conflicts for the young scientist. Scientific research is conducted largely behind closed doors, and the accuracy of any man's observations and the veracity of his reports depend ultimately upon his honesty. This honesty depends in turn upon maturity, upon some degree of security, and upon a sense of identification and fellowship with competitors. Under present conditions it is a tribute to scientists that violations of their code of honor are so rare that when lapses occur they become historic scandals. This issue is especially delicate in such fields of science as psychology, psychiatry, and psychoanalysis, in which it is difficult to repeat another man's observations for purposes of objective clinical or experimental or statistical confirmation. Consequently in these disciplines, reports of observations become themselves sources of controversy and suspicion. For many reasons I suspect (although I cannot prove this) that we may be seeing today the birth of a new psychosocial ailment among scientists, one which may not be wholly unrelated to the gangster tradition of dead-end kids.

Are we witnessing the development of a generation of hardened, cynical, amoral, embittered, disillusioned, young scientists? If so, for the present the fashioning of implements of destruction offers a convenient outlet for their destructive feelings; but the fault will be ours and not theirs if this tendency should increase through the coming years and should find even more disastrous channels of expression.

Certainly the idyllic picture of the innocent, childlike scientist who lives a life of simple, secure, peaceful, dignified contemplation has become an unreal fantasy. Instead, the emotional stresses of his career have increased to a point where only men of exceptional emotional maturity and stability can stand up to them for long, and remain clear-headed and generous-hearted under such psychologically unhygienic conditions. Thoughtful educators are beginning to realize that the socio-economic basis of the life of the scientist must be entirely overhauled; that the psychological setting of his life needs drastic revision; and that at the same time the emotional preparation for a life of research is at least as important as is the intellectual training.

I have offered a partial diagnosis with no ready corrective. For this I make no apologies. Medicine is proud of the fact that it has had the honesty and the humility to diagnose disease, sometimes for generations, before it could offer a cure. Students of educational, sociological, political, and economic problems would do well to emulate the medical sciences in this humility, instead of feeling that they must immediately suggest a cure, as soon as they recognize the fact that something is amiss.

It is also a fact that in this paper I have described the general nature of problems before they have been studied in detail. For this, as well, I offer no apology; because as I have indicated above, until educators awake to the importance of these problems, it will not be possible to finance, staff, or implement an adequate investigation of them. This paper will have served its major purpose, if it contributes to that awakening.

15 *The Image of the Scientist among High-School Students: A Pilot Study*

MARGARET MEAD AND RHODA MÉTRAUX

THIS STUDY is based on an analysis of a nation-wide sample of essays written by high-school students in response to uncompleted questions. The following explanation was read to all students by each administrator. "The American Association for the Advancement of Science,[1] a national organization of scientists having over 50,000 members, is interested in finding out confidentially what you think about science and scientists. Therefore, you are asked to write in your own words a statement which tells what you think. What you write is confidential. You are not to sign your name to it. When you have written your statement you are to seal it in an envelope and write the name of your school on the envelope. This is not a test in which any one of you will be compared with any other student, either at this school, or at another school. Students at more than 120 schools in the United States are also completing the statement and your answer and theirs will be considered together to really find out what all high-school students think as a group of people."

In general, the study shows that, while an official image of the scientist—that is, an image that is the correct answer to give when the student is asked to speak without personal career involvement—has been built up which is very positive, this is not so when the student's personal choices are involved. Science in general is represented as a good thing: without science we would still be living in caves; science is responsible for progress, is necessary for the defense of the country, is responsible for preserving more lives and for improving the health

---

Reprinted by permission from Science, *vol. 126, no. 3270 (Aug. 20, 1957), pp. 384–390.*

1. To control any possible influence which the wording of this statement might

and comfort of the population. However, when the question becomes one of personal contact with science, as a career choice or involving the choice of a husband, the image is overwhelmingly negative.

This is not a study of what proportion of high-school students are choosing, or will eventually choose, a scientific career. It is a study of the state of mind of the students among whom the occasional future scientist must go to school and of the atmosphere within which the science teacher must teach. It gives us a basis for reexamining the way in which science and the life of the scientist are being presented in the United States today.

Objectives

OUR specific objectives in this study were to learn the following.

1. When American secondary-school students are asked to discuss scientists in general, without specific reference to their own career choices or, among girls, to the career choices of their future husbands, what comes to their minds and how are their ideas expressed in images?

2. When American secondary-school students are asked to think of themselves as becoming scientists (boys and girls) or as married to a scientist (girls), what comes to their minds and how are their ideas expressed in images?

3. When the scientist is considered as a general figure and/or as someone the respondent (that is, the student writer) might like to be (or to marry), or, alternatively, might not like to be (or to marry), how do (*i*) the positive responses (that is, items or phrases, not answers) cluster, and (*ii*) the negative responses (that is, items or phrases) cluster?

4. When clusters of positive responses and clusters of negative responses are compared and analyzed, in what respects are the two types of clusters of responses (*i*) clearly distinguishable, and (*ii*) overlapping?

5. Is a generally positive attitude to the idea of science, an attitude which we are spending a great deal of money and effort to create, any guarantee of a positive attitude to the idea of science as a career?

Selection of Respondents

TWO separate samples of respondents were used in the study: sample A, a nation-wide sample of high schools, and sample B, a sample of high schools with widely different economic and educational characteristics.

have, part of sample B was collected without reference to the association. No difference in the formulation of the replies was found when the association was mentioned and when the association was not mentioned.

Sample A consisted of 132 public high schools (including one junior high school) that were selected from schools associated with the Traveling High-School Science Library Program sponsored by the National Science Foundation and administered by the American Association for the Advancement of Science. Of these, 118 were drawn from the high schools that participated in this program and an additional 14 from schools that qualified for the program but could not be included in it.

Sample B consisted of 13 special schools: four parochial schools, eight preparatory schools, and one public science high school. All these were from the eastern seaboard, selected to provide contrasts in educational and economic level to the smaller public high schools in the nation-wide sample (sample A). Sample B was collected after the homogeneity of the nation-wide sample had been ascertained.

The total enrollment of the schools participating in the study was 48,000. Schools with an enrollment of less than 300 students were asked to have each student complete one form; schools with an enrollment of more than 300 students were asked to complete 300 forms. The total sample (sample A and sample B) is drawn from the essays written by approximately 35,000 students, and the essays were kept together by the class, grade, and school from which the essays came.

The sample was randomized by drawing envelopes of these replies in groups that included three schools in one state, or three tenth grades, or all the separate classes in three schools, so that no essay was ever separated from the context in which it had been written.

Data-Gathering Instruments

WE ASKED each high-school student respondent to write a brief essay on a topic set by an incomplete sentence which was printed at the top of a page, on which provision was also made for giving the school, the grade, the class or section, the age and sex of the respondent.

Three different forms were constructed, each with a different incomplete sentence. Each of these three sentences was chosen to elicit one major aspect of the image of the scientist.

Only one form was used in any one school,[2] but the forms were so distributed that each form was used by at least one school in each state. These three forms are as follows.

FORM I

Complete the following statement in your own words. Write at least a full paragraph, but do not write more than a page.

When I think about a scientist, I think of

2. In six of the schools included in sample B, all three forms—to be used by different classes—were sent to the same school.

FORM II

If you are a *boy,* complete the following statement in your own words.

If I were going to be a scientist, I should like to be the kind of scientist who

If you are a *girl,* you may complete either the sentence above or this one.

If I were going to marry a scientist, I should like to marry the kind of scientist who

FORM III

If you are a *boy,* complete the following statement in your own words.

If I were going to be a scientist, I would not like to be the kind of scientist who

If you are a *girl,* you may complete either the sentence above or this one.

If I were going to marry a scientist, I would not like to marry the kind of scientist who

Use of the three forms made it possible to distinguish between answers giving official versions of the image of the scientist and those involving the respondents personally, and the use of two forms of the personal questions provided material on the links between negative and positive images, since many answers included responses relevant to both. Experience has shown that the way in which a question is phrased —that is, with a positive or with a negative emphasis—affects the phrasing of the answers by the respondents.

Analysis of Material and Problems of Validation

THIS study is based on qualitative data. The material reflects the way individuals feel and think about a subject as well as whether they will answer questions about the subject in the affirmative or the negative. The use of quantitative data, gathered primarily to count the number of individuals in any given group who will respond in one way or in another, is the more desirable technique when one is interested in whether individuals will agree or disagree with some stated opinion rather than how they feel or why they feel as they do. The check marks or brief responses gathered by quantitative studies are generally too sparse in the expression of feeling and imagery to permit the definition, or the redefinition, of shared attitudes; in such studies, attitudes which are assumed to exist are built into the questions.

The relative value of qualitative and quantitative studies has been debated in the behavioral sciences for some time. A resolution generally accepted at the present time is that the qualitative study is the method of choice for generating hypotheses, and the quantitative study the method of choice for testing hypotheses.[3] When the problem is one of delineating a shared aspect of a society-wide set of images—rather than of answering questions on which or how many students may be expected to respond in a given way—a qualitative study is preferable.

The identification of the pattern in any large sample of essays and of the cognitive and emotional processes which underlie the attitudes reported by individuals is best accomplished by trained behavioral scientists. Because any one analyst, no matter how well trained, may have some blind spots and biases, and because analysts differ in their types of disciplined perception, we had six different analysts work independently with six subsamples of the total sample drawn from different states. Because one kind of material may be more useful than another in outlining a given area, we used—in addition to the essay samples from the 35,000 students—a variety of other kinds of materials as well.

We are assured that we have identified important themes in the material by the multiplicity of independent analyses and by the use of a variety of data. We are assured of the validity of our conclusions by a comparison of the independent work of the analysts and by the agreement on materials from different parts of the country.

Stages in Analysis and Validation

THE stages in analysis and validation were as follows.

1. Sets of data were drawn from the main corpus by envelopes of answers, each set consisting of from 200 to 500 protocols, all from one state and including envelopes of answers to all three forms. Each of the six senior consultants was given a set of data. They worked in complete independence of one another until they met in conference to pool their results in discussion. This discussion was transcribed. The discussion indicated that the analysts were in agreement on the homogeneity of the attitudes found in the materials from different sections of the country. On the basis of this preliminary working of the material from sample A, further collections for sample B (including provision for a control on

3. There are a number of different quantitative studies of the broader subject under way: those directed by H. H. Remmers in the Division of Educational Reference at Purdue University, on high-school students' attitudes toward science; a study at the Survey Research Center at the University of Michigan on attitudes of the public toward science writing; two studies under the Science Manpower Project at Teachers College, Columbia University, one by Hugh Allen, on "Attitudes toward science and scientific careers: a research inventory for New Jersey high-school seniors," and a second by Frances Hall, on science teachers' attitudes toward science. The Interim Committee on Studies on the Social Perception of the Satellite Program and Personnel,

the use of the words *American Association for the Advancement of Science*) were planned and carried out.

2. A detailed pattern analysis was performed on 1000 essays, chosen to represent both the homogeneous nation-wide sample of public schools (sample A) and the highly diversified schools (sample B). This analysis of responses (that is, items, phrases) both checked on the patterns indentified by the senior analysts (among whom was included the analyst who made the detailed pattern analysis) and provided additional understanding of the patterns.

In making this analysis, essays from classes and schools were still kept together, so that each respondent could be placed and each essay could be placed within the major preoccupations of a class or a school. So some schools provide particularly clear material on the dichotomy between science as a subject for study and the personality of the scientist, or on ways in which an increasing sense of inadequacy was reflected in the rejection of science as a career. Everywhere it was possible to follow the divergent interests of boys and girls—as with the boys' interest in an active outdoor life and the girls' interest in the humanitarian aspects of medicine—but there were underlying assumptions shared by both sexes, such as the great importance of personal interests as a basis for career or marriage choice.

3. Fourteen graduate students were asked to report on smaller independent samples of essays. Graduate students were also enlisted to make collections of visual materials related to the image of the scientist in the culture of the United States today. Examples of this collection are illustrations from selected periodicals which present images of scientists, children's drawings made in response to the instruction "Draw a scientist," and the entire pictorial file from the public relations office of a pharmaceutical company.

4. Still another set of student essays from sample A was given to a seventh senior consultant, who had had no previous contact with any of the materials. Since she had not been involved in the earlier stages of the study, she could bring a fresh point of view to the final conference on the basis of which the report was written.

5. A final conference of the senior consultants was held, at which the preliminary findings were again reviewed, and the findings presented in this article were discussed in detail. There was general agreement that the findings effectively represented the data.[4]

under the chairmanship of Donald Michael of Dunlap Associates, Stamford, will also cover some overlapping areas.

4. For assistance in this study through the Institute for Intercultural Studies, which cooperated with the American Association for the Advancement of Science, we wish to thank Ruth Bunzel, Edith Cobb, Natalie Joffe, Martha Wolfenstein, Mark Zborowsky, also graduate students in Columbia University anthropology courses GS 271–2 and GS 198, and, for criticism of the report, Robert Weiss of the Survey Research Center, University of Michigan.

The Composite Image

IN READING the following composite statements, it is important to realize that they do not represent literary descriptions written by the analyst but rather composites of the responses made by the students in their essays, so that each "composite image" is to be understood as being something like a composite photograph which emerges from a very large number of superimposed photographs. Each phrase (response) both stands for a family of phrases (responses) which were used throughout the essays and is itself a recurrently used phrase (response). The phrases have been grouped in relation to themes, as they occur in the essays, but reference to the themes might occur in any order in the essays. It is important to realize that in organizing for presentation here the positive and the negative versions of the composite image of the scientists, the analyst has separated out from the answers the positive phrases (responses), on the one hand, and the negative phrases (responses), on the other hand, as an analytic device, whereas in the essays both occur—or may occur—together in a variety of combinations.

Before the image of the scientist is discussed, it will be useful to look at the way "science" appears in these essays. In the following composite statements, italics indicate the words and phrases (responses) used; detailed examples are given in parentheses, and explanatory notes in square brackets.

SCIENCE

Science is a very broad field which may be seen as a single unit (*science is very important, or I am not interested in science*), as a melange (*medicine and gas and electric appliances*), or as composed of entities (*biology* and *physics* and *chemistry* . . .) linked together by the personality of the *scientist*.

Science is natural science with little direct reference to man as a social being except as the products of science—*medicine* and *bombs*—affect his life. The subjects of science are *chemistry* and *physics* (*laboratories, test tubes, bunsen burners, experiments* and *explosions, atomic energy, laws* and *formulas* . . .), *biology-botany-zoology* (*plants* and *animals* [that is, as materials for laboratory work], *microscopes, dissection, the digestive system, creepy* and *crawly things* . . .), *astronomy* (*the moon, stars, planets, the solar system, outer space, astronomers, astrologers* [sic], *telescopes, space ships* . . .), *geology* (*the earth, rocks, mines* and *oil wells, out of doors* . . .), *medicine* (*cures for TB, cancer, heart disease,* and *polio, research, serums* . . .); *archeology* (*exploration, ancient cities, early man, fossils, digging* . . .). *Mathematics* is not a science but a tool and a measure of scientific aptitude.

The methods of science are *research* and *experimentation, invention, discovery, exploration, finding out new things* and *new ways of improving old ones*. Science means *doing* and *making: hard work*—not imagination—is the source of knowledge and the means of accomplishment.

The focus of science is upon the present. The past is important only as it is left behind (*without science we would still be living in caves*) and the future as a foreseeable goal (when we *find a cure for heart disease, see if there is life on Mars, discover new fuels. . .*). But as the past closes in behind us, the future opens to the curious (*there is still so much to discover*) into the yet unknown.

In thinking about science, different sorts of linked images occur which may be bracketed together when science is rejected or may be included when positive preference is expressed for one of a pair. So, science may be *theoretical* or *applied*, and either of this pair can be seen as more of a whole and be accepted (that is, the man in the laboratory is visualized as working through the whole problem; or the engineer can see the finished road), while the other is seen as partial and is rejected (that is, the engineer is visualized as working only on the end-product; or the man in the laboratory never sees the plan carried out). Likewise, science can be carried out *in the laboratory* or in a *far away place*; it may involve large-scale action (*traveling, digging, exploring, constructing, flying through space . . .*) or the skills of fine detail (*gazing through a telescope, poring over a microscope, dissecting, solving equations . . .*). The goals of science may be humanitarian (*working to better mankind, finding cures, making new products*, developing *programs for atoms for peace . . .*), or, in contrast, they may be either individualistic (*making money, gaining fame and glory . . .*) or destructive (*dissecting, destroying enemies, making explosives* that *threaten* the *home*, the *country*, or *all mankind . . .*).

Since, by implication, science is the source of unlimited power, its practitioners should have the highest and the most selfless motivations to use only its constructive possibilities—or its destructive possibilities only constructively—for the *welfare of their country* and the *betterment of people, the world*, and *all mankind*.

THE SCIENTIST: THE SHARED IMAGE[5]

The scientist is a man who wears a white coat and works in a laboratory. He is elderly or *middle aged and wears glasses. He is small,* sometimes *small and stout,* or *tall and thin. He* may be *bald.*

5. A few of the more mature students realize that this picture is stereotyped and incomplete. So, for instance, having described the scientist as the "man in the white coat," students continue: "On second thought—he might equally well be seining a small stream, feeding facts into an electronic computer, or injecting a radioactive fluid

He may *wear a beard,* may be *unshaven* and *unkempt. He* may be *stooped and tired.*

He is surrounded by equipment: test tubes, bunsen burners, flasks and bottles, a jungle gym of blown glass tubes and weird machines with dials. The sparkling white laboratory is full of sounds: the bubbling of liquids in test tubes and flasks, the squeaks and squeals of laboratory animals, the muttering voice of the scientist.

He spends his days doing experiments. He pours chemicals from one test tube into another. He peers raptly through miscroscopes. He scans the heavens through a telescope [*or a miscoscope!*]. *He experiments with plants and animals, cutting them apart, injecting serum into animals. He writes neatly in black notebooks.*

The image then diverges.

POSITIVE SIDE OF THE IMAGE OF THE SCIENTIST

He is a very intelligent man—a genius or *almost a genius. He has long years of expensive training—in high school, college,* or *technical school,* or perhaps even beyond—during which *he studied very hard. He is interested in his work and takes it seriously. He is careful, patient, devoted, courageous, open minded. He knows his subject. He records his experiments carefully, does not jump to conclusions, and stands up for his ideas even when attacked. He works for long hours in the laboratory, sometimes day and night, going without food and sleep. He is prepared to work for years without getting results and face the possibility of failure without discouragement; he will try again. He wants to know the answer. One day he may straighten up and shout: "I've found it! I've found it!"*

He is a dedicated man who works not for money or fame or self-glory, but—like Madam Curie, Einstein, Oppenheimer, Salk—for the benefit of mankind and the welfare of his country. Through his work people will be healthier and live longer, they will have new and better products to make life easier and pleasanter at home, and our country will be protected from enemies abroad. He will soon make possible travel to outer space.

The scientists is a truly wonderful man. Where would we be without him? The future rests on his shoulders.

into the veins of a monkey" (boy, 17, 12th grade). "I realize that there is more than microbiology [that is, the man in the white coat] to science. Therefore, I think of the atom, and somehow always of old men, working on various bombs and reactors. When I think of the use of atoms for peace, I think of young men working in offices. I don't know why" (girl, 14, 10th grade). "At the word *science,* I can imagine so much. The scope is unlimited and I sometimes do not connect the two words [*science* and *scientist*] any further than the laboratory. But if I could put the two together, a scientist would become more of an adventurer, a romanticist, than a figure who is nothing but a human IBM machine" (boy, 15, 10th grade).

NEGATIVE SIDE OF THE IMAGE OF THE SCIENTIST

The scientist is a brain. He spends his days indoors, sitting in a laboratory, pouring things from one test tube into another. His work is uninteresting, dull, monotonous, tedious, time consuming, and, though he works for years, he may see no results or may fail, and he is likely to receive neither adequate recompense nor recognition. He may live in a coldwater flat; his laboratory may be dingy.

If he works by himself, he is alone and has heavy expenses. If he works for a big company, he has to do as he is told, and his discoveries must be turned over to the company and may not be used; he is just a cog in a machine. If he works for the government, he has to keep dangerous secrets; he is endangered by what he does and by constant surveillance and by continual investigations. If he loses touch with people, he may lose the public's confidence—as did Oppenheimer. If he works for money or self-glory he may take credit for the work of others—as some tried to do to Salk. He may even sell secrets to the enemy.

His work may be dangerous. Chemicals may explode. He may be hurt by radiation, or may die. If he does medical research, he may bring home disease, or *may use himself as a guinea pig,* or may *even accidentally kill someone.*

He may not believe in God or *may lose his religion. His belief that man is descended from animals is disgusting.*[6]

He is a brain; he is so involved in his work that he doesn't know what is going on in the world. He has no other interests and neglects his body for his mind. He can only talk, eat, breathe, and sleep science.

He neglects his family—pays no attention to his wife, never plays with his children. He has no social life, no other intellectual interest, no hobbies or relaxations. He bores his wife, his children and their friends—for he has no friends of his own or *knows only other scientists—with incessant talk that no one can understand;* or *else he pays no attention* or *has secrets he cannot share. He is never home. He is always reading a book. He brings home work and also bugs and creepy things. He is always running off to his laboratory. He may force his children to become scientists also.*

6. When evolution is mentioned, it is mentioned negatively. It is impossible to tell what the absence of other than negative references to evolution means. In Remmers' study [H. H. Remmers and D. H. Radler, *The American Teenager* (Bobbs-Merrill, Indianapolis, Ind., 1957), p. 171] 40 percent of the teenagers checked "No," to the statement "Man has evolved from lower forms of animals"; another 24 percent checked "Don't know"; 35 percent checked "Yes."

A scientist should not marry. No one wants to be such a scientist or *to marry him.*

Discussion

THE "official" image of the scientist—the answer which will be given without personal involvement—which was evoked primarily in form I, but which recurs in the answers to all three forms, is a positive one.

The scientist is seen as being essential to our national life and to the world; he is a great, brilliant, dedicated human being, with powers far beyond those of ordinary men, whose patient researches without regard to money or fame lead to medical cures, provide for technical progress, and protect us from attack. We need him and we should be grateful for him.

Thus if no more than form I had been asked, it would have been possible to say that the attitude of American highschool students to science is all that might be desired.

But this image in all its aspects, the shared, the positive, and the negative, is one which is likely to invoke a negative attitude as far as personal career or marriage choice is concerned. While the rejection in the negative image is, of course, immediately clear, the positive image of very hard, only occasionally rewarding, very responsible work is also one which, while it is respected, has very little attraction for young Americans today.[7] They do not wish to commit themselves to long-time perspectives, to dedication, to single absorbing purposes, to an abnormal relationship to money, or to the risks of great responsibility. These requirements are seen as far too exacting. The present trend is toward earlier marriage, early parenthood, early enjoyment of an adult form of life, with the career choice of the man and the job choice of the woman, if any, subordinated to the main values of life—good human relations, expressed primarily in terms of the family and of being and associating with the kind of human being who easily relates to other people.

To the extent that any career—that of diplomat, lawyer, businessman, artist, aviator—is seen as antithetical to this contemporary set of values, it will repel male students as a career choice and girls as a career for their future husbands. But it is important to see also the particular ways in which the image of a scientific career conflicts with contemporary values. It divides girls and boys. The boys, when they react positively, include motives which do not appeal to the girls—adventure, space travel, delight in speed and propulsion; the girls, when they react positively, emphasize humanitarianism and self-sacrifice for humanity,

7. In this statement, we draw not only on the attitudes in this study but on a wide variety of other materials on the attitudes of contemporary young Americans.

which do not appeal to the boys. The girls reject science, both as a possible form of work for themselves, concerned with things rather than with people, with nonliving things (laboratory animals, not live animals; parts of anatomy, not living children), and for their husbands, because it will separate them, give their husbands absorbing interests which they do not share, and involve them in various kinds of danger. In earlier periods, when career choices and marriages occurred later, the girls' attitudes might not have mattered so much; they are very important today, on the one hand, because girls represent a principal untapped source of technical skill, and, on the other hand, because, with present adolescent social patterns, paired boys and girls spend a great deal of time discussing the style of their impending marriage and parenthood and the relationship of the boy's career choice to the kind of home they will have.

The image of the scientist's relationship to money also presents a problem, in a period of full employment, to young people who think that an adequate income is something that should be taken for granted. The scientist is seen as having an abnormal relationship to money. He is seen either as in danger of yielding to the temptation of "money and fame," or as starving and poor because of his integrity. The number of ways in which the image of the scientist contains extremes which appear to be contradictory—too much contact with money or too little; being bald or bearded; confined work indoors, or traveling far away; talking all the time in a boring way, or never talking at all—all represent deviations from the accepted way of life, from being a normal friendly human being, who lives like other people and gets along with other people.

Specific Indications about the Teaching of Science

FROM the standpoint of teaching, it is important to realize how the present image of scientific work lacks any sense of the delights of intellectual activity; the scientist works patiently and carefully for years, and only when he finds out something does he shout with joy. This lack of any sense that intellectual activity is rewarding in itself can be related to the lack of any mention of living things, plant, animal, or human, in the materials with which the scientist is believed to work. Plants and animals appear only as dead objects for dissection; the human body, as organs or systems studied in the laboratory and treated in medicine; whole human beings appear only as the dead denizens of dead and buried cities, and most of the scientists about whom they read are also dead. The lack of any sense of enjoyment can also be related to the central role given to mathematics as a tool, without any emphasis on the delights of observation, as in early natural history studies or in the perception of regularities and connections in the world around them, or between themselves and the world around them.

Because the materials were analyzed class by class and school by school, the study has also yielded, as a by-product, certain sidelights on science teaching: on the importance of particpation as opposed to passive watching, on the role which the personality of the teacher plays in attitudes toward science, on the effect on the rest of the class of the presence in it of one type of exceptionally gifted child.

One of the most recurrent responses is an expression of active boredom, the phrase, "I am not interested in science," or in a particular science course (chemistry or physics), followed occasionally by highly emotional expressions of fury and hatred of particular activities which are being demonstrated. "Interest" and "active enjoyment" seem to be so closely related that the student seated in a classroom who has to watch things being poured from one test tube to another or listen to a string of unrelated facts becomes permanently alienated. General science courses seem to be the ones in which this attitude toward science is characteristically invoked, except when a gifted teacher gives it some special emphasis. When mathematics is seen as the key ability on which all future scientific work is based, not liking and not being able to do mathematics become a specially weak point in the circle of the students' interests.

In contrast, other activities are defined as nonscientific because they are absorbingly interesting: "watching things grow that I have planted," or "working on my hot rod car."

The role of the teacher—as reflected in the comments of a whole class—is an exceedingly interesting one. The disliked teacher is personalized and vivid, but the teacher who has obviously been very successful and has caught the imagination and enthusiasm of the whole class does not emerge as a person at all but, instead, sinks into the background of good classroom conditions, together with "good laboratory equipment." Special aspects of the disliked teacher are commented on in detail. He may be described as an outsider, a stranger, with unusual habits of dress and manner, who does not know his subject well, who cannot talk about anything but his subject, who lives alone without the slightest tie to the community, who is "stuck-up and who is too busy for anyone but himself." It is easy to see how the only male teacher in the school presents special problems to the boys, if he himself is a figure they reject, and how easily the sphere of work for which he stands may be rejected also. So one boy writes, "Anyone who digs our teacher's gab is a square as well as being queer." Some of these consequences un-

8. The other side of this picture is sometimes seen in comments made by foreigners who have entered the sciences because Americans think they require less of a knowledge of the culture, and who because of their science training can get teaching positions in the schools. After a year or two of teaching in small-town schools, the

doubtedly flow from the convention in the United States that, ideally, science should be taught by men, with the result that men who might be more successful teachers in some other field are forced into teaching a subject which they dislike and in which they have no special competence. Similarly, foreigners and refugees—if male—may have a better chance to get positions as mathematics and science teachers than they have in other fields.[8]

The significance of the lack of particular mention of the good science teacher is equally important, for it is related to the lack of invocation of authority by the students, who state their opinions about science—even those obviously related to a particular teacher—as their own. Only when they disagree, when they wish to attack the current image of science as a good thing from a minority position—that is, from the viewpoint of some fundamentalist religious position which they accept—do they invoke authority. It is related also to the situation in American culture where, through generations, there has been a break between immigrant parent and native-born child. In this new setting, the European tendency for children to identify with the personality and occupation of the parents has been replaced by a tendency to follow the style set by members of one's own generation, especially those in one's own local school clique.

In the classroom, a disliked fellow-student who is regarded as a future scientist may also be described in some detail, as students say they do not want to be the kind of scientists who "go about with their noses in a book, looking superior." But in those classrooms where everyone has been committed to the joy of some experiment or project, no individuals emerge: it is impossible to say what is the sex, age, nationality, and personality of the teacher.

In summary, it may be said that where science teaching is successful, the teacher has created a situation in which his or her (one does not know which) personality sinks into the background, and in which no one student stands out as so especially gifted and preoccupied as to rouse annoyance in the class. Students and teacher appear to have worked as a group, accepting science as a part of *their* lives, preoccupied with no specific identified individuals.

Recommendations

MASS MEDIA

Straight across the country there is a reflection of the mass media image of the scientist, which shares with the school materials

foreign-born teacher flees back to the cities where he has friends or at least can live anonymously. (Based on life-history data from Chinese informants in the Chinese section of the Study Program in Human Health and the Ecology of Man, New York Hospital–Cornell Medical College, New York.)

the responsibility for the present image. Alterations in the mass media can have important consequences in correcting the present distorted image if such changes are related to real conditions. Attempts to alter the image, in which the public relations department of a particular company represents its research personnel with crew cuts and five children may improve the recruitment program of single companies, but do so only at the expense of intensifying the negative aspects of the image for the country as a whole.

What is needed in the mass media is more emphasis on the real, human rewards of science—on the way in which scientists today work in groups, share common problems, and are neither "cogs in a machine" nor "lonely" and "isolated." Pictures of scientific activities of groups, working together, drawing in people of different nations, of both sexes and all ages, people who take delight in their work, could do a great deal of good.

The mass media could also help to break down the sense of discontinuity between *the scientist* and other men, by showing science as a field of endeavor in which many skills, applied and pure, skills of observation and of patient, exact tabulation, flashes of insight, delight in the pure detail of handling a substance or a material, skills in orchestrating many talents and temperaments, are all important. This would help to bring about an understanding of science as a part of life, not divorced from it, a vineyard in which there is a place for many kinds of workers.

THE SCHOOLS

The material suggests the following changes which might be introduced in educational planning.

1. Encourage more participation and less passive watching in the classroom, less repeating of experiments the answers to which are known; give more chance to the students to feel that they are doing it themselves. A decrease in the passive type of experience found in many general science courses seems particularly necessary.

2. Begin in the kindergarten and elementary grades to open children's eyes to the wonder and delight in the natural world, which can then supply the motive power for enjoyment of intellectual life later. This would also establish the idea of science as concerned with living things and with immediate—as contrasted with distant—human values.

3. Teach mathematical principles much earlier, and throughout the teaching of mathematics emphasize nonverbal awareness:[9] let children

9. Studies of the College Entrance Examination Board Commission on Mathematics and the University of Illinois Committee on School Mathematics give promise of bringing about improvement in mathematics instruction. A study of junior-high-school mathematics will be undertaken at the University of Maryland this fall. There is more to be done, especially in the elementary grades, and state departments of edu-

have an opportunity to rediscover mathematical principles for themselves.

4. Emphasize group projects; let the students have an opportunity to see science as team work, where minds and skills of different sorts complement one another.

5. Emphasize the need for the teacher who enjoys and is proficient in science subjects, irrespective of that teacher's sex; this would mean that good women teachers could be enlisted instead of depending on men, irrespective of their proficiency. Since it would seem that the boys do not need to identify with an adult male as a teacher, this should leave us free to draw on women as a source of science teachers.

6. Change the teaching and counseling emphasis in schools which now discourages girls who are interested in science. This would have many diffuse effects: on the supply of women teachers and of women in engineering, on the attitudes of girls who are helping boys to choose careers, and on the attitudes of mothers who are educating their small children in ways which may make or mar their ability to deal with the world in scientific terms.

7. Deemphasize individual representatives of science, both outstanding individuals like Einstein—whose uniqueness simply convinces most students that they can never be scientists—and the occasional genius-type child in a class. (This type of child, who represents only one kind of future scientist and who is often in very special need of protection from the brutalities of his age mates, should probably be taken out of small, low-level schools, and placed in a more protected and intellectual environment.) Instead, emphasize the sciences as fields, and the history of science as a great adventure of mankind as a whole. (The monotonously recurrent statement "if it weren't for scientists we would still be living in caves" is an insult to the memory of millions of anonymous men who have—each in his way—made further advances possible.)

8. Avoid talking about *the scientist, science,* and *the scientific method.* Use instead the names of the sciences—biology, physics, physiology, psychology—and speak of what a biologist or a physicist does and what the many different methods of science are—observation, measurement, hypotheses-generating, hypotheses-testing, experiment.

9. Emphasize the life sciences and living things—not just laboratory animals, but also plants and animals in nature—and living human beings, contemporary peoples, living children—not the bones and dust of dead cities and records in crumbling manuscripts. Living things give

cation should be encouraged to establish state committees which can determine how work now in progress at the national level can be made effective in local schools. The Poloidiblocs, developed by Margaret Lowenfeld of the Institute of Child Psychology, 6 Pembridge Villas, London, W.11, are an important addition to the equipment for teaching young children mathematics.

an opportunity for wonder and humility, necessarily less present in the laboratory where students deal with the inanimate and the known, and contact with living things counteracts the troubling implication that the scientist is all powerful.

Conclusion

THIS report is not in any way a statement of the proportion of high-school students who will choose science as a career. It is a discussion of the state of mind of fellow-students, among whom the occasional future scientist must go to school, of the degree of personal motivation necessary to commit oneself to science, and of the atmosphere within which the science teacher must teach. Since most high-school students' attitudes closely reflect those of their parents, it is also an indication of the climate of opinion in which parents may be expected to back up their children in choosing science as a career, citizens may be expected to vote funds for new laboratories, and voters may be expected to judge Congressional appropriations for science education.

16 *The College-Student Image of the Scientist*

DAVID C. BEARDSLEE AND DONALD D. O'DOWD

THE IMAGE OF THE SCIENTIST among high school students has been studied in detail in recent years. Remmers and Radler[1] have reported on some beliefs of teen-agers about scientists, and Mead and Métraux[2] have summarized the image of the scientist revealed in essays produced by a large sample of high school students.

The beliefs of college students about the scientist are also of interest. Many students entering college seriously consider careers in science, and college students will eventually constitute an influential segment of the citizens whose views make up the public response to science.

Exploration of the college-student image of the scientist was initiated in a series of unstructured interviews with college undergraduates at Wesleyan University.[3] In these interviews, students described the scientist as being dedicated to his work and carrying it out with heroic devotion at the expense of concern with public affairs and even family responsibilities. The scientist was described as unsociable, introverted, and possessing few, if any, friends. Some students referred spontaneously to his high intelligence; others were more impressed by the pre-

---

Reprinted by permission from Science, *vol.* 133 (*March* 31, 1961), *pp.* 997–1001.

1. H. H. Remmers and D. H. Radler, *The American Teenager* (Bobbs-Merrill, Indianapolis, 1957).

2. M. Mead and R. Métraux, *Science*, 126, 384 (1957). [Reprinted as the previous chapter of this volume.]

3. E. W. Harbinger and A. LaCava, Wesleyan undergraduates, assisted us in this study. The research was carried out under a contract with the U.S. Office of Education, Department of Health, Education, and Welfare. Additional support was provided by the Faculty Research Committee of Wesleyan University.

cision of his thinking and the objectivity (that is, lack of emotional involvement) with which he handles most personal and professional problems. Two telling comments represent the common response of undergraduate men to the scientist. One student volunteered, "I wouldn't care to double-date with a scientist," and another student commented, "Maybe it's not a good idea for him [the scientist] to be married." A number of students were impressed by the scientist's apparent need to proceed in his work regardless of other demands on his time. In general, the college students revealed in these interviews beliefs similar to those found among high school students. The scientist, to use the student phrase, "is not well rounded."

In order to pursue further the subject of the student image of the scientist, a series of three successive questionnaires was designed and used in testing. A fourth version of the questionnaire was selected as the main instrument for an empirical study. It contained materials derived from the interviews and from standard questionnaires and scales developed in the earlier versions. In the questionnaire students were asked to indicate the appropriateness of a series of terms to each of 15 occupations, including that of scientist.[4] The terms were arranged in two-ended, seven-point rating scales of the following form:[5]

1. wealthy —:—:—:—:—:—:— not well-to-do
2. optimistic —:—:—:—:—:—:— pessimistic
3. excitable —:—:—:—:—:—:— calm

This design makes it possible not only to determine absolute values for characteristics attributed to the scientist but also to obtain an estimate of the standing of the scientist relative to individuals in other occupations.

The questionnaire was given to undergraduate men and women in four colleges in the northeastern United States: Wesleyan University, a second small and highly selective men's liberal arts college, a highly selective private women's college, and the college of arts and sciences of a state university.[6] At each college, probability samples of freshmen and seniors were chosen. Over 90 percent of the students selected at three of the four colleges returned completed questionnaires. At the second men's liberal arts college, all of the freshmen but only two-thirds of the seniors completed the questionnaire. Data from these seniors were not used in many of the following analyses. In all, about 1200 students were included in this phase of the study.[7]

4. The occupations that were studied are listed in Table 1.

5. The form of the questionnaire and some of the scales are taken from the work of C. E. Osgood [for example, C. E. Osgood, G. J. Suci, P. H. Tannenbaum, *The Measurement of Meaning* (Univ. of Illinois Press, Urbana, 1957)].

6. The data were collected during 1958 and 1959.

Image of the Scientist

IT IS possible to present a composite picture of the scientist from the responses obtained. Students from all of the colleges, both men and women, freshmen and seniors, were in sufficient agreement to justify a summary of the characteristics attributed to the scientist by all groups. There is clearly a well-defined stereotype of the scientist among college students as well as among high school students. In the following summary, the rating of the scientist relative to individuals in other occupations is considered.

The scientist, according to college students, is outstanding in several respects. Students see him most prominently as a highly intelligent person with a strong tendency to be both individualistic and radical in personal and social outlook. At the same time, the scientist is seen as socially withdrawn; he is indifferent to people, retiring, and somewhat depressed, and he rates low in social popularity. In over-all sociability the scientist rates lowest among individuals in the 15 high-level occupations. It is therefore not surprising that he is believed to have a relatively unhappy home life and a wife who is not pretty. There is an air of strangeness about him; he is hard to like and comprehend. He is respected for his great contribution to society, but he is not the kind of person one can easily get to know.

The scientist is believed to be highly intelligent but not interested in art. He is both self-sufficient and persevering. He focuses his powers in a rational and sensitive pursuit of answers to nature's mysteries. He is rated as reasonably successful and as having ample opportunity to advance in his field. At the same time he is seen as having only modest wealth. It appears that the scientist could exploit his situation to secure wealth and status, but he is so devoted to his work that he is satisfied with a modest income.

The scientist is moderately confident, optimistic, and realistic in his approach to life. He has power in public affairs yet is given only a moderately high score on responsibility. When combined with his radicalism, this finding suggests that there are grounds for an anxious public to become suspicious of his loyalty. After all, he has few friends, great determination, and an unusual set of values.

Rather surprisingly, the scientist is scored relatively low on stability, caution, and calmness. It appears that he has difficulty controlling his impulses. This is consistent with the picture of his radicalism. He is

7. A summary of the entire study appears in "College Student Images of a Selected Group of Professions and Occupations," *Final Report, Cooperative Research Project* No. 562, *U.S. Office of Education* (Wesleyan University, Middletown, Conn., 196c).

coldly intellectual in some spheres of his life—mainly in his work—and he is emotional in his response to social and political appeals.

The complexity of the scientist's nature must account for his being considered mildly interesting and colorful. He is thought to be very valuable to society and to derive very great personal satisfaction from his work. If one were to study his recreational habits one would find him most frequently at chess, rarely playing bridge, and never playing poker.

In summary, there emerges a picture of the scientist as a highly intelligent individual devoted to his studies and research at the expense of interest in art, friends, and even family. The scientist derives great personal satisfaction, a sense of success, reasonably high status in the community, and a modest income from his work. He serves mankind in a selfless way, almost unaware that he is doing so; he serves others by serving himself.

In public matters the scientist is influential, but he may be somewhat naive. He is extreme in his views on social matters, and he tends to become emotionally involved with issues outside his realm of professional competence. The scientist is coldly intellectual in his professional area but excitable in the public political sphere. He is clearly an intellectual, but unlike "eggheads" in the humanities, he is characterized by a vigorous and directed use of his intelligence. The image conveys a sense of strength of personality, but it is a little extreme, a little strange, somewhat contradictory, and, therefore, hard to comprehend.

Comparison with Images of the Nonscientist

AN ESTIMATE of the similarity of the scientist image with the images of individuals in 14 other occupations was obtained by correlating the mean scores obtained on 48 scales for the scientist and for people in these other occupations. The data from a subsample of the students tested were used to obtain the correlations presented in Table 1.

These data reveal that the scientist is believed to have much in common with the college professor. The similarity of ratings for the scientist and engineer was predictable, but the correlation with ratings for artist and school teacher had not been clearly foreseen. This correlation stems primarily from the students' grouping of all these roles as intellectual roles. It is clear that the students believe that scientists do not share many attributes with individuals in any of the business and industrial occupations.

Comparison of the image of the scientist with that of the college professor reveals some interesting differences between these roles that are often filled by the same person. Both occupations are entered by men of high intelligence with personality characteristics represented by high scores on *self-sufficient* and *persevering,* middle values on *strong, active, confident,* and *self-assertive,* and low scores on *stable* and *adaptable in habits.* Both professions are believed to attract men who are, to a high degree, radical and individualistic. Members of the two professions differ in that the scientist is thought to lack the artistic interest, good taste, and sensitivity of the college professor. The scientist is not a cultured intellectual, while the college professor attains the

TABLE 1. Correlation of the Profile of the Scientist with Profiles of Individuals in Other Occupations

Occupation	*Correlation*
College professor	+ 0.77
Engineer	+ 0.53
Artist	+ 0.51
School teacher	+ 0.49
Doctor	+ 0.44
Lawyer	+ 0.41
Social worker	+ 0.30
Accountant	– 0.03
Business executive	– 0.03
Industrial manager	– 0.03
Personnel director	– 0.18
Sales manager	– 0.25
Office supervisor	– 0.29
Retail store manager	– 0.29

highest score in this dimension. Moreover, the scientist is, to a striking degree, less interested in people and less sociable and popular than the college professor. The professor is interested in people and quite successful with them. The scientist is neither drawn to people nor socially attractive. Finally, the scientist is less interesting and colorful than the college professor. The scientist is scored above the college professor on two components of what might be called "material opportunity"—that is, wealth and the opportunity for advancement. The scientist has a more markedly active, persevering, and rational approach to life and work than the professor. In summary, the scientist has greater wealth and opportunity than the professor and a more forceful approach to intellectual problems. However, in the very important areas of social sophistication and esthetic interests the college professor leads the scientist by a wide margin.

When the full range of occupation profiles is considered, the scientist and the engineer have a good deal in common. In terms of strength and competence, as indicated by middle values on such items as *active, confident, strong, hard, self-assertive,* and *realistic about life,* they have very similar scores. Competence in either field connotes a reasonable degree of success, social status, and power in public affairs. The scientist differs from the engineer in that he is believed to be more intellectual and less conformist in personal behavior and political viewpoint. The scientist also is rated higher than the engineer in concern with esthetic matters, in spite of the relatively low rating of the scientist in the realm of cultural interests. The scientist is considered more persevering, self-sacrificing, and valuable to society, as well as more interesting and colorful. On the other hand, the engineer has two clear advantages over the scientist. First, the engineer is more concerned with people. He is a sociable, popular fellow as compared with the scientist. Secondly, the engineer is considerably wealthier, and he is a more "regular guy" than the scientist. This latter characteristic is indicated by the higher scores for the engineer on *clean cut, plays poker,* and *has good taste* (taste in clothes, house, car, and so on), and the engineer is believed more likely to have a pretty wife. In conclusion, then, the engineer is thought to be less of an "egghead" than the scientist. He is less intelligent, less nonconforming, less sensitive esthetically, and less valuable to society. At the same time, the engineer is a more normal, healthy American male, with somewhat the same traits of character as the scientist but with little of the scientist's tendency to go to extremes in behavior or emotional commitment. To summarize, engineers are "Simonized scientists," to bend a phrase recently reported in a national magazine.

Relation of Experience to Image

THE student responses were analyzed to determine whether the life experiences and current status of the students were associated with different beliefs about occupations. It was found in comparing the scientist image held by men with that held by women and the image held by students in private as against public colleges, by freshmen as against seniors, by students from different socio-economic backgrounds, by students from professional as against business families, and by students from different types of communities, that these groups do not differ in their beliefs about the scientist. This is clearly a stable image that is shared widely among college students with varied histories and experience.

In a study parallel to the one under consideration, 41 entering Wesleyan freshmen who indicated an intention to become scientists were compared with all the freshmen who planned to be active in other careers.[8] Those who intended to be scientists had a more favor-

8. This study was made by D. H. Bogart, a Wesleyan student.

able image of both the scientist and the engineer than the remainder of the newly arrived freshmen. The would-be scientists, as compared to the other freshmen, viewed the scientist as more colorful and interesting, of higher social status, more successful, more sensitive to art, and of a more sociable temperament. In absolute terms, the men wishing to enter the field of science rated the scientist quite high in material and social success and in esthetic interests, while they considered him moderately concerned with people. The scientist, as seen by these students, is interesting and colorful. Moreover, as compared with the non-science students, the science students had an image of the engineer that was closer to their image of the scientist. They viewed the engineer as more individualistic, persevering, and capable of deriving satisfaction from his work than did non-science students. In general, the engineer was seen as being more a man of parts by the pre-science students.

There is also evidence in the data that students on entering college have a more favorable view of the scientist than students who have already spent a semester in college. The new students have a more favorable view than second-semester freshmen of the intellectual ability, artistic concern, and success of the scientist.

Faculty Members' View

A GROUP of 27 college teachers of science at Wesleyan University were asked to respond to the same questionnaire that was given to the students. These men were a random sample of the science faculty. It is quite clear that the word *scientist* has similar connotations for them and for students. There was a correlation of +.91 between the average values attributed to the scientist by the Wesleyan students and by members of the science faculty on a group of 21 scales to which responses were made by both groups. The main differences between the two groups were, first, that the students attributed much more influence in public affairs to the scientist than the science teachers did, and, second, that members of the science faculty saw the scientist as more interested in art. Otherwise, the two groups were in close agreement.

Within the ranks of college teachers at Wesleyan, members of science and of social-science faculties are in almost complete agreement on the scientist image. On the other hand, faculty members in the humanities are more complimentary to the scientist than are the teachers of science or social science. A random sample of 23 teachers of the humanities rated the scientist quite high in material and social success and considered him more calm and more sociable than the science teachers did. The worldly success of the scientist seemed more impressive to teachers of the humanities than it did to teachers of the sciences.

Occupational Preferences

STUDENTS participating in the main study were asked to indicate the degree to which they would like to enter each of the 15 occupational fields if barriers related to expense, length of training, and native ability were removed. In other words, a male student was directed: "Rate each occupational position in terms of how much you would like to be in it if you could be in any occupation you wanted." The data revealed that a group of four occupations—those of college professor, lawyer, doctor, and business executive—were considered most desirable, in that order. The occupations of scientist and school teacher came next in order in a second grouping, at some distance from the first. A rather large gap appeared between this and the next grouping, the occupations of engineer and personnel director. When women were asked to estimate the attractiveness of these occupations for men, they also ranked the scientist in the fifth position. However, when college women were asked to name the single occupation for a future husband that would be most pleasing to them, only 3 percent indicated scientist. Approximately 20 percent of the women wished their husbands would be doctors, and another 20 percent selected the profession of lawyer.

Stereotypes of Specialized Scientists

IN STUDIES of the ranking in prestige of professions and occupations, the ranking of the term *scientist* differs from that of terms such as *chemist* or *biologist*, which describe scientists in specific fields.[9] In view of this finding, an exploratory study was designed to elicit the images of biologist, chemist, and physicist.[10] A small number of Wesleyan students were asked, in an interview, for their impressions of the personality, family life, status, social life, and motivations of men in each scientific field. Although the sample was small and unsystematically chosen, the agreement among students was so great as to suggest that the findings are of general significance. The stereotype of the specialized scientist in each case was more favorable than the image of scientist that was revealed in other interviews. According to these stereotypes the scientists in designated fields are more wealthy and successful, have richer social lives and more rewarding family lives, and are more pleasant and outgoing people than the "scientist" considered

9. National Opinion Research Center, in *Class, Status and Power*, R. Bendix and S. M. Lipset, Eds. (Free Press, New York, 1953).

apart from his field. The biologist is the most normal of the scientists in the sense that he approaches most closely the American ideal, and to the physicist are attributed many of the negative qualities that emerged in the interviews concerned with the generalized "scientist." The chemist falls between the two extremes.

Conclusions

THESE data suggest that there exists among college students a readiness to respond to the word *scientist* in a complex and differentiated manner. There is wide agreement concerning the image of the scientist among various classifications of men and women students in the Northeast. Members of one college faculty share this image with their students. The image is the same for freshmen as for seniors. It is safe to assume that the outlines of the image are the same for students at many colleges and for many college-educated adults. It is quite likely that the image is shifted somewhat in the first few months of a student's college career, but it is obviously not markedly changed. The image of the scientist among college students resembles in many ways the image held by high school students, as reported by Mead and Métraux.*

The specific features of the scientist image are important for several reasons. First, the image reveals the students' beliefs about the personality of the scientist and the style of life associated with a career in science. It means to the potential recruit that, if he selects science, he should have a certain set of personal qualities and can expect a particular kind of social life and certain types of personal associates, and it implies that the kind of life he will live is greatly limited by his work. If these features of the life of the scientist do not fit with the student's beliefs about himself or his hopes for the future, he is likely to be wary of committing himself to a career in science. At the same time, of course, the image influences the behavior of the student who has chosen science and leads him to develop those aspects of his character most in keeping with the stereotype of the scientist.

In short, the image has the effect of recruiting a certain type of person and discouraging others. This limits the range of people likely to consider the field, and it restricts the variety of basic talents available to science. Second, the public reaction to science, scientists, and the contributions of scientific research is likely to be colored by this image. This is particularly true in areas where arguments center around the generalized role of science. For example, the role of scientists in gov-

10. The interviews were conducted by E. W. Harbinger, of Wesleyan.
* See the previous chapter of this volume.

ernment or the advisability of admitting scientists to positions of high responsibility are issues frequently discussed in general terms. It may even be that the negative reaction of college students to courses in "general science" is attributable in part to the attitudes tapped by the word *science*.

The strong features of the image of the scientist are his high intelligence and his driving concern to extend knowledge and to discover truth. His work is of great value to mankind, and it brings him both a sense of satisfaction and a fair measure of success. The weaknesses in the image are many and disturbing. The scientist is seen as basically uninterested in people and unsuccessful with them. To the contemporary student, a person who does not care for people is suspiciously out of touch with life. The scientist is not interested in art—he has eschewed the life of the spirit that gives breadth and vitality to the life of the mind. Further, the scientist is a nonconformist and a radical, as well as a person with only moderate control of his impulses. These features suggest that college students possess beliefs that can easily be played upon to indict the scientist in times when loyalty is an issue of public concern. The undesirable aspects of this picture of the personal and intellectual life of the scientist make the role hard to accept in spite of the attractiveness of the work and the social contributions of the scientist.

The attractiveness of a scientific career in an abstract sense is clearly indicated by the high rank given it by men in statements concerning what they would like to be. Yet, surprisingly, few women wish to marry a scientist. It must be that, for men, the intellectual status, success, and material well-being of the scientist outweigh the many disadvantages of the scientist image. On the other hand, a woman married to a scientist must accept his personal qualities while benefiting very little in a direct way from the naure of his job.

Students clearly prefer the personality, social opportunities, and style of life of the college professor to those of the scientist. The scientist's only asset, by comparison with the professor, lies in the rewards associated with the work, and the differential is not great. The engineer and the scientist offer relatively interesting alternatives. The scientist is seen as an intellectual, with little capacity for social interchange; the engineer is a more normal "organization man," aiming at a nine-to-five existence, with an interest in good fellowship. It would seem that a student of science who could achieve the requisite training would be strongly drawn to college teaching with its richer, more humane connotations. On the other hand, the attractions of science and of engineering would seem to balance, with a person's view of himself playing an important role in his choice of one or the other.

It is interesting that students intending to pursue careers in science should have a more favorable image of the scientist than their colleagues who are planning other careers. It is not known whether commitment to a field changes the image or whether those with a more favorable image are drawn to the field. Probably both of these processes contribute to this difference.

It is comforting to find that scientists who are identified with their particular specialties are perceived as relatively normal people. These findings indicate that monolithic "science" is a source of concern to many sensitive citizens. On the other hand, men with professional specialties are considered more human, loyal, and comprehensible than "the scientist."

Science as a Way of Life

THE standard contemporary response to the finding that a product presents a "bad" image to the public is to turn for assistance to a team of public relations men who are instructed to change the image. To change an image as well developed and as widespread as the image of the scientist appears to be a most discouraging undertaking. This image is imbedded in a system of other stereotypes with which people, even highly educated people, structure their social world. To eliminate the unfavorable connotations from *scientist* would require a brilliantly conceived long-term campaign of confrontation through mass media and of educational innovation that is not likely to be undertaken. But is a massive campaign to alter this image appropriate? Scientists themselves, as well as their faculty colleagues, agree upon the essential features of the image. If it does represent, even in a distorted and exaggerated fashion, the characteristics of American scientists, it may be that to use publicity techniques would not only fail to hide the reality that lies behind the image but might also be dishonest.

Our studies give no data as to the actual (as distinguished from the perceived) characteristics of scientists. Yet C. P. Snow[11] has argued that indeed scientists *are* less interested than most educated men in esthetic matters and social affairs. Perhaps "the discipline" of science *does* narrow a man's interests, does create a group who do not meet the cultural ideal of the broadly educated man. If so, the "solution" is not to be found in an aping of Madison Avenue but, as Snow has also argued, in a more general appreciation on the part of the intellectual community of the demands the scientific mode of thought makes upon anyone, professional scientist or not, who seeks an objective understanding of the world around him. Perhaps, also, scientists have "over-conformed" to their own image of what a scientist is, and per-

11. C. P. Snow, *The Two Cultures and the Scientific Revolution* (Cambridge Univ. Press, New York, 1959).

haps the reality can change as more of them develop the broader interests and cultural appreciation constantly called for by liberal educators.

A final stance for the scientist consists in recognition of the possibility that to be a scientist is indeed to be different. The studies of Roe[12] and of Thorndike and Hagen[13] have shown that scientists tend to have characteristic developmental histories and personality structures. It may be that in order to do their work, recruits to scientific careers require some of the qualities which, in extreme form, appear in the stereotype of the scientist. If so, cannot the scientist accept this and get on with his work?

12. A. Roe, *The Psychology of Occupations* (Wiley, New York, 1956).

13. R. L. Thorndike and E. Hagen, *Ten Thousand Careers* (Wiley, New York, 1959).

17 *The Image of the Scientist in Science Fiction: A Content Analysis*

WALTER HIRSCH

In recent years sociologists have become increasingly concerned with the nature of "mass" or "popular" culture.[1] Paradoxically, in a "scientific age," science fiction has been neglected as a field of systematic study by sociologists and has been left largely to literary critics. The present paper aims to fill part of the gap between the stimulating but impressionistic approach of the humanist and the requirements of reliable knowledge which is the aim of the method of content analysis.

Science fiction should be of special interest to the sociologist. It concerns one of the basic focuses of contemporary culture—science—and it may serve as a vehicle for social criticism and for the construction of social utopias and counterutopias. In this paper we shall confine ourselves largely to a discussion of the image of the scientist presented in the genre and to the fluctuations of this image during the period 1926–50. Apart from its intrinsic interest, our analysis also has a "practical" relevance. There has been increasing concern about the shortage of scientists and engineers in the United States, especially vis-à-vis the apparent success of the Soviet Union in this field. Several recent studies of high-school students indicate that their attitudes

---

Reprinted from American Journal of Sociology, *vol.* 63 (1958), *pp.* 506–512.

Expanded version of papers read at the annual meetings of the Ohio Valley Sociological Society, April, 1957, and the American Sociological Society, August, 1957. The writer is indebted to Dr. Louis Schneider, Department of Sociology, Purdue University, for having suggested this study and for a number of hypotheses; to Mr. James Norton, of the Purdue University Statistical Laboratory, for aid in the sampling procedure and statistical analysis; and to Dr. Hanna Meissner and Lotte Hirsch for participating in the reliability analysis.

1. See, e.g., the May, 1957, issue of the *American Journal of Sociology*, "The Uses of Leisure," and Bernard Rosenberg and David M. White, *Mass Culture* (New York: Free Press, 1957).

toward a scientific career are often negative and their views of scientific work unrealistic.[2] The sources of these attitudes are still largely unexplored, but it is plausible that the reading of science fiction is such a source. Thus, Isaac Asimov, who is both a biochemist and a prolific science-fiction author, in an article entitled "The By-Products of Science Fiction" asserts:

> [In this genre] science and intelligence are . . . represented sympathetically. Scientific research is presented, almost invariably, as an exciting and thrilling process, its usual ends are both good in themselves and good for mankind; its heroes are intelligent people to be admired and respected.[3]

According to Asimov, the results of this will be that the readers of science fiction will be motivated to begin scientific careers. But there is no consensus on its content, even among its purveyors. Thus Philip Wylie, another writer of science fiction, challenges Asimov in his characteristically subdued manner:

> The science in science fiction is most commonly employed either ignorantly or for sadistic melodrama. Most writers' ignorance of human psychology is as abysmal as that of philosophy . . . and they but create a new and sinister folklore, in which the latest facts from Massachusetts Institute of Technology are superimposed on a human insight hardly more developed than that of bushmen.[4]

A glance through the welter of comments produced by practitioners and by friendly and hostile critics of the genre makes it obvious that there exists no consensus on the nature of its content. Before considering the putative effects on the readers, we must have relatively objective knowledge of what is presented to them. First, however, it is in order to give a brief description of the historic development of science fiction and of the people who produce and consume it.

Science Fiction as a Genre[5]

CONNOISSEURS often make a distinction between "science fiction" and "fantasy." The latter deals with themes of the weird and

2. See Purdue Opinion Panel Poll No. 45, "Physical Science Aptitude and Attitudes toward Occupations" (Division of Educational Reference, Purdue University, July, 1956), and Margaret Mead and Rhoda Métraux, "The Image of the Scientist among High School Students," *Science*, CXXV (August 30, 1957), 384–90. [Reprinted in this volume.]

3. *Chemical and Engineering News*, XXXIV (August 13, 1956), 3882–86.

4. "Science Fiction and Sanity in an Age of Crisis," in Reginald Bretnor (ed.), *Modern Science Fiction* (New York: Coward-McCann Co., 1954), p. 235.

5. For a more detailed treatment see Walter Hirsch, "American Science Fiction, 1926–1950: A Content Analysis" (unpublished Ph.D. dissertation, Department of

supernatural; the former, with phenomena explicable in "scientific" (or pseudo-scientific) terms. In practice, however, the distinction is rarely observed, and science-fiction magazines, so called, publish both types of stories, though some specialize in one or the other type. Specialization also extends to literary level and choice of themes, ranging from "space operas," with their black-and-white treatment of characters and stereotyped plots, to the highly sophisticated use of "psychological" themes and the unraveling of scientific puzzles.

The start of science fiction as a mass medium is usually associated with the founding of *Amazing Stories* in 1926 by Hugo Gernsback, an electrical engineer. Other periodicals soon came into existence, and by 1953 sixteen publishers put out thirty-five science-fiction magazines. (The number has since decreased considerably.) The "boom" also manifested itself in the publication of science-fiction novels and anthologies, both hard and soft covered, and in science-fiction films and radio and television programs.

The early issues of *Amazing Stories, Thrilling Wonder Stories,* and others had to rely largely on "classic" writers such as Jules Verne and H. G. Wells, whose stories they reprinted. Gradually a new, indigenous generation of writers arose, some of whom, like Ray Bradbury, became known to a larger and more diversified public than that which comprised the "typical" science-fiction readers of the 1920's and 1930's. There are no reliable data available on the quantity and makeup of science-fiction-magazine readership, but we do have some evidence that it shifted from a public with rather narrow "technical" and engineering interests to one with broader scientific as well as humanistic concerns.[6] The American readership was estimated at six million in 1954.[7] According to a survey made by John Campbell, Jr., the editor of *Astounding Science Fiction,* the typical reader of this magazine is male, a college student or graduate, under thirty-five, and engaged in the technical, professional, or managerial occupations.[8] It should be remembered, however, that this is not representative of the entire readership, since *Astounding Science Fiction,* together with the *Magazine of Fantasy and Science Fiction* and *Galaxy,* caters to a relatively highly sophisticated and educated audience.

Readers, writers, publishers, and critics of the genre communicate

Sociology, Northwestern University, 1957), chap. ii, "Science Fiction, Its Nature and Development." See also L. Sprague de Camp, *Science Fiction Handbook* (New York: Hermitage House, 1953), chap. iii, and August Derleth, "Contemporary Science Fiction," *English Journal,* XLI (1952), 1–8.

6. This evidence is partly inferred from content analysis (see Hirsch, *op. cit.,* chap. iv, "Time, Place and Characters").

7. See S. E. Finer, "A Profile of Science Fiction," *Sociological Review,* II (new ser., 1954), 239–55.

8. *Astounding Science Fiction,* July, 1949, pp. 161 ff.

with one another through the editorial and letter columns of the magazines and by way of more or less organized fan organizations which partake of the nature of a cult. The flames of organization and controversy are occasionally fanned by the monetary motives of publishers, but there appears to be an unusually high degree of *Gemeinschaft*-like interaction among the "fen" (as they refer to each other). Their emotions and creative aspirations are disseminated by numerous "fanzines," which are non-profit, amateur publications in contrast to the "prozines."

To what extent are the authors of science fiction scientists? Again, no extensive reliable data are avaliable, but it appears safe to say that, as a group, science-fiction writers contain a greater proportion with formal scientific or engineering training than is the case with authors of other types of fiction.[9] In view of this information, we assume that there is a "universe of discourse" involving a rather high degree of intercommunication between writers and readers. Consequently, our analysis of content is no mere academic exercise but should aid us in formulating hypotheses about the social functions of this type of literature.

Methodology

THE universe comprises all science-fiction stories published between 1926 and 1950, as listed in Donald Day's *Index to the Science Fiction Magazines, 1926–1950*.[10] The time period was divided into six subperiods, roughly corresponding to important historical events (Table 1).

TABLE 1

Subperiod	*Years*	*Characterization*
1	1926–29	Prosperity
2	1930–33	Depression
3	1934–37	New Deal
4	1938–41	Fascism and War in Europe
5	1942–45	World War II
6	1946–50	Postwar period

For each of the subperiods a random sample of fifty stories was chosen from the listing of titles in Day's *Index*, providing a total sample of three hundred stories. The writer analyzed and coded each

9. Of the "eighteen leading contemporary writers of science fiction" listed in De Camp (*op. cit.*, pp. 145 ff.), eight have a "scientific background" (i.e., they hold degrees in science or engineering or have done technical work or both); four of the eighteen are presently engaged in such work, and two hold Ph.D.'s in chemistry.

10. Portland, Ore.: Perril Press, 1952.

11. In order to provide some check on the reliability and validity of the analysis,

story on the basis of categories derived from about a dozen hypotheses.[11] Since the stories were read as they became available through loan or purchase rather than in chronological order of publication, an important source of possible cumulative biases was eliminated, so that they would not affect any generalizations based on time trends. The coded material was punched on Unisort cards and sorted manually for statistical and qualitative analysis.

Findings

1. SCIENTISTS AS MAJOR CHARACTERS

The proportion of scientists who are major characters in the stories declined steadily from 1926 to 1950. As indicated in Table 2, during the first (or prosperity) period, the proportion of *heroes* who were scientists was 44 per cent, but only 24 per cent during the postwar period (the difference is significant at the 5 per cent level).

Among the *villains* the scientists declined from 39 to 30 per cent of of the total (the difference is not significant). It is difficult to account for this unexpected finding in terms of "reflection" of reality; more likely the decline is due to the broadening of the readership in include those less exclusively interested in scientific and technical matters. It

TABLE 2. Percentage Occupations of Heroes

	Time Period					
*Occupation**	*1*	*2*	*3*	*4*	*5*	*6*
Physical scientist	44	34	36	26	23	20
Social scientist	—	2	2	—	4	4
Professional	11	17	12	10	13	10
Pilot, etc.	2	9	9	9	8	8
Military	3	5	2	5	11	10
Total *N* for each period = 100 per cent	61	58	56	58	47	50

* Other categories not shown.

may be argued that this long-range trend as well as the increase in heroic and the decrease in villainous scientists during the World War II period are attributable to the current involvement of scientists in nuclear and biological warfare. This is a plausible hypothesis, but a

two other analysts were given ten stories to code with the aid of detailed instructions. It was not possible to give them the extensive training necessary for a meaningful reliability check, but we were able to eliminate several of the apparently least reliable categories and become sensitized to the intrusion of biases after comparing and discussing the results. For a detailed discussion of the procedures used and of the problem of validity in content analysis see Hirsch, *op. cit.*, chap. iii, "Methodology."

qualitative analysis of the stories in our sample fails to support it. We had expected a substantial increase in the number of social scientists, but their proportion of the total number of major characters never exceeded 4 per cent. Patently, the popular conception of the scientist as one who deals with "natural" rather than "social" phenomena is also found in the image presented in science fiction. However, it is likely that the role of social scientists will become more prominent after 1950, judging from our impressionistic reading of the literature beyond the regular sample.

2. HEROES AND VILLAINS

Throughout the entire time period scientists occupy the most frequent occupational category of villains (except for the wartime period, when they are replaced in first rank by businessmen). Businessmen are generally in second rank, followed by politicians and criminals. Evidently the businessman is the bête noire of science fiction (Table 3).

We had expected this to be equally true of politicians and military men, based on the hypothesis that scientists will be portrayed as being frustrated by these other elites. The results do not bear this out, statistically speaking, though themes of conflict between scientists and other power groups in the "garrison state" appear increasingly during the postwar period. At any rate, it seems safe to state that the often vehement attacks on the business elite and its culture that we find in science fiction is not duplicated at present in any other domain of popular culture, certainly not in the mass media.[12] A more specific

TABLE 3. Distribution of Heroes and Villains by Occupation

		Heroes	*Villains*
	Scientists	109	53
	Businessmen	9	22
$N = 193$		Chi square $= 14.46$	$P < 0.001$

test of the value orientation of science-fiction writers is afforded by the analysis of the status of scientists in imaginary societies. The construction of such societies is a traditional device for expressing values which may not be safely stated in reference to actual societies. What is the status of scientists in either utopian or antiutopian societies depicted in our sample? Table 4 shows that, in general, scientists are represented as the "legitimate" elite (i.e., they are portrayed favorably) in comparison with non-scientists. We had expected these results, since it was our assumption that science-fiction writers tend to be spokes-

12. This finding is relevant to the proposition that science fiction is the major contemporary popular medium for the expression of social criticism and the building of social utopias and counterutopias. This aspect of our study will be treated elsewhere.

men for scientists and for the scientific ethos. The vigorous anti-scientific stance found among British authors like Aldous Huxley in *Brave New World,* George Orwell in 1984, or C. S. Lewis in the trilogy comprising *Out of the Silent Planet, Perelandra,* and *That Hideous*

TABLE 4. Authors' Treatment of Scientists and Non-Scientists in Imaginary Societies

	Favorable	*Unfavorable*
Scientists	24	9
Non-scientists	20	25
$N = 78$	Chi square = 5.10	$P < 0.05$

Strength is not typical of American science fiction in spite of the critical success of these authors and of American writers like Ray Bradbury. On the other hand, as we shall point out below, the naïve adulation of the omnipotent and omniscient scientist is no longer a feature of the genre as it had been in its childhood.

TABLE 5. Scientists Working in Independent and Bureaucratic Settings

Time Period	*Independent*	*Bureaucratic*
1–3	85	13
4–6	44	20
$N = 162$	Chi square = 6.65	$P < 0.01$

3. OCCUPATIONAL SETTING

Increasingly, from earlier to later stories, scientists are pictured in a "bureaucratic" setting rather than as "independent." (We categorized college professors as "independent," unless it was shown explicitly that they were involved in bureaucratic problems in their university.) Table 5 shows the relevant data. The "gentleman scientist," unencumbered by problems of "human relations," is slowly replaced by the scientist involved in a network of interpersonal and institutional pressures. In this instance science fiction tends to give a relatively realistic picture of the setting in which contemporary scientific activity takes place.[13]

4. THE SOCIAL ROLES OF SCIENTISTS

The image of the scientist is portrayed by a number of more or less conventional themes. The most frequent are the "Frankenstein" and the "Scientist as Savior" themes. Both these tend to decline during the

13. See, e.g., William H. Whyte, Jr., *The Organization Man* (Garden City, N.Y.: Doubleday & Co., 1957), chap. xvii, "The Bureaucratization of the Scientist."

later phase of the time period, while there is a relative increase in themes of Hubris—the scientist blasphemously attempting to attack natural or divine law—or the "Scientist as Martyr." What are the social problems depicted in the literature, and what part do scientists take in their solution?

The major social problems rank as follows in order of frequency. First, interpersonal relations, involving for the most part conventional "love interest," and, second, the effects of technology. These are basically of an unanticipated nature (e.g., biological mutations which have been produced experimentally and then get out of hand). The "Frankenstein" theme is, of course, associated typically with this type of problem. International conflict ranks third, followed by interplanetary conflict. The fluctuations indicate to some extent reflection of reality (e.g., there is an increase in problems of international conflict during Period 5), but reflection is not observable in all problem areas. Thus there is less concern with problems of poverty and unemployment during Period 2, the depression period, than in any other! Presumably, the "escapist" function of literature is operating in this case.

How are science and scientists involved in the solution of social problems? We had hypothesized that there would be a decline in the theme that utopia can be reached by the simple application of technology and natural science. Instead, human problems would be increasingly solved by (*a*) the application of social science (or natural and social science combined), (*b*) by the magical or charismatic powers of human beings, and (*c*) by the intervention of "aliens" from outer space and other nonhuman characters. The rationale for this hypothesis is as follows: Recent historical events have cast increasing doubts on the omnipotent role of technology and natural science. Tremendous strides have been made in the conquest of nature, yet we are still beset by wars, crime waves, depressions, mental illness, etc., none of which has been amenable to controls. To the extent that the scientific ethos still holds sway, the answer to the problem is "human engineering," or the application of the social sciences. But it seems equally possible that there has occurred a more radical disillusionment with any sort of rational scientific means and hence a recourse to "magical" solutions. Many observers of contemporary culture have attested to the rise of irrational tendencies in Europe and to the "loss of nerve" which has displaced the sanguine and facile optimism so characteristic of the America of "unlimited opportunities.[14]

The data in Table 6 bear out the first part of our hypothesis: there is a distinct decline in the use of technology and natural science as a means for solution of problems (the difference between Periods 1 and 6 is significant at the 5 per cent level). But there is no corresponding

14. *Ibid.*, chap. xx, "Society as the Hero."

TABLE 6. Percentage Means for Solution of Social Problems

*Type of Solution**	Time Period 1	2	3	4	5	6
Technology and natural science	46	28	38	26	28	26
Social science	3	—	4	3	6	2
Natural and social science	3	4	1	1	3	5
Courage of hero	15	34	20	29	20	2
Insight and ingenuity of hero	10	16	6	6	13	5
Human charisma	2	—	7	8	3	5
Intervention by "aliens"	7	5	11	11	7	28
Total *N* for each period = 100 per cent	61	92	71	70	67	43

* Other categories not shown.

increase in the use either of social science or of human magic and charisma; even the conventional attributes of heroes, such as courage and insight, do not fill the "vacuum." It is filled by the intervention of "aliens" (i.e., natives of other planets), especially during the postwar period (the difference between Periods 5 and 6 is significant at the 1 per cent level). Human actors, whether magicians or scientists, are being supplanted by "higher" powers.

Much could be said about these findings as a reflection of general cultural trends. However, our present concern is specifically with the role of science. If our analysis has any validity, it would seem that the public, which could be expected to view science as the obvious means for the solution of social problems, does not in fact view science in this way. The contemporary authors who may be considered spokesmen for the scientific ethos have lost the sanguine belief in the omnipotence of science held by their predecessors.

On the other hand, certain aspects of our analysis do not indicate a wave of resignation and irrational pessimism. The majority of the stories have "happy" endings, and there is no indication that contemporary science-fiction authors consider human nature essentially more evil or irrational than did earlier writers. Rather, there appears to be a greater recognition of the "reality principle," of the complexity of human behavior, of the "unanticipated consequences of purposive human action," as Merton has termed it. If the "aliens" represent any sort of principle or power, it would seem to be "chance" rather than some anthropomorphic deity. (In this connection it is of interest that in the stories we analyzed "chance" as a problem-solving device increases appreciably during the latest time period, while there is a notable absence of any "return to religion" in the conventional sense.) Scientists are no longer either supermen or stereotyped villains but real human beings who are facing moral dilemmas and who recognize that science alone is an inadequate guide for the choices they must make.

Conclusions

OUR analysis shows that the content of science fiction has undergone a number of significant changes. Some of these seem to indicate a reflection of actual historical and social trends, but the reflection is selective rather than mechanical. On the other hand, we notice both critical and utopian treatment of the present and future position of the scientist. The next problem to be dealt with concerns the perception of the content of the literature by its readers and the effect it has on their motivations for engaging in scientific activity or for choosing other occupations.*

* Another content analysis of fiction in "ordinary" mass media (*Collier's* and *The Saturday Evening Post*) has shown the paucity with which scientists are depicted as major characters in this genre, as well as the fact that scientific heroes are pictured as a rather bloodless lot, hardly distinguishable from "clean-cut" businessmen. These findings are also of interest in connection with the Mead and Métraux article reprinted in this volume. See Marilyn Kerster and Walter Hirsch, "Scientists in Popular Magazine Fiction," mimeographed, Department of Sociology, Purdue University.

18 *Scientists and Politicians*

HARRY S. HALL

UNTIL THE REVELATION of Hiroshima, Congressmen, like most laymen, had little reason to be much concerned with either science or scientists. The majority of them undoubtedly shared the popular conception of science as the well from which material benefits flowed in an endless stream symbolized by the familiar picture of the man in a white smock holding up a test tube to the light. As for scientists, most politicians seemed to view them either as useful tools for increasing the productive resources of industry, or as impractical visionaries and eccentric crackpots puttering with complicated equipment in their laboratories. As Senator Tydings candidly admitted to Dr. Szilard in 1945 during the hearings of the Special Senate Committee on Atomic Energy:

> Doctor, if in 1939 we had been conducting a hearing like we are conducting today, and men like yourself had come before our committee and projected the possible development of the bomb up to now with reasonable accuracy, I imagine they would have been called a lot of crackpots and . . . visionaries who were playing with theories. I certainly would not have had the receptivity that I have today to say the least.[1]

The atomic bomb changed this situation completely, forcibly thrusting science and scientists into the forefront of politicians' focus of attention. It demonstrated as never before the destructive possibilities

---

Reprinted, with permission, from the February, 1956, Bulletin of the Atomic Scientists, 935 *East Sixtieth Street, Chicago* 37. *Copyright* 1956 *by the Educational Foundation for Nuclear Science, Inc., Chicago* 37.

1. *Atomic Energy: Hearings before the Special Committee on Atomic Energy,* U.S. Senate, 79th Congress, 1st Session. U.S. Government Printing Office, Washington, D.C., 1945, p. 290.

of science. But, what was far more important, detonation of the bomb drove into people's consciousness the realization, hitherto understood by only a few laymen, that science was a major social force. Moreover, their preoccupation with the complex problems brought about by atomic energy necessarily involved Congressmen with scientists, even had the latter not voluntarily entered the arena of political action. For the first time, senators and representatives found themselves in considerable interaction with scientists. Whatever their previous conceptions of scientists might have been, politicians could not afford to ignore them. Whether Congressmen liked it or not, they had to take note of the crucial role of science and scientists in the atomic age.

How did politicians react to their enforced venture into strange new territory inhabited by equally strange men who spoke a foreign language and dealt with matters utterly beyond the experience of most Congressmen? What conceptions and attitudes regarding science and scientists did they form as a consequence of such entry?

Some answers to these questions may be obtained from the transcripts of Congressional hearings on matters in which scientists were interested in the immediate postwar years. Outstanding examples of such hearings were the ones on domestic control of atomic energy and establishment of a National Science Foundation. Most of the illustrative citations have been culled from these hearings,[2] although they are by no means exhaustive of the material. A few quotations from the 1953 Senate hearings into the Bureau of Standards rejection of the claims made for a battery additive indicate the contemporary nature of some conceptions and attitudes respecting scientists. Finally, it should be pointed out that while scientists were the principal witnesses of these hearings, the testimony of respected laymen, like General Groves and Morris L. Cooke, former head of the Rural Electrification Administration, about science and scientists undoubtedly helped shape Congressional thinking on such matters.

Scientists as Superior Beings

AFTER the astonishing news of Hiroshima and Nagasaki, overnight to many people, scientists became charismatic figures of a new era, if not a new world, in which science was the new religion and

2. *Atomic Energy: Hearings before the Special Committee on Atomic Energy,* U.S. Senate, 79th Congress, 1st Session.

Atomic Energy Act of 1946: Hearings before the Special Committee on Atomic Energy, U.S. Senate, 79th Cong., 2nd Session. U.S. Government Printing Office, Washington, D.C., 1946.

Science Legislation: Hearings before the Subcommittee of the Committee on Military Affairs, U.S. Senate, 79th Congress, 1st Session, U.S. Government Printing Office, Washington, D.C., 1945, 1946.

National Science Foundation: Hearings before the Committee on Interstate and scientists the new prophets. Like everyone else at that time, politicians, too, looked upon scientists with considerable awe and deference. Scientists appeared to them as superior beings who had gone far ahead of the rest of the human race in knowledge and power. Indeed, politicians seemed to regard scientists in much the same way that primitive men regard their magician-priests. That is to say, Congressmen perceived scientists as being in touch with a supernatural world of mysterious and awesome forces whose terrible power they alone could control. Their exclusive knowledge set scientists apart and made them tower far above other men. As Senator Tydings put it:

> There are a few men . . . or maybe several thousand in the world whose mental development in many lines—and particularly in the scientific line—is like comparing a mountain to a molehill when you compare them to the rest of us. (*Spec. Comm.*, 1945, p. 309)

Senator Russell expressed his feelings by remarking:

> My attitude toward scientists is . . . pretty much like the boy living in the country and going to the country doctor. He thinks the doctor can do anything. (*Spec. Comm.*, 1945, p. 170)

Senator Hickenlooper spontaneously referred to the magical quality of scientists in his comment to Dr. Bush on the release of atomic energy:

> We have got to the point where we have rubbed the lamp and the genie has come out and we cannot get him back into the lamp. (*Spec. Comm.*, 1945, p. 183)

The most explicit acknowledgment of scientists' superiority came from Senator Johnson of Colorado. In an exclamation that was both protest and plea, he burst out:

> I have one further observation to make, and that is that you scientists have gotten a long way ahead of human conduct, and until human conduct catches up with you, we are in a precarious way unless you scientists slow up a little and let us catch up. (*Spec. Comm.*, 1945, p. 141)

Senator Johnson's outburst also expressed the resentment and fear that

Foreign Commerce, House of Representatives, 80th Congress, 2nd Session, U.S. Government Printing Office, Washington, D.C., 1948.

National Science Foundation: Hearings before a Subcommittee of the Committee on Interstate and Foreign Commerce, House of Representatives, 81st Congress, 1st Session, U.S. Government Printing Office, Washington, D.C., 1949.

Battery AD-X2: Hearings before the Select Committee on Small Business, U.S. Senate, 83rd Congress, 1st Session, U.S. Government Printing Office, Washington, D.C., 1953.

ordinary men feel toward superior beings who are considered to be in a position to influence their destinies. Nor was Johnson the only Congressman who felt afraid. When Dr. Rabi complained to Senator Fulbright that the May-Johnson bill made scientists feel that politicians were treating them as a special class of citizens from whom other people had to be protected, Fulbright tacitly agreed, explaining:

> The reason for that is that you scientists scared us all to death with your atomic bomb and we are still very frightened about it. (*Sci. Leg.*, 1945, p. 992)

Earlier when Senator Magnuson asked Dr. Karl Compton:

> Doctor . . . After all, the reason that scientific hearings are now not only fashionable but interesting is because of the atomic bomb.

Senator Kilgore remarked:

> May I amend that and say we are all scared to death. (*Sci. Leg.*, 1945, p. 635)

An extremely significant dimension of politicians' perception of science as a major social force was the dynamic and progressively expanding character which they ascribed to it. Because of such a nature, science's role and impact would grow ever greater with time. In this view, science was likened to some sort of juggernaut rolling inexorably and invincibly forward. Its power and momentum were so irresistible that all obstacles in its path were crushed or swept aside.

Science as International

POLITICIANS' belief in the juggernaut character of science rested ultimately on two other perceptions (1) that scientists were an exclusive in-group, a tightly-knit fraternity of dedicated theoreticians and visionaries; (2) that science was international in scope in terms of its jurisdiction, activities, and membership. These two factors were responsible for science's invincible advance and for scientists' independence of control. The zeal with which scientific research was pursued plus its world-wide scope made it impossible to control either science or scientists. For, if repressed in one country, science would be carried on in another by different or even the same devotees.

Senator Hickenlooper expressed his view of the international character of science when he asked Szilard:

> Doctor, as a scientist and with what knowledge you have of the history and present activities of international science, do you be-

> lieve that science or even countries will stop the examination and exploration into the atomic energy field by any treaties or agreements? (*Spec. Comm.*, 1945, p. 272)

Senator Magnuson discussing his ideas on what the National Science Foundation should do, declared to the scientists:

> As you gentlemen so well know, science knows no geography. (*Sci. Leg.*, 1945, p. 566)

Their perception of science's international character and their distress, even resentment, of this fact was revealed in politicians' questions about the feasibility of international exchange of scientific information by itself, without an accompanying exchange also of other cultural material. Essentially, politicans were reluctant to admit that science was independent and self-sufficient enough *vis-à-vis* the rest of social affairs that its knowledge could successfully be exchanged without having at the same time other forms of international communication and interaction. Senator Millikin expressed his reluctance to Harlow Shapley in the question:

> Do you believe that the general exchange of scientific information between the nations and the exchange of scientists is feasible unless you have at the same time a general exchange of information and the permission for all kinds of people to cross borders? Do you think it can be confined to scientists? Can you isolate just one department of human knowledge and say, "Now after all this we will exchange information and personnel but we will not exchange anything else." Is it practical to do that? (*Spec. Comm.*, 1946, p. 164)

Senator Tydings, too, was concerned and reluctant to accept the fact that science could effectively be divorced and exist apart from the rest of society and culture. In a discussion with Vannevar Bush, he queried:

> Do you think it is possible to accomplish in the world a free exchange of scientific knowledge—in a world that does not permit the free exchange of religious knowledge?

Tydings indicated his unwillingness to accept Bush's affirmative reply by insisting:

> If we cannot do it in one field, what hopes may we have that we can do it in another field? If freedom of the mind is interdicted, so to speak, by the government in one of the oldest and strongest traditions, why is it reasonable to believe that the freedom of the mind will not be interdicted in this other field of science?

In return, Bush pointed out:

> Science, Senator, has always been more or less free; it has always had a flavor of internationalism, you know.

Tydings retorted,

> So has religion. (*Spec. Comm.*, 1945, p. 163)

The Scientific In-Group

CONGRESSIONAL reluctance to admit the independence of science from other areas of social life may be correlated with their perception of scientists as an exclusive little in-group sharply differentiated and set apart from the rest of society. This perception and its implications require more detailed analysis.

To begin with, politicians had a vague, inarticulate understanding that the scientists dealt with abstract symbols. Scientists and others, like General Groves, tried to explain the nature of basic scientific research. Yet, politicians couldn't really understand. Their dim perception of the abstract and symbolic nature of scientific work was expressed in an indirect way by the use of "theoretical," "visionary," etc., to characterize scientists. Thus when Bernard Baruch said:

> It has been my experience, having been in touch with the medical end of it, that there are two types of minds, one that will drift over into the applied sciences and another which is the roaming mind, what Edison told me once was the "source mind," the fellow who is interested in a particular subject and no money can get him off it. That fellow will develop that line—almost will make his own demarcation.

Kilgore rejoined:

> He is what we sometimes call the more visionary type. (*Sci. Leg.*, 1945, p. 919)

When the scientists tried to explain their orientation to abstract and remote symbols, it was clear that politicians did not understand—not even Senator Fulbright, who had an academic background himself. Thus, Drs. Wilson and Oppenheimer tried to explain the difference between fundamental and applied research as they conceived it. Wilson started off by saying:

> Programmatic (research) is when you just envisage research, when you can say that a man should do such-and-such a problem and measure such-and-such a quantity. That is programmatic work. In general when a man is really doing research, he is not able to tell you what he is doing. He is working on the frontier of knowledge. He does an experiment in an entirely exploratory manner, and then

> on the basis of a hunch he does another experiment. Finally, he sees something. He can't see it, but eventually it comes out. It is entirely creative, as opposed to most workaday physics that is in general done.

Oppenheimer tried to clarify in these words:

> Dr. Wilson said that programmatic research consists in the measurement of quantities which are known to exist. That may sound very funny, but half of the work of physics and most of the work of fundamental physics is not to measure the quantities that are known to exist, but to find out what quantities do exist. That is, what kind of language, what kind of concepts correspond to the realities of the world. What Dr. Wilson was saying—and I think quite rightly—was that if you really know what it is you have to do, if you can give it a name and say, "We don't know whether it is 10 or 20, but we would like to measure it," then you are doing something that increases knowledge very much but which doesn't qualitatively increase knowledge. When you try to find out whether it is possible to talk about such things as simultaneity and position . . . then you are going to make a contribution.

When Wilson tried to continue with "I think we can say that since 1940—" the listening senators interrupted to express their bewilderment and incomprehension. Magnuson said: "Dr. Fulbright would understand that"; but it was clear he did not for when Fulbright started to say: "That was a fine explanation—" he, in turn, was interrupted by Magnuson who commented, "He doesn't understand you either." (*Sci. Leg.*, 1945, p. 332)

Scientists as Dedicated Men

A SECOND characteristic of scientists as an exclusive in-group was their intense commitment to the ideals and pursuit of science. Their devotion seemed to resemble, in some respects, the fervor shown by the members of a religious sect who concentrate almost exclusively on realizing their goals to the neglect of everything else. Thus, Senator Young commented:

> Mr. Berne, I think you raised a very important point when you mentioned the salaries of scientists and so on. I have personal knowledge of several scientists who stayed in the work almost entirely because of the love of the work rather than the salary that they received. (*Sci. Leg.*, 1945, p. 845)

Senator Millikin made the same point in asking Compton:

Do you believe this is a correct statement that probably of all the professions in the world, the scientist is less interested in monetary gain—I am speaking of the pure scientist?

Compton replied:

I don't know of any other group that has less interest in monetary gain. (*Spec. Comm.*, 1946, p. 263)

The clearest expression of politicians' image of scientists as a band of dedicated men whose devotion to the ideals and pursuit of science has a religious quality was contained in politicians' conviction that scientists would continue developing the field of atomic energy—regardless of the consequences or attempts to prevent their work.

Senator Connally expressed his belief in the dedication of scientists during a discussion with Dr. Bush on other possible sources of atomic energy. Connally asked:

They will keep on trying, will they not, scientists will keep on experimenting, will they not?

When Bush replied: "Well, even scientists do the things they think can probably be done," Connally persisted in his image, saying:

That is true, but they will be trying other things, they will be working in other fields and among other elements. Maybe they think that they will fail, but they will keep on trying. (*Spec. Comm.*, 1945, p. 171)

Langmuir strengthened this image of scientists by putting their commitment to scientific ideals and research in an explicitly religious context. To Hickenlooper's question:

Having seen the proof, Doctor, that fission is possible on substantial scales, and having been admitted to this at least first dawning of this great field of energy, is there any reasonable possibility that science the world over could be prevented in any way from dabbling in this thing further and going on? In other words, science is inevitably bound to go forward with further experimentation in this field?

Langmuir's reply was framed in terms of one religious faith trying unsuccessfully to halt the progress of a competing religion; he asserted:

It is just like the Catholic Church trying to stop progress in science. It just can't be done, because the knowledge already is there to be found. It is in nature, and to prevent us discovering the facts of nature is a hopeless job.

Hickenlooper's rejoinder indicated his acceptance of Langmuir's statement on the futility of trying to control scientists in their work when he asked:

> And scientists will go on exploring the field despite what politicians or anyone else attempts to do?

Langmuir warned:

> If you regiment them here so they can not do it, they will go elsewhere. (*Spec. Comm.*, 1945, p. 127)

Politicians, then, were given to understand that scientists, like all sectarians, would and did disregard any attempt to obstruct their efforts to achieve the ideals they believed in. Their resistance and defiance extended even to the commands of legitimate authority which had moral justification for its orders. For example, Senator Hickenlooper asked Dr. Oppenheimer:

> Doctor, does the history of scientific development in the world at any time show that the scientists have ever discarded something that opens new ground for scientific investigation? In other words, can you by legislation, or even by intent of government, stop scientists from investigating or exploring new fields, especially when they have had success in a field, as is the case, the proven case, with atomic energy?

Oppenheimer's answer was:

> Well, enormous discouragements can be proposed and scientific advances can certainly be opposed. It was not easy during the days of the Renaissance for a man to investigate the laws of nature because he was likely to get himself into extreme personal trouble.

Hickenlooper's rejoinder indicated his belief that scientists would succeed in overcoming all obstacles because of their intense commitment to scientific ideals:

> But scientific development did go on. (*Spec. Comm.*, 1945, pp. 200–201)

The ethical imperatives of their scientific faith were perceived by politicians to take precedence over all other moral obligations and demands for scientists. Scientists pursued their work in complete indifference to its social and ethical consequences. This view of scientists was expressed by Senator Johnson when he exclaimed:

> Dr. Langmuir, how do you compose your two viewpoints? On the one hand you say that this country should appropriate $5,000,-

> 000,000 for scientific research; on the other hand, you say that this country should destroy $5,000,000,000 worth of the products of science. The scientists, according to your testimony, have made the world extremely insecure. Science has made, according to your statement just now, aggression inevitable and yet, at the same time, you say that we ought to keep pouring money into science.

When Langmuir argued: "Science is not a thing that we make; scientists don't create science in that way," Johnson retorted:

> But, you create atomic bombs, and now you want to go and throw them into the middle of the ocean because they have made the world insecure.

Langmuir explained:

> In order to get security—in other words, we buy something by that. It is a price. You cannot get security for yourself without giving it to other nations. We have no security now for the future, because we are now in stages one or two. We are now secure but we can foresee the case that this security is only temporary, that the time will come when not only we but no nation is secure, and we must do something about that. We must start now to do something about it, because otherwise disaster lies ahead, probably a worse disaster for us than for anyone else.

Johnson persisted in his conviction that all scientists wanted to do was to go on with their work under the financial support of the government with no concern for the social consequences of their research. He declared again:

> It looks to me as though you scientists have made the world extremely insecure, and now you are coming to the politicians and asking us to go about and make the world secure again by some sort of political agreement. At the same time, you are asking that the scientists who made the world insecure be given further appropriations to discover still another and more destructive element than atomic energy. (*Spec. Comm.*, 1945, pp. 120–21)

Scientists as "Uncommon" Men

THEIR orientation to the abstract and remote symbols of science and their devotion to scientific ideals and research set scientists apart from other men in politicians' eyes. Scientists were different. A simple but striking illustration of this view was contained in Congressmen's repeated use of the words "You scientists," or "you felllows." Politicians use of these words is taken here to mean they perceived scientists to be considerably removed from ordinary people. Scientists possessed distinctive attributes which put a wide gap between them and other persons. Hence, when Millikin remarked to Watson Davis, head of Science Service,

> I notice in the Orphan Annie strip they now have the atom bomb down to hand size and you can tote your own private demolition. But as a scientific man, you wouldn't know about the Little Orphan Annie strip. (*Spec. Comm.*, 1946, p. 192)

he was expressing his belief that scientists lacked the simple tastes and interests of the common man. Senator Thye's comment seven years later in the hearings on the Bureau of Standard's rejection of the claims made for a battery additive indicated the strength of this view. Thye stated his belief that scientists took very seriously trivial matters that no ordinary person would even worry about.

> Dr. Astin, since this question came up and for the sake of an amusing scientific argument, I saw a paper where two scientists had debated at great length on the question of whether boiled water would freeze quicker than unboiled water, and it was amusing to me to read the scientific language used in the debate on the simple question of whether water boiled would freeze more rapidly than water not boiled. (*Bat. AD-X2*, 1953, p. 230)

The image of scientists as a differentiated group that was set apart from other men was closely related to the further perception of its solidarity and exclusive nature. Politicians' resentment of this nature was expressed by Rep. Busbey in his complaint at the treatment one of his friends received:

> This friend of mine is, in my opinion, quite a scientist. He has many valuable inventions on the market. While he is a university graduate, he is not a Ph.D. He has one of the most modern laboratories, I think, in the United States. He has tried on numerous occasions to make his services and his ideas available down here to people and constantly he is given form after form to fill out. . . . He supports his ideas with certain research data and most of the time he cannot get replies to his letters. . . . Now I do not think that George Washington Carver had a Ph.D. degree, at least in his earlier years of research. I am inclined to agree with this friend of mine that it is possible that a man might have a good idea that does not have a Ph.D. Now, I would like to know why a man like that cannot get consideration. . . . I thought I had found the proper place for him to present that and they referred it back to the Chicago Office.

> They sent some young 23 year-old kid out there to look over his laboratory and discuss this thing with him who did not know the difference between the moon and a hunk of green cheese, and he went back and made an unfavorable report. I want to know who the man in authority is that a man of this caliber can discuss things of importance with down here. (*NSF*, 1948, pp. 33–34)

Congressman Beckworth also was concerned that once the scientists were in positions of power and responsibility—as in the National Science Foundation, for example—they would use their position to favor members of their own group, or those who were apprentices on their way to the status of full membership. Beckworth was afraid that men of equal, innate ability who lacked the necessary educational qualifications would be ignored and not even given the chance to try to compete. He also expressed his resentment and concern that, once an alienated group with a monopoly of knowledge such as the scientists possessed, was elevated to positions of power, it would proceed to put its own value system into effect and thereby violate democratic standards.

Politicians' perception of the alienated nature of scientists is considered here in terms of their image of scientists as withdrawn from the usual social relations and indifferent to concerns of ordinary people. The aloof and isolated nature of scientists was graphically presented to politicians by Morris L. Cooke, in his testimony on the National Science Foundation:

> The relationship which scientists—and engineers too—bear to the hurly-burly of American life, including its politics, is difficult for those outside these professional fields to understand. The scientists, as a matter of fact, are themselves only dimly aware of how detached they are individually and as a profession from the pulsating world around them. In the days when science was persecuted, the scientist was a recluse living as far as possible from the haunts of men. He still lives a life apart sharing almost not at all in our common activities and assuming no responsibility for the conduct of affairs outside the narrow confines of his own professional interest. Decisions dictated by slide-rule and test-tubes are his daily meat. The settlement of issues by the give and take involved in democratic compromise seems too crude in comparison with determinations reached by the two-plus-two-equals-four-method of mathematics. Scientists and engineers, with few exceptions, feel no responsibility whatever for the life of the community—the hospitals, the school system, the boys clubs, the forums for the discussion of public questions, the homes for the aged. (*Sci. Leg.*, 1945, p. 1003)

Politicians' acceptance of this characteristic of scientists, particularly with reference to their isolation from and rejection of political participation, was revealed by Sen. Fulbright's question. He asked Cooke:

> Why do you think people have such an aversion to politics? What is the basic reason that scientists and professional people do have such an aversion to it? (*Sci. Leg.*, 1945, p. 1003)

The fact of such isolation and rejection had been stated earlier by Dr. Curtis:

> But scientists don't care much, by and large, about political matters. Let somebody else take care of politics—we have our atoms to attend to. But the scientists have felt that they can and should make whatever contributions they could. (*Sci. Leg.*, 1945, p. 328)

International Orientation of Scientists

THE obvious corollary to the internationalism of science was the internationalist orientation of scientists. Such an orientation was explicitly affirmed by the scientists themselves, and, seemingly, accepted by politicians as one aspect of their image of scientists. Dr. Bowman gave one example of this orientation in urging governmental sponsorship and support of international scientific meetings. Similarly, Dr. Moulton declared:

> I want to assure you that the scientists are high-minded and often simple-minded in terms of political things but in international matters they have been high-minded; they are the ones who have maintained the highest level of international cooperation of any class of people. Dr. Shapley said that in the past few years . . . over 500 international conferences have been held. I have attended some of them. Scientists meet together with a minimum of friction. They speak the same language; they are interested in the same problems; they are looking forward to the same things. (*Sci. Leg.*, 1945, p. 80)

Even more impressive evidence of scientists' international orientations was presented to politicians by Dr. Meyerhoff. Citing the results of the poll of its membership by the American Association for the Advancement of Science, Meyerhoff stated:

> . . . the dissemination of information, both on a national and international basis, is favored by 76 and 74 per cent respectively. . . . And then the sort of thing which has been talked about by preceding speakers including Dr. Shapley this morning, that of international

collaboration, loomed rather large, to the extent of 64 per cent who felt that much more should be done in a more formal way. Incidentally, there was considerable interest expressed in the acquisition of information about research going on in other countries, so that we might be brought up to date and not lag behind as so commonly is the case.

Politicians indicated their perception and sensivity to scientists' internationalism when Magnuson asked:

> In other words, your 74 per cent for international dissemination ties right into the 64 per cent group that wanted more formal international collaboration?

Meyerhoff confirmed Magnuson's understanding:

> That is right. All people who voted for international collaboration voted for the exchange of information between countries. (*Sci. Leg.*, 1945, p. 88)

After having heard a number of scientists testify, Senator Millikin summed up his concept of their orientation in the following words:

> The scientist, as I see it and as I get it from the testimony that we have had, is an idealist. He is, in a true sense—and I am not using this in a disparaging sense—an internationalist. He is, because he is accustomed to interchange and meeting people from all over the world. His speciality is science. . . . (*Spec. Comm.*, 1945, p. 327)

The alienative aspect of scientists' international orientation was its supranational character. Scientists' alienation was perceived by politicians not merely in their withdrawal from affairs and their indifference to human concerns. More positively, it was seen in their adherence to ideals and interests that normally transcended those of the nation, and, under certain circumstances could conceivably be in conflict with them. Inevitably, therefore, politicians were led to the conclusion that scientists had attenuated sentiments of loyalty and attachments to national boundaries.

The ready mobility of scientists to go wherever they could do their work under the kind of conditions they required was seen as one index of such attenuation. Langmuir's statement that any attempt by authority to regiment and control scientists would result in their departure explicitly suppored this view. Additional confirmation was found in the emigration of scientists from Italy, Germany, and other countries before the Second World War. The continuing readiness of scientists to leave their countries of origin for more favorable conditions elsewhere was demonstrated to politicians by Dr. Schade's testimony. In a discussion with Schade, who had been sent to Germany by the U.S. Navy, Magnuson asked concerning German scientists' attitudes:

> Do many of the scientists with whom you have talked feel that they want to stay in Germany and help rebuild Germany or would they rather come to, say, America or Russia, or England, or other countries that might offer them refuge?

Schade answered:

> There is an astonishingly prevalent interest among the people of that kind that you talk to in the direction of wanting to come over here and work for us. Many of them feel that the future of Germany is nonexistent, or not very pleasant from their point of view, which is, of course, the truth. I think perhaps that is one of the reasons why such people have been astonishingly cooperative with people such as myself and my organization over there who were over there to find out what we could from the technical point of view as to what went on during the war that might be of interest to the Navy. . . . The Germans (i.e., scientists) were most cooperative and I suppose it was because many of them hoped to get a job over here some day. (*Sci. Leg.*, 1945, p. 272)

As politicians saw it, the mobility of scientists and the ease with which they could transfer from one country to another were mainly due to their having weaker sentiments of loyalty and citizenship. Scientists lacked that visceral feeling of rootedness, of being part of a land and a kinship group that true citizens had. As a consequence, they were able to leave their native land to go and work elsewhere without much feeling of regret or loss. They were never really citizens in the first place. Furthermore, scientists' mobility and lack of rootedness were intensified by the fact that their important possessions and necessities of life were abstract. Scientists carried them in their heads, hence were not hampered by valuing and owning material things that would make their movement more difficult. The role of scientists was thus inherently international and supranational. Nothing about it bound the scientists to any one country. He could live and work anywhere and would so long as he had the necessary conditions for carrying on his research. Transfer from one country to another could be accomplished without sustaining any significant and serious losses.

Their supranational orientation and attenuated feelings of loyalty and attachement to nation were explicitly affirmed by the scientists themselves. In a statement prepared for the hearings on the National Science Foundation, Shapley declared:

> Our American scientists and technologists at the present time have been derived from the adventurous pioneering stock of prac-

tically all the nations of the world. We call ourselves American by citizenship, but our blood is cosmopolitan. The scientists should, as rapidly as possible, call themselves citizens of the world and not the citizens of individual countries. (*Sci. Leg.*, 1954, p. 58)

Oppenheimer added his contribution to Shapley's clarion call when he said to Senator Fulbright:

> May I be very honest? Most scientists, because they are scientists, were certainly not happy with the absolute national sovereignty that prevailed ten years ago. They were not happy with the war. (*Sci. Leg.*, 1945, p. 313)

That politicians viewed scientists as having weak loyalties and attachments to nation was indicated by Senator Fulbright's observation:

> We have been told that they are very internationally minded people, that in peacetime they move back and forth between countries, and they have a great deal of resentment against restrictions on their development. . . . *I have always had the feeling that scientists did move about and are not so conscious of national sovereignty as lawyers or politicians or others.* (emphasis supplied) (*Sci. Leg.*, 1945, pp. 274–75)

While Fulbright saw no particular cause for alarm in this characteristic of scientists, it should be pointed out that he had come from an academic environment as professor and university president, and therefore could be sympathetic. But it seems hardly probable that most politicians who lacked such a background, shared his toleration. Rather, as Fulbright himself suggested, politicians elevate sentiments of loyalty and citizenship, attachment to nation and respect for the authority of government to the status of pre-eminent virtues. In their eyes, concern for the national interest and welfare should always be paramount.

Therefore, when Dr. Urey advocated destroying the stockpile of atomic bombs and all fissionable materials if necessary to convince the world of our peaceful intentions, politicians could see in such a proposal a failure on the part of scientists to put the national interest and security over all other concerns. Even more important, such advocacy suggested a lack of faith and trust in the government since it had said it would not use the bombs aggressively. Thus, Johnson could react to Urey's proposal in the following exclamation:

> Dr. Urey, the members of this committee, and insofar as I know, no one else unless it be General Groves and some of the people at the top know exactly the amount of the compound that we have that is ready to go into the development of bombs. It seems to me that your paper displays a lack of faith that others might have

> in us. Now, how are you going to convince the world, for instance, that we don't have a sizable amount of this compound stored away which can never be detected, never discovered, that we are holding back from the world? You do not seem to think that the other nations will have very much faith in the United States' good intentions. . . . (*Spec. Comm.*, 1945, p. 107)

The low intensity of scientists' feelings of loyalty and patriotism was also demonstrated to politicians by the fact that under the normal conditions of peacetime, the ideal of basic scientific research was considered by scientists to conflict with work for the government—even though their services were vital—and was given preference. An additional factor in their decision to return to private life was that their work for the government had never been held in high esteem scientifically. It violated their ideals of scientific research. Szilard, Urey, Langmuir, and others made this point in their testimony before Congressional committees.

Politicians, then, were given to understand that only an extraordinary situation such as war could intensify scientists' weak sentiments of loyalty and attachment to nation to the point of giving up temporarily their primary allegiance to the ideals of science and basic scientific research. Politicians' perception of the alienated nature of scientists was perhaps strengthened when they learned that, even under the crisis of war, the loyalties of Japanese scientists were suspect because they had trained abroad for years. The international orientation and sympathy, which they were presumed to have, raised doubts of their attachment to the nation and the government, to the point where their services were hardly used. As Dr. Compton testified:

> . . . The Japanese civilian scientists were apparently greatly distrusted by both the army and the navy, and we didn't find a single case of a university scientist who had been asked to do a war job who was given the information as to what the war job was to be—what was its military use. . . . The reasons for failure to trust the civilian scientists were several. One was that most of the Japanese top scientists have been trained in Europe or America and had been for four to seven years in residence, with frequent visits, and they were suspected by the Japanese military of having foreign connections or foreign sympathies, and that is one reason they were not trusted. . . .

Kilgore asked:

> Did you find this which was asserted 2½ years ago prevailed before the fall of France, that the military forces in France would not accept information or advice from the scientific personnel of

their universities or their leading scientists in solving problems? (*Sci. Leg.*, 1945, pp. 638–39)

Senator Magnuson reflected politicians' suspicion that scientists' primary allegiance was normally to the value system of science and scientific research and that such allegiance might conflict with the needs of government. To a witness's comment that every loyal American scientist would gladly join a proposed scientific reserve, Magnuson replied:

> And now, during the war we had no such Reserve, but the scientists voluntarily came down here and did the job. Do you think that while the scientists might do that again, the Reserve would probably be an anchor to windward? (*Sci. Leg.*, 1945, p. 176)

Another issue stemming from the same basic source of conflict that reinforced politicians' image of their attenuated loyalty and attachment to nation was the scientists' insistence on a more full and complete freedom to publish and exchange scientific information than politicians thought desirable. The vigor with which practically every scientist who came before the politicians insisted on their right to this article of their scientific faith was just one more proof of the dedicated nature of the scientific in-group. Politicians' concern was that granting this demand might well jeopardize the national security since vital secrets, whether inadvertently or deliberately, might be disclosed in the process.

The conflict on this issue was complicated by two factors. On the one hand, politicians feel that they are among the foremost champions and guardians of the national interest and security, especially as regards external sources of danger. As Congressman O'Hara said:

> Personally, I am a nationalist; first, last, and always. I think when we get to the point where we are thinking about everybody else in the world and forgetting our own national welfare and our own people, we are in rather bad shape. Maybe that is isolationalism. Call it what you will, that is the way I feel. (*NSF*, 1949, p. 90)

On the other hand, they were confronted with the fact that the subject matter with which scientists dealt was, and always would be, a mystery to them. Politicians were permanently barred from gaining access to and knowledge of scientific matters. Consequently, in this one respect, they looked upon scientists as a sort of secret society from which they were excluded

Politicians' perceptions of scientists as a secret society and their resentment of this barrier in their dealings with scientists was clearly voiced by Senator Tydings. After Alvin Weinberg had told Senator

Johnson he was wrong in believing that resistance to forward motion in water increased with the depth of the water, Tydings remarked rather bitterly:

> Apparently that is one of those scientific facts we are supposed to accept and not ask why. (*Spec. Comm.*, 1945, p. 337)

Senator Thye indicated that politicians still held the same view of scientists several years later in his comment during the hearings on the Bureau of Standards' rejection of the claims made for a battery additive. After Dr. Astin, the Director, had explained the technical grounds for the Bureau's decision, Thye exclaimed:

> That is where you have always got us as a scientist because you can get into that technical field and we are left behind in a daze: we are not sure whether we dare challenge you or not . . . (*Batt. AD-X2*, 1953, p. 304)

Politicians were not only frustrated by their inability to challenge scientists but also by their dependence on scientists in the new atomic age. Whether Congressmen liked it or not, their survival depended to a large extent upon trusting the scientists and admitting them to the public policymaking process. Senator Thye's outburst points up what is probably the single most important aspect of the relations of politicians and scientists: namely, the ability of men with such wide divergences in training and experience to achieve that degree of common agreement and mutual understanding that is required for cooperative effort in advancing the goals of a democratic society.

19 *The Elements of Identification with an Occupation*

HOWARD S. BECKER AND JAMES CARPER

ONE OF THE MAJOR PROBLEMS to which social psychologists are now addressing themselves is the process of identification and the nature and functioning of identity in conduct. These concepts are of strategic importance in any theory which attempts to relate the self and its workings to an ongoing social structure. As Foote[1] and Strauss[2] have pointed out, individuals identify themselves—answer the question "Who am I?"—in terms of the names and categories current in the groups in which they participate. By applying these labels to themselves they learn who they are and how they ought to behave, acquire a self and a set of perspectives in terms of which their conduct is shaped.

It appears theoretically useful to break the concept of identification down into its components, both for comparative purposes and in order to provide finer tools for the analysis of specific problems of social structure and personal development. This paper is an attempt to provide such a breakdown for one type of identification, that of a man with his work. Its purpose is to discover, by comparing three

---

Reprinted from the article by Howard S. Becker and James Carper in American Sociological Review, *vol.* 21 (1956), *pp.* 341–348.
This paper is based on work done while the authors were Ford Foundation Postdoctoral Fellows at the University of Illinois. We wish to thank the Committee on the Ford Grant of the University of Illinois which also provided funds for clerical assistance.

1. Nelson N. Foote, "Identification as the Basis for a Theory of Motivation," *American Sociological Review*, 16 (February, 1951), pp. 14–22.

2. Anselm Strauss, "Identification," unpublished manuscript.

3. Only men were interviewed, to avoid the complications introduced by sex differences in career patterns and ambitions. Foreign students were excluded to eliminate the difficulty of interpreting information relating to social systems about which

groups of persons about to enter the work world, some of the threads from which the fabric of occupational identification is woven.

The data on which this paper is based were gathered in the course of a study of the genesis of identification with an occupation in students doing graduate work in physiology, philosophy, and mechanical engineering. Graduate students were chosen for study not only because they were convenient but, more importantly, because of the central character of graduate school in developing professional identifications. Interviews lasting from one-half to two hours were tape recorded with students ranging from first year in graduate school to those about to receive the Ph.D.[3] While identifications are not so clearly defined in the first year as they become later, the consistency of our findings indicates that the process is already well started at that time.

The interviews were conducted informally. Questions were asked only to clarify points or to introduce some area in which information was desired that the interviewee had not spontaneously discussed in answer to the initial question: "How did you happen to get into. . . .?"

Comparison of the three groups suggested four major elements of work identification: (1) occupational title, and associated ideology; (2) commitment to task; (3) commitment to particular organizations or institutional positions; and (4) significance for one's position in the larger society. In what follows we present brief discussions of each of these variables and comparisons of the physiologists, philosophers, and mechanical engineers[4] to illustrate their dimensions and analytic utility. Illustrations are also included of the kind of theoretical use to which these concepts might be put.

Occupational Title and Ideology

KINDS of work tend to be named, to become well-defined occupations, and an important part of a person's work-based identity grows out of his relationship to his occupational title. These names carry a great deal of symbolic meaning, which tends to be incorporated into the identity. In the first place, they specify an area of endeavor

we knew little or nothing. Three philosophy students were excluded because they had no serious intentions of doing work in the field but were simply taking courses as a hobby. With these exceptions, we interviewed all the remaining students in philosophy (eleven) and mechanical engineering (twenty-two), and a randomly selected 50 per cent sample of those in physiology (eighteen), a total of fifty-one. The work was done at a large state university which may recruit from lower levels in the class structure. For this reason there may be important differences between our subjects and those studying in the same fields elsewhere.

4. Material on the process by which the identifications we describe as characteristic of each of these groups develop is reported in "The Development of Identification with an Occupation," *American Journal of Sociology*, 61 (January, 1956), pp. 289–298.

belonging to those bearing the name and locate this area in relation to similar kinds of activity in a broader field. Secondly, they imply a great deal about the characteristics of their bearers, and these meanings are often systematized into elaborate ideologies which itemize the qualities, interests, and capabilities of those so identified.[5]

These things implied by the occupational title are evaluated and are reacted to in terms of such evaluations. One may reject the specific work area the title specifies, preferring to be identified with some larger field; or he may eagerly claim the specific field, while minimizing the larger area; he may emphasize neither, or both. Similarly, the implicit statements about the person may be proudly claimed, whether these claims are recognized by others or not; or they may be as eagerly avoided, even though others attempt to impute them. The title, with its implications, may thus be an object of attachment or avoidance, and kinds of identification may fruitfully be compared in this regard.

The physiology students[6] feel themselves part of a larger group, devoted to building the edifice of science, and pride themselves on their participation in this endeavor and on the ultimate value of their work to society in the cure and prevention of disease. Nevertheless, they sharply differentiate their work from that of physicians and of other scientists involved in this enterprise. They feel that they make the important scientific discoveries on which medical practice is based, medicine itself being more empirical and superficial; one student put it metaphorically: "We write the music that the doctors play." Another stressed the fact that the scientist is free to pursue questions until he gets a real answer, while the M.D. must of necessity forego following up any particular problem intensively. In contrast to physicians, many saw themselves as men who would devote their lives to meeting the challenge of the unsolved problems of the field. They compare their work with that of other natural scientists—chemists, zoologists, and others—and conclude that theirs is the only science which really studies the problems of the living organism:

> Here you have living organisms, and there are certain rules that these organisms will follow. They don't hold fast; two and two isn't always four. It's up to you to interpret what happens, to be able to meet any emergency which arises. And you're working with something which is living and therefore responds to its environment. Whereas in chemistry, mathematics, there are certain reactions which occur and you can change them by doing little things

5. See, for example, Howard S. Becker, "The Professional Dance Musician in Chicago," *American Journal of Sociology*, 57 (September, 1951), pp. 136–144; and W. Fred Cottrell, *The Railroader*, Stanford, Calif.: Stanford University Press, 1941.

6. Of the eighteen physiology students interviewed, eleven were fully committed to the field; two were committed to closely allied fields of biological research; three

> but you yourself are the one that is producing these changes whereas in a living tissue, it itself is changing. . . . You're working with something which is alive just as you're alive and it changes and you actually can't control it completely, you just have to be able to work with it.

Others make a similar point in saying that physiology is not "cut and dried," as are the other sciences. In short, these men identify themselves as part of a discipline carrying on a peculiarly valuable kind of work, which no other group can do.

The engineers, like the physiologists, take great pride in their occupational title. Although in a few cases they feel equally identified with the titles of "research scientist" or "teacher," they all share the feeling that it is a good thing to be an engineer. Unlike the physiologists, the majority have no attachment to any particular part of their field; their specialty is the broad area of "technical work." They find the field desirable because of the remarkable skills and abilities engineering training is supposed to produce in them, abilities implied in the occupation's name. With few exceptions, these men are agreed that, as one put it:

> All our lives and all through our work we are being trained to think logically and to analyze. And if you can do these two, I don't think anything can stop you.

The ideology tells them that anyone called "engineer" has learned to reason so rationally and effectively that, even though this has been learned only with reference to technical problems, it operates in any line of endeavor, so that the engineer is equipped to solve any kind of problem in any area quickly and efficiently.

The philosophers, in marked contrast, have very little attachment to their occupational title, perhaps because of the august company in which it would place them:

> (It's all right to call you a philosopher, isn't it?) Well, I don't know. I do refer to myself every once in a while as a philosopher but I rather hesitate to because when I think of a philosopher I think of somebody like Plato or Aristotle.

The image they have of themselves is that of the "intellectual" whose interests cover the whole range of artistic, scientific, and cultural pursuits. Viewing their earlier specialization in particular fields as "too

were determined to become physicians; and two still had hopes of becoming physicians, but were well on the way to accepting physiology as an alternative. This discussion of aspects of identification describes all but the five interested in medicine and applies in large measure to the two undecided cases.

confining," they turn to philosophy which "does deal . . . with all crucial problems in one way or another."

> Frankly, I'm taking the viewpoint of a person who wants to know quite a bit about several things and I never want to give up my catholic interest, catholic meaning of course universal in this sense, and to specialize. Yet I realize that to know very much about anything I have to specialize. Philosophy is the best grab-bag for me. To do something in philosophy I don't have to go terribly deeply into a given discipline and stick with it all my life, so I can shift from one discipline to another. But at the same time to overcome the notion of it being grounded in nothing. Frankly, if somebody asked me what I mean by philosophy it would be very difficult for me to tell them what I mean. I'm just sort of in a big intellectual game and pursuit right now. It happens to go under that name and I think under the aegis of philosophy I'm more able to do this.

In short, they have chosen their occupational title simply as the least undesirable one available, since it will place them in the society's division of labor while allowing them to deal with a broad range of interests ordinarily divided between many specialties.

Commitment to Task

OCCUPATIONS may also be compared with reference to the degree to which their members feel identified with some specific kind of work. There may be a feeling that only some sharply limited set of work tasks, carried on in a particular way, is proper, all others being excluded, and that one is, among other things, the kind of person who does this kind of work. The opposite attitude may also exist: that there is no kind of task which is impossible. Again, a person may simply be vague on the matter, not really knowing what his work is or how he ought to go about it. The elements of attachment, or lack of it, to a specific set of tasks and ways of handling them, and of a feeling of capability to engage in such activities, thus also play an important part in identification with one's work.

The physiology students exemplify one extreme, identifying closely with a set of specific research tasks and a particular way of going about them. Although task and method may vary from individual to individual, each one has a fairly clearcut notion of what he is about. They see a limited range of problems to which their professional lives will be devoted, and a set of basic techniques in which they take great pride:

> You learn a little more about handling animals, doing regular surgery. After awhile, it becomes automatic. I think the first time I did it it took about twenty-five or thirty minutes. Now I can go into the throat of a dog, sew in the glass tube, isolate the artery there, put a glass tube in it and hook it to a pressure machine and have the whole thing recording in about five minutes.

Beyond this, they are committed to the notion of themselves as persons who do work which is precise, which can be reproduced by other investigators, which is theoretically sound and takes cognizance of existing knowledge available in the discipline's literature. There is no vagueness in this conception; they know what their specific problems are and how they will be handled, and they feel that they are qualified, by virtue of their technical training, to handle this kind of research successfully. They possess a concrete image of their professional future in terms of day-to-day activities which they will perform.

The engineers lie at the opposite pole, having almost no commitment to task—no kinds of work strictly theirs and beyond which they would neither dare nor care to go. Far from having a narrow conception of the engineer's work, eight of our twenty-two interviewees would be quite happy doing any kind of work our industrial system has to offer, as long as it is "interesting" and "challenging." Seven others stipulated simply that it be something technical while only five consciously limited their work to a particular technical specialty. They are quite ready to forget the specific kinds of work for which they have been trained and take on any kind of job which the title of engineer can win for them. This attitude is expressed in comments like this:

> (Now what did you have in mind, sort of, as a long-term goal in a thing like that?) Well, I think I have the same goal that probably every other kid fresh out of college has, that of going into some type of engineering work. With me, I think it would be production, as I've said. And eventually working up to higher management, I think. That's every young engineer's goal, whether he expresses it or not. (You mean non-engineering. . . .) Eventually ending up, using engineering as just a channel to go into management of some kind.

In addition to this kind of confident assertion that one is able to handle anything that comes up, the lack of commitment to task is found in the somewhat puzzled statements of younger engineers about what their work really is:

> I really didn't have a good idea of what an engineer does. And I still can't tell you. People asked me, "Well, what does a mechanical engineer do?" and I could give them examples, that's all. I

could go on and on and on in the examples, and that pertains to any engineer. All you have to do is just look at the placement records of engineers, and they go into everything.

(Six of the men interviewed included teaching, in combination with either specialized or general technical tasks, among the possible kinds of work for them.)

The philosophers present a third possibility. Lacking both the specific task attachment of the physiologists and the calm assurance of the engineers that all tasks are suitable for them, these students are not quite sure what they should be doing. Realizing that their future probably lies in the university, they accept teaching as a necessary task which is, however, not peculiarly theirs. The following is a typical answer to the question, "What does a philosopher do?"

> I suppose part of a philosopher's job is in telling people how much they don't know: It seems to be so old fashioned now to tell people how much they don't know. I suppose I will be teaching the various branches of philosophy. I'm interested in talking to students. In helping them with reading. In helping them with philosophical problems that come up. I'm sure that they can help me with some fresh ideas I hadn't considered heretofore. I'm interested in just learning as much as I can in my spare time. I might decide at some time to dabble in another profession. I just don't know. I'll never feel that there's any dearth of things to do. I don't know if I can ever categorize them. My job, my source of income, will involve taking so many hours of classes, teaching.

The clearest image of their work tends to center around the notion of continuing to learn and read in all areas of intellectual activity. Beyond this, they see all kinds of possibilities, ranging from semi-scientific research through journalism and artistic activity to such things as politics.

Organization and Institutional Position

AN OCCUPATIONAL identity tends to specify the kinds of organizations, and positions within them, in which one's future lies, the places in which it is appropriate, desirable, or likely that one will work. A person may see his professional future as tied to one organization, or to a very restricted range of organizations, or he may conceive of himself acting in his occupational role in a great many kinds of institutions. Again, he may feel tied to one particular kind of institutional position, or find it possible to conceive of holding a large variety of work statuses. These, with the further possibility of vagueness as to these matters, constitute continua along which various kinds of work identification may be located.

Participants in work institutions tend to see themselves in relation to those upon whom their success in these institutions depends. Research has demonstrated the importance of building connections with clients, colleagues, and others in the pursuit of success,[7] and identifications vary in the degree to which they reflect dependence on informal systems of sponsorship, recommendation, and control.

The physiology students see themselves as potential occupants of a few well-defined slots in a highly organized work world. There are only a few places in which they might do their kind of work: universities, where they would teach and do research; research foundations and pharmaceutical companies, where they would do only research; and government agencies, where they would engage in applied research. They do not consider themselves competent to handle positions of any other kind. They are unable to see beyond this narrow conception even to entertain the notion of becoming chairman of a Department of Physiology; this would mean moving out of the expected slot a little too far for comfort, involving as it would unfamiliar duties and responsibilities.

They expect such jobs to become available to them through the workings of a sponsorship system centered around their graduate-school professors. The initial job (the aspect of the career that looms largest at this stage) will come through the professor's contacts, and his recommendations will be of great importance. They feel quite dependent on this personal kind of sponsorship system and see no other way to get established professionally. They expect to progress through the hierarchy of university, industry, or government through careful research, knowledge of the field, and publication of important research.

In contrast, the engineers feel that their future lies somewhere in the country's industrial system, but do not think of any company (no matter what its specialty) or any position as impossible for them. Twelve of the twenty-two interviewees are prepared to work in any kind of industrial organization, while only six limit their possibilities to companies doing work in their technical specialty. (Three of the men expect to become teachers, and one wants to open his own business.) For the majority, any industrial firm in the country represents a possible employer.

Within this range of organizations they expect to compete for a broad range of positions. Of the eighteen who were considering in-

7. See, for example, Oswald Hall, "The Stages of a Medical Career," *American Journal of Sociology,* 53 (March, 1948), pp. 327–337; and Everett C. Hughes, *French Canada in Transition,* Chicago: University of Chicago Press, 1943, pp. 52–53.

dustry, only one would restrict himself to a position involving only his technical specialty. Nine are able to see themselves in any kind of technical position in industry, while eight are confident that they can compete successfully for any position, technical or managerial. Lacking any firm commitment to a particular task and armed with an ideology that stresses their universal ability, they see their futures in terms such as these:

> (What kind of job would that be that you would get into eventually?) It would be difficult to say. (Well, what are the possibilities?) Oh, assistant, for instance, a job that I would like to have. If I get back to X Co. I think I have a good chance, which would be a tremendous step forward, would be assistant to the general manager, for instance. While this would take me out of the technical field, it would be—It's a tremendous stepping stone, it's a big step forward as far as getting to the administrative end of it. I do not fancy myself as a research engineer who's just going to bury himself in his little office, content to work all his life compiling a set of tables, should we say, to take an illustrative example. A lot of people have done this. Or to investigate the natural laws. I think there's a—I don't fancy myself as doing this. I'd like to get ahead into a position where you are directing things, formulating policies, formulating the lines of the company. Do you follow me?
>
> (Yeah. Yeah. So you could conceivably end up as a general manager?) General manager of the plant, vice president. . . . (The sky's the limit, in other words.) It is. It really is. A good engineer can go any place these days.

Only a few feel that their future is in any way tied in with their "connections" with either prospective employers or with sponsors in the academic world. The majority felt quite independent in getting jobs, assuming that in the normal workings of the labor market in an economy becoming more and more "technical" they would be able to command a satisfactory position. This independence is reflected in the language they use to describe job-hunting: They are not "interviewed for a job," but rather "interview companies about jobs." This may, of course, be a temporary phenomenon associated with the present high demand for engineers.

The philosophers again suggest another dimension, having for the greatest part of their time in graduate school no clear notion of where they will work or what position they might hold. They think of themselves as intellectuals, and the term implies no specific relation to the occupational world. Late in their training they begin to realize that their futures are to be made in universities, and primarily as teachers rather than philosophers. By this time teaching, originally viewed as an important function of the philosopher, has become simply a way of earning a living and subsidizing the continuation of their intellectual pursuits. Since the state of the job market may limit opportunities to get such positions, they are ready to consider positions involving skills or experience acquired elsewhere. Any position which will allow continued intellectual activity, on or off the job, is considered suitable, even though it may have no relation to the professional organization of philosophy. If they do teach, they feel they are as likely to teach some other intellectual specialty as philosophy; anything within their cultural purview becomes a possible teaching subject.

They are vague about the ways in which jobs become available and professional success is achieved. Only two had a clear notion of the workings of academic sponsorship systems, although several believed that their professors might have some effect on their work future. They tend to be concerned about this, if at all, in a quite offhand manner:

> Lately I have begun to think that after all, in part philosophy is a business, so in part business ethics must apply to philosophy—and there are certain things you just have to do. (How about things like publishing, and so on . . . ?) Oh, that of course would be in your favor. But that again is something that I haven't thought about, haven't thought about writing any articles for journals. It's certainly . . . when I decided that I should realize that philosophy is partly a business, I also decided that I should think about writing for journals and I should take systematic notes on articles in journals, what kinds of articles are in there, what sort of thing they write about and how they write about it. (But you haven't done much of that yet?) No. It's certainly time to start, I would say.

Social Position

OCCUPATIONAL identities contain an implicit reference to the person's position in the larger society, tending to specify the positions appropriate for a person doing such work or which have become possible for him by virtue of his work. The most frequent reference is, of course, to social-class position and to the opportunities for class mobility opened up or closed off by entrance into the particular occupation. It is also possible for an identification to contain a statement of a particular relation of members of the occupation to the society, quite apart from class considerations.

The physiology students see themselves as achieving a desired move up in the class system. Twelve are men from the lower or lower-middle class who had hoped to become physicians, with the prestige of that profession playing a large role in their choice. This mobility hope has been wrecked on the reef of medical school entrance standards or abandoned during the tedious and trying voyage through "pre-med," and becoming a physiologist represents the salvage. Their parents, desiring to see their sons better themselves, figured importantly in the choice of medicine as a career, and these men remain sensitive to their parents' aspirations for them. Physiology as an occupation will get them some of the prestige and income they desired, although it is second-best; they will never approach the M.D. in these respects.

For four others (in the remaining two cases we did not get sufficient information to make a classification) physiology represents an escape into science from the mobility demands of their well-to-do families. They see physiology as an occupation giving them a respected position without necessitating the competitiveness of medicine or business. They are typically interested in academic positions, while many of the first group favor research positions in the drug industry, which they believe provide larger incomes.

For seventeen of the engineers, success in their profession spells successful social mobility. They are men whose fathers were skilled or unskilled laborers, farmers, or white-collar workers. Having entered engineering in many instances purely out of interest in "mechanical things," they are pleasantly surprised to find that it enables them to rise significantly (in social class terms) above their families and childhood friends. For the other five, a career in engineering is a means of continuing their families' solid middle-class status. They all expect to do well financially. At the least, they look forward to a very comfortable living, and eight expressed the desire to make "big money": "You can get your mansion on the hill." Being an engineer is a ticket to financial success and its accompanying social prestige.

The philosophy students' identification of themselves as "intellectuals" carried with it the implication that they are different in important respects from other members of the society. In every case they either consider themselves deviant or recognize that they are so considered by friends and relatives. Most importantly, these men of predominantly lower and lower-middle class origin have renounced the pursuit of class mobility in favor of the intellectual life. They have no concern with material success and tend to be proud of the meagerness of their financial future. It is expected that parents and others will be unable to understand these views, and they tend to break relations with people who would keep these interests before them. In contrast to the physiologists, parental aspirations play no part in the formation of their professional ambitions.

Discussion

THE dimensions of work identification detailed above suggest a number of problem areas in which they might be of use in further research. The question immediately arises, for example: To what degree do these dimensions constitute independent variables and to what degree are they functionally or causally related so that they will tend to appear, not randomly, but in relatively stable combinations or syndromes? The identifications of the three groups studied show a considerable degree of inner consistency, suggesting the existence of such relationships. The physiologists exhibit a congruent pattern of commitment to specific and restricted items: only certain limited kinds of tasks, organizational settings, and institutional positions are considered possible and acceptable, while they claim as their own only a small slice of the total pie of science; and these seem to fit naturally with the limited social-class mobility they expect their work to provide for them. One might reasonably assume that these limitations tend to reinforce each other, both psychologically and in terms of movement within social structures. The engineers' identifications are equally consistent in the other direction, with commitments to a broad area of work, a wide range of possible tasks, organizations, and positions, and the expectation of great social mobility. Again the hypothesis of mutual reinforcement seems appropriate.

While these cases suggest the likelihood of relations of functional interdependence between the elements of identification, that of the philosophers points more to the possibility of causal relationship, since the chief characteristics of their identification—lack of commitment to any organizations or positions, to specific tasks or mobility aspirations—seem all deducible from their basic commitment to the intellectual life as they conceive it. Both detailed genetic studies of the development of identification and cross-sectional studies of the relationship of these attributes seem indicated for the solution of this problem.

The use of such distinctions as these would also provide variables for the intensive analysis of the problem of the development of identification, and for those problems centering around the functioning of identifications in society; for example, the problem of the relation of variation in elements of identification to variation in occupational role behavior, and that of the way in which the identifications of individuals function within the organizations in which they work.

Let us consider, as an instance of the latter class of problem, the way in which variations in these elements of identification affect the relative ease of an individual's mobility through occupational institutions, keeping in mind the effect which differences in mobility potential have on the organizations in which these people work. If, in identifying himself occupationally, an individual exhibits an intense identification with a particular institutional position or a particular set of tasks or with both of these, movement to some other position, or movement which involves a shift in the actual job done, becomes more difficult. The physiologists exemplify this tendency. Tied to their particular research problems and techniques, they are unable to envision themselves occupying any but the few positions they know of in which they can pursue these problems in the way they know best. Even the minor move to department chairman worries them. Neither the engineers nor philosophers have any such commitment to task or institutional position, and movement is much more possible for them, should it become an alternative available to them. The engineers, for example, are ready to move into any kind of position in the country's industrial system, while the philosophers are able to consider with equanimity many kinds of positions involving a variety of skills, as long as they allow for continued intellectual activity. The physiologists' limited view allows them to fit easily into the limited mobility pattern of the university when they actually take their positions in it, just as the engineers' high mobility potential allows them to meet with ease the personnel needs of the expanding industrial economy into which they are moving. In this sense, identifications have functional consequences for institutions.

The Organization of Scientific Work and Communication among Scientists

4

THE eight selections in Part 4 deal with important aspects of the organization of scientific work and communication among scientists. As science in our time has become increasingly large and increasingly differentiated by specialization and type of skill, the coordinative problems of organization and communication have also had to increase. We must remember what Professor Derek Price has pointed out, that 90 per cent of all the scientists who have ever lived are still living. This fact indicates the recent rate of increase in the size and differentiation of science.

In an unusual comparative historical study, which is our first selection, Professor Ben-David shows how organizational changes in German and American science were responsible for higher rates of medical discovery in those countries in the nineteenth century than in France and Britain, where medical science failed to make these organizational shifts. Germany and America, he argues, more quickly recognized the need for specialized scientific disciplines, more quickly created specialized scientific jobs and facilities for research, and more quickly established large-scale training organizations. In Professor Ben-David's selection we are again reminded of the usefulness of historical data for the sociology of science.

The next selection is by Professor Morris Stein, a clinical psychologist who has found it necessary to understand the nature of scientific organization in order to understand the psychological problems of scientific creativity that are his primary interest. Although he emphasizes the personality element in his research, Professor Stein's work is genuinely social-psychological because he relates personality possibilities and variations to organizational role structures.

One of the very large new areas for scientific research is the industrial laboratory. Professor Shepard's paper is a survey of the organizational and personal problems that are typical of this new environment of scientific research. Many of these problems also occur in other research environments, such as the government and the university. It should be noted that Shepard's paper and the two following ones by Pelz and Kaplan were all published in the scholarly journal, Administrative Science Quarterly. This journal is an important focus of a new and growing field of specialized social science theory and research: the field of organization and administration. Efforts to understand the organizational problems of science can borrow in considerable measure from the generalized knowledge that this new social science specialty has accumulated, as it can also contribute something to that knowledge. Some of the organizational problems of science are common to organizations in many other fields, though of course science has some organizational problems of its own.

Another one of the very large new areas for scientific research is the government laboratory. Many scientists are now working in the whole set of laboratories that make up an essential part of the National Institutes of Health, the American government's great and recent investment in medical research. With his colleagues from the Institute for Social Research, University of Michigan, Professor Pelz has studied some of the organizational factors that affect research performance in one of these N.I.H. laboratories. Again, some of the problems are special to the type of research organization under study, but others are common to research organizations in other areas.

A comparative study of governmental medical research laboratories in the U.S.S.R., laboratories quite similar in many respects to those of our National Institutes of Health, is provided in Professor Kaplan's paper. Kaplan is here especially interested in the role of the "research administrator," who, though not directly responsible for the work of the scientists in his organization, can affect them indirectly.

In the following selection, by A. H. Cottrell, who is Professor of Metallurgy in Cambridge University, we see how the wise and observant natural scientist can sometimes himself be a good student of scientific organizations and scientific communication. His remarks on the solitary scientist and on the scientist in organizations bring out clearly the changing patterns of scientific research. And his account of how international scientific conferences may sometimes be of the most fruitful consequence for scientific discovery indicates the importance of these and other conferences in the network of formal and informal communication among scientists, a network without which they cannot carry on their best work.

Especially during the last twenty years, there has been a very great increase in the scale of financial support for scientific research from the governments of nearly all modern industrial nations. Science is now of the essence of national power and well-being. Inevitably, a large part of this support has gone to what has long been, and remains, the center of scientific research in such societies, their universities. This changing relationship between the government and the universities has stirred much concern for the effects of government support on the autonomy of the universities and their ability to carry on their basic responsibilities to fundamental research and teaching. The selection by Dr. Kidd is the summary and concluding chapter of his extremely wise and judicious book on the relationships between American universities and their government. While recognizing problems that need further adjustment, he thinks that on the whole these relationships are mutually beneficial. Dr. Kidd's views are based on lengthy personal experience in government and on further study that has included interviews with the principal persons involved, both in the universities and in government.

Finally, in the last selection, Dr. Menzel of the Bureau of Applied Social Research, Columbia University, reports on his survey of the communication patterns, needs, and problems of scientists. This survey builds on the whole previous program of research in communications carried on by the Bureau of Applied Social Research over the last twenty years. We see again how a problem that may seem special to the sociology of science is in fact a problem that has its more general aspects, its common features with other social areas. The sociology of science and other special sociologies can best grow together.

20 *Scientific Productivity and Academic Organization in Nineteenth-Century Medicine*

JOSEPH BEN-DAVID

THE PURPOSE of this paper is to describe and explain differences as well as fluctuations in the productivity of the medical sciences in Germany, France, Britain, and the United States, from 1800 to about the time of World War I. Scientific productivity as defined here does not comprise any evaluation of the greatness or depth of various scientific ideas, or of the "efficiency" of scientific production as measured by some input-output ratio. It refers only to two gross quantities: the number of scientific discoveries (including scientifically important technical inventions), and the numbers of people making such discoveries. Provided that these numbers are not a fixed proportion of the general population or some other general quantity, they are a measure of the active interest in science existing in a society at a certain point of time.

The two suggested indexes of productivity—the numbers of discoveries and of discoverers—have not precisely the same meaning and there are obvious objections to both. It can be argued that since scientific discoveries are disparate units of unequal significance, it is meaningless to count them.[1] The first part of the claim is true, but not the

---

Reprinted from the article by Joseph Ben-David in American Sociological Review, *vol.* 25 (1960), pp. 828–843.

A preliminary draft of this paper was written while the author was a fellow at the Center for Advanced Study in the Behavioral Sciences, Stanford, California. He is indebted to Harry Alpert, S. N. Eisenstadt, Jacob Katz, Morris Janowitz, Robert K. Merton, D. Patinkin, Dr. George G. Reader, and the late Dr. J. Seide for comments on the manuscript or discussion of its subject matter, and to A. Zloczower for his help with the research.

1. The method is applied and discussed by T. J. Rainoff, "Wave-like Fluctuations of Creative Productivity in the Development of West-European Physics in the Eighteenth and Nineteenth Centuries," *Isis,* 12 (1929), pp. 291–292. See also S. C. Gilfillan, *The Sociology of Invention,* Chicago: Follet, 1935, pp. 29–32; Joseph Schneider, "The Cultural Situation as a Condition for the Achievement of Fame," *Ameri-*

deduction from it. It has been shown time and again that "great" discoveries had been preceded by intensive activity manifested in numerous "small" discoveries, often leading to the simultaneous finding of the final solution by more than one person.[2] Similarly, one of the signs of a great discovery is that it leads to a greater number of smaller discoveries based on the newly discovered principle.[3] Therefore, viewing science as a flow of constant activity, great discoveries appear as waves built up gradually by the ant-like work of predecessors, leading first to an upsurge of activity by followers and disciples and then diminishing into routine when the potentialities of the great idea have been (or seem to be) exhausted. Thus there is no need to weight the individual discoveries. The weighting is done automatically by the clustering of discoveries around the significant event. This is not to deny that there are lone discoveries, neither expected beforehand nor understood after they are made. For the historian who sits in judgment of individual greatness and stupidity, these are important events that prove the absurdity of our method of counting. But if one's purpose is to gauge the extent to which various social systems induce people to scientific productivity, then the relatively negligible weight accorded to the lone discovery is a good index of the relative lack of inducement to engage in research in that society.

The use of the number of discoverers (not students or graduates) as an index of scientific activity can be justified by similar reasoning. Such men as Newton, Lavoisier, and Einstein did not spring up in scientific deserts but in environments of intensive scientific interest, and their work inspired disciples and followers. So we can expect a general correspondence between this index and the previous one. Yet, there are numerous problems involved in the use of this index. In principle, the same numbers of discoveries can be made by quite different numbers of people, so that there may be no relationship between the two counts. In fact, however, the variation is quite limited, because the accomplishment of even a single scientific discovery demands as a rule considerable investment of time and training: one can assume that discoveries will be made by persons with special characteristics ("discoverers") and not randomly either by them or others. Thus we take this figure too as a good index of the social inducement to engage in research. No more than general correspondence between the two sets of data is expected, however, because, first, there may be variations due to institutional circumstances in the length of the creative period of discoverers, and in the chances of "outsiders" for making

can Sociological Review, 2 (August, 1937), pp. 480–491; Frank R. Cowell, *History, Civilization and Culture: An Introduction to the Historical and Social Philosophy of Pitirim A. Sorokin*, London: Black, 1952, pp. 90–106; and especially the methodological comments of Robert K. Merton, "Fluctuations in the Rate of Industrial Invention," *The Quarterly Journal of Economics*, 59 (May, 1935), p. 456.

discoveries; and, second, even if these things were constant, the shape of the two curves would still differ because each discovery is a single event counted only once, at the time of its occurrence, while discoverers must be counted over a period of time or at an arbitrarily fixed point of time (such as their age at the beginning of the professional career). For these reasons we expect this second index to correspond with the first only in registering relatively long-term and gross changes. But in such details as the exact time of the changes and short term fluctuations no correspondence between the two indices can be expected.

A second problem requiring preliminary clarification is the definition of medical sciences. We have adopted the criteria of our sources, which include all discoveries that eventually became part of the medical tradition. Undoubtedly this implies the inclusion of some non-medical discoveries and discoverers; therefore, from the viewpoint of the history of scientific ideas, this may not be too meaningful a category. However, in a study of scientific activity one needs data reflecting activity in more or less homogeneous institutional frameworks, irrespective of whether they do or do not relate to a logically coherent system of ideas. On this score, medicine in the nineteenth century seems to be a good choice. Through most of the century it was closely interwoven with the natural sciences. It had been the first profession based on the study of natural sciences, and medical faculties were the first university departments to teach them. For many years the only large-scale and permanent organizations where research was systematically conducted were the teaching hospitals. Also, the art of the apothecary and the science of chemistry were often connected until the early nineteenth century. Thus the sciences associated with medicine have formed a complex of scientific activity which has been related to well defined social structures since the eighteenth century, whereas most of the basic sciences were the professional concern of only a few individuals in any country well into the second half of the nineteenth century. The medical sciences, therefore, appear to be well suited for discerning the effect of structural changes upon scientific creativity during the period under consideration.

The Questions to Be Explained

TABLE 1 is based on a count of medical discoveries made in the countries here surveyed from 1800 to 1926, according to a "Chronology

2. *Cf.* William F. Ogburn, *Social Change,* New York: Huebsch, 1922, pp. 90–122; Bernhard J. Stern, *Society and Medical Progress,* Princeton: Princeton University Press, 1941, pp. 41–44.
3. Merton, *op. cit.,* pp. 464–465.

of Medicine and Public Hygiene."[4] The data reveal two different trends.

First, between 1810 and 1819 a rise in the number of discoveries in France and Britain begins, followed in Germany in the next decade. By 1840, the rise has passed its peak in France and Britain and a decline sets in lasting until the 1870's. Second, an upsurge starts simul-

TABLE 1. Number of Discoveries in the Medical Sciences by Nations, 1800–1926

Year	*U.S.A.*	*England*	*France*	*Germany*	*Other*	*Unknown*	*Total*
1800–09	2	8	9	5	2	1	27
1810–19	3	14	19	6	2	3	47
1820–29	1	12	26	12	5	1	57
1830–39	4	20	18	25	3	1	71
1840–49	6	14	13	28	7	—	68
1850–59	7	12	11	32	4	3	69
1860–69	5	5	10	33	7	2	62
1870–79	5	7	7	37	6	1	63
1880–89	18	12	19	74	19	5	147
1890–99	26	13	18	44	24	11	136
1900–09	28	18	13	61	20	8	148
1910–19	40	13	8	20	11	7	99
1920–26	27	3	3	7	2	2	44

SOURCE: F. H. Garrison, *An Introduction to the History of Medicine*, 4th edition, Philadelphia and London: Saunders, 1929.

taneously in all these three nations and in the United States in 1880. These parallel movements reflect the story of the convergence of chemical, anatomical, physiological, and pathological discoveries in the first half of the nineteenth century, and the spate of bacteriological and surgical innovations which followed the work of Pasteur, Lister, and Koch in the last quarter of the century. Both waves show only that certain fruitful ideas had been simultaneously, or nearly simultaneously, exploited in Western European countries beginning from the early nineteenth century, and in the United States as well from the end of the century. Apart from indicating that scientific communication among these countries was well established by that time and that therefore the phenomena reflect the course of scientific ideas, they call for no sociological explanation. What needs to be explained is the conspicuous change in the relative shares of the countries during this period. French supremacy in the beginning of the century with Britain a close second gave way to an overwhelming preponderance of German discoveries through the second half of the last century. The American share was

4. Published in F. H. Garrison, *An Introduction to the History of Medicine*, 4th edition, Philadelphia and London: Saunders, 1929.

rapidly increasing from the 1880s and became the largest by 1910–1919. Since this was the time of World War I, comparison with the European countries may seem of doubtful validity; but the relative decline of the European countries started prior to the war and lasted well into the twenties, so that it should not be attributed entirely to the war. Figure 1 shows the proportion of the total discoveries in each nation

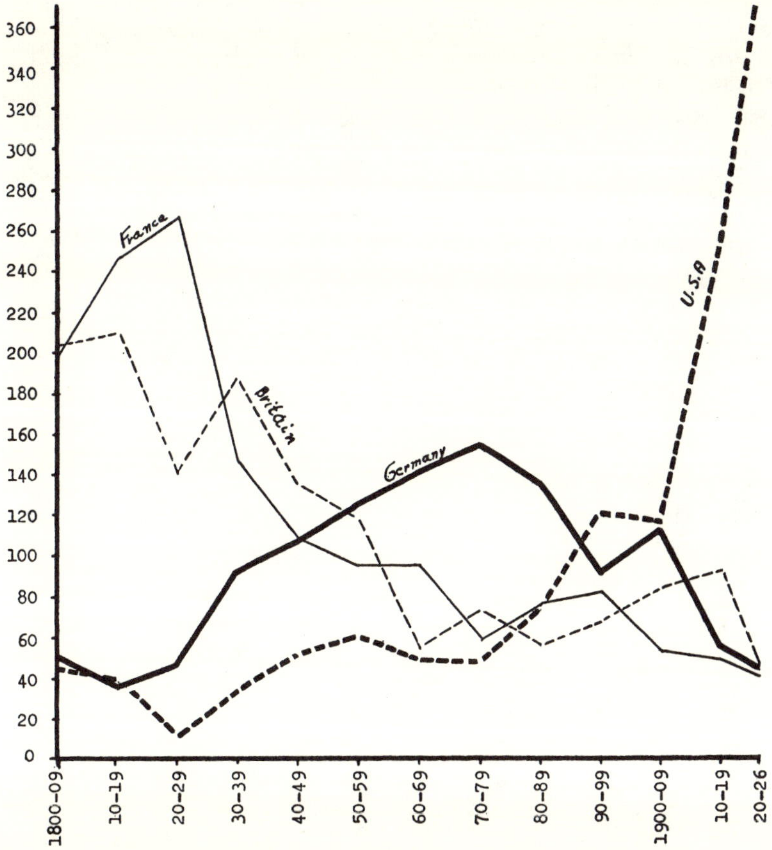

FIGURE 1

Changes in the relative share of medical discoveries in selected countries, 1800–1926.

during each period as a proportion of the country's relative share over the whole period:

$$y = 100 \frac{\text{country's share in decade (\%)}}{\text{country's share over whole period (\%)}}$$

A significant aspect of this change of relative positions is that it is connected with an atypical growth in the curve of discoveries in the country which is gaining the largest share. Thus the number of German discoveries continually increases through the middle of the nineteenth century in a period of decline in France and Britain. A similar deviation marks the change in the relative position of the United States at the beginning of the twentieth century.

A similar pattern marks the number of discoverers. Table 2 shows the "productivity" of the various countries in terms of scientists.[5] France and Britain, with the largest numbers at the beginning of the century, fall behind Germany starting about 1835. While the number of German scientists entering upon their careers increases regularly, with only one considerable drop until 1885–1890, there are fluctuations and a generally downward slope in France and England through the middle of the century. The American trend, like the German, shows much less fluctuation. Thus, with respect to major trends, the two indexes validate each other.[6]

Two questions emerge: What explains the change of scientific leadership from France to Germany to the United States? And what explains the "deviant" nature of the development in Germany during the middle and in the United States toward the end of the nineteenth century, as manifested in (1) the continuous rise in the number of discoveries during periods of relatively low creativity in the other countries; and (2) the relatively smaller fluctuations in the number of people embarking upon scientific careers in these two countries compared with the others?

Hypothesis: The Organizational Factor

NEITHER the changes in scientific leadership nor the deviant nature of the German and American developments can be manifestations of differences in the scientific ideas in the various countries. This could be the case only if international communication had been deficient, so that new ideas in one country would have no effect upon the work of scientists in the others. This was by no means the case, as

5. Based on W. A. Newman Dorland, *The American Illustrated Medical Dictionary*, 20th edition, Philadelphia and London: Saunders, 1946.

6. The pattern which emerges from these indexes parallels the qualitative descriptions of up-to-date histories of medicine and science. See, e.g., Arturo Castiglioni, *A History of Medicine*, New York: Knopf, 1947; Richard H. Shryock, *The Development of Modern Medicine*, New York: Knopf, 1947; H. T. Pledge, *Science Since 1500*, London: Philosophical Library, 1940. Rather than simply referring to such sources, I prefer to present the numerical indexes in detail for two reasons: (1) They contain some information not sufficiently emphasized—or even blurred—in those sources. Thus the small amount of medical research in Britain is blurred in the qualitative descriptions by the dazzling brilliance of England's few scientist-intellectuals

TABLE 2. Discoveries in the Medical Sciences at the Age of Entering Their Professions (Age 25) in Various Countries, 1800–1910

Year	*U.S.A.*	*England*	*France*	*Germany*	*Other*
1800	1	7	8	7	4
1805	1	8	5	8	2
1810	3	11	6	6	2
1815	2	12	12	7	3
1820	3	11	23	18	2
1825	2	17	15	18	6
1830	8	12	25	10	6
1835	11	13	26	29	7
1840	5	24	22	35	12
1845	5	14	13	33	5
1850	10	18	21	37	10
1855	15	16	20	49	27
1860	16	23	13	61	23
1865	25	15	36	71	26
1870	25	15	31	83	41
1875	40	31	23	84	46
1880	48	17	40	75	50
1885	52	16	34	97	52
1890	43	11	23	74	41
1895	47	9	27	78	29
1900	32	9	17	53	30
1905	28	4	4	34	25
1910	23	6	7	23	18

SOURCE: Dorland's Medical Dictionary (20th ed.).

demonstrated by the parallel upward movements of the curves of discoveries in all the countries in periods of crucial scientific advance. Independently from this fact, whatever barriers to scientific communication had existed between France and Germany during the first decades of the nineteenth century had disappeared by the beginning of the fourth decade. By about the same time, the British too established contacts with continental science, from which they had become isolated

and by the glamor of the British medical profession. Also, the different patterns of growth of scientific personnel (discoverers) is a subject not sufficiently emphasized in the histories of medical science. (2) What is called here scientific productivity is only one aspect of the development of science; in terms of the interrelationships of scientific ideas it is perhaps a peripheral one. Since traditionally the history of science is an history of ideas, even the few historians interested in such sociological phenomena as differences in the scientific development of various countries are not very explicit about the bases of their judgments, nor do they sufficiently differentiate between the various aspects of science as a social activity. It is important, therefore, to present explicitly the quantitative basis of the historians' judgment and clearly delimit the particular aspect of scientific activity dealt with here from others.

with decreasing splendor during the eighteenth century, as did the Americans.[7] Therefore nothing immanent to science as a body of ideas explains the observed differences and changes. The explanation has to be sought in external circumstances.

Among the possible external causes there are some general and obvious ones, such as population growth and the growth of national income. A few unrefined attempts to assess the population factor suggested that this is not a promising line of approach. The introduction of this factor does flatten out the curves somewhat, but does not eliminate the characteristic waves of development, and it hardly affects the changes in the relative position of the nations.[8]

Nor do differences in national or personal income seem to be relevant. The indexes of national income in all the countries here surveyed show a fairly gradual and constant rise through the whole period without such ups and downs and such extensive changes in the relative positions of the countries as indicated by our data. Moreover, the United States and Britain were the richest of these countries, at least since the middle of the nineteenth century (and no doubt earlier in the case of Britain). Yet, as to medical discoveries, these countries were relatively backward during much of the period.[9] None of these factors, therefore, seems to be directly and consistently related to the differences in the growth of discoveries in the various nations.

Thus, it is assumed that the conditions determining the differences are to be sought in the *organization* of science. But this is a complex phenomenon: we still must seek the particular organizational factor which reasonably answers our questions. It is proposed to isolate this factor by comparing the main aspects of the organization of science in France and Germany during the first half and the middle of the nineteenth century and those same aspects in Britain and the United States during the three decades preceding World War I. This particular pairing is selected because France and Germany maintained a publicly supported network of scientific instruction and research from the early nineteenth century, while Britain and the United States did not begin to develop their systems until the second half of the century. There were short-lived experiments in Britain during the first half of the century, but these were overshadowed by the archaic nature of the most

7. *Cf.* Shryock, *op. cit.*, pp. 193–196; Paul Diepgen, *Geschichte der Medizin*, Berlin: Gruyter, 1955, Vol. II/1, pp. 204–207; Charles Newman, *The Evolution of Medical Education in the Nineteenth Century*, London: Oxford University Press, 1957, pp. 265–269.

8. The sources used for population data were *La Population Française: Rapport du Haut Comité Consultatif de la Population et de la Famille*, Paris: Presses Universitaires de France, 1955, p. 19; Michel Huber, Henri Bunle, et Fernand Boverat, *La Population de la France*, Paris: Librairie Hachette, 1943, p. 19; W. S. and E. S.

important universities. If it is possible to isolate a theoretically relevant condition common to the organization of science in Germany and the United States, but absent in France and Britain, that condition may reasonably be taken as the cause of the observed differences.

France and Germany

THREE conditions are mentioned in the literature in explanation of German scientific superiority in the nineteenth century: (1) the relative excellence of laboratory and hospital facilities for research and the faster recognition of the importance of new fields of research, especially physiology; (2) the clear recognition of the aim of the university as a seat of original research, and efficient organizational devices to achieve that aim, such as far-reaching academic self government, the freedom of the teacher regarding the content of his courses, the freedom of the student in the choice of his courses and his teachers (including easy transfer from one university to another), the requirement of submitting theses based on research for attainment of academic degrees, and, above all, the institution of *Habilitation,* that is, the submission of a high level scientific work based on original research as a precondition of academic appointment; (3) the existence of a large number of academic institutions which made possible the mobility of teachers and students, and resulted in an atmosphere of scientific competition that did not exist elsewhere.[10] The superiority of the German scientific facilities from about the middle of the century is an undeniable fact. But instead of explaining the differences in creativity, it is itself a phenomenon that needs explanation.

The pioneering country in the establishment of modern scientific facilities was France. Founded in 1794, the Polytechnique had been the model academic organization in the natural sciences. Among other new features, it possessed the first academic research laboratories (in chemistry). The physiological laboratory at the Collège de France, where Magendie and Claude Bernard conducted their studies, was considered most inadequate by the middle of the nineteenth century. Yet it was there that modern experimental physiology began. The idea of studying illness as a natural phenomenon, not necessarily for the sake of cure, was

Woytinsky, *World Population and Production,* New York: Twentieth Century Fund, 1953.

9. For national income data, see, e.g., Colin Clark, *The Conditions of Economic Progress,* 2nd edition, London: Macmillan, 1951.

10. *Cf.* Abraham Flexner, *Universities: American, English, German,* Oxford: Oxford University Press, 1930, pp. 317–327; Donald S. L. Cardwell, *The Organization of Science in England,* London: Heinemann, 1957, pp. 22–25; H. E. Guerlac, "Science and French National Strength" in E. M. Earle, editor, *Modern France,* Princeton: Princeton University Press, 1951, pp. 85–88.

first conceived in Paris, and the beginnings of systematic clinical research in medicine were made in the hospitals of that city.[11]

Until the 1830s German medical research and natural science research in general was backward compared with the French, and probably with the British too. The famous network of modern German universities already existed from the time when, following tentative beginnings at Halle, Goettingen and Jena, the University of Berlin was established in 1809.[12] But the universities, rather than promoting, retarded the development of empirical science. They regarded philosophy as the queen of sciences, and usually disparaged empirical research. The biological sciences in particular were under the sway of *Naturphilosophie,* which stimulated much imaginative writing but little research.[13]

Only around 1830 did this atmosphere change under foreign influence. Liebig, who had studied in Paris, established in 1825 the first chemical laboratory at the small university of Giessen. A few years later Johannes Mueller, the central figure of German physiology, abandoned his early attachment to *Naturphilosophie* and became converted to the empirical method by studying the works of the Swedish chemist, Berzelius. About the same time the Vienna school of clinicians adopted the methods of investigation initiated by the Paris clinicians, and various learned journals began to propagate the new scientific approach in the medical sciences.[14]

Thus the French showed at least as much understanding of the value and the needs of scientific research as the Germans. It should not be assumed that this understanding suddenly declined around the middle of the century. The influentials of French science at that time, such as Dumas, and later Pasteur, Claude Bernard, and Victor Duruy, were certainly not less enlightened and brilliant than their German counterparts. In fact, they may have been more sympathetic to the needs of scientific research than German academic policy makers, since obscurantism was rather prevalent within both the faculties of the German universities and the governmental offices in charge of higher education.[15]

11. *Cf.* Shryock, *op. cit.*, 70–71, 151–169; Newman, *op. cit.*, p. 48; Guerlac, *op. cit.*, pp. 81–105.

12. *Cf.* Flexner, *op. cit.*, pp. 311–315; R. H. Samuel and R. Hinton Thomas, *Education and Society in Modern Germany*, London: Routledge & Kegan Paul, 1949, pp. 111–113; Jacob Barion, *Universitas und Universitaet*, Bonn: Rörscheid, 1954, pp. 14–20.

13. *Cf.* Shryock, *op. cit.*, pp. 192–201; Diepgen, *op. cit.*, Vol. II/1, pp. 23–28.

14. *Cf.* Cardwell, *op. cit.*, pp. 22–75; Shryock, *op. cit.*, pp. 188, 195; Garrison, *op. cit.*, pp. 451–452.

15. See Guerlac, *op. cit.*, pp. 85–88 on France. On the relative backwardness of German academic administration, see Ervin H. Ackerknecht, *Rudolf Virchow: Doctor, Statesman, Anthropologist,* Madison: University of Wisconsin Press, 1953, pp. 139–140; Samuel and Thomas, *op. cit.*, pp. 114–130; Max Weber, *Jugendbriefe,* Tübingen: Mohr, n.d., pp. 151–152. In order to realize the amount of obscurantism and intolerance in German universities at the time it is useful to read the otherwise

The greater expansion of German scientific facilties and the prompter recognition of new fields are therefore as much in need of explanation as the continuous growth in German discoveries.

The second condition—the presumably peculiar values and organization of the German university—is also a very doubtful explanation. The idea of academic freedom notwithstanding, atheists, Jews, and socialists were often kept out of academic careers in Germany. Academic self-government was not necessarily enlightened: liberal scientists in the 1840s regarded it as an essentially retrograde arrangement. In fact, some of the most beneficial academic decisions—with relation to the growth of science—were taken by civil servants, most notably Friedrich Althoff, who interfered with academic self-government. Even the *Habilitationsschriften* were often rather mediocre pieces of research, and there was nothing in the constitution of the universities efficiently to prevent mediocre professors from confirming inferior theses.[16]

At the same time, the ideas as well as some of the arrangements said to be characteristic of the German universities also existed in France. Freedom of teaching already formed the core of the tradition at the Collège de France before the Revolution and was carried further than in the German universities. The ideals of pure research were formulated in French scientific ideology at least as clearly as in German and they were practiced and encouraged in a great many ways.[17] There is no proof that the lack of the paraphernalia of academic self-government interfered with the research of French scientists more than in Germany. It is true that, compared with the *Habilitation*, the French *aggrégation* and the system of open examinations seem to be inefficient ways of selecting people for academic careers. But there is little evidence that this irrelevant hurdle actually prevented potentially creative people from entering scientific careers. Moreover, there were other means, such as numerous prizes and public honors, which encouraged original research in France.[18]

Decentralization has been written about much less than the first two conditions, partly because it was an unintended circumstance, and partly

shallow work of Richard Graf du Moulin Eckart, *Geschichte der deutschen Universitaeten*, Stuttgart: Enke, 1929.

16. *Cf.* Flexner, *op. cit.*, pp. 317–327; Samuel and Thomas, *loc. cit.*

17. See Claude Bernard, *Morceaux choisis*, dirigé et préfacé par Jean Rostand, Paris: Gallimard, 1938, pp. 16–18, for one of the most beautiful descriptions of the traditions of the freedom of teaching and research as it was practiced at the Collège de France. See also Ernest Lavissé, *Histoire de France*, Paris: Librairie Hachette, n.d., Vol. IX/1, p. 301, on the pioneering beginnings of the teaching of pure sciences in the same institution in the 1770s and 1780s.

18. For a good description of how the French system of examinations actually worked, see René Leriche, *Am Ende meines Lebens*, Bern und Stuttgart: Huber, 1957, pp. 53–55. Leriche, like others, attributes the lack of originality of French medicine to the examinations. But his own account shows that the problem was rather the lack of career opportunities for young medical scientists (*ibid.*, p. 34).

because its effect upon research is less immediately evident. The decentralization of the German academic system was the result of the political dismemberment of the German-speaking people. There were 19 independent universities in Germany proper, maintained by the princes of the numerous small states constituting Germany in the eighteenth and early nineteenth centuries, as well as German language universities in Switzerland, Austria (including the Czech provinces), and Dorpat in the Baltic Sea Provinces of Russia.[19] At the same time the French boasted a unified academic system, most of it situated in Paris. Although some of the features of this centralization introduced by Napoleon were deplored, the central administration of science and academic institutions generally was considered to be desirable by French politicians of science.[20]

Nevertheless, decentralization seems to have been the decisive factor in determining the differences in the scientific creativity of the two countries. It gave rise to academic competition, and competition forced upon the individual institutions decisions which would not have been made otherwise, or at least not made at that time. In all areas crucial to the development of the medical sciences German policies turned out to be in the long run more farsighted and bold than French policies, although the first initiative was often taken by the French.—What, then, was the actual competition and how did it influence the decisions about the crucial problems of academic policy?

The Crucial Decisions

GIVEN the situation of the medical sciences (and perhaps of the sciences in general) at the beginning of the last century, the problem faced by the French and the German systems (and not confronted by Britain and the United States until later) was to find adequate criteria for the evaluation and support of science. The governments, and increasingly the people too (especially in France), believed in the value and usefulness of science. Academies, universities, and other institutions were set up everywhere, or rejuvenated where they existed before, in order to promote research and to disseminate knowledge. One of the aims of these institutions was to enable a selected few scientists, who had already proved their greatness, to devote all of their time to financially supported scientific research. But it was not intended to create in these institutions academic careers which one entered as in any other

19. With the addition of Strassburg in 1872 there were 20 universities in Germany. *Cf.* Christian v. Ferber, *Die Entwicklung des Lehrkörpers der deutschen Universitaeten und Hochschulen 1864–1954*, Vol. III of Helmuth Plessner, editor, *Untersuchungen zur Lage der deutschen Hochschullehrer*, Göttingen: Vandenhoeck & Ruprecht, 1956, pp. 37–38. The German-language universities of Switzerland were Zürich, Bern, and Basel; and of Austria, Vienna, Prague, Graz, and Innsbruck.

profession. The large majority of the scientists had independent means or a lucrative profession (very often medical practice, even in sciences not connected with medicine), and pursued their scientific interest in their free time, often at a considerable personal cost. This idealistic pattern seemed to fit perfectly that sacred pursuit of truth which was science. Academic appointments therefore were regarded as honors rather than careers, and turning science into an occupation would have seemed something like a sacrilege.

A corollary, in this amateur stage of science, was the absence of specialization. The great names of the early nineteenth century were those of generalists who were creative in more than one field. And the new scientific disciplines developed from their work. While it was increasingly believed that the new disciplines required specialists, the fact that they were opened up by generalists seemed to indicate that specialization was not really necessary. Moreover, there persisted the reluctance to abandon the conception of general science which explains to the adept all the secrets of nature. Thus there was considerable disinclination to substitute for the *savant* such narrow specialists as chemists, physiologists, and the like. And there was even more reluctance to redefine such a traditionally unified field as medicine into a number of subspecialties.

The second problem was the development of criteria for the support of research. Today it is still difficult of course to decide what constitutes adequate and sufficient support of research, but at least budgets can be drawn for determined purposes. At that time even this was impossible, since research was an unpredictable, erratic process, and important discoveries were made as often outside as inside the laboratories.

Finally, there was the question of training scientists. Until the second half and particularly the last quarter of the nineteenth century, science had few practical applications. Most of it was pure science benefitting no practice. Under these circumstances, to train every medical student, would-be chemist, and engineer in scientific research was about as justified as it would be today to teach every concertgoer advanced musical composition.[21]

These problems existed in both countries and were approached in France and Germany with the same concepts. Yet, to repeat, the long-term decisions made in France concerning all three problems were the opposite of those made in Germany.

20. *Cf.* Guerlac, *op. cit.*, pp. 87–88.

21. On the state of science in the early nineteenth century, see Pledge, *op. cit.*, pp. 115–151. On scientists in the same period, see Elie Halévy, *History of England in 1815*, London: Pelican, 1938, Vol. 2, pp. 187–200; René J. Dubos, *Louis Pasteur: Franctireur de la science*, Paris: Presses Universitaires de France, 1955, pp. 3–4; and Diepgen, *op. cit.*, Vol. II/1, pp. 2–5, 66–69, 152–153.

SCIENTIFIC CAREERS AND SPECULATION

The creation of regular careers in science and the recognition of specialized disciplines were closely connected problems. Both may be illustrated with the case of physiology, the most decisive science for the development of medicine in the nineteenth century.

As a systematic discipline, physiology emerged at the beginning of that century. François Magendie, considered to be the founder of experimental physiology, was professor of medicine. He established the new specialty and could follow it undisturbed (though practically unsupported) at the Collège de France, because of the full degree of academic freedom prevailing in that institution. But his disciple, Claude Bernard, who became the most outstanding representative of the new field around the middle of the century, for many years had to use his private laboratory and private means to pursue his research. At last, against the opposition of those who regarded the new discipline as merely a branch of anatomy, a special chair was created for Bernard at the Sorbonne in 1854. Soon thereafter he also fell heir to Magendie's chair at the Collège de France and held both appointments until 1868; he then transferred his work to the Museum of Natural History, relinquishing the post at the Sorbonne to his disciple, Paul Bert.

The recognition of the discipline of physiology, however, did not create opportunities for purely scientific careers in the traditional field of medicine. In this connection, the only change was that after the retirement of the chair's incumbent a single successor would have to be found. This was not a prospect on the basis of which one could realistically take up research as a career. Therefore, potential scientists first had to build up a practice, and engaged in research as a part-time activity.[22]

Thus, the academic career changed very little in France through the nineteenth century. Appointments were made from an undifferentiated group of practitioners—amateur scientists—and usually at a fairly advanced age. Even academically successful persons did not become full-time scientists before they reached their forties or fifties, and since the

22. *Cf.* Bernard, *op. cit.*, pp. 154–157, 263–285; J. M. D. Olmsted, *Claude Bernard: Physiologist*, New York: Harper, 1938, pp. 51–89. For the situation at the beginning of the twentieth century, see Edouard Rist, *25 Portraits des médicins français, 1900–1950*, Paris: Masson, 1955, pp. 29–40.

23. See Rist, *op. cit.*, pp. 97–104, on the career of S. A. Sicard, who seems to have been a relatively lucky and successful scientist. When at the age of 51 he was appointed as professor he had to abandon his life-long interest and research in neurology because the vacant chair was designated for internal pathology, and course preparation in the new field required a great effort.

24. *Cf.* K. E. Rothschuh, *Geschichte der Physiologie*, Berlin-Göttingen-Heidelberg: Springer, 1953, pp. 93, 112–118.

25. George Rosen and Beate Caspari-Rosen, *400 Years of a Doctor's Life*, New York: Schuman, 1947, pp. 248–250; Ernst Gagliardi, Hans Nabholz and Jean Strohl,

chair to be vacated was not known they had to maintain as broad interests and activities as possible. But in the second half of the century there was increasingly less chance for non-specialists to make important discoveries. French scientific productivity therefore declined even in fields pioneered by Frenchmen. Whenever a discipline reached the stage of development where its efficient pursuit required specialists, there was little chance that the French system would produce such scientists.[23]

Physiology as a science was received with more sympathy in Germany than in France, but its recognition as an academic specialty there also ran into difficulties. The man who did most for the introduction of the discipline to Germany, Johannes Mueller, was a generalist who taught, in addition to physiology, anatomy, ophthalmology, and surgery.[24] His eventual successor in Berlin, Du Bois-Reymond, had been refused one professorial chair after another because he was considered a mere specialist.[25] The early creation of a separate chair in physiology (for Purkinje in Breslau, 1839) had no general effect, and for some years physiology and anatomy continued to be taught by the same person in all other German universities. But pressure for the separation of the disciplines by the younger generation of scientists continued, and those with some bargaining power raised the demand when they were offered university chairs. Thus, when Carl Ludwig was offered a professorship at Zürich in 1849, he accepted it only on the condition that a separate teacher be appointed for anatomy;[26] thereafter the recognition of the new discipline proceeded rapidly. No university could afford to neglect the new field, so that by 1864 there were already 15 full professors of physiology in Germany and several others in the wider system of German-language universities.[27] The separation of physiology from anatomy at this stage became the official policy of university administration. In some cases, where traditionally-minded incumbents were reluctant to abandon one of the disciplines, the separation was forced upon them by administrative pressure.[28]

All of this led to a complete transformation of the scientific career in Germany. In spite of the strictures against narrowness and of the

Die Universitaet Zürich und ihre Vorlaeufer 1833–1933, Zürich: Erziehungsdirektion, 1938, pp. 548–549.

26. *Ibid.*, pp. 539–548. Virchow, who was also offered the chair, refused to accept it on the ground that he wished a chair for pathological anatomy exclusively (without teaching responsibilities in either surgical anatomy or physiology). In Ludwig's time the nominal unity of physiology and anatomy was still maintained; the separate teacher in anatomy was only an extraordinary professor. But when Ludwig left Zürich in 1855 and the position was offered to Koelliker, the chairs were finally separated upon the latter's suggestion (although Koelliker himself refused the job). For similar instances of creating new specialties at the same university in order to attract or retain teachers, see *ibid.*, pp. 562, 879.

27. *Cf.* Von Ferber, *op. cit.*, p. 204.

28. For example, Valentin in Bern, in 1865; see Bruno Kisch, *Forgotten Leaders in Modern Medicine*, Philadelphia: American Philosophical Society, 1954, pp. 174–175.

continuing lip-service paid to the image of the scientist who works because of devotion, science became a specialized and regularized occupation. As we have seen, success, fame, or even sheer enterprise had a good chance for reward. Once a new and fruitful field was recognized in one university, strong pressure led other universities to follow suit, thereby creating more opportunities for those willing to work in the new field. Therefore, it was possible—and for the very able also worthwhile—to concentrate after graduation on one well defined and promising field of research with the definite aim of a scientific career. Not only was it unnecessary first to build up a practice and to retain as general interest as possible, but if one had taken such a course his academic prospects would have been negligible in competition with the full-time specialists. Thus specialized science became a career, and the amateur general scientist disappeared in Germany.[29] This difference in career possibilities, not the distinction between *Habilitation* and *aggrégation*, explains the greater research orientation of German than of French science.

The same mechanisms which explain the development of scientific roles also explain the development of facilities for research, and the introduction of scientific methods into the training of physicians. The creation of new facilities was part and parcel of the bargaining between universities and scientists. Facilities (laboratories, assistants, and so on) were offered to attract desirable candidates or to prevent scientists from moving elsewhere. The extension of facilities made possible, and to some extent made necessary, the training of a growing number of persons capable of doing research. Since not all such individuals could be given academic appointments in the basic medical sciences or otherwise, they used their research skills and interests to transform clinical medicine into an exact science. These processes and their results may be briefly illustrated.

RESEARCH FACILITIES

As has been pointed out, the French were the first to establish modern institutions for scientific training and research. But the facilities

29. Max Weber, writing in 1918, regarded science as a most risky career; see his "Science as a Vocation," in H. H. Gerth and C. Wright Mills, *From Max Weber: Essays in Sociology*, London: Oxford University Press, 1947, pp. 132–134. But, it should be realized that Weber was referring to a crisis situation in an already established discipline; the circumstances were much more hopeful in the middle of the nineteenth century. Of those who took their *Habilitation* between 1850 and 1859; 85 per cent received full-time academic appointments, while for those who received their *Habilitation* between 1900 and 1909 only 62 per cent received such posts. The corresponding proportions in medicine are 84 and 48 per cent, respectively. (This does not necessarily mean a relatively greater decline of research opportunities in medicine, because there were good research opportunities outside the universities in public hospitals.) See Von Ferber, *op. cit.*, pp. 81–82, for the statistical data; and Adolf Struempell, *Aus dem Leben eines deutschen Klinikers*, Leipzig: Vogel, 1925, on the *Habilitation* as a preparation for a hospital career.

and arrangements established in France about 1800, considered to be ideal for their time, were hardly extended or changed until World War I or later. The Pasteur Institute, established in 1888, was the first independent research institute of the world. Again, it remained the only one in its field in France at least until World War I.[30]

Thus in France a new type of organization was apt to remain a single show-piece for 50 years, while in Germany such novelties became routine features of the organization of research in a much shorter time. By the 1840s there were apparently more and better chemical laboratories in Germany than in France, and by the sixties the contrast was extreme. At a time when it was an achievement for Pasteur to obtain any (and most inadequate) laboratory facilities, the Prussian government built new laboratories at Bonn and Berlin (the Bonn laboratory, for example, could accommodate more than 60 students) equipped with the most up-to-date facilities, and the older ones probably were also more adequate than anything that existed elsewhere. And there were good laboratories at other universities in Germany.[31]

There were similar differences between Germany and France in the development of facilities for medical research and of specialized research institutions. The New Vienna School of clinical research began in the thirties, and its facilities seem to have been modest even until midcentury. But there were gradual improvements in one place after another, and by the sixties there evolved fairly uniform standards which made it possible to conduct clinical research with the aid of adequate laboratory facilities in a number of places.[32] Finally, the establishment of specialized research institutions became a matter of routine in Germany soon after the beginnings made in France. They became a tool regularly used by the universities' administrations, the governments, and local bodies to encourage and develop the work of famous scientists.[33]

SCIENTIFIC TRAINING

The differences in the development of medical training were no less conspicuous. The fact that until about the 1880s all the great advances in the basic medical sciences contributed little to the cure of illness largely explains the persistent and overwhelming emphasis on the

30. The ideal arrangements of French medical schools in 1798 are noted in Newman, *op. cit.*, p. 48. Concerning the quite different picture presented by French academic medicine early in this century, see Abraham Flexner, *Medical Education in Europe*, New York: The Carnegie Foundation for the Advancement of Teaching, 1912, pp. 221–223; and Leriche, *op. cit.*, p. 34. On the Pasteur Institute, see Guerlac, *op. cit.*, p. 88.

31. *Cf.* Cardwell, *op. cit.*, p. 80; and Dubos, *op. cit.*, pp. 34, 78–79.

32. *Cf.* Diepgen, *op. cit.*, Vol. II/1, pp. 207–209; on the situation in the 1860s, see Theodor Billroth, *The Medical Sciences in the German Universities*, New York: Macmillan, 1924, p. 27; and at the turn of the century, Flexner, *op. cit.*, pp. 145–166.

33. *Cf.* Flexner, *op. cit.*, 1930, pp. 31–35.

practical art of medicine rather than on its few scientific bases in the training of the student-physician. Indeed, apprenticeship and bedside demonstrations were the most important parts of medical training in France, England, and the United States.[34]

Only in Germany did the training of the doctor become a privilege of scientists. By the 1860s even clinical chairs were given exclusively to people with attainment in research rather than to outstanding practitioners. And from the middle of the century, even public hospitals were increasingly staffed by doctors both interested and trained in research. Thus much earlier than elsewhere (possibly prematurely), medicine in Germany became an applied science.[35] As a result, when the great opportunities for clinical research arose, following the discovery of the bacteriological causation of illness and the perfection of anesthesia and aseptic surgery, there were in Germany enough doctors trained in research to take full advantage of the opportunity, and to transform public (even non-teaching) hospitals into veritable institutions of applied medical science.[36]

Decentralization and Competition

THUS, regarding all three crucial decisions—developing scientific facilities, creating scientific roles, and training larger numbers of research personnel than were justified by existing practical needs—the German system "behaved" with uncanny foresight. It has been shown that this foresight was not the result of greater individual wisdom. It was the result of competition due to the unintended decentralization of the German system.

"Competition" in this paper refers to the general condition underlying all the processes described above: it is a situation in which no single institution is able to lay down standards for the system of institutions within which people (in this case students and teachers) are relatively free to move from one place to another. Under such circumstances, university administrators required neither exceptional boldness nor foresight for continually expanding facilities and training, and for creating new scientific jobs. There was little if any need for fateful individual decisions. Improvements and innovations had to be made from time to time in order to attract famous men or keep them from leaving. In this way, laboratories and institutions were founded, as-

34. *Cf.* Diepgen, *op. cit.*, Vol. II/1, pp. 212–214; Vol. II/2, pp. 154–155, 286–288; Abraham Flexner, *Medical Education: A Comparative Study*, New York: Macmillan, 1925, pp. 211–212, 241, 248.

35. Diepgen, *op. cit.*, Vol. II/1, pp. 152–153. See also Theodor Billroth, *loc. cit.*; Bernhard Naunyn, *Erinnerungen, Gedanken und Meinungen*, Munich: Bergmann, 1925, pp. 375–376.

36. There was a parallel development in chemistry. There too the availability of

sistantships provided, new disciplines recognized, and scientific jobs created. These innovations were repeated throughout the system because of pressure from scientists and students in general, irrespective of practical needs and of what a few scientific influentials thought.

If competition inevitably brought about the adoption of fruitful innovations in the universities, it also forced them to correct mistakes and to eliminate traditions which retarded scientific development. This process has been shown in the case of the separation of physiology from anatomy and the introduction of scientific criteria in clinical training in Germany.

Britain and the United States

THE similarities, differences, and differential effects observed in the cases of France and Germany were, in essentials, repeated in the cases of Britain and the United States.

From the middle of the nineteenth century, British—and soon after, American educators—scientists, and administrators displayed increasing interest in the organization of science in Germany. Scientists and intellectuals who visited Germany returned home enthusiastic about German academic life, and soon German university training became a standard preparation for scientific careers among British scientists.[37]

Consequently British universities, though retaining certain traditions, introduced measures to bring themselves in line with German standards and practices. Oxford and Cambridge, which until the 1860s were training centers primarily for the rich and the clergy, began to emerge as institutions of empirical science and positive scholarship pursued in an atmosphere of academic freedom and autonomy. The newer University of London and the universities in the provinces imitated the German pattern even more closely and were imbued, from the beginning, with the spirit of empirical science.

The rapid growth of the modern academic system also began in the United States in the 1860's. The Land Grant Act passed in 1862 and other circumstances brought about a large increase in the number of American colleges and universities between the sixties and the eighties.[38] In the present context, the most important events were the rise of the graduate schools in the seventies, and in the following decade the establishment of Johns Hopkins Medical School which was directly

relatively large numbers of trained chemists afforded Germany the opportunity to build up within a short time a chemical industry based on applied science, after the discovery of the aniline dyes made the practical application of science a permanent possibility; *cf.* Cardwell, *op. cit.*, pp. 134–137, 186–187.

37. *Ibid.*, p. 50.

38. *Ibid.*, p. 80.

influenced by the German example.[39] Eventually older institutions such as Harvard also abandoned certain traditions derived from pre-nineteenth century England and adopted new methods in imitation of the German.[40]

In this development of a system of up-to-date institutions for medical research and training Britain had most of the advantages over the United States, similar to those possessed by France over Germany at the beginning of the century. The British began the adoption of the German patterns earlier, and they began from a higher level than did the Americans.[41] Nevertheless, while the effect of the academic reform on British science was slow and partial, in America it produced a conspicuous rise in scientific creativity.

That the social mechanisms at work in these cases were similar to those involved in our first pair of comparisons can be illustrated best by the organization of clinical research and the creation of clinical chairs. Attempts to copy the Germans by making hospital departments into virtual research establishments and filling the clinical chairs according to criteria of scientific achievement ran into serious opposition in both countries. They seemed like an infringement on the rights of the profession, whose members had run the teaching hospitals independently of the universities, and it also seemed to be endangering the charitable purpose of the hospitals. Therefore, when Oxford and Cambridge decided to overhaul their medical training programs along German lines, they confined themselves to the basic departments and sent their students to continue their clinical studies in the hospital medical schools in London. This division was a decision in favor of preserving the traditions of the professional fraternities attached to the various public hospitals and of the philanthropic bodies which governed these hospitals. Of course, it could also have been justified by the aim to keep apart pure research and professional practice.[42] However, *a priori* reasons for incorporating the teaching hospitals in the universities and staffing them on the basis of attainments in research might have been advanced. As

39. *Cf.* Flexner, *Universities . . .* , *op. cit.*, p. 73; and Abraham Flexner, *I Remember*, New York: Simon and Schuster, 1940, pp. 63–64.

40. *Cf.* Edward D. Churchill, *To Work in the Vineyards of Surgery: The Reminiscences of J. Collins Warren* (1842–1927), Cambridge: Harvard University Press, pp. 193–197, 257–271.

41. *Cf.* Newman, *op. cit.*, pp. 269, 276; Cardwell, *op. cit.*, pp. 46–51, 80, 103–107, 110–114, 118–119, 134–137, and *passim*. Flexner, *Universities . . .* , *op. cit.*, pp. 46–65; Richard H. Shryock, *American Medical Research: Past and Present*, New York: New York Academy of Medicine, 1947, pp. 106–108, 118–119.

42. *Cf.* "The First Hundred Years: Notes on the History of the Association," extracts from Ernest M. Little, "History of the Association," *British Medical Journal*, 1932, 1, pp. 672–676; A. M. Carr-Saunders and P. A. Wilson, *The Professions*, Oxford: Clarendon Press, 1933, p. 87; Flexner, *Medical Education . . .* , *op. cit.*, p. 28; Newman, *op. cit.*, pp. 49–50, 133 ff.

shown above, this was one of the problems which could not at that time be decided on *a priori* logical grounds; only future experience could indicate the effective choice.

The conditions for acquiring the needed experience existed in England, since there were approximations of a proper university hospital and university clinical departments in the London University College Medical School (founded as early as 1836), and similar opportunities arose when the provincial universities were established.[43] Yet, instead of representing competing alternatives, none of these departments ventured further than the model established by the Oxford-Cambridge-London triangle; that is, their clinical departments were run by local practitioners as practical training centers rather than being organized as university departments engaged in research and staffed by persons selected on the basis of scientific eminence. This was quite different from the situation in Germany, where, for example, the little University of Giessen successfully pioneered in establishing its chemical laboratory, imitated later by universities of much greater prestige. It also differed from the innovation of the Johns Hopkins Medical School, where a full-scale medical faculty that included basic as well as clinical departments was established—a pattern that was followed by other universities and led to a rapid transformation of American medicine reminiscent of German, notwithstanding the strength of a professional and philanthropic tradition similar to that of Britain.[44]

All this shows unequivocally that the British system was not competitive. Yet seemingly it was decentralized, since universities and public hospitals were private institutions financed and governed in a variety of ways, as in the United States. In fact, however, Britain also had a centralized system, though centralized in a somewhat different way than that of the French. The provincial universities did not begin to confer degrees until 1880 (with the exception of Durham, established in 1831) and their status, as well as the status of London University, never reached that of the two ancient universities. The system was totally overshadowed by the Oxford-Cambridge duopoly, which, in spite of differences in matters of religion and politics, represented basically similar educational philosophies and academic policies.[45] The special

43. *Cf.* Flexner, *Universities . . . , op. cit.*, pp. 242–244.

44. *Cf.* Donald H. Fleming, *William Welch and the Rise of Modern Medicine*, Boston: Little, Brown, 1954, pp. 173 ff. On competition in American academic life in general, see Logan Wilson, *The Academic Man*, London: Oxford University Press, 1942, pp. 157–174, 186–191, 195–214; and Theodore Caplow and Reece J. McGee, *The Academic Marketplace*, New York: Basic Books, 1958.

45. *Cf.* Flexner, *Universities . . . , op. cit.*, p. 249; Bruce Truscot, *Red Brick University*, Harmondsworth: Penguin Books, 1951, pp. 19–29. See also R. K. Kelsall, *Higher Civil Servants in Britain*, London: Routledge & Kegan Paul, 1956, p. 137, on the preservation of the educational duopoly in another field; as late as 1950, 47.3 per cent of British civil servants in the ranks of Assistant Secretary and above had attended Oxford or Cambridge.

position of these two institutions was maintained in large part by their unwritten exclusive right of educating the political, administrative, ecclesiastical, and professional elite of the nation. In the case of medicine, the two universities were, as we have seen, connected with the leading medical corporations of London, whose members traditionally received their pre-professional education in "Oxbridge." Thus the centralization of academic life, which in France was the result of administrative design, was achieved in England through the more subtle functions of a class system, in which academic institutions like people "were kept in their place" through internalized traditions and networks of semi-formalized bonds among persons, groups, and independent organizations.

The United States, then, provides a case similar to the German, where competition within a decentralized system encouraged the establishment of specialized research roles and facilities. The usefulness and the necessity of such roles and facilities in the clinical field were not yet generally recognized at the turn of the twentieth century (in spite of the already existent German examples), and there was strong resistance against them in Britain as well as in the United States. At this time, like medical scientists (or natural scientists) in general, clinicians were still conceived as primarily practitioners and only secondarily as scientists. Thus the problem of transforming the practitioner-amateur scientist role into a scientific career in the clinical field was similar to the earlier problem of the creation of scientific roles in general. At this stage as in the previous one, competition was the decisive factor in the emergence of the new career.

Conclusion

THE continuous growth in the curves of German discoveries during the middle decades of the nineteenth century and in the American curves starting from the 1880s is thus attributed to the extent to which these societies exploited, through enterprise and organizational measures, the possibilities inherent in the state of science. They were quicker than France and Britain in the recognition of new disciplines, the creation of specialized scientific jobs and facilities for research, and the introduction of large-scale systematic training for research. They were also quicker to abandon traditional notions which had lost their usefulness. None of these conditions alone could have sustained scientific growth for a long period of time. It was no coincidence, however, that they went together,

46. This is the subject matter of A. Zloczower, "Career Opportunities and Scientific Growth in Nineteenth Century Germany with Special Reference to the Development of Physiology," unpublished M.A. thesis, Hebrew University, Jerusalem, Israel, 1960.

since a common underlying factor, competition, determined the crucial decisions concerning all of these conditions in the two decentralized systems. Successful scientists were rewarded with university chairs and facilities. Their success encouraged others to take up science and, incidentally, transformed the pursuit of science into a regular professional career; it created pressure for further expansion of facilities and training, and exposed the inadequacies of out-of-date traditions.

This interpretation of the curve of scientific discoveries, according to which their growth was due to increased opportunities for entering research careers (and not, for example, to better selection of scientists), is also consistent with the differences between the countries shown in the second index based on the numbers of discoverers. As pointed out earlier, beginning in 1835 in Germany and in 1860 in the United States, the growth in the numbers of those entering upon scientific careers became continuous, while in France and Britain there were fluctuations over the whole period. Continuous growth represents a situation in which research becomes a regular career; fluctuations, a situation in which research to a large extent is a spontaneous activity engaged in by people as the spirit moves them.

In conclusion, some of the implications and problems raised by the existence of a positive relationship between scientific productivity and academic competition may be noted. According to the present explanation, this relationship is due to the impetus provided by competition for entering promising but undeveloped fields of research. This, however, suggests that the growth of discoveries in any field may be limited by the capacity for expansion of the institutional framework (jobs and facilities), a suggestion which seems to be worth further exploration.[46]

Another question concerns the *quality* of the impetus given to science by competition. The present hypothesis suggests that competition increases the gross amount of discoveries of all kinds through the thorough exploitation of potentially fruitful fields of research. It says nothing about the conditions conducive to the creation of fundamentally new ideas, and it is quite possible that the social conditions that stimulate basic innovations differ from those that facilitate the exploitation of fruitful ideas already discovered.[47]

Finally, nothing has been said about the conditions that maintain scientific competition. Political decentralization gave rise to competition in Germany, and political decentralization enhanced by private financing and administration of higher education led to competition in the Uinted States. It is not argued, however, that competition is the only possible outcome of any state of decentralization, or that com-

47. *Cf.* Joseph Ben-David, "Roles and Innovations in Medicine," *American Journal of Sociology,* 65 (May, 1960), pp. 557–568.

petition, once established, is self-maintaining. Decentralization may lead to collusion or mutual isolation as well as to competition; and competition may be replaced by either of these alternatives. Determination of the general conditions that ensure competition, therefore, is another problem which needs further study.

21 *Creativity and the Scientist*

MORRIS I. STEIN

FOR THE PAST FOUR YEARS my staff and I at the University of Chicago have been investigating the relationships of both psychological and sociological factors on the creativity of industrial research chemists. During the course of this period the research has been supported by the Research Division of Armour and Company, the United States Public Health Service, and the Sub-Committee on Research Personnel of the Industrial Research Institute.[1] We have studied three large industrial research organizations, have obtained data on several hundred scientists, and are in the process of completing intensive studies of forty-six individual scientists who consitute our present population. The data that I shall present later in this paper are based primarily on this last group of men. The purpose of the research is to conduct a basic study of the problem of creativity in an effort to increase our knowledge and understanding of creativity and so, in the long run, to facilitate the selection and recruitment of scientists and to eliminate some of the problems in administering research.

The purpose of this paper is threefold. First I shall present a discussion of the theoretical assumptions that have guided our research. This will be followed by a sociological analysis of the roles that the scientist in industry is expected to fulfill, and then I shall conclude with a brief discussion of the psychological factors that we have found to be related to creativity in our subject.

Before starting on the body of my paper, I should like to caution you about two specific factors. First, the sociological and psychological data obtained in one culture need not necessarily apply to the prob-

1. The research is presently continued under Dr. Stein's direction at the Research Center for Human Relations, New York University.

lems of another culture. While we in the United States may share a great many things with people in England, there are also a good many ways in which we differ from each other. Consequently, neither all my techniques nor all my results may apply to English scientists. I, for one, am curious to learn just where they do and where they do not overlap. Secondly, the data that I shall present are based on chemists. We have not studied physicists, mathematicians, biologists, or other scientists, and within chemistry we have not studied all its branches. I make mention of this because we do know that there are differences between the organic and the physical chemists and we believe that there will be differences among the other scientific groups also. The scientific specialty of the subjects we studied, therefore, limits the extent to which the specific results may be applied to other scientific groups. Nevertheless, I am certain that the variables I shall discuss will be among the crucial dimensions in any study of scientific creativity. With these two cautions in mind, I shall now turn to the body of my paper.

Theoretical Background

METHODOLOGICALLY, creativity has been studied from two major viewpoints—sociological and psychological. The sociologists tend to argue from such data as the simultaneity of invention,[2] the value system of the culture,[3] and patterns of cultural growth,[4] that there are factors in the environment that facilitate or obstruct creative developments.[5] In this frame of reference the individual and his psychological characteristics are de-emphasized, for it is assumed that the social forces have broad enough tolerance limits to allow for individual differences. In contrast to the sociological viewpoint, the psychologist looks to forces within the individual; he concentrates on such factors as intelligence, personality, and attitudes and studies their relationship to creativity. An implicit assumption in the psychological framework is that the individual is an "alloplastic" organism, that it can alter its environment and that it can actualize its own needs and potentialities. Consequently, psychologists working in this area tend to overlook the social milieu in which the individual creates, although they may utilize such factors in interpreting their results.

Both the sociological and psychological approaches have each yielded

2. W. F. Ogburn and D. Thomas, "Are Inventions Inevitable?," *Political Science Quarterly*, 1922, 37, 83–98.

3. H. G. Barnett, *Innovation: The Basis of Cultural Change*, New York: McGraw-Hill, 1953.

4. A. L. Kroeber, *Configurations of Culture Growth*. Berkeley: University of California Press, 1944.

significant information for an understanding of the creative process. But few, if any, studies have utilized both approaches in the same research, so that our knowledge of the relative contributions of both sociological and psychological data in a single situation is quite limited. It is our hope that by the time our research is completed we shall be able to suggest some answers on the relative contribution of both sociological and psychological factors to scientific creativity, since we study the environments in which our men work as well as their individual personalities.[6]

On the basis of our present knowledge of the social sciences we have selected three assumptions that are basic to our undertaking. These are:

1. Creativity is the resultant of *processes* that occur within the individual. In general one tends to judge the creativity of others in terms of the "products" that they have produced, or, stated differently, in terms of the "distance" between what they have produced and the status of the field before they came on the scene. Such an orientation causes us to overlook the fact that creativity is a process. It is a process of hypothesis formation, hypothesis testing, and the communication of results. Creativity may be manifest in any one or all of the aspects of this process. Some people are "creative idea men"; others may not be able to generate the ideas but they are quite creative in developing the means for testing them; finally, still others are creative in the manner in which they present ideas or findings to others. And, to be sure, there are individuals who are "high" in all aspects of the process.

2. Creativity is the resultant of *processes* of social transaction. Individuals affect and are affected by the environments in which they live. They do not interact with their environments without changes occurring in both directions. The early childhood family environment transaction pre-disposes the individual to creativity or sets up intrapsychic barriers to creativity. Later, adult environment transactions similarly encourage or inhibit creativity.

3. For purposes of empirical research our definition of creativity is as follows: Creativity is that process which results in "a novel work that is accepted as tenable or useful or satisfying by a group at some point in time."[7] By virtue of this definition we limit ourselves to studying individuals who are regarded as creative by significant others in their environment. Some of you may regard this definition as "too social," but I submit that almost any criterion in this area has its roots

5. These factors have been discussed at greater length in my paper entitled "The Cultural Context of Creativity" (mimeographed).

6. A further elaboration of this approach may be found in G. Stern, M. I. Stein, and B. Bloom, *Methods in Personality Assessment*, New York: The Free Press, 1956.

7. M. I. Stein, "Creativity and Culture," *Journal of Psychology*, 1953, vol. 36, 311.

in the judgments of others.[8] With these assumptions in mind, let us now turn to the environments in which our subjects work.

The Industrial Research Chemist's Environment and His Roles

OUR subjects are industrial research chemists. We regard the companies in which they are employed as subcultures or subsystems within the broader culture of systems of our society. Each of the subcultures has specific goals to accomplish, a prescribed status system, a value system, a system of rewards, etc. With regard to some of these factors each subculture may be similar to others, but in other respects it may differ and manifest its uniqueness and individuality. The constellation of that which it shares and that which it holds as unique constitutes its "attraction value" for individual scientists seeking employment. Thus, individuals may be attracted to certain companies because of their scientific prestige, because of the opportunities they make available for creative research, because of their salaries or social security policies, because they are located in areas in which the individuals wish to live, or for a variety of other factors. And, by the same token, certain companies may possess "negative" or "avoidance" value because of their organization policies.

In addition to its attraction value, each company may be said to have a specific "selective orientation"[9] in terms of which it selects and recruits its scientists. This is manifest when companies seek individuals with special scientific or social backgrounds that they regard as necessary qualifications for employment. It is quite critical that a company determine for itself just what its "selective orientation" is and just what its "attraction value" is. In other words, a company should spell out for itself just what its goals are, what it wants to achieve, and how it strikes others. I can well imagine that you regard this point as quite obvious, but it has been my experience that this is rarely done over a sufficiently long time span so that recruitment of scientific personnel for creative research can occur in a systematic fashion. At the present time there seems to be something of a fad to look for creative individuals, but relatively little attention is being paid to developing crea-

8. In our research we speak of "manifest creativity" and "potential creativity." The former refers to that which is regarded by the significant others as creative. The latter refers to those individuals who are not now regarded as creative but who on our various psychological tests appear very similar to those who are regarded as creative. For these individuals, we assume that there may be certain factors in their environments that may account for this. This we hope to test in the future.

9. T. Parsons and E. Shils, *Toward a General Theory of Action*. Cambridge: Harvard University Press, 1951.

10. This area has been discussed at greater length in my paper entitled "On the

tive environments. If word gets around among prospective scientists that Company X does not permit its scientists the freedom of selecting their own projects, then it will have low attraction value for top-level persons. If the company has analyzed its needs and knows precisely what it is looking for, then, when the right man comes along, he can feel that he is going to make a significant contribution to the company's future and he will be motivated to do creative research. Therefore, my suggestion to you is that you analyze your own situation carefully so that you have a clear picture of the kind of man you want before you start looking for creativity.

Once the scientist is within the employ of the company, he has several roles to fulfill. For the industrial research chemist there are four roles—the scientist, the professional, the employee, and the social role.[10] I shall discuss each of them in an effort to specify the demands that are made upon our men.

THE SCIENTIST ROLE

As the "scientist" the industrial research chemist, like all scientists, is expected to "discover, systematize, and communicate knowledge about some order of phenomena."[11] In his role as scientist the individual undertakes activities not because they will be of benefit to anyone who may be considered his client, but because they will result in more knowledge. "Scientists, in the purest case, do not have clients."

In fulfilling his role as scientist, the individual conforms to the ethos of modern science, which, as described by Merton, involves the following four institutional imperatives or constraints:

1. *Universalism.* The source and claims for truth are to be subjected to "pre-established impersonal criteria."[12]

2. *Communism.* This refers to the fact that "the substantive findings of science are a product of social collaboration and are assigned to the community. They constitute a common heritage in which the equity of the individual producer is severely limited.[13]

3. *Disinterestedness.* Science demands objectivity and has no place for the personal and subjective motivations of the individual.

4. *Organized Scepticism.* This last institutional imperative involves "the suspension of judgment until 'the facts are at hand' and the detached scrutiny of beliefs in terms of empirical and logical criteria. . . ."[14]

Rôle of the Industrial Research Chemist and Its Relationship to the Problem of Creativity" (mimeographed).

11. E. C. Hughes, "Psychology: Science and/or Profession," *American Psychologist,* August, 1952, 7.

12. R. K. Merton, *Social Theory and Social Structure,* New York: The Free Press of Glencoe, 1949, 1st ed., p. 309.

13. *Ibid.,* p. 312.

14. *Ibid.,* p. 315.

THE PROFESSIONAL ROLE

Overlapping with the scientist role is the professional role. As a professional the industrial research chemist has been trained in a specific tradition and "only members of the profession are treated as qualified to interpret the tradition authoritatively and, if it admits of this, to develop and improve it."[15] This statement holds true for both the professional and the scientist. What distinguishes them is that while the latter is concerned primarily with increasing knowledge and in communicating with his colleagues, the professional earns his livelihood by giving what Hughes has called "esoteric service" to a client.[16] The client for the industrial researcher is "the company." By accepting a position with a company, a researcher both implicitly and explicitly accepts the task of working on problems related to the products that the company produces. But the company is not only the researcher's client, it is also his patron in that it provides him with the financial security, the equipment, the personnel, etc. to carry out his work. It is this client-patron role, the company vis-à-vis the researcher, that puts certain restraints on the fulfillment of the scientist's role for the industrial researcher and is manifest in the following:

1. *Limited Communism.* While the scientist role demands that procedures and results should be shared with the scientific fraternity, the professional role demands that they are to be shared only with certain selected individuals whose number may vary from none to many but never outside the company. To be sure, this is a function of time, since once a company has secured the patent rights to a process or product, it may then permit its employees to discuss it. Until such a time, even papers to be presented at scientific meetings often have to be "cleared" by the company's patent office, to protect the company's interests.

2. *Focused Truth.* Following the institutional imperatives of the scientist role, the individual is free as well as obligated to pursue the problems and unknowns that arise in the course of his work. He need not encumber himself with the artificialities or practicalities of the differentiation between pure and applied research. In the industrial system, however, the goal of each man's work is to be focused on the product or products that can be produced and sold by the company to the consumer. Furthermore, the best possible product need not be developed at one time, since there are always possibilities for "new and improved" products.

3. *Selflessness.* The researcher who has many ideas has to be capable of yielding them to others. This decision is often arrived at by both the administrator of research and the researcher himself, although it is pos-

15. T. Parsons, "A Sociologist Looks at the Legal Profession," in *Essays on Sociological Theory* (rev. ed.). New York: The Free Press, 1954, p. 372.

sible that the administrator alone may assume such responsibility. At such times it is often necessary that the researcher withdraw any self-involvement in his ideas, even though, if they do not work out, his reputation may suffer when the people to whom his ideas are assigned make mistakes that he might not have made and invalidate his ideas. The need for selflessness also exists when a project or problem has been completed. At such times the products of a man's labor are sent to the pilot plant and additional alterations may be made.

4. *Communication with Lay Personnel.* The industrial researcher, by virtue of his client-patron relationship with nonscientific personnel who are in decision-making positions, must be able to communicate with them in nontechnical terms. Emphasis on this type of communication may occur early in the research process where the researcher need to convince management of the value of his ideas, as well as at the end of the research when his efforts must be condensed into the ubiquitous "one page or less" so that management thinks that its investment has been a wise one.

5. *Vested Interest.* The professional role demands that the industrial researcher be loyal to and maintain the interests of his company, his division, department, section, or work group. At scientific meetings or in contacts with customers, the industrial researcher is expected to support the vested interest of his company. He must keep his eyes and ears open to see how the company can be of greater value to its customers and how it can maintain its position among and/or surpass its competitors.

THE EMPLOYEE ROLE

The third role for the industrial researcher is the employee role. It is to be distinguished from the two previous roles in that *their* adequate fulfilment adds *new* information to the system, while the performance of the employee role pertains to the flow of already existing information and to the maintenance of the system as an ongoing enterprise. The industrial researcher shares this role with others in the company. Some of the factors involved are:

1. *Consistent Productivity.* The man on the job must produce with some degree of consistency. To be sure, it is expected that the consistency will be a function of the difficulty of the problems that the man is investigating. But even on the most difficult problems he is expected to show progress in the course of his work.

2. *Financial Awareness.* From the planning stage through the production stage, research and development cost money. The time it costs to sit and think is budgeted and charged for like equipment. The researcher must always be aware of the costs of his activities, as well as

16. Hughes, *op. cit.*

be concerned with whether or not his results will "ring the cash register."

3. *Efficiency.* Since time, equipment, and personnel are costly, the industrial research chemist is expected to be quite efficient in all his undertakings. The best idea and working procedure is one that results in a novel product that requires a minimum of retooling and reallocation of personnel and functions. Those ideas which require much shifting about of personnel may generate morale problems and those ideas which require the expenditure of large amounts of money may involve many groups of individuals in the decision-making process who must give their approval on the problem.

4. *Accepting Status Position and Adjusting to Authority.* While in his research efforts the industrial researcher may be iconoclastic and even defiant of authority—i.e., explore areas in which the authorities in his field say that certain things are impossible—the employee role demands that he accept the limitations and circumscribed power assigned to his status position. He has to go through channels and work through others in more powerful positions in order to get what he wants. If he is openly defiant of those above him, he may well jeopardize his position. He has to learn to adjust to them or to get around them without too frequent or open conflicts.

5. *Regularity and Flexibility.* Although the industrial researcher may work independently on his own research and set his own rules in this area, he is nevertheless part of a working community and he must abide by the rules and regulations that affect the total working community. He must attend his job regularly and be there for the prescribed working hours. He may not have the opportunity to pursue a "hot idea" after "closing time," for to do so may require special permission and clearance from the safety engineer and the night watchman. The researcher is required to keep accurate records of his research efforts, because of patent office requirements, and also for cost-accounting purposes so that the total cost of the development of a product may be calculated. Within this emphasis on regularity there is the emphasis on flexibility. The research man himself may come up against a problem on which he requires the efforts of another person, or someone else may come to him for aid, or a problem may arise in production that involves a previous problem of his; and for these and other interferences the researcher has to be in a position where he stops what he is doing for a reasonable amount of time to help them.

THE SOCIAL ROLE

The social role refers to the behavior patterns that an individual is expected to manifest in his interpersonal relationships with superiors, colleagues, and subordinates. The individual's social role varies as a function of his position in the company's status hierarchy. The higher the status position, the more immunities and privileges accrue to him; and he might even be able to alter the role so that it is more congruent with his own personality. At lower levels in the status hierarchy, however, the individual may feel that it is impossible to alter the role to suit his needs.

The social role differs in one very critical aspect from the roles considered previously. For the scientist, professional, and employee roles there are usually either written or verbalized codes and regulations with which the individual may acquaint himself. But for the social role the prescriptions are not codified and not verbalized. One learns about them through personal experience, or the individual may be informed about them by close friends. At times when they are verbalized in a professional discussion they may be denied, for the social role includes the "irrational" factors in the social process with which the scientist does not want to concern himself, especially since he may not be too adept at fulfilling them. Yet fulfilling the social role adequately is a prerequisite for establishing smoothly functioning communication networks that facilitate one's work and often gain for him the *opportunity to be creative*—a factor that has often been overlooked in research in this area.

The discussion of the social role that follows is not intended to be all-inclusive, nor does it attempt to account for all the nuances and variations that occur as a function of time in the company or status position. Indeed, the frame of reference for this analysis is the hypothetical individual who "succeeds"—i.e., rises in the administrative hierarchy. While no single individual may fulfill all the requirements, the more successful ones are able to fulfill most of them.

The expectations with regard to social behavior are:

1. The industrial researcher is to be assertive without being hostile or aggressive.

2. He is to be aware of his superiors, colleagues, and subordinates as persons; but he is not to get too involved with them as persons.

3. He may be a lone wolf on the job; but he is not to be isolated, withdrawn, or uncommunicative. If he is any of these, he had best be creative so that his work speaks for itself.

4. On the job he is expected to be congenial but not sociable.

5. Off the job he is expected to be sociable but not intimate.

6. With superiors he is expected to "know his place" without being timid, obsequious, submissive, or acquiescent.

7. But he is also expected to "speak his mind" without being domineering.

8. As he tries to gain a point, more funds, or more personnel, he can be subtle but not cunning.

9. In all relationships he is expected to be sincere, honest, purposeful, and diplomatic, but not unwilling to accept "shortcuts," be inflexible, and Machiavellian.

10. Finally, in the intellectual area he is to be broad without spreading himself thin, deep without being pedantic, and "sharp" without being overcritical.

The scientist, the professional, the employee, and the social roles are, then, four of the roles of the industrial research chemist as we have observed them.[17] There are differences both between and within companies in the extent to which any one or a combination of these roles are emphasized. Unless we are aware of the character of the roles that our subjects are expected to fulfill, significant factors that contribute both the necessary and sufficient conditions for understanding and predicting creativity are apt to be overlooked. For example, if a company is one in which a high value is placed on "basic" or "pure" research and a man is expected to fulfill scientific-professional roles, then the constellation of factors that would predict creativity in this environment would have greater weights placed on values for theoretical factors, his capacity to deal in abstractions, and his independence than would be the case in a company where there is greater emphasis on professional-social roles. I wish to emphasize this point, for while our research has indicated that there are many similarities in the personalities of the creative men who are employed in different companies, there are also crucial differences. Therefore, the combination of tests that predict creative performance in one company may be different from that which predicts creative performance in a second company. With this in mind, I shall now turn to some of the psychological factors that we have found to differentiate the more from the lesser creative chemists.

Psychological Factors

UP TO this point in my discussion I have considered only the environmental side of the creativity issue. I should now like to turn to a consideration of the psychological factors that are related to creativity. To bridge the gap between what I have said above and what is to follow, let me state the issue that confronts us. Assuming that an individual is in a situation where the emphasis is on scientific creativity and he has already achieved the opportunity to be creative,

17. A fifth role, the administrative role, which would likely complete the list, is not considered here.

what are the necessary psychological characteristics that he should possess in order to be manifestly creative?

To answer this question, we concern ourselves both with the man's past history and his present psychological status. Some of our hypotheses in this area are:

1. The creative individual may be characterized as a person who is oriented to growth. He is an autonomous person who seeks self-expression and is not inhibited either by submissiveness to authority or by internal conflicts that limit his inquiries and inhibit his capacity to determine the correct solutions to his problems.

2. The creative individual has experienced complex interpersonal relationships early in life, which he has resolved by detaching himself from others, by relying on his own capacities, and by solving his problems through the use of his intellectual faculties.

3. The creative individual is a more differentiated person whose behavior is determined by his own value-hierarchy.

4. The creative person has a more positive picture of himself and is capable of tolerating the ambiguity that is involved in long-term research.

Now let us turn to the empirical work and results. Our subjects, as I stated, are industrial research chemists. They were selected for us by their superiors, who reliably ranked them on creativity as I defined it previously. These rankings were corroborated by ratings from their colleagues and subordinates. Each man was studied with a variety of psychological tests that were either designed or selected to yield information on the variables with which we were concerned. Sixteen men were studied in one company, and thirty men were studied in another. Approximately two days were spent with each man either in individual or group testing procedures. Our data are broken down in terms of the results obtained from the "more" and "less" creative individuals within each company, and then the findings are collapsed so that we can study both company (subculture) differences as well as "more" and "less" creative differences independent of companies. The results that I shall present are those which are independent of differences that are obtained both within and between companies.

BIOGRAPHICAL RESULTS

The past histories of our men were studied with a series of biographical questionnaires. In one of these questionnaires the men were asked to recall their attitudes to a variety of persons during childhood. The results indicate that the more creative individuals felt more distant from their fathers, mothers, and adults in general than did the less creative individuals. They also say that their mothers were less consistent in their behavior toward them, and there is a trend that suggests that adults in general were similarly distant. The more creative subjects say they identified less; that is, they wanted to be less like their mothers than did the less creative individuals, and there is a trend for a similar result to hold up with respect to fathers. In the area of group activities, the more creative men say they enjoyed group activities less than did the less creative men, and there is a trend that suggests that they enjoyed solitary activities more and enjoyed competitive and cooperative games less than their less creative colleagues. Finally, the less creative subjects tended to play more with friends who were younger than themselves than the more creative subjects did. These findings suggest that the more creative individuals were more detached from their immediate families and more isolated in their social relationships. It is assumed that they found self-satisfaction by turning to the area of inanimate objects and abstract relationships such as one finds in science.

These results are samples of the biographical data we have obtained. Let us now turn to data related to our subjects' present psychological status.

RESULTS RELATED TO PRESENT PSYCHOLOGICAL STATUS

Intelligence

An individual's intellectual capacity has often been associated with the capacity to do creative work. The test that we used here was one that is based on the capacity to deal with verbal analogies. For example, the man is asked to complete an item such as: fruit: orange–animal:———.The results on this test indicate that there is a trend for the more creative men to do better on this test of verbal intelligence than the less creative men. But the difference is not statistically significant.

Energy Level

To investigate certain aspects of our subjects' energy level, they were asked to write a brief phrase at their normal pace and then to write the same phrase as rapidly as possible. The data indicate that there is a trend for the more creative subjects to have a more rapid normal pace than is true of the less creative subjects. When both groups were asked to write as rapidly as possible, there were no significant differences. It is interesting to note, however, that there is a significant difference between the two groups when one compares the differences between the two conditions. The less creative men improve more than do the more creative men. The result here suggests the possibility that even under normal conditions the more creative men work closer to their peak pace than is true of the less creative men.

Reactions to Authoritarianism

From the observations of scientists generally and from an analysis of their case studies, it may be hypothesized that it is characteristic of the scientist not to be submissive of authority, nor is he one to follow tradition blindly. To test this hypothesis, the "F-Scale," a test measuring authoritarian attitudes was administered to our subjects; and the data indicated that the more creative men were less authoritarian. The lesser creative men score higher on this test and therefore may be said to be more submissive to authority and more acceptant of tradition.

Other Personality Characteristics

Studies of the histories of creative men suggest that they may differ from their less creative colleagues in terms of three major characteristics. They are devoted to their goals and make inordinate sacrifices in order to achieve them; they integrate complex situations into simplified and meaningful new developments; and they are dynamic in that they strive for distant goals. These three characteristics might be subsumed under the single psychological variable *autonomy.* In addition to the characteristics just considered, it might be suggested that the more creative person, because of his independence (or nonstereotypy), is likely to see himself as possessing attitudes that are different from others—specifically his work group and the population at large. To test these hypotheses, a questionnaire testing attitudes related to these psychological factors was constructed, and the men were asked to indicate what their attitudes in a variety of situations were and what they thought the attitudes of their colleagues, and of people in the United States as a whole, might be to these very same factors. The data indicated that the more creative men differ from the less creative men in that they are more integrative, more dynamic, more autonomous, and see themselves as differing in their attitudes from their work groups and the general population. The one variable on which the groups do not differ significantly is devotion. The reasons for this, whether they lie in poorly constructed items or whether the variable does not really discriminate between the groups, are still not clear.

Rewards

The last area from which I shall sample our results is the area of rewards. Our subjects were asked to assume that they had just developed a major product for their company. The company had a series of twelve rewards that it wanted to give to the man, and its problem was the order in which to give them to him. He was presented with a list of these twelve rewards—four of them were in the area of science (i.e., they would enhance the man's scientific development), four of them were in the area of administration, and four of them were financial rewards. He was asked to list the order in which he would like to receive each of them. There was a fair amount of homogeneity on the order in which the men preferred their rewards. Using this rank-order procedure, both the more and the less creative men differed significantly from each other on only two rewards. The more creative men ranked the receiving of a cash bonus higher than did the less creative men, while the latter ranked the opportunity to go to scientific meetings at company expense higher. In view of the homogeneity of response in the two groups, you may be interested in the average rank order, since they may be of value in programs that are designed to motivate for creativity. The order is:

1. Make you assistant to the research director.
2. Send you to an executive training program.
3. Give you a substantial increase in salary.
4. Give you a percentage of the profits from your research.
5. Give you administrative experience in other divisions within the company.
6. Permit you to choose your own problems.
7. Give you more people to carry out your ideas.
8. Make you assistant coordinator of research, production, and sales.
9. Send you to all professional meetings, expenses paid.
10. Give you a flat cash bonus.
11. Increase your laboratory space and facilitate the purchase of equipment.
12. Guarantee your job for five more years.

The surprising thing to us in this list of rewards was the fact that scientific factors were ranked lower than we had anticipated. Administrative and monetary rewards were ranked higher than we had anticipated. On reviewing the nature of the situations in which our subjects worked, the reason for our erroneous expectation became obvious. Our expectation that scientific rewards would be more preferred was based on observations of scientists in academic institutions or nonprofit research laboratories, from which we inferred that scientific rewards were most important. When we think of the fact that our scientists selected industrial situations for their rewards it is reasonable that their preferences for certain rewards should differ from the hypotheses we suggested initially.

There is one additional point that I should like to add in this area of rewards. The data that I reported above were obtained by asking the men to rank-order the various rewards. We also obtained their rank order of rewards in another fashion. We presented each reward matched against every other reward, and the man was to select the one of the two he preferred. On the basis of this paired comparisons technique a rank order could also be established. The rank order so obtained was correlated with the rank order obtained from the check-list technique described above. Here there are differences between our more and our less creative men. The correlation between the two rank orders are +0.88 and +0.72 for the more and the less creative men, respectively. This difference is statistically significant and suggests that the more creative men are more consistent in their reward preferences than are the less creative men.

Summary and Conclusion

THE purpose of this paper has been to present a survey of the theoretical rationale, research design, and results presently underway in a study of the problem of creativity in industrial research chemists. The research is oriented to investigating both the sociological and psychological factors related to creativity in the aforementioned population. An analysis of the environments in which our men work and the roles they are expected to fulfill was presented, and the results relevant to the psychological status of the subjects were discussed. As we draw closer to a complete understanding of our subjects through the analysis of our tests and results, we hope that we will come still closer to our goal of understanding the processes involved in creativity and so facilitate the selection of creative scientists and the administration of creative research.

22 *Nine Dilemmas in Industrial Research*

HERBERT A. SHEPARD

PROBABLY NO OTHER BRANCH of industrial activity causes as much managerial unrest and uncertainty as does research and development. The concern is due largely to the enormous difficulty of evaluating the laboratory's actual and potential economic contribution to the firm. The inapplicability of traditional methods of calculation creates an interest in discovering other ways of judging whether performance is good or poor. Through this door the never-never land of opinion, myth, and prejudice is entered—a land which breeds unrest in manager and scientist alike.

It used to be said that the way to do industrial research was to hire good scientists and leave them alone. Certainly no such simple formula can be taken seriously today. But neither have we arrived at the point where discovery of the secrets of successful research and development management can be claimed. Perhaps the best that can be done at present is to impose some order in the never-never land—to identify some of the problems that make for unrest and some of the issues that need to be resolved.

The Laboratory: Department of Today or Department of Tomorrow?

THE needs of the firm as interpreted by top management exert a powerful influence on the laboratory. For public relations purposes —and it is generally held that a number of laboratories have been

---

Reprinted from Herbert A. Shepard, "Nine Dilemmas in Industrial Research," Administrative Science Quarterly, *vol.* 1 (1956), *pp.* 295–309.

1. Warren G. Bennis, "Role Conflict and Market Structure in Two Physical Sci-

founded primarily for the sake of institutional advertising—the laboratory is likely to be referred to as the "Department of Tomorrow." In many companies, however, "today" is the principal concern of the laboratory, which may be quite incapable of bringing about a different tomorrow. The potential of the laboratory may not be understood, because of inadequate representation in top management. Even where it is well represented, its scope is likely to be limited to questions of technological feasibility; hence its power as an innovative force is less than it might be.

While certain kinds of new products and processes are undoubtedly desired by most companies engaged in research and development, other requirements are likely to have high priority. Many, and in some cases nearly all, of the laboratory's activities contribute relatively little to the development of new products and processes. For example, technical assistance and "trouble-shooting" in other parts of the company take up much laboratory time. Some laboratories are required to supervise quality control activities. Defensive research must be carried on; i.e., limited resources must be devoted to keeping abreast of certain fields, not with a view to innovation, but to ensure that the company would not be far behind its competitors should one of them take an innovative step. Similarly, much effort may be devoted to imitative and substitutive work, to match competitors' products. Patents may be produced for trading purposes only. In industries where technology is complex, the laboratory may be regarded as a training ground for managers and specialists needed in other parts of the company.

Thus in industries which are highly competitive, where no firm has a large cushion of monopoly, the laboratory is needed to meet immediate competitive threats, and few resources are available for expensive innovation or long-term research. The consequences for the scientific staff of the laboratory are likely to be frustration and disillusionment.[1]

A "research-minded" top management in a company able to make the high-risk investment involved in long-term research and development provides a different environment for the scientific staff. Many companies have established research centers with campuslike surroundings and an "academic" atmosphere. The term "research-minded" is a common one in laboratories and seems to refer not only to an interest in taking a risk of this kind, and hence a willingness to make a long-term investment, but also to an understanding of and sympathy with the needs and desires of research and development personnel. Sometimes the term seems to mean an interest in research for its own sake.

Many other sets of managerial expectations also have an influence

ence Research Laboratories" (Unpublished manuscript, Massachusetts Institute of Technology, 1953). See also Bennis, "The Effect on Academic Goods of Their Market," *American Journal of Sociology,* 62 (July 1956), 28–33.

on the laboratory. Company policies and procedures, beliefs about proper methods of organization, stereotypes about research and scientists, and comparisons between the conduct of laboratory personnel and other company personnel draw certain patterns of response from the research and development staff. All such outside forces influence the internal structure of the laboratory organization and affect the attitudes and activities of research personnel and the contribution of the laboratory to the firm.

The Scientist: Local or Cosmopolitan?[2]

COMPANY policies and practices are usually intended to ensure a prior interest in the company's welfare on the part of every employee, to lead him to identify his own success with the company's. Salaries, fringe benefits, personnel programs, and the implied threat of punishment for disloyalty are employed to this end. The scientist may, however, identify himself primarily with his professional group. Professional and scientific values emphasize the importance of contributing to knowledge, and prestige is granted on this basis. The scientist is supposed to be guided by intellectual curiosity. His motto is, "How much do we know about this?" whereas the businessman's motto is, "What is the value of this to the company?"

These conflicting forces are disruptive not only of laboratory-company relations, but also of relations within the laboratory. The research staff itself is likely to be divided into what Robert Merton calls the "cosmopolitans" and the "locals."[3] The former are oriented toward success as members of their profession, and their interest in the company is limited to its adequacy as a provider of facilities for them to pursue their professional work. Since they are productive, they may be valuable to the company, but such value is an almost accidental by-product of their work. The locals are good company men, but their interest is likely to be less in their work than in their advancement in the company. They may therefore be of less value to the company than the cosmopolitans. This state of affairs creates a dilemma for management. The locals are likely to be more observant of company policies and procedures, create fewer "human relations problems," and in all disciplinary respects be good employees. The cosmopolitan is likely to be a

2. The following four sections draw upon a number of sources, principally Lowell W. Steele, "Problems of Integrating Scientific Research and Industry" (unpublished doctoral thesis, Massachusetts Institute of Technology, 1952); Warren G. Bennis, "The Social Science Research Organization: A Study of the Institutional Practices and Values in Interdisciplinary Research" (unpublished doctoral thesis, Massachusetts Institute of Technology, 1955); and R. K. Merton, *Mass Persuasion* (New York, 1946).

3. Robert Merton, "Patterns of Influence: A Study of Interpersonal Influence of

"problem person," without whom the organization would function more smoothly.

The Results of Research: Guarded Secrets or Advertised Achievements?

THE company's interest in a research program depends in part on the prospect of profit resulting from the lag between itself and its competitors in the introduction of a new product or on the prospect of a more permanent monopoly. In companies which owe their past success in part to the careful protection of "trade secrets," there is usually a desire to keep the laboratory activities under wraps. The scientist's professional standing depends, however, on publication of his achievements. Even where secrecy is not the issue, management may deplore the cost in professional time and laboratory expense required to prepare material for professional publication or for presentation at meetings of professional societies. Despite C. F. Kettering's dictum to the effect that locking the laboratory door excludes more than it keeps in, and despite the importance of establishing a good scientific reputation for the laboratory as a recruiting device, relatively few managements actively encourage publication of research results.*

The Budget: Company Time or Research Time?

COMPANIES operate on annual budgets, but research and development projects are likely to go on for years without producing useful results. It may take a long time to investigate a research problem even to the point of estimating the probabilities of eventual solution, and this is a source of strain between scientist and businessman. Management is frustrated by its inability to determine whether progress on a project is as good as could be expected; the research staff is frustrated by management's inability to understand the nature of the technical problem. This communication gap can lead to demoralization of the research staff and to loss of confidence on the part of management. Or it may lead to more complicated states of affairs. For example, the laboratory may report only on those projects in which it can show the kind of

Communications Behavior in a Local Community," in Paul F. Lazarsfeld and Frank N. Stanton, *Communications Research,* 1948–1949 (New York, 1949). Pelz and associates use the terms "institution-oriented" and "science-oriented" to make a very similar distinction. See Donald C. Pelz, Glen D. Mellinger, and Robert C. Davis, *Human Relations in a Research Organization,* I and II (Ann Arbor, Mich., 1953).

* For a recent survey of managerial practices, see *Publications of Basic Research Findings in Industry,* 1957–59, National Science Foundation, Washington, D.C., 1961. [Eds.]

progress that management understands, other research activities being "smuggled."

In another way as well, difference in time perspective produces conflict. To the company, a new competitive threat, a delay in production, a dissatisfied customer are problems demanding immediate attention. To the research worker with his heart in a long-term project, such problems are irritating interruptions, and if he is frequently called upon to "put out fires" he is likely to become resentful and suspect the sincerity of management's support of his work.

Authority: Delegated or Shared?

TWO organizational traditions, the professional and the industrial, meet in the laboratory. Internal organization of the laboratory will be discussed in greater detail below, but certain conflicts between the two traditions can be pointed out here. Industry respects certain kinds of orderliness in productive organization, and takes them as evidence of efficiency. Thus centralizing services, controlling hours of work, budgeting time, controlling expenditures and decision making by graduated delegation of authority from top to bottom, and many other practices are regarded as elementary principles of good organization.

The scientific and professional organizational traditions are based on assumptions that are different from those of industry in some respects. In the first place, power is supposed to be exercised not from top to bottom, but sideways. Achievement is evaluated by "the weight of scientific opinion." Unless a colleague's competence or honesty becomes suspect, he is expected to make all decisions relating to his work for himself. The idea of a boss is an anathema, as are the other external controls imposed by industrial methods of organization—in fact, they are held to be inconsistent with the basic tenets of professionalism. "Professional people work as 'senior' and 'junior' rather than as 'boss' and 'subordinate' . . . the professional thinks of [a project] as a group project, with each member of the team having independent responsibility vested in him."[4]

Most laboratory managements make certain gestures in recognition of the scientist's "rights" as a professional. For example, time clocks may be omitted, or the scientist may be allowed to incur expenditures up to a certain amount on his own authority. Nonetheless, there is continual pressure from management for more conformity to its organizational ideals, even including such matters as neatness of dress, regular

4. Peter F. Drucker, "Management and the Professional Employee," *Harvard Business Review*, 30 (March–April 1952), 86.

5. Paula Brown, "Bureaucracy in a Government Laboratory," *Social Forces*, 32 (March 1954), 259–260.

hours, and a more systematic approach to the coffee break. At the same time there is continual pressure from the research staff for more "freedom."

Organization: Project Groups or Functional Groups?

THERE is a strong tendency for laboratories to become organized "functionally" into permanent specialist groups—a process whose consequences for morale and productivity are hard to assess.[5] There is some evidence that it has a depressing effect on creativity, enthusiasm, and interpersonal and intergroup cooperation.

Many forces operate so as to divide the laboratory into specialist groups. Efficient processing of common classes of problems suggests the desirability of placing together those who have expert knowledge of a given class. The principle of putting like specialists together is quite in accord with professional organizational traditions. Just as this arrangement makes for understanding by one's colleagues, it makes for objectivity in personnel evaluation by the supervisor, by enabling him to compare his men's performance on similar tasks. Similarly, it facilitates observance of such matters as the submission of technical reports and other rules and regulations. Moreover, it permits each man to be employed at the work he does best and allows him to develop as a professional specialist. It provides him with a secure status in the company, and a clearly defined path of advancement. It makes for good communication between the laboratory and other parts of the firm, through the group's permanency and the rich background of experience it obtains on the class of problems assigned to it. It makes for reliability in budgetary forecasting; with experience, the specialist group becomes able to predict its requirements.

Functional organization has its disadvantages, however. Among the shortcomings mentioned by laboratory managers are resistance to crossing specialist boundaries and cooperating with other groups; a tendency to work on ever more specialized problems of ever decreasing significance; lack of creativity and responsiveness to challenge—the scientist becomes a specialist in knowing what cannot be done and is afraid to venture into unfamiliar fields. The permanency of groups in a functional organization may be detrimental; the author and his associates found that colleague, self, and management ratings of research and development groups fell off directly with the length of time members had been associated with one another.[6]

6. Herbert A. Shepard *et al., Some Social Attributes of Industrial Research and Development Groups* (Progress Report, Massachusetts Institute of Technology, 1954).

An alternative to functional organization is organization by *ad hoc* project teams, with specialists from various fields as members, so that each team is equipped to solve a particular complex problem requiring knowledge from several disciplines. Since many problems are of this character, the project-team approach has an immediate and obvious advantage over the functional approach. The team members' energies and interests are all concentrated on a single complex problem. Such teams often turn out to be enthusiastic, creative, and dramatically successful.

Unfortunately, organization by this method presents serious administrative and personnel problems. Project teams tend to perpetuate themselves and become specialist groups, there is often strong resistance to project termination and transfer to new groups. As team members, research workers do not achieve the high degree of specialized knowledge the laboratory requires to keep abreast of advancing technology. The project tends to rely too much on its own resources, and resists consulting the literature or other specialists. Teams sometimes ignore rules and regulations and are hard to discipline. Intergroup jealousies are frequent. Too much depends upon the leadership and technical competence of the team's supervisor; men with the necessary qualities are rare. The path of promotion is not clear; supervisors have to evaluate representatives of several disciplines, and may lack the knowledge to do this objectively; project termination causes insecurity. Incompatibility of team members can have a disastrous effect on performance.

In most laboratories, some combination of functional organization and *ad hoc team* organization is used. Each appears to have its advantages and disadvantages from the point of view of administration, productivity, creativity, and satisfaction in work. Choosing the most appropriate combination of functional and project modes of organization is not the only organizational problem in the laboratory, however, nor does its solution automatically take care of all human relations problems.

Research Management: Fulfillment or Regression?

STUDIES by Donald C. Pelz and associates[7] have emphasized the importance of leadership style for the job satisfaction and effectiveness

7. Pelz *et al., op. cit.*
8. Communication from C. D. Orth, III, in seminar, Massachusetts Institute of Technology, 1954.
9. Communication from M. I. Stein, in seminar, University of Michigan, 1955.
10. Anne Roe, *Making of a Scientist* (New York, 1953).
11. H. B. Moore and S. J. Levy, "Artful Contrivers: A Study of Engineers," *Personnel*, 28 (Sept. 1951), 152.

of research workers. Pelz's results indicate that dominating supervision produces apathy and resistance, and that laissez-faire or inactive leadership produces dissatisfaction and low productivity. An active stimulating leadership role is associated with good performance and high morale. Studies by C. D. Orth and Ralph Hower[8] point in more detail to the various roles the research supervisor has to take. He must understand the needs of groups and individuals in his organization and other parts of the company. He must be interpreter of the group's needs and achievements to higher levels of management and act as interpreter of managements' requirements to his group. He must protect the group from demands and pressures which he believes would adversely affect it. He is a source of technical guidance and judgment, of inspiration and stimulation. He must coordinate the ideas and resources of his group. Studies by Shepard and associates show the successful research group leader to be a more active social being than the unsuccessful group leader. He has many friends in the laboratory, centers his social life around the laboratory. Studies by M. I. Stein[9] suggest that the successful research group leader has certain charismatic attributes: subordinates tend to identify with him and model themselves on his pattern.

All of these observations point in the same direction: the effective research group leader is a creative, dynamic, enthusiastic person who relates easily to others. Such men are said to be rare among engineers and scientists. According to Anne Roe,[10] the successful natural scientist typically avoids interpersonal intimacy. According to H. B. Moore and S. J. Levy,[11] the engineer is usually an authoritarian who attempts to handle all problems by impersonal regulations and procedures.

David Moore and Richard Renck find that scientists and engineers typically "view the organization as confused and ill-conceived; they view management as confused and ill-advised."[12] The contempt for administrative and other nontechnical occupations is reminiscent of Thorstein Veblen and the Technocrats. Engineers have a strong desire for "status and recognition of their individual accomplishments" and also "desire to be considered an integral part of an organization."[13] The engineer "is tense. Emotions which rise are ignored, suppressed and repressed unless they are so intense that they overwhelm his self-control (which is usually pretty strong). Feelings are regarded as invalid phenomena. . . . Irritability is one of the most common expressions of the tension in engineers."[14]

12. David G. Moore and Richard Renck, "The Professional Employee in Industry," *Journal of Business,* 28 (Jan. 1955), 62.

13. G. H. Metz, "Management and the Professional Employee," *Addresses on Industrial Relations* (University of Michigan, Bureau of Industrial Relations Bulletin no. 22, 1954).

14. Moore and Levy, *op. cit.,* pp. 152–153.

This mixture of feelings and attitudes implies that the lot of the research administrator is not a happy one. The professional employee resents encroachments on his autonomy. Moreover, the emphasis on individual achievement makes teamwork hazardous. Jealousies, antagonisms, and feelings of injustice are likely to arise over such matters as authorship of reports, patents, and papers. Jealousies are likely to be accompanied by disparagement of the professional competence of the others involved, with disastrous effects on collaboration. The phenomena associated with these needs and interpersonal difficulties are the basis for labeling the professional as a "prima donna."

On the other hand, some observers report that the professional works well on a team.[15] Paula Brown stresses the research workers' earnest attempts to govern their behavior rationally.[16] Shepard stresses the free exchange of task-oriented information in the determination of status and the maintenance of cooperative relations.[17] Moore and Levy note that their engineers "seek good relations, smooth interchange and workable friendships with others, and will push their feelings back into unconsciousness if these appear."[18] The major implication of these observations appears to be that the reward system should be sensitive to and accurate in response to individual task-achievement and that the control system should permit a high degree of individual and group autonomy in technical decision making.

Rewarding research workers for good performance presents a difficult administrative problem. In American industrial society, success is measured largely by the size of the organization one controls. But not all successful engineers and scientists are temperamentally suited for supervisory or administrative work, nor would they, as creative research workers, be most profitably employed in such work. Scientists and engineers are themselves somewhat ambivalent about organizational advancement. On the one hand, it may be regarded as the only generally accepted evidence of their value. On the other hand, they fear loss of professional status and competence. Even laboratory officials who have been many years away from the bench are likely to "keep their hand in" by working on some technical project. While this indulgence may be to some extent sentimental, it is also a means of preserving professional self-respect.

Many laboratory managements have attempted to solve their own dilemma and the successful scientist's dilemma by establishing a "dual hierarchy." One way up is promotion through the ranks of supervision and management. The other way up is through special privileges, au-

15. Drucker, *op. cit.*
16. Brown, *op. cit.*
17. H. A. Shepard, "The Value System of a University Research Group," *American Sociological Review*, 19 (Aug. 1954), 460.

tonomy, high salaries, and impressive titles (research associate, senior scientist, and so on) which can be obtained without incurring supervisory responsibilities. Unfortunately the dual hierarchy rarely succeeds in achieving its goal. Sometimes management cannot avoid the temptation to "shelve" someone whose ability to contribute has waned by giving him such a title—thus ruining the title for others. Even under the best of circumstances, the promotion to "senior scientist" is at once a symbol of achievement, and of managerial incompetence.

Nonprofessional Workers: Caste or Class?

IN MUCH that has been said about the scientists and engineers in the foregoing pages, a certain snobbery is implied. Frustrated in his desires for recognition, enforced to conform to regulations that he considers inappropriate for a professional, the scientist or engineer is quite likely to be defensive about his status. Moore and Renck find that "negativism and a tendency to strike out almost indiscriminately at all aspects of the work environment"[19] are associated with the low morale they find to be characteristic of scientific and engineering staffs. One aspect of the environment against which engineers and scientists rebel is the requirement that they do work which they consider "low-level" or "nonprofessional." Many regard participation in activities involving the drafting board, "hardware," and glassware as somewhat degrading, or, as they put it, "an uneconomical use of professional time." This attitude does not endear them to the nonprofessional workers. And given this attitude, any sign of incompetence in these matters on the part of the engineer wins him the enmity of nonprofessional personnel. Such enmities appear to flourish in functionally organized laboratories, where interpersonal difficulties can quickly develop into intergroup conflicts. In project organizations, on the other hand, there often develops an intense loyalty between the professional and nonprofessional members of the team. The professional sometimes regards himself as teacher and protector, the nonprofessional, as student and apprentice.

Professional Workers: Elite or Proletariat?

TO MOST of us, the term "profession" connotes service or ministry rendered out of specialized knowledge, high standards of conduct (professional ethics), and a kind of democracy as the means of social control within the profession. Altruism is the keynote. In the lay mind, and to some extent in the professional, professional values appear to stand in opposition to wealth- or power-seeking. Certainly for the members of

18. Moore and Levy, *op. cit.*, p. 152.
19. Moore and Renck, *op. cit.*, p. 60.

some professions—notably clergy and teachers—the worldly rewards have been small, and throughout the professions "commercialism" is frowned upon.

While these values are in a formal sense characteristic of professional groups, there is a good deal of variety among the professions in this respect. Moreover, there is in every profession a difference between the announced values and the values that govern in practice. The popular model of the professional visualizes a client to be ministered to, as in medicine and law. The model has to be distorted almost out of recognition if it is to include the members of the engineering department of a large firm. A kind of equalitarianism among the individualistic members of the profession is also part of the popular model: the law partnership, the medical clinic, the university faculty, and the professional society being representative organizational forms. Again, this aspect of the model is abandoned in the engineering department of a firm.

Such departures from the generally accepted model of a profession reduce the meaningfulness of the word. The criteria for professional status became blurred. A host of new "professions" have arisen: public relations, purchasing, cost accounting, personnel administration, and so forth. Even the idea of management as a profession is not unpopular, though few go as far as Mary Parker Follett when she asserts that "the business man has opportunities to lead the world in an enlarged conception of the expressions 'professional honor,' 'professional integrity.' "[20]

In fact, the professional model is at such variance with the state of affairs in industry that a discussion of professionalism would seem irrelevant here if it were not for the fact that a large proportion of scientists and engineers in industry think of themselves as "professionals." As has already been indicated, this self-definition has a number of important consequences through its effect on attitudes, expectations, and reactions to the demands of the industrial environment.

Concern about the implications of professionalism has been especially keen in relation to the issue of unionism. Professional associations in science and engineering have, with few exceptions, opposed organization of their members for collective bargaining. Accordingly, growth of unions in these professions has been slow. Many industrial scientists and engineers feel that unionization constitutes a rejection of professional ideals and responsibilities and a downgrading in their status to the ranks of labor.

But powerful economic and social forces create a pressure toward unionization. The individual engineer or scientist is no longer an influential member of his company, with a monopoly on specialized knowledge needed by the firm. He is one of a hundred or a thousand such

20. Henry C. Metcalf and L. Urwick, eds., *Dynamic Administration: The Collected Papers of Mary Parker Follett* (New York, 1942), p. 143.

specialists. The individual engineer or scientist is peripheral and replaceable. The parallel with craft unionism is striking. The highly skilled craftsman was once a proud individual, his own master or an influential member of his organization. Just as the guild was a satisfactory association for craftsmen, so the professional society has met the needs of the modern scientist and engineer. Just as the guilds were the predecessors of the unions, though not often their parents, so the professional society may eventually give way to, or at least be supplemented by, the professional union. The means used by professional associations in the past for protecting the economic welfare of their members—primarily control over entry to the profession—may not be adequate to meet the challenge of mass industrial employment.

Conclusion

THE reader can infer, and any laboratory manager knows, that imposing order on the never-never land of ambiguous relations between science and industry is a relatively easy matter if the author undertakes no more than a classification of areas of conflict and uncertainty. The really important and challenging job is to determine what can be done to improve the situation. To date, however, most of the formal studies of research organization have necessarily been of the kinds that describe how things are and draw inferences about how they came to be that way. Statistical analysis of data resulting from studies of current practice, or even from more selective observation of things as they are, is often an inefficient path of knowledge. Kurt Lewin once remarked that the theory of gravity was neither conceived nor tested by such methods—and added that the best way to learn how anything works is to try to change it.

There is a crying need for a more action-oriented approach to the study of research organization. Organizational experiments as carefully conceived and carried out as the physical experiments that are conducted in the laboratory are the obvious answer—painfully obvious, surely, to persons who make a career of theory and experiment. Every laboratory in the country unwittingly produces quantities of information bearing on organizational problems—and pours it all down the drain, through a lack of interest and skill in the design, conduct, and analysis of social experiments. There is no shortage of promising hypotheses. But to date there has been a shortage of research laboratories with a will to experiment. Perhaps the same forces that result in a limitation of the laboratory's innovative potential by defining its contribution as "technical" inhibit systematic study of the impact of personal, social, and organizational factors on the conduct of research itself.

23 *Some Social Factors Related to Performance in a Research Organization*

DONALD C. PELZ

THIS REPORT will summarize some of the highlights from a study conducted in a large government organization for medical research.[1] The organization employs some three hundred investigators who conduct laboratory research on the frontiers of medical knowledge. The organization is regarded by its employees as having an atmosphere similar to that of a large university. The research is generally of a fundamental rather than applied character, and considerable freedom is allowed the individual investigator.

As part of the study, judgments of individuals' scientific performance were made by panels of investigators. The latter were experienced scientists, nonsupervisory as well as supervisory, who were familiar with the work of the others. Where possible, each scientist was evaluated twice, once in comparison with others in his own laboratory or division, and again in comparison with others in the same scientific discipline. A particular individual might be judged by as many as fourteen other scientists, only one or two of whom would have supervisory authority over him. Essentially, this was an evaluation by one's peers. The assessors' judgments were assigned numerical scores from 1 to 9, the

---

Reprinted from Donald C. Pelz, "Some Social Factors Related to Performance in a Research Organization," Administrative Science Quarterly, *vol.* 1 (1956), *pp.* 310–325.

1. The study was financed by the National Institutes of Health, U.S. Public Health Service, and the Department of Health, Education, and Welfare. Additional analysis and writing were made possible by a grant from the Foundation for Research on Human Rehavior, Ann Arbor, Michigan. Assisting in the study were Glen Mellinger, Robert C. Davis, and Howard Baumgartel.

2. D. Marvick, *Career Perspectives in a Bureaucratic Setting* (Ann Arbor: Institute of Public Administration, University of Michigan, 1954).

scores from different judges were averaged, and where a scientist received both a laboratory and a discipline score the two results were combined.

In our analysis we wanted to study the effects of the social environment on the individual's performance. For this purpose we wanted to reduce variations in performance caused by individual factors such as basic ability, quality of training, or type of experience. It seemed reasonable to assume that such factors would be reflected in the scientist's amount of education and in his job grade. The data showed that with increase in grade, there was a definite increase in average performance; and non-Ph.D.'s scored lower than Ph.D.'s of equal rank. By adding or subtracting an appropriate constant, an "adjusted" score for each individual was obtained such that the residual variation still associated with grade or doctoral degree was no greater than 0.4 per cent. In this adjusted measure it seems likely that differences owing to individual factors have been reduced.

Science versus Institutional Values

IF SOCIAL factors can influence a person's scientific achievement, they may do so in part by affecting his motivation toward the work. To study such effects it was first important to determine whether motivational factors have any association with performance.

In a previous study of a government organization administering research grants, Dwaine Marvick[2] demonstrated the significance of certain general motivations or values. He isolated two syndromes labeled the "specialist" and "institutional" orientations. The specialist seeks the approval of a larger circle of peers wherever they may be found; the institutionalist seeks rewards within a localized institution. We wanted to know whether such orientations would relate to scientific performance.

In the present study each scientist filled out a questionnaire, one part of which listed nine factors that might be important in a job. Each person indicated how much personal importance he attached to each factor. A correlational analysis located three items which seemed to represent the specialist or (as we named it) the *science* orientation. Another cluster of items seemed to indicate the *institutional* orientation.[3]

3. The three science-oriented items were: stress on using present abilities or knowledge, freedom to carry out original ideas, and chance to contribute to basic scientific knowledge. The three institutional-oriented items were: stress on having an important job, association with high-level persons having important responsibilities, and sense of belonging to an organization with prestige in the lay community. The latter cluster might also be labeled a "prestige" orientation. The two indexes were neither positively nor negatively correlated; a person high on one might be either high or low on the other.

In an analysis of these data, R. C. Davis[4] found that the index of science orientation was significantly related to scientific performance, whereas the index of institutional orientation was not. Furthermore, it appeared that a strong science orientation went with high performance mainly when the institutional orientation was weak; strength of science orientation was not significantly related to performance when the institutional orientation was strong.

Similar results were obtained with data from another study conducted at the Institute for Social Research by S. Lieberman and L. Meltzer[5] as part of the Survey of Physiological Sciences sponsored by the American Physiological Society. In a questionnaire sent to a nationwide population of physiologists, similar questions about the personal importance of various job factors were asked. Scientific performance was measured in this instance by the number of times each person had been cited in the *Annual Review of Physiology* for the preceding three years. When a parallel index of science orientation was formed, a significant relationship was again found with performance. No relationship was found between the number of citations and the scientist's motivation toward prestige items such as advancement in his profession.

These results raise some provocative questions for research administrators grappling with knotty problems of salaries, titles, and promotion policies. The point to be stressed here, though, is that the index of science orientation does relate to actual performance and thus points toward a way in which performance might be increased. Some results to be given later suggest that science motivations may be affected by the type of leadership in the organization. It may also be possible to raise the level of science motivation through the selection of new personnel or through the kinds of recognition given.

Contact with Scientific Colleagues

A second series of analyses approached the question of whether the style of communication with one's colleagues can raise or depress the level of performance. If so, what is the optimal pattern of such contact? Should a scientist rub elbows daily with his peers, or should he be assigned a cubbyhole where he can work without distraction? If frequent contact proves worth while, should it be mainly with others of similar background, with whom he can talk the same language, or would

4. R. C. Davis, "Factors Related to Scientific Research Performance," in *Interpersonal Factors in Research, Part I* (Ann Arbor: Institute for Social Research, University of Michigan, 1954). See also, R. C. Davis, "Commitment to Professional Values as Related to the Role Performance of Research Scientists" (unpublished doctor's dissertation, University of Michigan, 1956).

5. S. Lieberman and L. Meltzer, *The Attitudes and Activities of Physiologists* (Ann Arbor: Survey Research Center, University of Michigan, 1954).

he benefit more by the stimulation of ideas from persons of different backgrounds?

To study these questions G. Mellinger[6] used nonsupervisory scientists at a senior level who were mature investigators at civil service grades of GS–12 and higher, with several years of experience.[7] On their questionnaires scientists named up to fifteen people in the organization with whom "some contact is of greatest significance to you in your work," and indicated the frequency of contact with each.

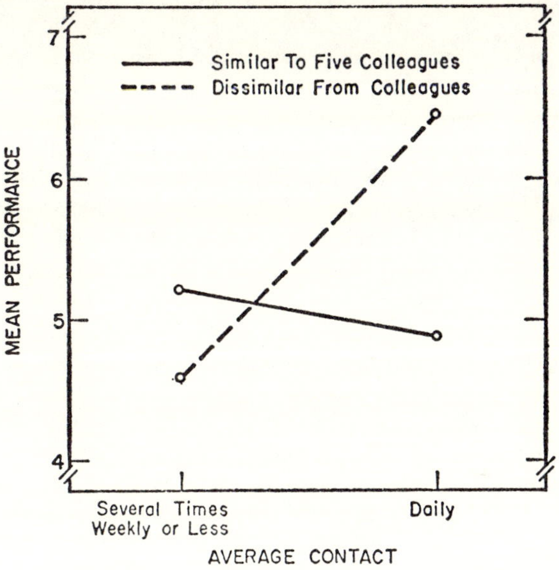

FIGURE 1

Scientific performance related to similarity to five colleagues in prior employment and to frequency of contact (at senior level).

For each respondent we selected the five scientific colleagues he name as most significant, excluding supervisors above him, technical assistants, and administrative personnel. For each colleague we computed a measure of similarity on the science and institutional orientations described above; we averaged these for the five colleagues, and also computed the mean frequency of contact. Figure 1 shows the

6. G. Mellinger, manuscript prepared for *Interpersonal Factors in Research, Part II* (in process; Ann Arbor: Institute for Social Research, University of Michigan, 1956).

7. In the following data a distinction will be made between senior- and junior-level scientists. They differ in experience and in autonomy and may require different kinds of social environment for high performance.

results when these measures are jointly related to scientific performance.

Among the four possible combinations of frequency and similarity, highest performance is found when scientists have *frequent* contact (more than several times weekly) with colleagues who are on the average *dissimilar* from themselves in values.

Increase in contact is significantly related to performance only if colleagues are on the whole dissimilar from themselves. When colleagues have similar values, more frequent contact is not accompanied by higher performance; there is, in fact, a slight decline.[8]

These findings suggest that scientists benefit by frequent opportunities to exchange ideas with persons having different values. Since the institutionally oriented persons in our study tend to lay stress on improving the nation's health, the results might mean that those who emphasize basic research can benefit from contact with those who stress applied research, and vice versa. The interpretation suggested here is that frequent contact with dissimilar colleagues stimulates higher performance. It might also be argued that abler scientists seek out those of different values. The second interpretation seems to be preferred by some scientists who do not feel that social factors can influence level of performance. The facts are, however (from data shown here), that the colleagues of abler scientists are on the average just as similar to these scientists as the colleagues of less able scientists are to the less able scientists. It does not look as though the more competent investigators deliberately seek contacts with colleagues of dissimilar values.

Another result in a parallel vein was obtained with a different measure of similarity. On the questionnaire scientists were asked about the kinds of situations in which they had previously worked; most frequently named were government, university, and hospital (including private practice). Depending on the number of such situations that they shared in common, each pair received a high to a low similarity score; and the average similarity between a scientist and his five main colleagues was computed.

The results are shown in Figure 2. We note that highest performance occurs when the senior scientist has frequent contact with five colleagues markedly dissimilar from himself in type of previous employment.

Frequency of contact is associated with significantly higher performance only for those scientists whose colleagues are dissimilar from themselves. When colleagues come from similar backgrounds, increase in contact from weekly to daily tends to go with slightly lower performance.

8. This does not mean that among this group "the fewer contacts the better." Most scientists report average contacts weekly or more often with their colleagues. Among the few isolates who report less than weekly contacts, performance tends to be low.

On the other hand, among moderately isolated scientists (who see their colleagues weekly) dissimilarity in previous employment seems to be a handicap; performance is significantly lower in comparison to those having similar colleagues.

It is likely that those who have worked previously in government, university, and hospital situations will develop different views as to the best way of approaching a research problem. Whatever the nature of

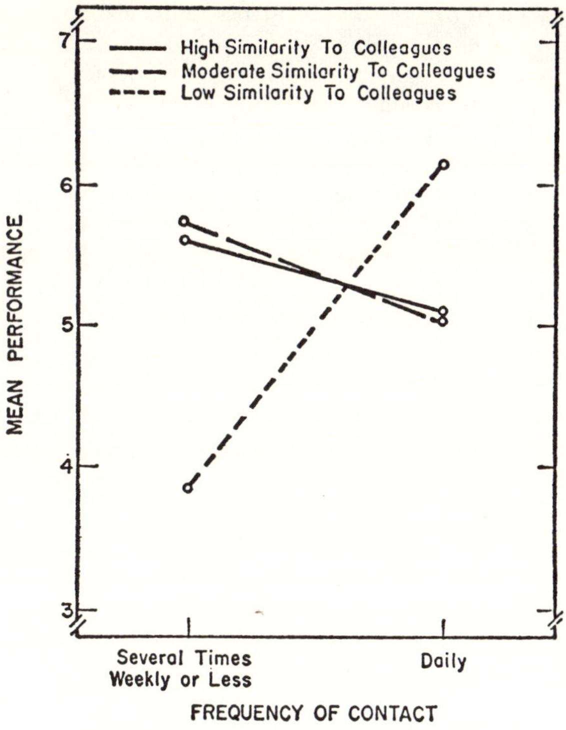

FIGURE 2

Scientific performance related to similarity in values and to frequency of contact with five main colleagues (at senior level).

these differences, it appears that they can encourage high scientific performance, provided there is ample interchance among the individuals. But diverse viewpoints may be a hindrance if they constitute the only contacts of a relatively isolated individual.

Is the similarity illustrated in Figures 1 and 2 the same thing or two different things? The correlation between the two similarity measures is in fact slightly negative (—0.15) and not statistically significant.

It appears that we are dealing with two independent kinds of similarity. What their precise nature is remains a question for future research.

Such results raise the question of scientific discipline. Will scientists benefit from close contact with others in different disciplines? Does organization of laboratories along interdisciplinary "project team" lines give better results than organization into single-discipline groups?

In our study we classified each scientist according to the major and minor disciplines he was currently using in his work. These classifications were based in part on the disciplines he named on his questionnaire and in part on the groupings established to assess performance. Similarity scores were assigned, depending on whether scientists had major fields in common, minor fields in common, or no fields in common.

The results of this analysis are not as clear-cut as the previous ones, but they point in a similar direction. In general, senior scientists whose five main colleagues differ from themselves in field of work to perform somewhat better; and this tendency is more marked with frequent contact.

The tendencies (according to data not shown here) are also sharper when the colleagues are not equally dissimilar, but vary in their degree of dissimilarity. Thus it appears that for high performance the scientist requires variety in his daily fare. Additional support of this idea will be given below.

Contacts with Single Individuals

THE foregoing analyses raised further questions. We have studied contacts with five colleagues; will similar results obtain with the scientist's one most important colleague? And what about his chief—should the scientist work under a supervisor who is in a different field? To explore such questions we again used frequency of contact and several measures of similarity. The present analysis was extended to junior-level scientists (in civil service grades GS–9 and GS–11, most of whom have doctoral degrees but limited experience) as well as the senior level used previously.

One particularly surprising result emerged, as shown in Figure 3. The measures used are similar to those in Figure 1 except that they refer to the one most significant colleague rather than to an average for five colleagues. The figure demonstrates that high scientific performance is associated with daily contact provided that the scientist's major colleague has *similar* values. By contrast, Figure 1 showed that high performance goes with frequent contact provided that contact is mainly with five colleagues who have *dissimilar* values.

How to reconcile these results? The following view seems plausible: For maximum performance it is helpful to have at least one close colleague with a similar orientation—someone who "talks the same language," with whom the scientist can air his problems and get a sympathetic hearing. But one or two such individuals are enough. To provide the stimulation of new ideas, it is important that the remaining contacts be with people of dissimilar orientation. In short, one kind of environment for high performance is frequent contact with a variety of viewpoints, a few similar, but most of them different.

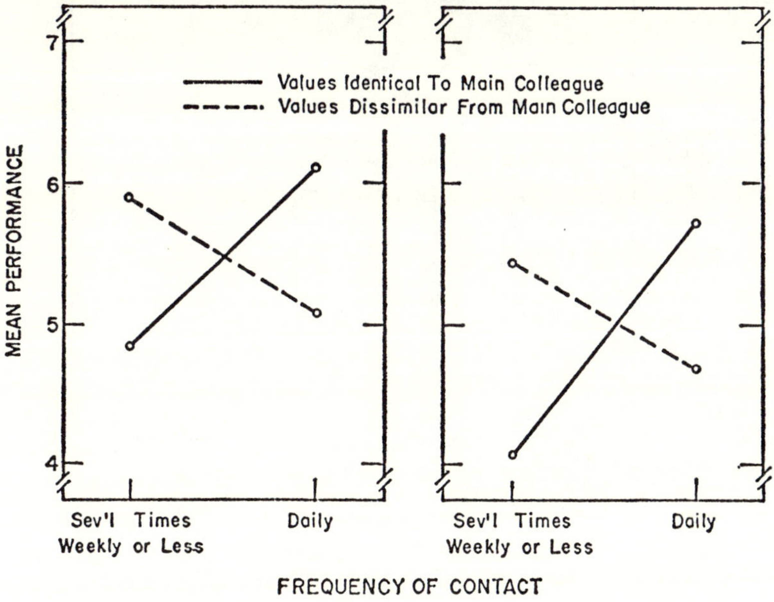

FIGURE 3

Scientific performance related to similarity in values and to frequency of contact with main colleague. (Junior level, left; Senior level, right.)

Other conditions for achieving variety are possible. Consider the relatively isolated scientist who sees his colleagues only once or twice a week. The data from Figures 1 and 3 suggest that in this case his major colleagues should possess different values and the remaining colleagues similar ones. Again the principle of variety is indicated.

The same principle emerges when we examine the relationships of the scientist to both his chief and his major colleague in terms of scientific field. The data in Figure 4 show that scientific performance tends to be higher if the scientist's chief and major colleague are heterogeneous in scientific field—one similar and the other dissimilar. Lower performance occurs when the two are homogeneous—both dissimilar to the scientist or (for senior investigators) both similar.

It does not seem to matter whether the chief plays the role of "confidant" and the colleague that of "stimulator," or vice versa. The important thing is that both roles be provided, especially for senior scientists. The key lies in variety.

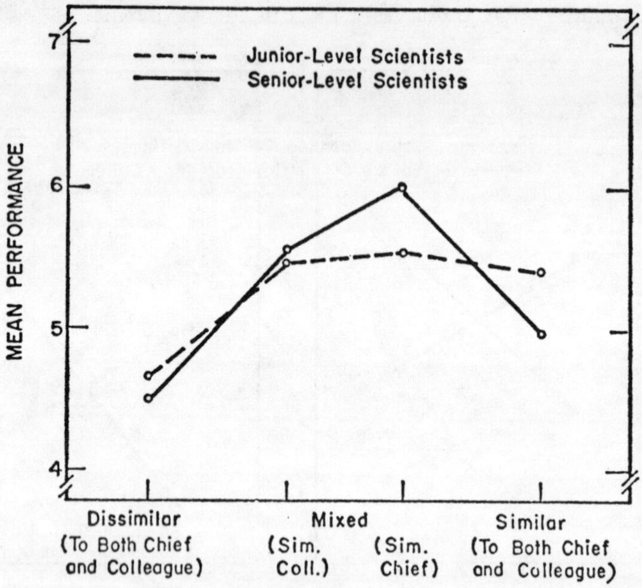

FIGURE 4

Scientific performance related to similarity of field to that of chief and main colleague.

Leadership Methods of the Immediate Chief

WHAT sort of supervision do scientific personnel require for maximum performance? One view commonly held is that the scientific supervisor shoud do little except keep out of the way of his subordinates. The soundest way to encourage high achievement, according to this philosophy, is to secure good people, give them good equipment and assistants, and then leave them alone. This view arises in part as a protest against the continental tradition of the *Herr Geheimrat,* the professor directing his students, the master his disciples.

Many scientists conceive of leadership in this either/or manner, and do not recognize the existence of a middle group between domination and isolation. Does such a middle ground of scientific leadership exist, and, if so, how does it affect performance as compared with the extremes?

In the following analyses, measures of supervisory behavior were obtained from the average reports of two or more subordinates. This procedure helps to guard against the possibility that the supervisor is simply adjusting his methods of leadership to the abilities of a particular subordinate.

In this section we shall deal with "small working groups"—two or more investigators who name the same person as their immediate chief. (In terms of the administrative structure, such a group might consist of a unit, a section, or even a small laboratory.)

A major supervisory variable we attempted to measure may be called independence or autonomy from the chief. Two items were used: (1) the percentage of scientists in each group who report that in selecting work problems or in interpreting results they make their *own decisions;* (2) the extent to which they feel the chief's activities and decisions can *influence their work.* With these two items, four patterns of leadership were defined, ranging from "dependence" (subordinates make few decisions, and the chief has considerable influence) to "independence" (subordinates make many decisions on their own, and the chief has little influence). The intermediate categories were "mutual influence" (both subordinates and the chief have much to say) and "separation" (neither subordinates nor the chief have much to say—apparently decisions depend on the nature of the task or on higher chiefs).

The results (not illustrated here) show that performance is slightly higher under an intermediate degree of independence, rather than under full dependence or full independence. Junior scientists (civil service grades GS–9 and GS–11) benefit somewhat more from "mutual influence" between chief and subordinates; senior scientists (GS grades 12 and up) benefit somewhat more from "separation."

It seems plausible that too much independence may deprive the subordinate of the stimulation that a competent chief can provide. On the other hand, too close dependence on the chief may stifle individual initiative. By this line of reasoning, highest performance should result if we can combine the benefits of frequent stimulation with the assurance of freedom for initiative. To test this hypothesis we analyzed not only independence but also amount of contact with the chief.

The results in Figure 5 support the hypothesis that at the junior level, performance is highest when independence from the chief is combined with frequent contact with him—when the individual has frequent interaction with the chief, but also has considerable voice in the final decisions. A similar but less striking tendency exists for senior-level scientists (data not shown).

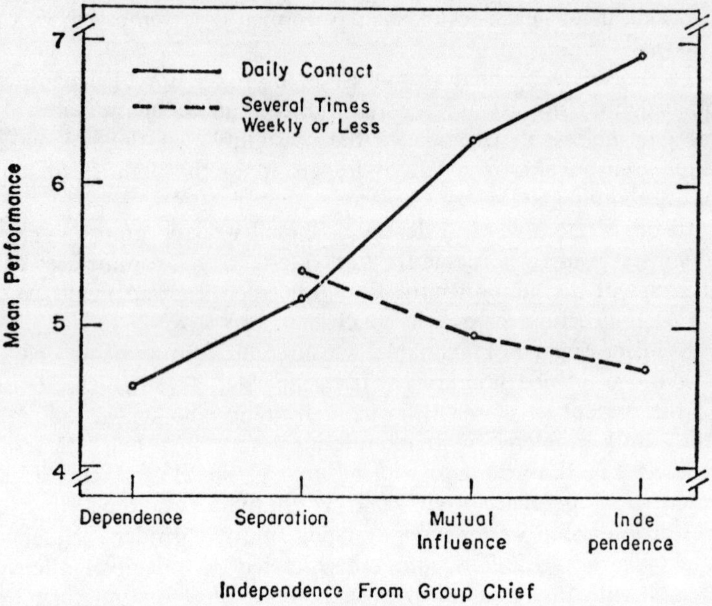

FIGURE 5
Scientific performance related to independence from group chief and to contact with chief (at junior level).

Leadership Factors in the Laboratory Chief

THE final set of analyses to be reported were conducted by Howard Baumgartel.[9] Studies in industry have indicated that a department head can influence the productivity and morale of employees at several levels under him. Does the same hold true in a research organization? To study this question, Baumgartel analyzed certain measures of the chiefs of twenty laboratories. The latter were basic administrative units ranging in size from six to thirty-three professional investigators, and as many subprofessional assistants. A typical laboratory was subdivided into three or four sections.

For this analysis it was not possible to use the measure of individual performance as a criterion, since the measure was originally obtained in part by comparing individuals with others in the same laboratory, thus reducing interlaboratory differences. But we have seen above that questionnaire items such as science-oriented values do relate to per-

9. H. Baumgartel, "Leadership, Motivation and Attitudes in Twenty Research Laboratories" (unpublished doctor's dissertation, University of Michigan, 1955); "Leadership, Motivation and Attitudes in Twenty Research Laboratories" (paper read

formance; these and other attitude items were therefore used as criteria.

Baumgartel developed three measures of the laboratory chief. One was an index of the chief's "scientific performance and motivation," based on several intercorrelated items including his own score on performance and his motivation toward the science values. This index measures the extent to which a chief combines both high technical qualifications and strong motivation toward scientific goals.

A second index described patterns of supervision and was similar to the variable of "independence" discussed above. (The latter, in fact, was suggested by Baumgartel's analysis.) Examination of several items revealed three clusters or types: a "directive" type (characterized by high influence by the chief, little freedom for subordinates to make decisions, and moderate frequency of contact between them); a "participatory" type (high influence by the chief plus considerable freedom for subordinates, and frequent contact); and a "laissez-faire" pattern (little contact, little influence by the chief, and considerable freedom of decision by subordinates).

A third leadership factor was the extent to which the leader's actual methods of making decisions corresponded to the methods preferred by his subordinates. The discrepancy between these two measures constituted an index of "role conformity."

It was predicted that subordinates' motivations toward science values, their sense of progress toward these values, and their attitudes toward leadership would be higher under (a) chiefs with a high index of performance-and-motivation in comparison to a low index; (b) chiefs who used participatory leadership in comparison to those using directive; (c) chiefs who used participatory leadership in comparison to laissez-faire; (d) chiefs who conformed to the role expectations of subordinates in comparison to those not conforming.

A summary of results is given in Table 1. A "positive" relationship is one which conforms to the prediction; a "negative" relationship is one showing no difference, or a difference opposite to prediction.

Of forty-five tests where the direction of relationship was predicted, forty are positive, and sixteen of these are statistically significant at the 5-per-cent level of confidence (using a one-tailed test). Only five results are negative. Thus the data in general support the hypotheses.

Some specific features deserve attention. First, the laboratory chief's own index of performance-and-motivation is more significantly related to his subordinates' motivation and sense of progress than is his pattern of supervisory behavior. In this organization, successful leadership depends not simply on administrative skill but also depends heavily

at American Sociological Society meetings, 1955); and Leadership, Motivations, and Attitudes in Research Laboratories, *Journal of Social Issues,* 12 (1956), no. 2 (in press).

on the leader's personal qualifications and his own motivation toward the task.

TABLE 1. Measures of Scientists' Motivations and Attitudes Related to Factors in Laboratory Chief's Leadership*

	Leadership Factors			
Measures from Scientists	*Performance-and-Motivation*	*Partic., not Directive*	*Partic., not Laissez-faire*	*Role Conformity*
Science motivations (based on 4 measures)				
Significantly positive	3	1	0	No differences expected
Positive	1	2	3	
Negative	0	1	1	
Sense of progress toward scientific goals (based on 3 measures)				
Significantly positive	2	1	0	No differences expected
Positive	1	2	3	
Negative	0	0	0	
Attitudes toward leadership (based on 8 measures)				
Significantly positive	No differences expected	3	3	3
Positive		5	2	5
Negative		0	3	0

* This table summarizes the results of forty-five tests of relationship between leadership factors and scientists' motives and attitudes. (Science motivations were measured in four ways; sense of progress in three ways; and attitudes toward leadership in eight ways.) It was predicted that these relationships would be positive. The table shows the number of tests in which the relationships were positive and negative. A "significantly positive" relationship is a difference in the predicted direction which is large enough to attain the 0.05 level of confidence (using a one-tailed *t* test).

Second, there is fairly clear evidence that participatory leadership is more effective than directive leadership (and also slightly better than the laissez-faire pattern).

Summary

IN A large government organization conducting basic medical research, the level of individuals' scientific performance is found to be higher under the following conditions:

1. Strong personal emphasis placed upon science-oriented values of

using one's abilities, having freedom to pursue original ideas, and making contributions to basic scientific knowledge.

2. Frequent (daily) contact with several scientific colleagues who on the average have been employed in settings different from one's own, who stress values different from one's own, and who tend to work in scientific fields different from one's own.

3. At the same time, frequent contact with at least one important colleague who has similar professional values.

4. A chief and a major colleague one of whom is in the same scientific discipline and the other in a different one, rather than both similar or both dissimilar.

5. A chief who gives neither complete autonomy nor excessive direction, but who frequently interacts with subordinates and who also gives them the opportunity to make their own decisions.

In addition to these findings on scientific performance, analyses of scientists' motivations and attitudes in twenty laboratories show that:

6. Motivation and sense of progress toward scientific goals are stronger under laboratory chiefs who themselves are highly competent and motivated individuals.

7. Motivations and attitudes also tend to be stronger under laboratory chiefs who employ participatory rather than directive or laissez-faire leadership.

To summarize these findings in a few words, it appears that many scientists may benefit from (a) close colleagues who represent a *variety* of values, experiences, and disciplines, and (b) supervisors who avoid both isolation and domination and who provide frequent stimulation combined with autonomy of action.

24 *Research Administration and the Administrator: U.S.S.R. and U.S.*

NORMAN KAPLAN

IN A RECENT PAPER I described the newly emerging role of the research administrator in the U.S. and tried to analyze a number of conflicting definitions and problems that stem primarily from the organizational structure in which the role is embedded.[1] During the summer of 1959, an attempt was made to compare these findings on the American research administrator with the situation of the Soviet research administrator. Some preliminary results of this comparative study are reported in this paper.[2]

After a brief description of the study in the U.S.S.R., I will outline a typical large-scale Soviet medical research organization. The administrator is located in this structure and his role is then described and compared with that of his counterpart in American research organizations. Although there are many similarities between these two

---

Reprinted from Norman Kaplan, "Research Administration and the Administrator: U.S.S.R. and U.S.," Administrative Science Quarterly, *vol.* 6 (1961), *pp.* 51–72.

1. Norman Kaplan, "The Role of the Research Administrator," *Administrative Science Quarterly*, 4 (1959), 20–42.

2. Revision and extension of a paper read at the 127th annual meeting of the American Association for the Advancement of Science, December, 1960. Some of the ideas were initially developed in a lecture on "Comparative Research Organization," delivered at the Fifth Institute on Research and Development Administration, American University, Washington, D.C., April, 1960. This investigation is part of a larger series of studies on the organization of scientific research. Grateful acknowledgment is made for the support of these studies by a Public Health Service research grant (RG 5289), from the National Institutes of Health, Division of Research Grants, U.S. Public Health Service.

3. Most of the Soviet institutes visited were selected prior to my arrival in the U.S.S.R. on the basis of available knowledge here concerning their focus on medical research generally, and on cancer problems in particular. I am particularly grateful for the advice and suggestions offered by the late Dr. C. P. Rhoads, director of the Sloan-Kettering Institute, and Dr. John R. Heller, then director of the National Cancer

roles in the two societies, some basic differences emerge, which are of potential significance for both the concrete study of research organization and for organizational theory generally. In a later section of the paper, some of the factors that may account for this basic difference are explored. Finally, some implications of this analysis are discussed.

Description of the Study

ONE of the most important objectives of the study in the U.S.S.R. was to obtain data on the organizational structures and practices in research institutes that would permit comparisons with results previously found in the U.S. The study was therefore restricted to research institutes in the medical field, and especially those concentrating on cancer research, so as to examine roughly similar types of organizations engaged in roughly similar activities in both the U.S.S.R. and the U.S.

In all, I interviewed the director or deputy director, as well as a number of department heads and other scientists, in thirteen medical institutes located in Moscow, Leningrad, and elsewhere in the U.S.S.R.[3] Interviews were frequently conducted in a mixture of English, French, and German, as well as Russian. Sometimes we relied on interpreters almost entirely, and in general, either lay or scientific interpreters were almost always available. A qualitative interview guide was used, and on the whole the cooperation in answering questions very specifically was exemplary. Most of the interviews lasted a minimum of two hours, and many were much longer. In a few instances it was possible to conduct several interviews with the same person on successive days.

Most of the organizations visited were under the jurisdiction of the Academy of Medical Sciences, and the majority of these were concerned primarily with cancer research.[4] The smallest institute had over

Institute, and now president of the Memorial Sloan-Kettering Cancer Center. The selection of institutes, as well as initial contact with their directors prior to my arrival in the U.S.S.R., was greatly facilitated by the availability of an excellent document compiled by David P. Gelfand, *A Directory of Medical and Biological Research Institutes of the U.S.S.R.* (U.S. Public Health Service Publication No. 587; Washington, 1958). Finally, Mrs. Galina V. Zarechnak, of the National Library of Medicine, very kindly made available a prepublication draft of her study of the history and organization of the Soviet Academy of Medical Sciences, which provided valuable background helpful in the selection procedure as well as in the subsequent interviews with Soviet medical scientists.

4. I am pleased to record my gratitude to the institute directors, vice-directors, and other Soviet scientists who helped me to explore some of these problems of research organization. I am especially grateful to Professor S. A. Sarkisov, a member of the Presidium of the Academy of Medical Sciences, and Professor N. N. Blokhin, the director of the Institute of Experimental Pathology and Therapy of Cancer in Moscow (Dr. Blokhin has since become the President of the Academy of Medical Sciences), for their help in facilitating my visits and interviews, and in general, for enhancing my welcome at the various medical institutes in the U.S.S.R.

two hundred people while the largest had over a thousand research workers including auxiliary staff. In size, scope, and nature of specific research activities, these institutes were not unlike many to be found in many parts of the U.S.

Structure of a Research Institute

AS ONE might expect where most of the institutes studied are under the jurisdiction of a single organization, namely the Academy of Medical Sciences, the basic structure tends to be the same in most of the institutes.[5] Differences were, of course, encountered but these appear to be related primarily to differences in size and especially to differences in emphasis with respect to clinical operations. In this section the basic outline of the structure encountered in most research institutes is described in general terms. No claim is made that this structure is typical of all medical research institutes in the U.S.S.R., let alone all scientific research institutes. My interviews lead me to believe, however, that the deviations and differences which may exist in other research institutes are not basic ones. This will necessarily be an exploratory account, since the primary purpose here is to locate the role of research administration and the administrator.

The director is the chief executive of the research institute and has over-all responsibility for the conduct of the research program and the maintenance of the research institute and its staff. He is appointed by the Presidium of the Academy of Medical Sciences for a three-year term which is renewable indefinitely. Directly below him in the organizational hierarchy is the deputy director or vice-director and typically the title contains the phrase "for research." He assists the director, acts for the institute in his absence, and has primary responsibility for the conduct and co-ordination of the scientific program of the institute. Below the vice-director are the departments into which the institute is divided with the department heads or chiefs reporting directly to the vice-director. The number of departments as well as their composition depends upon the size of the institute and the scope of its program. Below the department heads, one is likely to find a number of laboratories with the laboratory chiefs reporting directly to the department heads.

The basic outline of this type of structure is very familiar and certainly resembles that of most larger medical research institutes in the U.S. and many European countries. Parenthetically, it might be noted that I saw only one organization chart at all the institutes visited,

5. For a general, and somewhat critical, review of the history and organization of the Academy of Medical Sciences based primarily on Soviet documentary sources, see: Galina V. Zarechnak, *Academy of Medical Sciences of the U.S.S.R.; History and Or-*

although most of the directors with whom I talked were quite willing to help me draw one up.

Two other elements are always present in the organizational structure and should be described in some detail. The first is the Scientific Council (*Soviet*) which is nominally responsible for the over-all research plan of the institute, evaluating progress of the institute and of individuals, and in general dealing with any organizational or scientific problems that may arise. The director of the institute is the chairman of the Council which is made up of all or most of the department heads. The Party is represented formally on this Council by the secretary of the local Trade Union of Scientific Workers who is normally one of the regular scientists on the staff. Senior scientists who may not be department heads may also be on the Council. In addition, at least two eminent scientists, usually in related fields, but always from other institutes, are also members of this Council. The total number of members varies, of course, according to the size of the institute, and most of the ones about which information is available vary from about twelve to about thirty-five members. The frequency of meetings varies from institute to institute, but in general there are regularly scheduled meetings once or twice a month although they may occur as often as once a week.

The Council appears to combine in a single group the functions normally incorporated in two separate groups in most scientific research institutes in the U.S. One function is that of executive committee for the institute as a whole, which in the U.S. would be composed typically of department heads, the vice-director for research, and the director as chairman, as in the U.S.S.R. The second function is typically carried out by a separate group in many institutes in the U.S. and is called a scientific council made up of scientists who are not regular members of the organization, but who are invited once or twice a year (or perhaps more frequently) to evaluate the scientific work of the institute. This scientific council in most U.S. institutes has no operating functions. It is difficult to know whether the scientific council in the U.S.S.R. institute would appear above the director's box on an organization chart, or whether it would more appropriately be on the same level as that of the director, with a dotted line denoting a primarily advisory function.

Finally, we turn to the position of the research administrator. Every institute visited has such a person and the title is usually a variant of "vice-director for administration" or simply "director of administration." As in most U.S. organizations, he has primary responsibility for finances,

ganization, 1944–1949 (Public Health Monograph No. 63; Washington, 1960). See especially her charts and descriptions of Soviet research institutes, pp. 12 ff.

supplies, apparatus, equipment, furniture, repairs, maintenance, and other such service activities. The size of his staff tends to vary with the size of the institute as a whole and in some of the larger institutes the administrator may have a staff of over thirty persons working in a number of separate departments. The administrator's position in the organizational hierarchy is also difficult to locate precisely. He reports to the director of the institute but he has very little if anything to do with any other scientists. Although he reports directly to the chief executive, he is not a part of the executive committee nor is he typically considered a member of the executive hierarchy. Interestingly enough, I did not meet him personally at most of the institutes visited, with one or two exceptions, when the director wanted a precise figure or fact I had asked about and he consulted the administrator. With this background, it is now possible to examine the role of the research administrator in more detail.

Role of the Research Administrator

AS ALREADY noted, the administrator reports directly to the director of the research institute and may have a fairly large staff. Furthermore, he is responsible for more or less the same kinds of activities as is his counterpart in the U.S. Some differences begin to appear as we note that the Soviet administrator is typically trained in what would be the American equivalent of business accounting and business procedures. It is not considered essential, or even desirable (as it is frequently considered here in the U.S.), that he have a scientific background or that he come from the ranks of the scientists. This difference becomes somewhat accentuated when we note his absence in greeting a foreign visitor, where the analogous situation in an American institution would find the administrator one of the more important men present a such a meeting. This is particularly to be expected when that visitor is more interested in problems of organization than in the substantive content of the work of the institute.

It is at first surprising to hear him referred to as the "bookkeeper" and his job described essentially as a bookkeeping one with few if any policy-making responsibilities. This term as used there implies more than simply the keeping of the financial books, referring also to "keeping the books" on maintenance, equipment, and so on. In many respects, we find that he occupies a position sometimes designated in

6. These are 1959 rubles. It is difficult to translate these earnings into terms which permit suitable comparisons with the U.S. Furthermore, it is unnecessary to do so for our purposes here since the object is to show that the administrator's salary tends to be much closer to that of technician or beginning scientist, and not, as in the

American organizations as that of chief clerk. He has administrative responsibility for the clerks who work under him but has no other decision-making functions.

It is not surprising, therefore, to find that typically he is paid considerably less than most of the research scientists—normally only somewhat above the research technician with no advanced training. While laboratory technicians may earn approximately twelve to fifteen hundred rubles per month and the director of a research institute may have a base salary of at least five to six thousand rubles a month, the chief administrator earns approximately twelve hundred to two thousand rubles per month.[6] The range for the administrator indicates primarily the differences in size of organization and length of experience. A researcher starting out with the first advanced degree probably earns about eighteen hundred to two thousand rubles a month. In short, there can be little doubt that the chief administrator, who is referred to as the bookkeeper and whose duties correspond to those of a chief clerk, is in fact paid as one would expect a bookkeeper or chief clerk to be paid compared with the more technically trained research scientists in the research hierarchy.

For the research administrator there is little or no conflict concerning authority and control over science and scientists. These are exercised by the scientists themselves and not by a lay administrator. The Soviet administrator, when compared to his American counterpart, occupies quite a subordinate position in the research organization, despite the fact that the two have essentially the same titles and many of the same functions in a research organization.

The American research administrator is paid a good deal more than most American scientists in the same research organization and frequently is paid nearly as much as many senior scientists. In the organizational hierarchy, he is always at or near the top of the organizational structure. Although his duties may correspond very closely to those described for the Soviet administrator on a formal basis, the American administrator has many decision-making functions, overtly or covertly.[7] Many of these, incidentally, seem to stem from the unwillingness of the research director to make the decisions himself. The American research director often feels that he has little time for purely administrative decisions, and furthermore, the administrator is often thought to be better equipped to make them. In the U.S. it is often considered desirable for an administrator to have a scientific background, and not

U.S., closer to that of the senior scientists, associate directors, or even department heads.

7. The observations on the role and status of the American research administrator are drawn largely from an earlier paper. Cf. Norman Kaplan, *op. cit.*

infrequently chief administrators in research organizations are recruited from the ranks of scientists. The American administrator is definitely a public figure and in fact serves to save the research director's time in public relations. He frequently exercises authority over scientists with regard to the kind of equipment they can get, space allotment, and adherence to budgets, although much of his authority is exercised indirectly, frequently with the budget or some such impersonal instrument as the indirect mechanism employed.[8]

Finally, when the American administrator is not a scientist, it is not very likely that he can move up much higher in the scientific research institute. This is, of course, similar to the Soviet situation. But frequently the American administrator, even without scientific training, who moves out of the scientific realm whether in the same organization (e.g., an industrial firm or the government) or whether from an organization in one institutional sphere to another, can move up very high in the organizational hierarchy by virtue of his *expertise* as an administrator.

In sum, the American research administrator is better paid, compared with his Soviet counterpart and with scientists in the research organization. He enjoys much higher prestige in America and, of course, he is the source of many more conflicts and problems in a research organization.[9]

This brief description indicates some vital differences in the role of the research administrator in the U.S. and the U.S.S.R. Since we are dealing with essentially similar types of organizational structures and with organizations concerned with roughly similar problems, handled in approximately the same way, and whose over-all size is roughly comparable, we are faced with the question: Why is the role of the Soviet administrator so very different? We are amazed that the Soviet administrator occupies such a subordinate position in the research organization compared with the American administrator. Of course, we could with equal validity ask the question: Why does the American administrator occupy such a superordinate position in the research organization relative to his Soviet counterpart? Asking the

8. Kaplan, *ibid.*, p. 33.

9. *Ibid.*; for other evidence see E. Orowan, "Our Universities and Scientific Creativity," *Bulletin of Atomic Scientists*, 15 (1959), 237–238; L. Kowarski, "Psychology and Structure of Large Scale Physical Research," *ibid.*, 5 (1949); A. M. Brues, "The New Emotionalism in Research," *ibid.*, 11 (1935).

10. See, for example, the discussion by Herbert Simon, where he questions the distinction (attributed to Frank J. Goodnow) between policy and administrative processes, "Recent Advances in Organization Theory," in *Research Frontiers in Politics and Government* (Washington, 1955), esp. pp. 24–26. This kind of distinction has been emphasized by many political scientists commenting on the alleged stability and resilience of the civil service apparatus in Great Britain, France, and other nations in the face of marked changes in the political leadership of the state. This thesis is explicitly challenged in a brilliant analysis of the Nazi Germany case by Frederic S.

question both ways raises interesting subsidiary problems, some of which are considered in the remaining sections of this paper.

Possible Explanatory Factors

AS AN American commenting on the Soviet scene, it seems to make sense to try to amplify the question in terms of Soviet experiences first. The first obvious question is what happens to all the administrative tasks? Obviously, the American administrator and his staff have much to keep them busy; in fact, they always seem overburdened with a variety of administrative problems. Who takes care of these problems in the Soviet research institute?

This seemingly simple and obvious question turns out, of course, to be fairly complex upon closer examination. For one thing, we must ask whether there is the same "amount" of administrative work and detail in the Soviet and American research institutes. We must also inquire whether the Soviet administrator has approximately the same kinds of duties but simply a lower status, or whether he has lower-status duties and a lower status as a consequence.

We are almost forced to start with the notion that the Soviet scientists, as compared with their American counterparts, tend to view the content and the boundaries of research administration differently. The Soviet view of the research administrator essentially restricts him to a bookkeeping function and in terms of administrative theory, might be labeled the pure execution of policy.[10] The American view of the chief administrator is often much broader. The hypothetical distinction between the execution of policy and the formulation of policy often does not work out in practice. Furthermore, the American scientist's tendency to delegate any problem that he considers essentially nonscientific results in a concept of the chief administrator's role as essentially residual—it becomes in effect all things and all functions which the director or the other top scientists are unwilling, unable, or reluctant to do themselves. In return for the alleged freedom resulting from a broadly conceived view of administration, the American scientist-

Burin, "Bureaucracy and National Socialism: A Reconsideration of Weberian Theory," in Robert K. Merton *et al.*, eds., *Reader in Bureaucracy* (New York, 1952), pp. 33–47.

To our knowledge, the distinction between the formulation and the (mere) execution of policy has been confined almost exclusively to the political sphere. It has not been studied adequately in other kinds of large nongovernmental organizations. Is there, for example, a "neutral" apparatus in large corporations which remains essentially intact in the face of sharp changes in the leadership and control of the company? Our analysis here points to the possibility that the Soviets effectively avoid the problems which may arise if the distinction is recognized insofar as the scientists keep administrative policy-making functions for themselves rather than delegating these and by downgrading the administrator to the level of a chief clerk.

director must also give up some of the areas of decision making which at the same time he continues to feel are still his prerogative; hence the almost continual underlying conflict between the administrator and the scientists in many American research organizations.

In the U.S.S.R., and for that matter in most of the rest of Europe, the scientist and the director of scientific research organizations appear to be much less reluctant than their American counterparts to assume administrative duties which have a bearing on the conduct of the research.[11] They cheerfully delegate keeping the books and other financial, personnel records, and similar bookkeeping-type operations to a chief clerk, who is called a research administrator. But most other so-called nonresearch duties the Soviet scientist, as well as the European scientist generally, seems more willing to do himself. In general, it may be said that scientists at all levels, from the laboratory head to the director, are more willing to involve themselves in the nonclerical aspects of administration—and especially anything which is viewed as connected with the effective conduct of the research itself.

It is not simply as a matter of prestige that the American scientist-director argues in favor of sloughing off administrative duties. Far more important in the eyes of most scientists is the opportunity to concentrate on the conduct of research without being diverted by what seem extraneous organizational and administrative responsibilities. If it is true then, as we have asserted, that the Soviet scientist is far more willing to engage in administrative tasks than his American counterpart, does he in fact spend less time on research, since presumably he has to spend more time on administration?

11. Published evidence for this statement is admittedly scanty. However, it was strongly supported by my own observations and interview data. In some German laboratories, for example, the Director explicitly provides "on the job training" in administrative duties for his young postdoctoral research assistants. Usually, the young man is given responsibility for "helping" with purchasing activities for a six-month period, and then may be shifted to equipment maintenance for a similar period, and so on. This is viewed, in part, as a continuation of the traditional apprenticeship pattern to ensure that the young man will have gained the experience necessary to qualify him for a more senior post ultimately. Another consequence, of course, is that the director's own total administrative load is lightened considerably by being shared with subordinates. But, significantly, the director delegates some administrative responsibility to other *scientists*, and not to professional full-time administrators. To some extent, I suspect that this pattern is less a deliberately considered policy and more an extension of the traditional patterns of the small research institutes to the much larger organization which is becoming more prevalent today. The absence of this kind of strong tradition in the U.S. is perhaps partly responsible for the greater reliance on professional administrators here.

12. The stereotype of excessive red tape and bureaucracy in the U.S.S.R. is widely supported in the literature and is generally shared by most foreigners visiting the Soviet Union. How much of this stereotype can be attributed to the facts of the case, and how much to preconceived ideas coupled with inadequate comparative analyses is difficult to determine. To our knowledge there have been no studies of bureaucratic tendencies and administrative proliferation in the research institutes of the U.S.S.R.

The answer is paradoxical indeed. Most directors of American research institutes seem to have little, if any, time for their own research. The Soviet director, on the other hand, asserts that he spends most of his day on the conduct of his research and that this is in fact his first duty. When asked for an estimate of how much time a director in the Soviet medical institute had to spend on administrative duties, he typically answered that it was an average of about an hour a day. This increased, of coure, at certain times of the year when new budgets had to be in, but generally the time reported spent at the institute not devoted to research was extremely low. We then have the apparent paradox of the Soviet director more willing and more likely to engage in administrative duties than his American counterpart and yet being able to spend considerably more time on his own research than his American counterpart.

Dismissing some of the more tenuous kinds of answers, we can only suggest one rather startling possibility. There is simply less administration. This is exceptional on two counts. We could expect that, given the same type of activity and the same size in comparable organizations, the administrative duties (not counting the purely routine ones, which are handled by clerks in both situations) would be roughly the same in order to meet the requirements of maintaining the organization. It might even be expected by some that, given comparable organizations, the level of administrative duties in the Soviet organization would be considerably higher because of the nature of Soviet society with its greater emphasis on centralization and its general bureaucratic tendencies.[12] But I must conclude tentatively that there is probably

However, medical scientists have commented on the "medical bureaucracy" in the Soviet clinical practices and in the hospitals. Cf. the comments in the U.S. Public Health Service, *The Report of the United States Public Health Mission to the Union of Soviet Socialistic Republics* (Public Health Service Publication No. 649; Washington, 1959), especially p. 25.

Much has been written on the bureaucratic facets of Soviet industrial organization, but even here, this notion has been sharply criticized. See, for example, David Granick, *Management of the Industrial Firm in the U.S.S.R.* (New York, 1954), especially the concluding chapter in which Granick makes an explicit attempt to compare the extent of bureaucratization in Soviet and non-Soviet industrial organization. He notes, for example, "It appears an open question whether Soviet industry is not . . . less bureaucratic than are most giant firms in capitalist society" (p. 262). Granick attributes the fact that many Western observers see so much bureaucracy in the U.S.S.R. to their treatment of planned and centralized control over the economy as being synonymous with "bureaucracy." It should also be noted that there has been increasing concern with the growing bureaucratization of private business organizations in the U.S. A study by Seymour Melman of this problem over a fifty-year period in the U.S. cites an increase of 87 per cent among productive workers compared with a 244 per cent in administrative personnel in American manufacturing industries in the period 1900–1940 ("The Rise of Administrative Overhead in the Manufacturing Industries of the United States 1899–1947," *Oxford Economic Papers,* 3, N.S. [Feb. 1951], 62).

less administrative detail and bureaucratic red tape in the Soviet medical institute.

Why Does This Difference Exist?

IT MIGHT be concluded that less administration and red tape would be possible in the U.S.S.R. because of the relative simplicity of the financial support structure. One of the obvious reasons for a complicated and large administrative staff in many U.S. medical institutes is the complexity of the financial structure and the necessity to keep track of the dozens and sometimes hundreds of different grants from different agencies with differing termination dates, differing rules concerning permissible practices, differing requirements for progress reports, renewal procedures, and so on. In the Soviet medical institutes, which are under the jurisdiction of the Academy of Medical Sciences, the budget stems from that single source.

Is it very likely that this difference in the financial support structure accounts for differing administrative loads in similar medical institutes in the two societies? The answer, perhaps strangely, is that this explanation is not very likely because when we examine organizations in the U.S. essentially similar to the Soviet ones with respect to financing, we do not find this to be the case. One of the best examples of such a comparison would be one of the National Institutes of Health, which also has a single source for its budget, namely, the Department of Health, Education, and Welfare and ultimately Congress. Despite this single source, or perhaps because of the characteristic federal accounting and auditing regulations, the reporting procedures and the administrative load generally are probably not very different from that found in most other U.S. medical organizations of similar size and scope. In fact, it may be suspected that the administrative load is at least as heavy, if not heavier, in such an organization.[13] It is probably the case then that the particular kind of financial structure is not of central significance in this context, although as pointed out in the previous paper, it may influence an already existing level of administration.[14]

If it is probably not the relative simplicity of the financial structure, what other possible factors might account for the hypothesized lower level of administrative activity in the Soviet medical research institute?

13. It must be emphasized that this comparative evaluation is purely impressionistic. It is based largely on available documentary sources and talks with scientists and administrators at the National Institutes of Health. It would certainly be desirable and worth while to check this further in a more precise quantitative fashion.

It should also be emphasized that our impressionistic comparison is between the seven institutes of the National Institute of Health and their intramural research organization with somewhat similar types of institutes in the U.S.S.R. under the central administration of the Academy of Medical Sciences. This comparison is not intended as a reflection of the effectiveness or policies of the National Institute of Health ad-

Perhaps the Soviet government is willing to require fewer formal controls which in turn reduces the amount of administrative activity simply because they tend to trust their scientists more than we do. This is an intriguing hypothesis because it is probably true that the average American feels that the Soviet government trusts none of its citizens at all, while the American government, industrial firm, or scientific research institute under private auspices would seem more likely to trust their scientists. Unfortunately, it was not possible to obtain any data which would either confirm or deny this hypothesis. If, however, there is less administration in otherwise comparable organizations, then a factor such as this may play an important role.

There is relatively little disagreement that the scientist is accorded considerably more prestige and is relatively better paid and rewarded in the material sense in the U.S.S.R. than he is in the U.S. Conversely, the administrator, at least in the medical research institute, enjoys far less prestige and material reward than the administrator in the U.S. To the extent that the prestige accorded, as well as the material rewards, reflect an evaluation of the relative importance of the activities carried on by scientists and administrators, we have perhaps an additional small bit of evidence in support of the hypothesis that the Soviet scientist is trusted somewhat more.

Another factor of potentially great significance is the nature of higher authority over the organization. The director of the medical institute is responsible to the Academy of Medical Sciences and specifically to the scientists who make up the Presidium of the Academy. He is thus responsible directly to other scientists and not to government administrators or "politicians." His American counterpart is typically responsible to a board composed of laymen who are not often very familiar with the nature of science. Being unfamiliar, they are much more likely to require reports, statistics, and data, which they can understand and which in turn require the services of, and enhance the importance of, the administrator.

Returning to an earlier injunction that the question must necessarily be asked both ways, namely, why does the Soviet institute seem to have "less" administration and why does the American institute have "more" administration, we are led to inquire into some of the

ministration structure or its administrators. In fact, its administration, as a whole and at the institute level, seems to be highly regarded by the National Institute of Health as well as by other scientists and research officials who have any familiarity with it. In my own experience, these institutes, when compared with *others of the same size and scope in the United States*, are consistently highly rated in this regard. The comparison with institutes of the Academy of Medical Science, however, highlights the importance of the external environment and the demands stemming from it, which may affect administrative requirements within the organization.

14. Kaplan, *op. cit.*, p. 32.

consequences of administrative decisions and programs. Administration, in the American sense of the term as defined here, is necessary in order to accomplish a minimum of co-ordination, communication, and control in an organization. But presumably these should be the same in the U.S.S.R. and the U.S. given similarity in organization and its activity. Part of the problem, however, is co-ordination, control, communication for whom and for what purposes? At its simplest, these are necessary for the director; he must be able to exercise control functions and may need help for this. But it becomes more complicated when the director is in turn responsible to other authorities and must provide certain information to them, primarily for purposes of control. As already noted, the complex multiplicity of research budgets in many American institutes may require the exercise of control to meet the differing criteria of a large number of organizations, all of which have provided funds for part of the larger program.

The American research director's having to account for the activities and expenditures of his research organization to a board of trustees or directors—to laymen in general (at least with respect to the intricacies of scientific research)—tends to force the director to provide certain types of nontechnical reports and information. Since these board members may have little technical knowledge of the substance of the research, and since they tend to have a great deal of knowledge concerning the operation of large organizations, both they and the director of organizations responsible to them feel that certain types of reports are most desirable to indicate proper control and reasonable progress although they may have little intrinsic value for the conduct of the research. All of these inevitably increase the administrative load and, in fact, make it very difficult for the director to spend much time on co-ordinating the research itself, let alone doing any of his own.

The additional problem of raising funds, not at all unimportant in most American research institutes, also consumes a good deal of the time and energy of a director and administrators to whom such functions can be, and frequently are, delegated. In the U.S.S.R., on the other hand, whatever the problems concerning the amounts and scope of the financial support, it is a single body of *scientists* to whom the director must go for his financial support for the following year. The men of the Academy of Medical Sciences are presumed to have a fairly intimate knowledge of the scientific character of the work and are less likely to require reports which we might consider normal for boards of trustees here.

Finally, two other far more speculative factors which may affect

15. Some confirmation of the importance of administrative personnel in American industry may be found in Melman, *op. cit.*

16. For a comparative study of scientific personnel systems, see: Edward McCren-

research administration in the two countries should be mentioned. The first has to do with the contemporary origins of the large-scale research institute in the two societies. In the U.S.S.R., it is apparently the case that the university institute, following the old European tradition, was expanded into a large-scale organization under the Academy. In the process, the high prestige and the relative autonomy of the scientist (with some notable exceptions of political incursions) was maintained. In the U.S., on the other hand, there was little tradition for the relatively autonomous institute, whether attached to the university or not, and the scientist in general enjoyed relatively little prestige or autonomy. By the time research in the U.S. was expanded in the university and outside, and the complexity of the research organization grew with this expansion, the organizational model which many felt worth imitating was the successful big business enterprise. Moreover, the business organization model was borrowed at a time when the organizational specialist—the administrator—was becoming increasingly important.[15]

The other major factor has to do with the diversity of not only our financial support structure but also the occupational structure for scientists in the U.S. Titles vary from institution to institution, salary scales vary from one institutional sphere to another (industry versus government etc.), and in general there is diversity with respect to most aspects of the employment, supervision, and evaluation of the scientist. This necessitates the collection of a good deal of information to provide some basis for the evaluation of scientists and institutes.

In the U.S.S.R., on the other hand, there is a single system, with many subdivisions to be sure, defining salary scales in different types of institutes, employment grades related primarily to educational attainment, and other more or less fixed criteria. Thus large areas open to discretion in the U.S. are fixed in the U.S.S.R. and require relatively little administrative action.[16] There are, of course, numerous formal and informal ways of by-passing this otherwise inflexible structure which need not be considered in any detail here. The point to be stressed is that having this centralized and generalized system of promotion policies, grades of employment, salary schedules, etc., may actually reduce the administrative load as well as the amount of discretion that can be exercised in any specific institute. Whether the perceived disadvantages of this centralization outweigh this particular advantage is yet another question.

In closing this section, it must be emphasized again that we are primarily concerned with exploring several significant aspects of the administration of research institutes. Obviously, neither the short period

sky, *Scientific Manpower in Europe* (New York, 1958). Chapter vii is particularly relevant inasmuch as it contains a discussion of Soviet practices compared with others.

of time spent in the U.S.S.R., nor the preliminary nature of my inquiries permit anything other than a very tentative analysis. It should also be obvious that the various possibilities, theoretical and otherwise, which may account for the apparently sharp differences in the administration of medical research encountered in the U.S.S.R. and in the U.S. have hardly been exhausted. In subsequent studies of this problem, these are among the hypotheses deserving of futher exploration. In the final section which follows, I turn to an examination of some of the implications of my observations and the hypotheses just noted.

Summary and Conclusions

THE observations, that the character of research administration and the role of the research administrator in roughly similar types of medical research organizations in the U.S.S.R. and the U.S. are different, call for an explanation. We want to know why this is so and how these differences operate, as well as how this affects the conduct of medical research.

How this is accomplished is possibly easier to describe, and the main points previously made can be summarized briefly here. The primary difference revolves around the definition of the chief administrator. In the medical research institutes of the U.S.S.R., he is defined primarily as a chief clerk. In the U.S. there is no single clear-cut definition, but in general he tends to be defined as something much more than a chief clerk, varying from general business manager to a general manager of an organization. In the U.S., the chief administrator normally has some decision-making functions while in the U.S.S.R. he appears to have practically none. This difference in definition leads to obvious differences in recruitment patterns as well as in the rewards involved in the job.

For the Russians, there is little or no problem concerning the type of person to be recruited for this job. He does not require any advanced education. He must be a competent keeper of books and records (financial and others), and, to be a chief administrator in a fairly large institute, he must be able to supervise the activities of a number of subordinate clerks. For the American research organization, on the other hand, the character of the desirable recruit for chief administrator tends to vary. Some believe the best sort of person for this position is a man who knows how to run and manage an organization. An underlying assumption is that most large organizations, irrespective of their particular activities, are essentially alike with respect to organizational problems, and consequently the best type of man for this position is a specialist in administration who is, with respect to organizations, a generalist. That is, he can move fairly easily from running a research organization to running a soap factory. Another school of thought, however, believes that there is something fairly unique about the management of the scientific research organization and tends to favor a former scientist or at least a man with scientific background who has administrative experience or at least displays a flair for administration. Involved in such a flair is the ability to deal with people and to talk with scientists, in particular to understand their problems as well as their general antipathy toward large bureaucratic organizations.

Given the Soviet requirements and definition, the man recruited need not be paid a very high salary relative to others in the research organization. He is, in effect, a fairly low level, white-collar worker among considerably better trained and more advanced personnel in the various scientific fields. In the U.S., on the other hand, the man recruited must be paid a fairly high salary relative to other scientists because he too has advanced training, and what is most important, his market includes other types of large organizations where he commands a high salary.

We should certainly expect that the differences built into these two conceptions of the chief administrator should manifest themselves in other ways in the research organization. As already noted, we can entertain one of the two major possibilities: Either the amount and character of administration (management control, etc.) is roughly equivalent in the Soviet and American medical institute of the same size and character, in which case we should expect that the functions of the administrator in the U.S. setting are carried out by one or more functional substitutes in the organization; or, it is possible that the amount and character of general administration is quantitatively and qualitatively different in the Soviet institutions and hence few, if any, functional substitutes may be necessary. Our tentative analysis appears to favor the latter possibility although some questions and modifications must be considered.

First, it has been suggested that there is in fact "less" administration in the Soviet institutions and that, furthermore, the scientist himself, and in particular, laboratory and department heads as well as the scientific director of the institute, appear to be more willing to carry on some so-called administrative duties, which tend to be shunned by their American counterparts. Most important, these Soviet scientists report that such duties do not infringe on their research time and, in fact, are far more likely to report that they do their own research. This suggests the hypothesis that given a reduction of administrative requirements, and an adequate division of labor with respect to the remaining requirements among the scientists, it is possible to have a more effective organization in which the primary goal of the pursuit of scientific research is not diminished significantly.

In fact, it might be argued that the apparent saving of time in delegating many management activities *bearing directly* on research is in the long run a myth. The structure becomes far more cumbersome, cleavages and antipathy may rise between the research people and the administrative people, and the administrator is forced to make decisions in situations where scientific competence and intimate knowledge of the scientific research is necessary. This results in additional mechanisms in the organization to reduce cleavage and to communicate information, which may be far more cumbersome than an ordinary division of labor among the scientists themselves. If the scientist is willing to accept some minimum amount of administrative duty as part of his job, and as part of the price he must pay for the benefits derived from working in a large complex organization, then the net results in terms of what he can accomplish scientifically may be far greater than if he delegates many of these management functions to specialists in management. Such a step would be extremely difficult in many American research organizations because, among other things, it would necessitate the reduction in status, prestige, and monetary rewards of the chief administrator as he is now defined.

It is unfortunately not possible to discuss relative differences in the effectiveness of the conduct of research in the U.S.S.R., and the U.S. medical research institutions.[17] This is so for many obvious reasons, in-

17. In recent years there have been numerous reports evaluating the "quality" and other characteristics of medical research in the U.S.S.R. by American and other Western medical scientists who have visited the Soviet Union. It would obviously be presumptuous of me, a layman with respect to the medical sciences, to give my own evaluation. However, my impression, from reading many of these reports and from talking with some of the medical scientists who have been there, is that Soviet medical research is generally viewed as competent, and in particular subfields, as quite outstanding. The growing program of translation of Soviet medical and scientific journals must also be viewed as evidence of the importance attached to Soviet research.

An extremely useful selected and annotated list of references has been compiled by Elizabeth Koenig of the National Institutes of Health Library: *Medical Research in the U.S.S.R.* (Public Health Service Publication No. 710; Washington, 1960). Among the most relevant reports in terms of the institutes I visited are the following: J. R. Paul, "American Medical Mission to the Soviet Union," *Scientific Monthly*, 85 (1957), 150–156; M. B. Shimkin, "Oncology in the Soviet Union," in *Year Book of Cancer*, 1957–58 (Chicago, 1958), pp. 506–510; M. B. Shimkin and R. E. Shope, "Some Observations on Cancer Research in the Soviet Union," *Cancer Research*, 16 (1956), 915–917; J. Turkevich, "Soviet Science in the Post-Stalin Era," *Annals American Academy Political Social Sciences*, 303 (1956), 139–151; H. Hamperl, "Pathologie in USSR" (Pathology in the U.S.S.R.), *Deutsche Medizinische Wochenschrift*, 82 (1957), 416–419; C. W. Scull, M. Nance, F. Grant, and G. F. Roll, "Some General Observations on Medical and Pharmaceutical Research in the Soviet Union," *Journal American Medical Association*, 167 (1958), 2120–2123; *The Report of the United States Public Health Mission to the Union of Soviet Socialist Republics, Including Impressions of Medicine and Public Health in Several Soviet Republics* (Public Health Service Publication No. 649; Washington, 1959); *U.S. Public Health Service, United States–U.S.S.R. Medical Exchange Missions*, 1956; *Microbiology and Epidemiology* (Public Health Service Publication No. 536; Public Health Monograph No. 50; Washington, 1957).

cluding our lack of adequate measures, but also because of differences in emphases, relative time devoted to the attack on different sorts of problems, and a host of ordinary but complicated problems of assessing the effectiveness of any kind of organization. One point which has some implication for general organizational theory must, however, be stressed. In general, our observation and analysis force us to ask how much administration is necessary in a complex organization. We have tended to assume, perhaps without sufficient evidence, that the level of administrative activity in research organizations (as well as in others) is at, or very near, the minimum necessary for co-ordination, control, and communication considered adequate to maintain the organization. The findings tend to throw some doubt on the validity of this assumption, at least for medical research institutes, and in a very speculative way possibly for most other types of complex organizations as well.

In summary, it seems highly possible that the Russians really do use much less formal administration in scientific organizations than we have thought possible. I have tried to suggest some of the factors that may contribute to this and, in particular, would stress the strategic role of the larger society as well as differences in approach toward large-scale complex organizations. The nature of the financial structure, the kinds of controls exercised by higher authority external to any given organization, as well as the general prestige level of scientists relative to administrators and others seem to affect the situation. It is hoped that additional empirical research can be conducted inside the U.S.S.R., as well as further comparative research in other countries and in other types of organizations in the U.S., to test some of the assumptions and hypotheses suggested here as well as to move closer toward a theory of complex organizations.

25 *Scientists: Solo or Concerted?*

A. H. COTTRELL

THE SOLITARY SCIENTIST, withdrawn from the world: this is a popular image and it is easy to see why. Here is Newton, alone in a country orchard. There goes Henry Cavendish, terrified of the company of women, escaping back to the austere beauty of his electrical experiments. And here is Einstein on the river bridge, seeing in the water the reflection of curved space itself, while the crowds pass by.

Science attracts the solitary mind. Its creations are like great mountain peaks newly discovered in some abstract world of the intellect. They can be seen and their full magnificence enjoyed only by making the long, difficult journey oneself. Creation itself is a solitary business. No committee ever corporately created a new idea any more than an orchestra wrote a symphony. The idea has to start in one mind and then spread outwards. And the struggle to create is mainly a struggle of will against one's own weaknesses, like the struggle the long-distance runner has to make, withdrawing into himself to drive his protesting body along.

Even so, this image of the solitary scientist is mostly a misleading one. He may be at odds with the world but, if he is to be effective, he can never be right out of it: certainly not the world of his own scientific contemporaries. Science, particularly modern science, is not like that. It cannot sustain a complete withdrawal. However deeply you go in you soon have to come up for air; otherwise, scientifically, you begin to die.

Chasing the Atom Down the Centuries

THERE are people who delight in digging up the origins of scientific discoveries and inventions. Science to them is a field for arch-

Reprinted from The Listener, *Sept.* 15, 1960, *pp.* 411–412 aelogical exploration. They chase the atom down the centuries back to the ancient Greeks; they track the theory of heat back to Lucretius and beyond; the quantum hypothesis back to Newton's light moving in fits and starts. They discern pale anticipations of natural selection before Darwin and Wallace. They find that Einstein's theory of relativity flowered, not alone in a desert, but as the brightest bloom in a flower garden. They prove, in fact, that science is not really revolutionary after all: that there is nothing new under the sun. The scientist, however remote he may seem, is always bound closely to the scientific life around him. He cannot work in a vacuum. He has to take the ideas and problems as they exist among his fellows, transmute them in his own personal way, and then bring them back as offerings to his community. He both takes and gives, in the scientific currency of his time.

This is because science is at heart a progressive, evolutionary subject. Nothing good or sound is ever thrown away. It all goes into the foundations. Those who come after, whether they call themselves pure or applied scientists, are all science-users in that they use all that has gone before them. At any instant there is a narrow frontier dividing what has gone before from the problems still to be solved. This frontier occasionally gets stuck at places, where it runs up against one of the big problems of science, but for the most part it is steadily moving forward, recognizably changing position even from day to day.

It is not surprising that this should be so. Countless hands now push it along. Keeping in touch with its position from day to day, even on a narrow sector, has become surprisingly difficult. It is in fact becoming the main problem that the modern scientist has to face in his own work. It is a strikingly narrow frontier: such width as it appears to have is usually due to his own inability to find it exactly. Yet finding it exactly is half the battle in modern research. If the scientist does not find it, he is in danger either of being too timid and wasting his talents on a stale problem, or of being too bold and futilely attempting the impossible. On the other hand, if he can pinpoint its position really well, deciding the next step forward and how to make it is often surprisingly easy. This almost looks like a new principle in scientific research. Certainly, as scientists become more and more aware of it, it is changing the pattern of scientific research itself. Keeping in touch is the thing, and that means meeting as many people working in your own field as possible and meeting them as often as possible. If you can meet them every day, in a great laboratory, so much the better.

Rash of Conferences

FAILING that, there is the scientific conference. These conferences have broken out all over the world like a rash. By going perhaps to New York and Madrid, back to London and Paris, then on to

Moscow, calling at Zurich and Prague before going on next to Tokyo, a man can fill his days completely with scientific conferences on his own particular branch of science. If an important new idea, discovery, or technique, is brought forward on the first day of such a conference, it will be minutely scrutinized, turned inside out, taken apart and put together again so many times during the next three days, and so many thousands of words will pour over it round the coffee tables, that already, before the conference is ended, it will have become historical; a classical well-worn foothold for those trying to make the next leap forward.

This is an age of research teams, big laboratories, international research projects. New ways of estimating scientific capability appear. Mutterings are heard about the "non-viability" of those unfortunate research groups in outlying places, too isolated to keep in touch with the frontier and too small in numbers to make any part of it their own. The large laboratory, the big project, the stacks of research reports, the breathless letter-to-the-editor, the international conference; all this is bustling modern "city-science," something far removed from the quiet contemplation in a country orchard. We may not like it, and some of the rarest spirits shy away from it altogether, preferring to find some quiet, unfashionable corner of science. But it is undoubtedly here to stay: it is proving to be such an effective way of creating science.

A few very great scientists can, of course, stand apart from all this. They are strong enough to create their own frontier wherever they are and have no worries about isolation. They can leap far ahead without losing their balance or can stand back and find great new problems right under everyone's noses, in the most familiar places. The way they choose to work is then entirely a matter for their own temperament. Some never seek solitude. Rutherford never did. The research students and visiting scientists flocked round him in huge numbers and he loved it. Even for the giants, standing apart can be precarious. People are still wondering what Einstein might have done in his later years if he had not decided to let quantum physics go on ahead without him: or Eddington, if he had not tried to jump quite so far ahead in the physics of fundamental particles.

A False Idea

IT IS sometimes said that pure science is for the solitary individual and applied science for the crowds. This idea springs partly from history; partly also from the immense prestige of pure science in this country, which lifts up its outstanding names into public figures. But it is false: the opportunity for brilliant invention and the need for patient teamwork are just as strong in both kinds of science. I wonder what Rutherford's research students, enjoying together the most exciting company in the world of pure science, would have thought of the idea? Or Shockley and Bardeen with their transistor? Or Whittle with his jet engine? Or Kroll, with his process for making titanium, a man who delights in playing David to the Goliaths of the scientific world?

An even worse suggestion is that pure science is adventurous and applied science dull. This false idea has, it seems to me, done much harm to applied science in Britain, frightening away many of the best brains. Adventurousness, conservatism, originality, dullness: these are human qualities that run through the fabric of all human endeavours. They belong to the people doing the work and not to the work itself, certainly not in such a freely creative subject as science. Some of the most exciting pure science at present, the determination of the structure of living molecules, requires extreme patience in repeated chemical extractions and *x*-ray examinations. The adventure is in trying to achieve the goal, not in the day-to-day work itself. There are daring projects in applied science that enthral their research teams with a sense of high adventure. Satellites and rockets; Zeta; making real diamonds from coal; or, looking backwards, Bessemer audaciously making steel by lifting a pool of molten iron up on a bed of air: these are all scientific adventures. To stake everything on a bold enterprise—that is the way to adventure in anything.

Big Projects and Great Laboratories

THE big projects and great laboratories are gradually leading to a new kind of creativity in science. By pooling their individual talents and efforts in a common purpose the members of such organizations are able to attempt new ventures with a boldness and on a scale far beyond the scope of the individual. It is a collective creativity, rather like that involved in, for example, producing an opera or a ballet. In these one needs not only the Stravinsky's and Benjamin Brittens but also all the various orchestralists, soloists, chorus, designers, costumiers, and producers. They all create as individuals, but to the common purpose; and it is the result of all their efforts, the collective creation, that is the finished work of art. Many modern laboratories are striving to achieve the same sort of thing in science.

In some parts of science it has not yet become necessary to work in these large groups. In much of chemistry and biology, for example, and in applied sciences such as metallurgy and electronics, it is still possible to do highly original work on a small laboratory bench with a microscope and a few pieces of glassware. There are certainly advantages in the large laboratory for such small-scale researches; for example, in keeping in touch with the frontier and in the stimulating company of one's colleagues; but great size is not an overriding necessity with them. On the other hand, in other parts of science—for example, in nuclear physics, radio-astronomy, atomic energy, and aeronautical research—there is no alternative. The apparatus is so big.

Organizing scientific research on the scale of big operatic and theatrical productions is still something new in science, and we hardly know yet how to bring it off. Science is behind the arts here. It has nothing equivalent to the centuries-old traditions of the renowned theatrical companies and symphony orchestras, who have learned the secret of holding together large groups of highly individualistic creative artists so as to combine their resources in great single productions. This way of working still feels new and strange to scientists, particularly those from universities where the solitary genius is still the ideal image of the scientist.

The arts undoubtedly have something to teach science here. We have learnt how to train the scientific equivalent of the individual composer, painter, and writer—even the *prima donna*—but have little idea beyond this. How to train the chorus, the orchestra, the *corps de ballet,* how to encourage that delicate balance between individual creative research and collective teamwork, is still far from clear. The armies of Ph.D.s that now march out from the universities each year are our best attempt so far at that problem.

Even worse, we have no idea at all of how to train the people who stage the great productions; the equivalents of Diaghilev, Garrick, Eisenstein, and D'Oyly Carte. The answer may well be that they are not trainable. But at least we ought to have some basis on which to recognize the right qualities and to select whom we should encourage to go in that particular direction. At present we just have to hope that those in possession of such gifts will recognize the fact for themselves. Some do, and the laboratories they build become famous for the sustained brilliance of their work. But it often fails. There are monumentally dull laboratories as well as dazzlingly brilliant ones. Without a tradition of creative leadership to guide us it is inevitable that this should be so. Does a great scientist make the best director of a large laboratory? If so, what sort of man should he be? The solitary man is obviously unsuitable. On the other hand, if he goes too far the other way he may so dominate everyone with his brilliance and personality that he turns his staff into technicians, stifling their creativities by capping every idea with a better one of his own. How does one in fact lead a team of individualists?

Day-to-Day Administration

THEN there is the problem of day-to-day administration. Should the weight of this fall on the shoulders of the director? If so, he has to remain the creative prime mover of the whole thing and yet deal with all the problems of finance, accommodation, staff appointments, and the thousand-and-one cares of any large group of people. If the management goes over to professional administartors, can these remain sufficiently unobstrusive and not spoil everything by tidying it all up, bringing in efficiency, rules, and red tape?

One striking feature of the large research laboratories is that, with a few exceptions, they rarely live long: usefully, that is. They mostly shine brilliantly for about ten years or so and then suddenly appear to burn themselves out. Their dynamism, self-confidence, and sense of direction all seem to disappear. Why does this happen? Is it inevitable? Do the good people leave, or become stale, or does the administration eventually become top-heavy? We do not know.

The people at the laboratory bench also have their problems. Except when you are fortunate enough to be riding a wave of success, full-time research day in and day out can be a nerve-racking business. There is a tendency to feel that you have not justified your existence when your experiments fail to come off and your theories are proved wrong. Worrying about such things takes the joy out of research, and in the end spoils the research by forcing it into timid channels where a mediocre success of sorts is made certain.

The universities have an elegant solution to this problem. There, a man is expected to do both teaching and research, about half his time at each. He makes his reputation by research but makes his living by teaching. The solid, reliable load of his teaching duties assures him of his value to the community and liberates him from the anxieties that can chill the work of the full-time professional research scientist. How is the problem to be solved in the large research laboratory? Part-time teaching might be possible for those near large centres of population but, at best, this could only be a stop-gap solution. Perhaps the old ideas about the solitary genius still haunt the corridors of the modern laboratory too much. Perhaps we shall have to look to the theatrical arts to exorcize them.

26 *American Universities and Federal Research*

CHARLES V. KIDD

UNIVERSITIES NEED SUPPORT for science from government, and government needs knowledge obtainable only by university research. As a result, the two have been placed in a state of unprecedented mutual dependence. The nature and degree of this relation, and the effects of federal research funds on universities and on the federal government, are what this book is about. The preceding chapters have dealth with an aspect of one of the most important questions of our time—the relations between science and society.

Science and Society

FEDERAL funds are a force outside science which organizes and directs science. They are one important way through which society exerts pressure on science. Whether these forces can be exerted without reducing the freedom and productivity of science and of the universities is an intricate question which lies at the root of most of the questions discussed earlier.

Some philosophers of science have maintained that any external forces tending to direct the substance of science must be pernicious. Polanyi, a philosophical liberal to whom the freedom of the individual scientist is an article of faith, takes this position:

> No committee of scientists, however distinguished, could forecast the further progress of science except for the routine extension of the existing system . . . The pursuit of science can be organized,

Reprinted by permission of the publishers from Charles Vincent Kidd, American Universities and Federal Research *(Cambridge, Mass.: The Belknap Press of Harvard University Press, 1959), Chapter 12, "Conclusion," pp. 206–231.*

therefore, in no other manner than by granting complete independence to all mature scientists . . . The function of public authorities is not to plan research, but only to provide opportunities for its pursuit . . . To do less is to neglect the progress of science; to do more is to cultivate mediocrity and waste public money.[1]

If he is right, our federal government is undertaking to influence the direction of science in a way which inherently cultivates mediocrity and wastes public money.

Polanyi's view is not universal. The most extreme dissent is expressed by those who see science as the servant of a materialistically determined state. Bernal and Hogben have stated this view in the context of a consistent philosophy. Bernal, for example, wrote that:

The relevance of Marxism to science is that it removes it from its position of complete detachment and shows it as a part, but a critically important part, of economic and social development . . . Science should provide a continuous series of unpredictable radical changes in the techniques themselves. Whether these changes fit in or fail to fit in with human and social needs is a measure of how far science has been adjusted to its social function.[2]

Hogben has stated a variation on the same theme:

To get the fullest opportunities for doing the kind of work which is worthwhile to themselves scientific workers must participate in their responsibilities as citizens. Among other things, this includes refraining from the arrogant pretence that their own preferences are a sufficient justification for the support which they need. This pretence, put forward as the plea that science should be encouraged for its own sake, is a survival of Platonism and of the city-state tradition of slave ownership. Science thrives by its applications. To justify it as an end in itself is a policy of defeat.[3]

George Sarton has stated a middle position which seems to me esthetically and logically more satisfactory:

Great social events cast their shadows before and after upon science as well as upon other human activities; and however alive and independent science may ever be, it never develops in a political vacuum. Yet each scientific question suggests irresistibly new ques-

1. M. Polanyi, *The Logic of Liberty* (Chicago: University of Chicago Press, 1951), p. 90.

2. J. D. Bernal, *The Social Function of Science* (New York: Macmillan, 1939; London: Routledge & Kegan Paul Ltd., 1939), p. 415.

3. L. Hogben, *Science for the Citizen* (New York: W. W. Norton, 1951; London: George Allen & Unwin Ltd., 1951), p. 741.

tions connected with it by no bounds but the bounds of logic. Each new discovery exerts as it were a pressure in a new direction, and causes the growth of a new branch of science, or at least of a new twig. The whole fabric of science seems thus to be growing like a tree; in both cases the dependence upon the environment is obvious enough, yet the main cause of growth—the growth pressure, the urge to grow—is inside the tree, not outside. Thus science is as it were independent of particular people, though it may be affected at sundry times by each of them.[4]

The question that must be asked is—to pursue Sarton's analogy—whether the environment for science created by federal funds is healthy or unhealthy.

Causes and Effects

THOSE in the physical sciences have difficulty in describing situations fully, and in establishing ultimate cause and effect relations. It should therefore be taken for granted that the effects of federal research funds on universities cannot be determined objectively and precisely.

CULTURAL PRESSURES

The primary reason for uncertainty is that neither the federal government nor universities exist independent of the culture of which they are parts. The omnipresent force of the values of men affect both universities and federal agencies.

One of these forces has been our persisting predilection for the practical, in the sense of the immediately useful. This trait, as de Tocqueville pointed out in 1831, has fostered applied as contrasted with fundamental science:

> In America, the purely practical part of science is admirably understood, and careful attention is paid to the theoretical portion which is immediately requisite to application. On this head, the Americans always display a clear, free, original and inventive power of mind.
>
> Men who live in democratic communities not only seldom indulge in meditation but they naturally entertain very little esteem for it . . .
>
> Possessing education and freedom, men living in democratic ages cannot fail to improve the industrial part of science . . . The local

4. G. Sarton, *The History of Science and the New Humanism* (New York: Braziller, 1956), p. 177.

5. A. de Tocqueville, *Democracy in America* (New York: A. A. Knopf, 1946), vol. 2, pp. 41–47.

conditions and the institutions of democracy prepare scientists to seek the immediate and useful practical results of the sciences.[5]

This national characteristic affects universities as well as federal agencies. In some universities good applied work is valued more highly than good fundamental work. Many faculty members prefer to conduct applied research although funds are available for basic research and the government exerts no pressure on them to do applied work; they are trained for and happy with applied work. For these reasons, the emphasis of university research would not shift markedly towards the basic end of the spectrum if unlimited federal funds were available without strings.

Another force that operates upon both universities and federal agencies is a national tendency to seek bigness as an end in itself. Speaking of university scientists, Paul Klopsteg, the Associate Director of the National Science Foundation, has noted that.

> In their feeling of need for financial aid and their desire to initiate team work, they did not discriminate between the military developments on which they had worked and their own pre-war basic research. The new pattern of large operations had been extended, in their minds, to include their own work. They would be greatly handicapped unless they had a great deal of money and many assistants. In many instances, their university administrations were easily persuaded to adopt the same view.[6]

Since World War II, faith in the power of research has become a national characteristic. The federal government contributed heavily to this development, notably through the development of the atomic bomb. However, widespread acceptance of the efficacy of research, coupled with a mistrust of scientists, has deeper roots. Faith in the power of science has strongly influenced both federal agencies and universities in the postwar years. Medical schools, for example, are conducting research on a scale out of balance with their teaching and community service responsibilities. Since about half of the research is financed by federal agencies, they are often held responsible for creating this imbalance; but the federal agencies and the medical schools are the expression of a national faith in the power of research.

Since pressures for growth are exerted by general cultural pressures, the federal funds have "caused" the growth of research in universities only in a limited sense.

FORCES WITHIN UNIVERSITIES

Forces generated within universities themselves have conditioned the effects of federal research funds. For example, the tradition

6. P. E. Klopsteg, "University Responsibilities and Government Money," *Science*, 124, 920 (1956).

of service to government is strong, particularly among the tax-supported institutions. They feel an obligation to make their faculties available for government research.

Federal funds have provided the means by which universities could caricature themselves if they so desired. Some universities which have wanted to establish large applied research enterprises have been able to finance the expansion with federal funds. Conversely, those universities which have held to rigorous standards of scholarship and intellectual excellence have been able through self-discipline to avoid untoward effects of federal funds, and to use them productively.

The most significant factor affecting university research programs has not been the federal government but the standards of excellence and discrimination maintained by the intangible social pressures of the faculty. The most important effect of the federal funds has therefore been to provide momentum in directions set by cultural values and by forces within universities.

FEDERAL AUTHORITY

The executive and legislative branches of government do, however, decide some important matters affecting university research. By law Congress controls appropriations, an important power. But the executive agencies also have a great deal to say about the level of research appropriations, despite the power of Congress to appropriate. They also have wide latitude to determine the kind of research that will be supported and where it will be done. For example, they have largely independent authority to decide whether federal funds will be used to support research in universities or elsewhere. Sometimes this authority is exercised not only by making funds available but also by placing direct pressure on universities to undertake specific kinds of research. This is the case, for example, when federal officials, primarily those in the Department of Defense, have pressed universities to undertake research not entirely congenial or to assume management responsibilities they would prefer not to have. Finally, the agencies have virtually unlimited discretion to determine the terms and conditions under which research funds will be made available. All in all, the agencies are not entirely passive in the face of cultural forces and forces from within universities.

Productive Growth

WITH all of the reservations that must apply in undertaking to assess causes and effects in a situation as complicated as government-university relations in research, some conclusions seem warranted.

Before stating these, however, the criterion by which the effects of federal funds can be measured should be explained. On can undertake to weigh the effects of the funds against what the situation might have been if the money had not been provided. This is the standard for assessment in this section of the chapter. In the final section the present state of affairs is assessed against a hypothetical ideal; there the future, rather than developments to date, is the center of attention.

University research in virtually all fields is more extensive, and in most fields of higher average quality, as a result of federal funds. The size and complexity of the total research effort have not stifled the individual.

Faculty members of universities are receiving a great deal of federal money for basic research—basic in the sense that the investigations are not limited to attaining an immediate practical end, and that the scientists are free to pursue their own lines of investigation in a congenial working environment. Certainly, a substantial proportion of the most competent scientists in the country work with the aid of federal funds which permit them to do what they want to do. Basic research is not starved for lack of support; in fact, it is better financed in this country than ever before.

The federal funds have improved the quality of graduate training in the physical and biological sciences. They have also tended to increase the number of graduate students, but other forces have been much more important. Graduate students of exceptional intellectual attainments are probably better trained than ever before, in part because of the existence of federal funds. A sharp cut in these funds would cause the number of graduate students in all physical and biological sciences to decline unless a large federal fellowship program or some other compensatory measure was adopted.

On the other hand, the funds may have tended to decrease the quality of undergraduate instruction, but this is by no means certain.

Federal research funds have beyond any doubt been the most important influence in making the university laboratories of the United States very well equipped, with markedly favorable effects on both the quantity and the quality of research. On the other hand, such gadgeteering and over-concentration on collection of data as may have resulted have been attributable in part, but not entirely, to the federal research funds.

Finally, it seems to me that not only are universities and individual scientists not controlled by the federal government but that their freedom has in many respects been extended by the federal research funds. While the federal government may have, in the words of one observer, "threatened constantly to distort universities by giving more emphasis to applied or developmental than to basic research,"[7] a remarkable process of mutual adaptation has kept the threat a potential one.

While these views on the effects of federal funds are subjective, there is wide agreement that research, and universities, are substantially better off with the money than they would have been without it. The dangers of federal domination of science and of pressures towards mediocrity, predicted when federal funds were first made available on a large scale, have for the most part remained potential. When one considers the dangers which have been averted, the financing of university research by the federal government must be judged a success from the viewpoint of both parties.

So far as the future is concerned, the system is imperfect, and potential dangers exist. There is, indeed, no reason to suppose that the slow process of solving the problems created by the federal-university relationships will ever be completed. The government has purposes which are not those of universities. Universities are not in the business of governing, and both government and universities continue to change.

The Means of Reconciliation

IN RETROSPECT, a number of underlying considerations have contributed to a successful solution of the problems generated by the federal funds. These, rather than the direct, current effects of the funds are likely to provide the most useful guides to policy in the future.

THE NATURE OF RESEARCH

The nature of research tends to make the solution of problems in this field less difficult than in some other areas, such as educational policy. As science is international, local traditions, customs, and attitudes are not expected to influence the process and the substance of research. In consequence, there is no tradition that it should be financed and controlled locally. On the contrary, there is a long and strong tradition of conduct and support by the federal government, and a weak tradition of support by state and local governments.[8] For this reason, discussions by university and federal officials are not colored by a general assumption that the federal government should get out and stay out of the business of supporting research.

Research is by nature dynamic. The rate of advance of the substance of science has implications for federal policy—economic, social,

7. J. D. Millett, *Financing Higher Education in the United States* (New York: Columbia University Press, 1952), p. 355.
8. A. H. Dupree, *Science in the Federal Government* (Cambridge, Mass.: Harvard University Press, 1957).

and political—that cannot be ignored. Stresses are brought urgently to the surface. They must then be dealt with as concrete, limited, operating problems by the universities and by government.

BREADTH OF FEDERAL AUTHORITY

The breadth of the missions of federal agencies and the amount of money available to finance research in universities are in many respects a protection to them. Money has been provided for virtually every kind of research. This has tended to reduce the possibility of control by federal agencies. The pressures exerted by individual federal agencies to increase the quantity of research of interest to them have been in large part nullified by the number and variety of pressures from all agencies combined. Similarly, the large number of federal agencies which support work—often work of the same kind—protects science, scientists, and universities. After a period of wavering, the principle of diversity for research support has been adopted as a general policy by the federal government. This principle of operation perpetuates what is unkindly called duplication and overlapping and more charitably called competition in ideas and simultaneous verification. This kind of duplication is abhorrent, however, only to those who value symmetry above productivity. As an English scientist has tartly observed, and with some truth:

> It is urgent, for the sake of the welfare of science, that people who worship mere tidiness should occupy themselves in some suitable and congenial occupation and not strive to impose impossible conditions of work on the original scientific investigator.[9]

RESTRICTED USE OF AUTHORITY

So far as the statutes are concerned, the federal agencies could have imposed virtually any conditions upon universities as a prerequisite to receipt of research funds. The federal agencies had to decide in providing research funds to universities whether the terms and conditions accompanying the money would relate solely to research, or whether an effort would be made to use the funds as a means of securing wise general university policies.

The philosophy that the federal government should decide what is good for universities and then require them to act in the approved manner, or forbid them to act in a disapproved manner, has been expressed in relation to a number of federal research policies. On the ground that secret research has no place in a university, it has been proposed that the federal government should refuse to place contracts

9. J. R. Baker, *Science and the Planned State* (New York: Macmillan, 1945; London: George Allen & Unwin Ltd., 1945), p. 53.

involving secret work with them. On the ground that universities should not depend too heavily on the federal government, it has been argued that they should be required to pay part of the cost of research financed by the federal government. On the ground that the loyalties of faculty members will be divided if their salaries are paid by federal agencies, a general policy forbidding the payment of the salaries of faculty members from federal research funds has been advocated.

Even though the policies suggested might be good ones, the federal government should not impose them as a condition for receipt of research funds. Universities have different missions, and what would be good for one might not be desirable for another. The process of taking into protective custody the judgment of universities has no logical end, and removal of the power of decision and ultimate responsibility would weaken them. Finally, the volume of federal funds is so large that any general assumption as to how the universities should act would take on the character of a self-fulfilling prophecy. Universities would tend to become what the executive branch of the federal government says they should be.

The freedom left to universities creates problems for the universities and for the federal government. Some universities may well adopt policies they will later regret. The federal government will, in taking a rather passive role, probably be called to task for failing to see what its shortsighted policies have done, just as it is now common to hear complaints that the government has caused universities to overemphasize applied research. But such problems can be solved, and they are a small price to pay for freedom. On balance the refusal to require universities to adopt general policies that seem appropriate to federal administrators has been one of the most potent reasons underlying the generally satisfactory relations between federal agencies and universities. In this connection, it is useful to recall that the only serious break between the federal agencies and universities developed when the agencies imposed conditions not related to effective performance of research. This was when scientists engaged in nonsecret research had to meet vague "loyalty" and "security" requirements.

PRODUCTIVE GUIDANCE

How has it been possible to shift the distribution of research effort across the various spectra—basic to applied, the range of scientific disciplines and the array of scientifically interesting problems—without doing more harm than good?

When the consensus among competent people is that it is desirable to alter research effort, a deliberate attempt to change research patterns can be productive. For example, informed people in this country believe that greater concentration toward the basic end of the research spectrum would be wise at this time. Deliberate efforts to achieve this concentration constitute productive guidance, not control. Conversely, the federal government has not forced a heavy investment in fields before, in I. B. Cohen's apt phrase, "the total scientific situation is ripe."[10] The danger of premature research drives has existed in the form, for example, of demands to set up Manhattan Districts to discover a cure for cancer, and it may exist again. But almost without exception the strong, effective external pressures have been for research in "ripe" fields.

Pressures to shift the emphasis of research must be considered not only in relation to science as an abstraction, but also to scientists as human beings. Sometimes investigators turn from what they would otherwise be doing to undertake a task set for them by a federal agency. Such redirection of effort can be disruptive. A scientist generally has an area of work which absorbs him, which he has grasped, within which he is at home and can wander freely, confidently, and productively. Such persons often have strong intellectual and emotional attachments to their fields, and may well work less effectively if they move in response to external factors. On the other hand, some scientists are stimulated by changing their specialties, or by working on a new problem in concert with congenial persons with acute minds. Moreover, a large volume of federal research funds has been provided under conditions which leave the initiative in seeking funds with the investigator and which permit the direction of research to alter in response to investigators' ideas. All in all, the elements of disruption and of productive stimulus may well balance each other.

STRUCTURAL ADAPTATIONS

Structural adaptations in both universities and the federal government have not been mere administrative changes, but an important means of adjusting to the new problems and relationships generated by research.

Within the federal government, new organizations have been created as parts of existing departments or as independent agencies. These organizations reflect recognition that research calls for special administrative structures, procedures and policies. The organizations have been staffed by a new kind of professional federal administrator concerned not with the operation of federal laboratories, but with the provision of research funds to universities. These people typically understand the goals and problems of universities, and of faculty members. In fact, many of them come from universities, and some of them return. The cumulative effort of this rapport between federal and uni-

10. I. B. Cohen, *Science, Servant of Man* (Boston: Little, Brown, 1948), p. 32.

versity people on the decision-making process is substantial. One effect is to move discussions from the level of abstraction where doctrine must be invoked as the basis for establishing broad policies. The working relation is based on a series of limited problems to be solved by persons with common professional backgrounds.

The existence of a network of personal relations at various levels in universities and government may be the means of establishing mutual trust and understanding which could further improve communication between them. There appears to be emerging for the first time, a relation between government and universities which parallels the broader rapport existing in Great Britain between the University Grants Committee and the universities. In explaining the singularly smooth and effective operation of this remarkable administrative device, former President Dodds of Princeton pointed out:

> The conclusion is that the success of the University Grants Committee rests fundamentally upon unwritten conventions and personal and social relations of a homogeneous community of university men, in and out of government, who share common tastes and a common outlook unmatched by any similar relationships in the United States.[11]

In this country, similar rapport may be developing not through wearers of the old school tie, but through common professional interests. It would not be wise to idealize the beneficent effect of the relations which now exist. Some representatives of universities condescend to their colleagues in government. Some governmental representatives do not understand the functions and operations of universities. Nevertheless, the relations are becoming more extensive, closer, and more productive.

Not only have the federal agencies changed in structure and function, but a new type of organization—the research center—has been invented to undertake work for which no other laboratories are wholly suitable. These are not normal government laboratories, even though the physical facilities are owned by the federal government and all operating funds are federal. They are not university laboratories in the usual sense, although many of them are associated with universities. They are not industrial laboratories, although many of them are organized and operated like industrial laboratories.

Universities, on the other hand, have adapted themselves structurally to handle the new and extensive tasks created by federal funds. Complicated business affairs have made it necessary to establish special organizations, ranging from sections of existing business offices to re-

11. H. W. Dodds, L. M. Hacker, and L. Rogers, *Government Assistance to Universities in Great Britain* (New York: Columbia University Press, 1952), p. 113.

search institutes which are in large part autonomous. These offices have been staffed with people who know both university and government business practices. In addition, universities with extensive federal grants and contracts have had to establish means of reviewing at some central point the research that faculty members propose to undertake with federal funds.

In short, both universities and federal agencies have adjusted structurally and functionally to rapid and extensive changes. This remarkable adaptability, demonstrated in the face of complex and rapidly changing problems far outweighs the deficiencies of the system and the unsolved problems. It is, in fact, probably the most important single reason why the system has operated reasonably well.

EFFECTIVE COMMUNICATION

Pressures from society will be exerted on science and they must be exerted without wrecking science. In this vitally important and delicate process of containing social pressures on science, while at the same time stimulating and guiding it towards the service of society, the importance of communication between the worlds of science and politics cannot be overstressed. The persistence of social pressures on science are perhaps the main reason why a sensitive communication between the worlds of science and politics is needed. If scientists and universities do not participate in shaping and directing these pressures, decisions—perhaps unwise ones—will be made for them.

Those acting in a political capacity must impress on the world of science the urgency of social problems to which research can contribute. Conversely, those functioning in a political capacity must be told when fields of study are or are not "ripe," and they must be willing to leave unripe fields fallow. Some of those functioning as scientists must be willing to accommodate to social pressures, and to assist in a realignment of scientific effort when important social goals appear attainable through science.

The administrator and his advisers are in a key position between the world of science and the wider world of social and political needs. One of the most significant tasks of the federal and the university administrators is to adapt science to social pressures while at the same time protecting science against unhealthy social pressures.

Attainment of a workable compromise would be impossible without a sensitive communication system. The problems arising in the search for a workable relation could be solved neither by the agencies, the universities, nor the National Research Council, and the machinery invented and adapted for the task is the scientific advisory group.

The advisory groups, made up largely of scientists from universities, are a new means for deciding whose research will be financed in universities. Many decisions formerly made within single universities have been transferred to faculty members from many universities regrouped by discipline or by research problem. This method of arriving at decisions, which was almost a prerequisite to the successful operation of the system, has shifted important powers of decision from federal administrators and from universities. Both parties have accepted the transfer of power because both have gained much more than they lost.

Scientists throughout the country are well satisfied with the advisory system. Most scientists of stature have some advisory or consulting relation with a federal agency, and consider the function worth while. It gives them, and the scientific disciplines which they represent, a powerful voice in the use of large amounts of research money. The sense of participation felt among university scientists is remarkable.

Apart from the parties immediately concerned, the nation profits from this system. The advisory groups are a means of making the state responsive to the needs of science, rather than the master of science.

Guides to the Future

IN TURNING from the past to the future, I assume that the applications of science to practical problems are so powerful and so pervasive that scientists will not be completely independent of the needs of the community. Large amounts of federal money will probably be available for research in universities for the indefinite future. Given this probability, we should not be immobilized by staring fixedly into the eye of the danger of federal control, or other dire consequences of federal support. The wiser course is to consider how the imperative demands of society can be harnessed for the good of science and society, and how an environment can be created to meet the needs of science.

CONTINUITY OF POLICY

Most of the policies and mechanisms developed over the past decade to deal with federal research in universities are sound. The gradual extension of the degree of freedom given to investigators should continue. The tendency to provide funds in large blocks for research described in general terms should continue. The provision of funds for both basic and applied research from many federal agencies should continue.

The continuation of a sound relation between the federal government and universities, and the removal of some existing difficulties,

12. D. Price, *Government and Science* (New York: New York University Press, 1954), p. 84.

will be made easier if both parties continue to recognize that their objectives are in part different and that the important questions will be resolved through political processes rather than through technical discussions. The important problems relate to the basic role of the federal government in science, the functions of universities and the functions of university, industrial, and federal laboratories. Not only are important questions of goals and values involved, but a great deal of money is at stake. In this connection, Don Price, who has observed at first hand the operation of the research programs of the Department of Defense, wisely noted that: "Whenever any program requires so much money and involves the fate of so many competing institutions —industrial as well as academic—that program is certain to become loaded with political views."[12]

Many problems will be resolved through mutual accommodation, adjustment to the pressures that each can and will bring to bear to ensure the attainment and vital institutional goals, and adherence by each party to what it considers essential.

ADAPTABILITY

Maintenance of the adaptability that has characterized government-university research relations over the past decade is of prime importance because the substance of research will shift and national needs that can be met by research will change. Over long periods of time, the relative urgency of stimulating fundamental as contrasted with basic research may change in this country as it has in others. A recent British report asserted, for example, that the chronic economic difficulties of Great Britain are caused "particularly by the failure to exploit the results of scientific research. While we more than hold our own in fields of pure science, we lag behind other countries in the application of science to the development of new processes."[13] Not only will research change, but the broader question of the role of the federal government in relation to all of the functions of universities will come to the fore. In this connection, the lack of balance between resources for teaching and for research will have to be dealt with, and the role of the federal government in redressing the balance will be a central problem over the years ahead.

RESEARCH OUTSIDE UNIVERSITIES

The research centers that were so prominent during and after World War II should be continued and expanded. The centers are uniquely able to carry out certain tasks. More significantly from the

13. *Sixth Annual Report of the Advisory Council on Scientific Policy* (1952–53), report on the exploitation of science by industry (London: Her Majesty's Stationery Office, 1954), Cmnd. 8874, p. 1.

point of view of universities, they are a means of avoiding mere bigness. President Stratton of the Massachusetts Institute of Technology has pointed out the implications for universities of the sheer bulk of much modern research:

> New patterns of research are emerging that will be difficult to reconcile with the true spirit of a university. To breach the barriers of knowledge, we have learned the effectiveness of team action backed up by adequate modern equipment. . . . If, now, we have discovered a method that is effective, then it is inevitable that it should be developed and expanded. If, in fact, the results of fundamental research are as vital to the national welfare as we have claimed, then we would be lacking in our responsibility if we were to fail to push the organization of scientific research to its ultimate conclusion. This road apparently leads to larger laboratories, a growing costliness of facilities, and an increasing need for the planning of programs.
>
> There is a basic incompatibility between the true spirit of a university and those elements of management which tend to creep into the organization of projects, the planning of programs, and the utilization of costly facilities. One must ask whether the universities can by themselves satisfy the need for all the fundamental research that appears necessary in this country and whether it is wise for them to attempt to do so. . . . One must recognize that there may be an ultimate need to establish central institutions to supplement the universities in fundamental research. . . . If we strive to contain the widening scope of research entirely within our large universities, we shall end by changing their character and purpose. In so doing, we shall render the greatest possible disservice to the cause of research itself.[14]

In the interest of universities, an increasing proportion of the nation's research, including basic research, should in the future be conducted outside universities in order to sustain the freedom and flexibility of the university environment. For this reason, laboratories outside universities should not be regarded as competitors. They should be viewed as a means whereby the volume of research required in response to irresistible economic, political, and humanitarian forces can be undertaken without forcing universities to undertake inappropriate work.

Unresolved Problems

WHILE a reasonably successful reconciliation of the goals of the federal government and of universities has been achieved, and while

14. J. A. Stratton, "Research and the University," *Chemical and Engineering News*, 31, 2582 (June 22, 1953).

clear guides to the future exist on some questions, difficult unresolved problems remain. Many questions relating to administrative structures for science are unresolved. More significant problems relate to the effects of federal activities upon the substance of research and on the strength of the nation's educational system.

STRUCTURES FOR RESEARCH

Despite the remarkable adaptation of structural forms within universities and government to meet the demands of a rapidly expanding national research effort, the lag typical of the adjustment of organizational forms to the tasks imposed upon them exists in universities and in government.

Most universities are still evolving policies and structures which will reconcile the desires and needs of individuals with institutional goals.

Even though the research centers are a means by which pressures on universities to conduct inappropriate large-scale research can be reduced, the centers absorb scientists. Most of the scientists come from universities, and rapid growth of centers can impair the capacity of universities to train the next generation of scientists. For this reason, the rate of growth of centers, their relations to universities, and the nature and extent of their teaching responsibilities comprise one of the most complex and significant problems facing government and universities over the years ahead.

Within the federal government, the fact that there has been no point within the executive branch at which individual universities could present their total needs and resources poses a problem.

More significantly, the executive branch of the federal government has not been organized in a manner which has permitted an effective, continuing examination of the total effects of the research activities of all federal agencies upon higher education. Most of the federal agencies which distribute funds to universities have a research and not a training mission. Even those with a training mission, such as the National Institutes of Health, the Atomic Energy Commission, and the Office of Naval Research, have no statutory authority to concern themselves with the long-run effect of federal research expenditures on higher education in this country. They may do so out of a sense of responsibility, within limits set by law, but the National Science Foundation is the only agency with a clear mandate. The agencies—such as the Department of Defense, the Department of Agriculture, the Atomic Energy Commission, and the National Institutes of Health—which have made the decisions having the greatest effect upon universities are not the places where general federal research policies relating to universities can be formulated.

The National Science Foundation has from its inception recognized the significance of these problems. In 1958, the Foundation produced the most thoughtful report, *Government University Relationships*, yet issued on these matters. The questions need even more sustained and close inquiry.

Apart from studies, the Foundation is already playing a unique and important role in bringing to light and resolving problems relating to education for the sciences at the high school through the graduate school levels. These activities are being extended to make the Foundation an even more effective catalytic agent for bringing individual universities and federal agencies into a relation permitting review of the total effects of the federal programs on the university. Finally, the development of a strong network of advisory groups interested in this problem, including persons in both scientific and administrative positions in universities, could help the Foundation to deal with a major weakness of the entire federal research relation with universities.

At the top, it is important that the preservation of a strong, diversified, free system of universities be viewed as a goal as important as the attainment of the operating objectives of Federal agencies. This attitude should be diffused throughout the agencies and reflected in their actions. At the same time, concern for the vigor and stability of universities should be combined with and tempered by other considerations of equal importance, such as the proper division of federal research among federal laboratories, industrial laboratories, and universities, the relations between strategic military plans and scientific effort, and science in relation to foreign policy.

By moving to the White House responsibility for those aspects of national science policy that are effectively dealt with centrally and by emphasizing the responsibilities of the National Science Foundation for the promotion of basic research and for education in the sciences (Executive Order 10,807, March 13, 1959), the President strengthened the structure for dealing with the effects of federal research on universities. The structure was further strengthened by reconstituting the President's Science Advisory Committee as a group of nongovernmental advisers reporting directly to him. The appointment of the Special Assistant to the President for Science and Technology was another important step. Finally, the problem of securing coherent policies among the agencies was attacked by the establishment of the Federal

15. *Science and Technology Acts of* 1958, analysis and summary prepared by the staff and submitted to the Senate Committee on Government Operations on S.3126: A Bill to Create a Department of Science and Technology (Washington: Government Printing Office, 1958). While the staff arguments for creation of the Department seem strained, the document contains comprehensive references to the question of a Federal Department of Science.

In contrast, Lloyd Berkner's proposal, "Government Sponsorship of Scientific Research," *Science*, 129, 817 (1959), for a Department of Science whose function would

Council for Science and Technology. Significantly, an answer was not sought through recommending the establishment of a Department of Science, a proposal made periodically since 1884, and most recently in 1958 as S.3126, a Bill to Create a Department of Science and Technology.[15] Science should not be centralized, but should continue to remain diffused throughout the federal establishment. Moreover, problems of policy relating to the whole executive branch can be dealt with most effectively not in departments but in the Executive Office of the President.

While the federal structure continues to evolve toward a stronger, more solidly based and logically articulated machine, I think that additional steps will eventually prove desirable. The establishment of administrative devices through executive order lacks the sanction of Congress, and those characteristics of the Executive Office of the President which are most enduring are those established by statute. The next step in the evolution of a fully adequate structure for science policy at the top of the executive branch might therefore be the establishment by statute of a science and technology office in the Executive Office of the President. This organization might well be analogous in structure and function to the Council of Economic Advisers. Such a group would have the proper status, and it would be an enduring part of the Executive Office of the President. The establishment of a Congressional Joint Committee on Science and Technology, analogous to the Joint Committee on Atomic Energy, would further strengthen the total federal structure for dealing with science.

Since a statutory Council for Science and Technology would in a certain sense be a part of the executive branch of government, the need for vigorous, independent criticism of the scientific policies of the Federal government would become even more important. This role is one which might be played most effectively by the National Academy of Sciences—National Research Council.

The federal structure for science policy is much stronger than it was a decade ago, and adaptation is proceeding at a reasonable rate. The years since World War II may well be viewed in the future as a period when activities of the government relating to research in universities grew at an extremely rapid rate, and somewhat at random. Over the next decade the affairs of federal agencies and those of uni-

not be the formulation of science policy at the top but rather strengthening "governmental scientific and technical services not principally involved in attaining existing department objectives or strongly related in the organic sense to the functions of a single federal department but of the utmost importance to the Government and the people as a whole" is sound and should in my opinion be adopted. The only significant objection to the proposal, which Berkner pointed out, is that "Congress would end by dumping all the scientific activities of the departments and agencies into such a department," a move that "would, of course, be little short of catastrophic."

versities will become more closely intertwined. The adaptation of structure will have to continue if the problems generated thereby are to be solved reasonably well.

SUBSTANCE OF RESEARCH

Structures should be designed for the effective performance of functions, and the most significant unresolved problems relate to the substance of research and education rather than to the administrative forms.

While basic research is now better supported than ever before, gaps exist because some important investigations are outside the area of interest of any agency except the National Science Foundation. For example, investigators of high competence in such areas as bioluminescence, archaeology, meteorology, pure mathematics, linguistics, astronomy and radio astronomy, field studies in biology, and geology have received inadequate support. In addition, science has evolved over the past few years in a manner emphasizing needs for larger and expensive physical facilities for basic research. Not only in nuclear physics, but in such fields as oceanography, meteorology, and materials research, large new research equipment is needed. There is little prospect of filling these glaring gaps without substantial aid from the federal government.

An important task of the National Science Foundation in supporting research is to provide money to promising scientists and promising ideas in neglected fields of science, and those fields whose relevance to the solution of practical problems is not obvious are most likely to be neglected. This task is not solely support of general-purpose basic research, even if the term were operationally usable, but the wider one of filling any kind of gap that may appear in basic science. For example, analytical engineering—the sophisticated scrutiny of available knowledge to determine what kinds of "things" can be built and what can be invented with existing knowledge—has been inadequately supported.

Looking ahead, extension of the postwar progress in securing support for basic research is important. In this connection, the National Science Foundation's budgets for support of research have been skimpy until fiscal year 1959, when the budget of the Foundation was almost tripled to a level of $140 million. Over the period 1953 through 1957 the Foundation has been able to support only 28 percent of the $140 million requested.[16] While such statistics must be viewed cautiously, work in some fields has suffered because of inadequate financial support.

At the same time, it should be borne in mind that lack of money

16. National Science Foundation. *Seventh Annual Report, 1957* (Washington: Government Printing Office, 1958), p. 56.

is not now the most important barrier to the establishment of greater national strength in basic research. The more acute problems relate to the training of manpower, the effects of heavy research support upon universities as educational institutions, and the terms and conditions under which federal funds are distributed to them.

With respect to all kinds of research, the terms and conditions under which federal funds are provided to university scientists should be further improved. Funds are still dependent upon year-to-year appropriations which may fluctuate sharply. A productive next step would be to place research and development appropriations on a two- to five-year instead of a one-year basis. Much federal money is still provided for research that sets narrow research tasks and requires specific performance from the investigators. Here, it is useful to recall that the obligation of the federal agencies is not to meet the full wishes of the universities. A degree of incompatibility between the objectives of the agencies and universities is built into the system because the agencies have functions that are not necessarily best discharged by acting in a way most congenial to universities. They may adapt their policies to the extent that maintenance of strong universities is a governmental concern compatible with attainment of the direct governmental objectives. Within these limits, the agencies do have an obligation, thus far imperfectly met, to concern themselves with the strength of universities.

Education and Research

THE interrelated tasks of training and using manpower, strengthening of our total educational structure for the production of scientists, and fostering the development of highly talented students are emerging as functions that are as significant to the nation as support of research.

The critically important problems of training enough scientists well and of using them well are far from resolved. The support of scientific training provided as a by-product of support of research has been useful, but fortuitous and not entirely satisfactory. A much wider federal scholarship and fellowship program is needed to ensure that no talent remains untrained. This need was met in part by enactment of the National Defense Education Act in 1958. In addition, the status of the large group of professional nonfaculty scientists loosely attached to universities should be improved. But there appears to be no way of using this talented group more effectively until universities have enough money to hire them as faculty members.

The federal research funds have affected not only research in universities, but all their functions. For the long run, the effects of the funds on the educational functions of universities may be more significant than their effects on research. Such unresolved questions as these remain:

How are universities to exert countervailing forces against the pressure to expand the physical sciences?

What are the implications of an expanding volume of research for the capacity of universities to teach the impending wave of students?

How are universities to attract and keep teachers with research talent in the face of competition from industry, financed in substantial part by federal money?

How are universities to withstand the temptation to accept federal funds which would over the long run divert them from their central functions?

If federal research funds pose a threat to the freedom of universities, the primary threat is that universities will not find satisfactory answers to questions such as these, rather than that bureaucrats will exercise arbitrary authority.

There is, finally, the difficult question of general federal aid for the financing of universities. This is related to federal aid to research for two reasons. First, such aid is already being provided indirectly and in unmeasurable amounts from research funds through such channels as provision of equipment inevitably and properly used for teaching, provision of income (often including partial or total payment of salaries) to teachers, and payment of the full costs of research. Since the role of the federal government in supporting higher education is traditionally a subject for heated debate, it is remarkable that the significant Federal payments to higher education derived from research funds have not been more vigorously argued. The discussion has been muted because the support has been piecemeal, dispersed among a number of Federal agencies, and a by-product of the less debatable function of aiding research. Second, the question of general aid to universities arises in a more significant form in relation to the need of the nation not only for a strong system for producing scientists but also for a system able to educate citizens for a democracy in times of unprecedented change and challenge.

So far as capital expenditures are concerned, the nation's entire system of higher education is underbuilt and underequipped. The Special Assistant to the President for Science and Technology, Dr. James R. Killian, pointed out early in 1959 that:

> Of the 700 colleges and universities in the United States which offer graduate degrees, something less than 300 offer graduate degrees in scientific, medical, and engineering subjects, and not enough of these can claim real distinction in any one scientific field. The few top-quality departments are in danger of becoming overloaded, and

> the peak in graduate enrollment is still several years off. It is especially important that we build more first-rate graduate schools of engineering and that they be developed in close association with excellent departments of science.[17]

It will require hundreds of millions of dollars for capital construction to meet existing and emerging educational needs. Part of the needs will be met by private gifts and by state appropriations. But unless trends in state support and private giving are sharply reversed, some form of general federal aid for construction will also be required if the problem is to be solved in time. Apart from fiscal questions, which are not to be ignored, there is ample precedent in the activities of a number of federal agencies for the policies which are necessary to provide federal funds for construction of educational and research facilities without impairing the freedom of universities.

Whether federal aid for the general operating costs of universities will be required is not relevant here, but the research experience is relevant to the question whether it would be possible to provide federal funds for education without introducing federal control. The threats to the freedom of universities encountered in providing general aid would be much the same as those encountered in providing research aid, for teaching and research are inextricably interwoven. The research experience leads me to the conclusion that, if general federal aid for higher education will be required, the possibility that funds can be provided without restricting the freedom of universities is bright. The general federal research policies point the way. These include the development of a sound set of guiding principles with extensive help from nongovernmental advisory groups, a pragmatic administrative approach, maintenance of diverse sources of support, and powerful sources of independent, responsible criticism.

Values

UNDERLYING all of the specific unresolved problems relating to government-university relations are the cultural values and pressures that have played such an important role since World War II. These will continue to exercise a dominant influence on the policies of both government and universities. Those facets of our national culture which have in the past militated against intellectual excellence as a paramount goal will probably change slowly because they are so deeply ingrained. Until learning of all kinds and exploration of the unknown in all fields are more widely respected as ends in themselves, social

17. J. R. Killian, "Science and Public Policy," *Science,* 129, 134 (January 16, 1959).

pressures will be exerted away from rather than toward intellectual excellence. The attitudes which together comprise a national judgment that intellectual excellence is not highly admirable are firmly embedded. They affect the tone of government and every level of education. They can be altered only slowly, and only by extraordinarily strong forces.

Attitudes toward secondary education must change. This involves general willingness to do the things—such as raising salaries and lifting the intellectual standards required of teachers—that are necessary to attract a higher proportion of highly intelligent people as secondary school teachers, and to spend the required billions. Changes of this kind are not easily brought about. As Archibald MacLeish has pointed out, "a radical change in educational policy cannot be ordered as an automobile manufacturer orders a new model. When it comes, it will come out of a change in the community's conception of itself."[18]

Widespread attitudes toward the functions of higher education must also change. In this connection a perceptive report of the Carnegie Foundation for the Advancement of Teaching has observed that:

> One would like to believe that Congressmen, the heads of Federal agencies, and others concerned with Federal action in higher education will be guided by something approaching a philosophy of higher education and a sense of its role in American society. This could occur only through the vigorous and enlightened action of America's leaders in higher education . . . Now the truth is that measured against this challenge, public understanding of American higher education leaves much to be desired. And it must be admitted that this is partly the fault of our educational leaders.[19]

Part of the problem is that some universities, reflecting important aspects of our society, do not establish an environment where intellectual excellence is valued most highly. A university system closer to the ideal of a company of scholars engaged in creation, transmission, and preservation of knowledge would involve a change in the tone, the values, the rigor, and the intellectual discipline of many universities.

The most important prerequisite to the establishment of a stronger university system and to the achievement of other national goals in education and science is the strength of the national will to give first place to education at all levels, and to the fostering of excellence in all fields.

18. A. MacLeish, "What Is a True University?" *Saturday Review* (January 31, 1959), p. 13.

19. Carnegie Foundation for the Advancement of Teaching, *Annual Report for 1956–57:* Summary of a discussion by the trustees on federal programs for higher education (New York: Carnegie Foundation for the Advancement of Teaching, 1957), p. 16.

27 *Planned and Unplanned Scientific Communication*

HERBERT MENZEL

UNDER A GRANT from the National Science Foundation of Washington, D.C., the Bureau of Applied Social Research of Columbia University has undertaken to explore ways in which communication research by interview survey methods can contribute to an understanding of the needs and means of scientific information-exchange. On the basis of such an understanding, proposals to improve scientific communication might be generated and evaluated. As a first step, it was decided to study the information-exchanging behavior of the biochemists, chemists, and zoologists on the faculty of a single academic institution—a prominent American university.[1] This paper reports selected results. A more complete account is on deposit with the National Science Foundation.

The objectives of the research that is ultimately envisaged had been defined as follows:

Reprinted from Proceedings of the International Conference on Scientific Information, 1959, *pp.* 189–212.
This paper is Publication A-259 of the Bureau of Applied Social Research, Columbia University. It is based on a pilot study carried out at the Bureau under the supervision of Charles Y. Glock. William A. Glaser and Robert H. Somers collaborated with the author in the execution of the work.
The Bureau operated under a grant from the Office of Scientific Information of the National Science Foundation. The encouragement given this work by Helen L. Brownson and Harry Alpert of the Foundation is gratefully acknowledged. Special thanks are due the biochemists, chemists, and zoologists whose generous contributions of interviewing time and attention made this work possible. Their visible interest in the matters discussed was a source of continuous stimulation.

1. The 77 scientists whose interviews are analyzed here include all but 8 of the following: teaching faculty in biochemistry; teaching faculty and research associates in chemistry and zoology; provided they were in residence on the campus of the university during the spring of 1957. (Four biochemists, one chemist, and one zoologist refused to be interviewed, or to complete an interrupted interview. Two zoologists were interviewed for background information only.)

1. To distinguish the types of informational *needs* which scientists have, and to determine in what respects they remain unsatisfied.

2. To examine the *means* and occasions of scientific information-exchange, in order to single out the features which make them more or less able to meet the scientist's several needs.

3. To analyze characteristics of the scientist's specialty, his institution, and his outlook as possible *conditions* which influence his needs for information, his opportunities for satisfying them, and, hence, his information-gathering habits and felt satisfactions.

The exploratory study was intended to define problems, categories, and procedures for more systematic investigation. Although this report contains numerous frequency counts based on interview responses, they are to be regarded as illustrations of the possible outcome of further work and not as reliable findings. They may not even reliably describe the three academic departments studied, since the interview schedule was continuously modified and developed as the work proceeded.[2]

A SYSTEMIC VIEW

While the population of the initial study is small, it was decided to cast a broad conceptual net: to consider *all* the channels through which scientists exchange and gather information, and *all* the functions which scientific communication facilities are called upon to perform. In fact, the functions which the facilities must serve were made the organizing principle of the study. Rather than to ask, at the outset, "How well does this journal perform? How much does this meeting accomplish? What is wrong with that indexing system?" it was decided to ask as a first set of questions: "What are the functions of the scientific communication system? What mechanisms are now available for performing them? What are the inadequacies in the present performance of each function?"

This decision was founded on the belief that specific topics for investigation can be wisely selected and defined only after the broader context has been scanned. Furthermore, studies of communication processes among nonscientific publics[3] had shown that different communication functions are often performed, at their best, through different channels, and that the diverse channels may supplement one another in intricate ways.

Accordingly, we include in our purview not only the scientific literature and its manifold storing, abstracting, and indexing appendages; not only the formally established meetings and conferences, but also

2. The schedule contained both structured and unstructured questions. A copy of the most recently used version is appended. The average interview took 1.9 hours.

3. Many such studies are reviewed in Elihu Katz and Paul F. Lazarsfeld's *Personal*

the informal, person-to-person modes of communication like correspondence, visits, and corridor conversations.

Secondly, we conceive of scientific communication as not necessarily limited to simple transactions between an individual scientist and a source of information. Communication includes more complex processes: several different channels of communication may have to interact to complete a transaction; one or more persons may serve as relays between the source of a message and its ultimate consumer; and contacts at each intervening step may be initiated now by the receiver, now by the bringer of the message. For these reasons we shall speak, somewhat loosely, of the "scientific communication system," meaning the totality of all publications, facilities, occasions, institutional arrangements, and customs which affect the direct or indirect transmission of scientific messages among scientists.

Thirdly, we believe that policies to improve the scientific communication system must be planned in terms of the entire range of its contributions to scientific progress, and not only in terms of the most obviously necessary informational services. At the present time, most plans are quite naturally directed at maximizing the efficiency of the system in the performance of its two most obvious functions: that of bringing scientists the available answers to specific questions, and that of keeping them abreast of current developments in given areas. Yet the policies that recommend themselves for these purposes may not be adequate to assure the fulfillment of other functions of the scientific communication system; and it is just possible that some of these same policies, say the shortening of papers at meetings or the streamlining of periodicals, may be detrimental to the system's other functions, which are not so obvious, but nevertheless important.

MULTIPLE FUNCTIONS OF THE SCIENTIFIC COMMUNICATION SYSTEM

The scientific communication system serves the progress of science not merely through the reference services it performs and through keeping scientists up to date in their chosen areas of attention. It serves in a variety of other ways as well: by enabling scientists to brush up on past work in additional areas; to verify the reliability of one source of information through the testimony of another; to ascertain the current demand for research on given topics; to locate rare materials; and so on. In fact it would be a mistake to think that the functions of scientific communication for the progress of science are limited to satisfying the informational needs of which each scientist is aware; that the only im-

Influence: The Part Played by People in the Flow of Mass Communication, New York: The Free Press, 1955, and in Elihu Katz' "The Two-Step Flow of Communication," *Public Opinion Quarterly,* 21, 1957, pp. 61–78.

portant job of the scientific communication system is, so to speak, to give each scientist what he wants, and knows he wants.

One important function of scientific communication which transcends the informational requirements each scientist can define for himself is that of directing the scientist's attention to new topics beyond those with which he has "kept up" in the past. Another is to assure the eliciting of suggestions and criticisms from fellow scientists. These and other rarely considered functions of the scientific communication system, and some of the mechanisms by which they are satisfied, are discussed in the report which has been deposited with the National Science Foundation.

Keeping Scientists Up to Date

YET even the performance of functions of which everyone is well aware requires more than the prompt appearance of information in the official channels (journals, meetings, etc.) and more than painless access to these media. It is, for example, not these formal media alone that keep scientists informed of current developments in their chosen areas of attention, in spite of the prodigious amounts of planned effort devoted to this communication function by individual scientists as well as by the professional organizations and publishers.

In fact, the news which comes to the attention of scientists is not restricted to the information obtained when they intentionally "gather information," as it is called. Fortunately so! For a good deal of the news comes to their attention in unplanned and unexpected ways, during activities undertaken and on occasions sought out for quite different purposes, proves to be of considerable significance to them. At least this was a frequent experience among the scientists we interviewed, in spite of the fact that their intentional activities for gathering information about current developments ranged all the way from the assiduous perusal of current periodicals to the button-holing of colleagues who had returned from conferences.

It was thought that it would be instructive to examine instances of significant scientific news coming to the attention of scientists through other ways than those which they systematically employ to "keep up." This line of investigation was included in our study not only to learn about the operation of communication through informal and personal channels; by implications, it was thought, this approach would also throw light on possible inadequacies in the formally established methods of bringing current news to the scientist. Under an ideally functioning communication system, it was thought, the routinized and regular

4. Unless otherwise indicated, indented matter is quoted from scientists' statements during the interviews. Matter in brackets [ ] paraphrases or supplements

methods of gathering information would convey to each man all the scientific news that is pertinent to his work. Any pertinent information that actually reaches a man in some extra-routine or accidental way would therefore indicate a service not adequately performed through the routine system. As will be seen below, however, this form of interrogatory also taught some unanticipated lessons which are, perhaps, even more important.

In order to obtain records of instances of useful scientific news obtained during activities not undertaken for this purpose, the following question was asked:

> Have there been any instances where some unlooked-for piece of information came your way that turned out to have bearing on your work?
> (If Yes) Tell me about the last time this happened.

Supplementary questions were asked in order to obtain complete accounts of the experiences. Not all the replies given proved pertinent to the present topic. Excluded from the list finally used were all accounts of information obtained in the course of routinely scanning the literature, attending meetings, or engaging in any other activity which was explicitly designed to find out what is new. Also excluded were episodes of information learned in the course of ordinary intercourse with departmental colleagues. Thirty-five usable accounts were obtained.

UNPLANNED MECHANISMS

The extra-routine mechanisms by which these messages reached the interviewed scientists are of four basic types:

1. The scientist *searches the literature for one particular item of information, and in the process stumbles across another* which proves useful to him. This, of course, is in addition to the countless times when a scientist comes upon some useful information which he had not anticipated in the course of his routine perusal of journals, or in the course of listening to the program of meetings which he regularly attends. What is meant here is rather a scientist searching the literature in order to find the answer to some specific question, and coming across pertinent information of another sort, information which he would probably not have seen had it not been for the accident of his search for the first topic. Thus a paleontologist reported:[4]

> This morning my assistant wanted information on the geology of Southern Britain at a certain time. The same journal happened to contain information which will be interesting to our formal analysis

scientists' statements. Matter in parentheses ( ) quotes what the interviewer said to the scientist.

problem. . . . (Do you think you would have seen this particular item otherwise?) It is doubtful.

2. The second kind of situation which leads to the unexpected acquisition of information pertinent to one's work occurs when a scientist, in the course of contact for another purpose, *informs a colleague of his current work* or of some obstacle which preoccupies him at the moment, *and is rewarded with an item of information* that becomes important to his progress. A zoologist made an unexpected find in this way:

> I learned in this way the whole technique for solving the problem of the recording technique for . . . . This had baffled scientists for twenty years. I went to a man at . . . Institute to buy some wire—Dr. . . . . , in the . . . Laboratory. He is an able and imaginative fellow. In conversation, we talked about my research and problems, and he dropped the hint which enabled me to solve the problem of the recording technique.

3. Sometimes a scientist hears about new developments from *a colleague* who *volunteers the information* while they are thrown *together for another* purpose. (We exclude here information gleaned during corridor conversations at scientific meetings, or any other occasion attended for the explicit purpose of gathering news.) This may happen frequently during informal visits by one scientist to the laboratory of another. For example:

> I went to the . . . Institute two months ago to give a talk. I stopped to see Dr. A. He is working in a different field. He found that a certain substance crystallized under certain conditions. We are interested in finding many different kinds of crystals. We will try using his experimental methods here. What Dr. A. found may not ever be published by him—it was a side effect, as far as he was concerned.

4. There is a fourth manner in which information of immediate relevance comes to the attention of scientists by what appears to be accidental routes. Frequently a colleague will deliberately seek out a scientist whom he knows to be interested in the matter, *in order to convey to him some information* that he happens to have heard. Thus, for example, a biochemist:

> . . . started a new project because I heard that someone in Germany had positive results in a related field. He published it one-half year later. . . . (How did you hear about the German scientist?) He had sent his unpublished results to another man in America who knew my interests and told me.

And a chemist gives a very detailed and instructive account of such an incident:

> One of the problems in our work is to do a certain chemical separation. Recently a friend of mine had been in Europe. He met a young German who was developing a new technique. So we now try to apply this to our problems. Neither my friend nor I knew about the existence of this procedure before the encounter. The young German had invented this. My friend had not been looking for it—he was going through Europe, visiting labs and drinking beer with the people at the various labs. This technique was of no particular interest to my friend, but he knew it would interest me, and told me when he got back.

Table 1 shows how many accounts of experiences of each of these four kinds were reported by the interviewed scientists. It also shows that about half of the messages transmitted in these personal ways were actually in print at the time.

TABLE 1. Useful Information Obtained "Accidentally": How Obtained *

Manner in Which Message Reached the Interviewed Scientist	Number of Messages		
	Total	*Published*	*Not Published*
Found in the literature while searching for another topic	4	4	—
Contributed by a fellow scientist upon being informed of colleague's current work	13	8	5
Spontaneously mentioned by a colleague while together for another purpose	4	—	4
Specifically addressed to the interviewed scientist by a colleague	9	2	7
Other, or not specified	5	2	3
Total	35	16	19

* Exclusive of information learned during ordinary intercourse with departmental colleagues, while scanning the literature, while attending meetings, or while engaging in any other activity explicitly designed to find out what is new.

CONTENT OF INFORMATION OBTAINED

What was the content of these messages which reached their consumers in such unexpected ways? Ten of the messages informed the scientist of new findings or principles (a biological mutant described, an archaeological find reported, a chemical reaction performed, etc.); eight informed him of the existence of new techniques, procedures, or apparatus; four furnished him details on the performance or adaptation of a technique; five told him who was doing work on a given topic or from whom a particular material could be obtained (Table 2).

TABLE 2. Useful Information Obtained "Accidentally"; Content *

	Number of Messages		
Content of Message	*Total*	*Published*	*Not Published*
New findings or principles	10	6	4
New procedure or apparatus	8	3	5
Details on procedure	4	2	2
Who does what; where to obtain material	5	1	4
Not indicated	8	4	4
Total	35	16	19

* Exclusive of information learned through ordinary intercourse with departmental colleagues, while scanning the literature, while attending meetings, or while engaging in any other activity explicitly designed to find out what is new.

The figures in Tables 1 and 2, as most of the figures throughout this report, are to be regarded as no more than suggestive and illustrative of possible findings of more systematic studies, as mentioned earlier. What they do suggest is that the apparently accidental obtaining of pertinent information plays a large role in the work of the interviewed researchers. Examples of it come up again and again in our interview protocols.

INDIVIDUAL ACCIDENT—AGGREGATE REGULARITY?

Why should this manner of learning of new developments be so prevalent? Part of the reason must be sought in the nature of specialization among basic researchers at the top level. They not only specialize to a high degree, but they also delineate their specialties in highly individual and original ways; often no more than a small handful will be specializing in precisely the same area. All the possible ways of classifying content cannot possibly be taken into account in the organization of journals, in the indexing and abstracting services, or even in the selection of titles for papers. Any given researcher is likely to find that the way of classifying reports which would be most relevant for his purposes has not been used. Within the confines of a narrow field, he attempts to scan everything that comes out; but beyond that he must depend largely on friends who work in the adjoining specialties, yet know what is of interest to him, to flag the pertinent material for him.

> You see what happens around here [says a biochemist]. Everyone knows what problems you're working on. Whenever you come

across something which might be of interest to another you make a note of it. This way the individual is able to be acquainted with a lot more than he would be if he didn't have the others on the look-out too.

If this is true, it becomes imperative to consider the information network as a system, and not merely as an aggregate of information-dispensing or information-consuming individuals. What is little better than an accident from the point of view of the individual may well emerge as an expected occurrence from a larger point of view. For while there is only a small likelihood that any accidentally obtained piece of information will be of use to the individual scientist who obtained it, the likelihood that it will be of interest to at least one of his departmental colleagues is much larger. And if enough members of a given department or research group are plugged into branches of the professional grapevine through consultantships, secondary appointments at other institutions, committee services, and personal correspondence and visits, they may collectively be able to assure each of them a good share of the news about work in progress that interests him.

The formal and organized means of communication—especially the periodical literature, including its voluminous abstracting services and review publications—serve the scientist most efficiently when he knows precisely what he is looking for, when he needs the answer to a specific question. When it comes to bringing scientists together with information the significance of which to their own work they have not anticipated; when it comes to pushing out the frontiers, it may be that the system of informal and "accidental" means of communication, inefficient though it may be, is as reliable a mechanism as one can get. In fact, the very frequency of the "chance" occurrences of information transmittal, which was illustrated in the preceding pages, suggests that they may not be altogether accidental; perhaps, if knowledge of a particular item hadn't come to the researcher one way, it would have come through another, although a little later.

One chemist told us of an experience which seems to bear this out rather dramatically. He had done some experimental work in 1955 and had published a report without fully realizing the relevance of his work to the chemical theory of a certain reaction mechanism. Between 1955 and 1957, he was led to earlier literature which suggested this significance of his experiment to him. During the same period, this fact was also brought home to him through three contacts with other scientists which had ensued from his work in three quite independent ways. To what extent one can depend on these apparently fortuitous mechanisms of communication to bring the right combination of scientist and information together is, of course, not known. It is, however, worthwhile to consider the totality of information exchanges among scientists as a system, to accept what appears as "accidental" communication as part of the system, and to examine the ways in which the system, including its unorganized components, may be made to operate more efficiently and more reliably.

Furnishing Answers to Specific Questions

THESE pages make it fairly clear that the functions of the scientific communication system extend beyond the bounds of enabling the scientist to get the information which he knows he wants. Yet even to get the scientist the information which he knowingly seeks takes more than the means officially established for this purpose. This can be seen when one examines the ways in which scientists secure the available answers to specific questions. Yet no other function of the scientific communication system has received more solicitous care through formal arrangements than this "reference function." Copious amounts of planned effort and many specially designed devices—indexes, abstracts, card files, compendia, handbooks, loose-leaf services, and what not—are employed in its service. Analysts study the completeness of coverage of these facilities, the time lags involved in their preparation, the suitability and logical structure of the categories they employ. Scientists and engineers streamline the existing facilities and devise new ones—microfilm libraries, new cataloguing systems, mechanical retrieval systems. These are also the topics which occupy the largest single portion of the program of the present International Conference.

SEARCHES OUTSIDE THE LITERATURE

Here, once again, our exploratory study sought to gain insight by examining the reverse side of the coin. Instances when scientists secured answers to specific questions in other ways than those deliberately designed for this purpose were recorded. As before, it was hoped that this procedure would, on the one hand, illumine the operation of the informal avenues of communication, and that it would, on the other hand, point up the services which the formal reference facilities fail to perform. Eventully, such knowledge may suggest ways of having the formal facilities do more adequately the job they presently fail to do. Or, depending on the circumstances, it might be found more practicable to improve the operation of informal avenues of communication. More of this later.

The following question was included in the interview:

> . . . Can you tell me about the last time you used another channel than just the literature to find the answer to some question that arose in connection with your work?

CONTENT OF INFORMATION SOUGHT

In examining the replies to this question, it is striking how intimately the content of the information sought is tied up with the reason for seeking it outside of the regular channels of the literature search. For in two-thirds of the reported cases, the nature of the information sought either made it improbable that it would appear in the literature at all, or made it seem very difficult to track down, even if published. Most of these searches were for practical details to supplement basic knowledge which was already at hand: unpublished minor details of already published findings; information about the use of techniques and the adaptation of apparatus; quests for the fruits of experience and know-how. For example:

> I have two former Ph.D.'s at . . . Institute. We have some equipment that was developed there, and I called them up for questions about it not long ago.

Or:

> Last week there was a conference . . . in the city. . . . There were specific questions that were troubling me, about modifications in our instruments. I made it my business to have lunch with some people who were working at other laboratories in the country, and found out if . . . they had any experience with specific devices that are mentioned in the literature. E.g., . . . I asked someone, "Have you used this? Does it work as well as the article reports?"

In the remaining one-third of the episodes, the information was secured through personal channels although the nature of its content would not seem to have barred it from appearing in print, or from being traced if published. But in half of these cases, the information sought had not, in fact, appeared in print at the time it was secured by the interviewed scientists. For example,

> I now intend to write someone in Chicago and he will answer me. . . . A number of things are not available to me [otherwise], and they have it in Chicago . . . [It is on] the purely theoretical calculations of electronic structure of molecules.

In the other instances the information is known to have been available in print at the time it was secured. Personal channels were used in the following ways: obtaining citations from students; having a friend at a pharmaceutical company arrange for the searching, excerpting, and collating of literature; securing in conversation with a local fellow zoologist the published background information about an organism.

Table 3 summarizes the types of specific information which the interviewed scientists reported having secured through informal channels rather than through a literature search.

TABLE 3. Answers to Specific Questions Sought through Personal Channels: Nature of the Information Sought

	Number of Episodes Related by:			
Nature of the Information Sought	*Total*	*Biochemists*	*Chemists*	*Zoologists*
Publication or indexing unlikely				
Facts to be newly established	2	—	1	1
Practical details on:				
Materials	2	1	—	1
Apparatus	2	—	2	—
Techniques	10	1	2	7
Findings	2	—	2	—
Publication, indexing not unlikely				
Techniques	5	2	1	2
Findings	5	2	2	1
Total	28	6	10	12

PERSONS CONSULTED AS SOURCES

Given that information was sought through personal channels, how did these scientists know to whom to address their questions? How did they decide whom to ask, when more than one possible source existed? Did they address inquiries "cold," or did they tend to seek out colleagues with whom they had already established some sort of relationship? Answers to these questions may shed light on the importance which various forms of access to personal communication hold for scientists. They may also give some insight into what is, perhaps, lacking in the communication picture of scientists in less favored positions, where opportunities of access to experts are not so ample.

Nearly half the inquiries here recorded were addressed to scientists who were easily identified as the ones most qualified to answer the particular question: they were the authors of publications on which clarification was sought, the developers of instruments or techniques regarding which counsel was needed, or the recognized leaders in a specialty.

A few of the inquiries were addressed to individuals uniquely qualified to answer them, although their expertise in the matter was not generally known at the time. How, then, did the inquiring scientists know that it was these particular colleagues who could answer their questions? In each case, it was a more or less fortuitous circumstance which established the contact. One illustrative case is that of a chemist:

> I wrote to Professor X, of the physics department at . . . University, for some information on . . . about four months ago. It was factual

> information. I knew the answer was not in the usual literature, but to salve my conscience, I looked in the most recent *Physical Review.* I did know that Professor X had done an experiment which would have given him this information as a by-product. . . . (How had you known that Professor X had done this experiment?) Because he had done it at . . . [summer laboratory] two summers ago, when he had lived next door to me.

Perhaps the most extreme example of the fortuity in learning who has vital information is the following odyssey of a zoologist.

> We had begun using a new technique for measuring the amount of . . . in the blood, a technique we had first heard of through the literature. But we had problems with it, so we wrote to the man who had developed it—and even called him on the [long distance] telephone.

Up to this point, information about a procedure was sought from its publicly known author. But this attempt to gain the desired knowledge proved unsuccessful, and the report continues:

> But that didn't really help, and we heard that he was having trouble with it, too. It's a notorious technique in that it is very difficult to get consistent results. We might have been able to solve the problem if we had kept working on it. But then I went to a meeting in . . . [a European city], and saw from the meeting abstracts that a man was there to present a paper who had also used this technique. It turned out that he hadn't done the work [himself], but that he had used a modification which had been devised by a biochemist at NYU. So he gave me this man's name at NYU and when I got back we got in touch with him and solved the problem.

It is to be hoped that most searches for information lead to their goal less circuitously, at least when the goal is so close at hand. It is worth noting that the informal source of information, once located, proved very effective. This episode indicates that much would be gained if *finding* the right personal source of information could be made more efficient. It may well be possible to make better *formal* arrangements for locating such *informal* sources of information, even where it would not be practical or economical to have the actual information carried in the formal media.

A third and final category of the inquiries was addressed to scientists who, although qualified to answer, were not the oustanding experts on the subject. They were, however, previously known to the inquiring scientists, and it was, apparently, this accessibility which determined their choice. Thus questions were, variously, addressed to men "who had worked with me here for a year," "who were known to me from previous conferences," "who had been my fellow-students," "with whom I had worked in the same lab previously," and the like.

The various personal sources of information which were used by the interviewed scientists to secure the answers to specific questions are summarized in Table 4.

TABLE 4. Answers to Specific Questions Sought through Personal Channels: Source Used

Source to Whom Inquiry Was Addressed	Number of Episodes Related by:			
	Total	*Biochemists*	*Chemists*	*Zoologists*
Publicly known top experts	12	3	4	5
Privately known top experts	3	—	2	1
Accessible, not top experts	14	3	4	7
Total	29	6	10	13*

* One zoologist counted twice because he used both a publicly and a privately known top expert in pursuit of the same question.

Some Questions for Future Research

THE reader has already been warned that he must not regard the figures in this table, or elsewhere in this paper, as reliably representing the communication behavior among some defined population of scientists, not even that of the 77 biochemists, chemists, and zoologists who were interviewed at one American university. The purpose of this exploration was to formulate problems and procedures for further investigation; a large part of the schedule of questions was developed and modified as the interviewing progressed, rather than uniformly applied; and even as finally used, some of the questions are not regarded as satisfactory.

But even if there were no question of the reliability and representativeness of what is reported here, it is quite obvious that only a scant beginning of an understanding of some problems and potentialities of scientific communication has been achieved. Even the fuller report deposited with the National Science Foundation does no more than to set the stage for more systematic investigation, as, indeed, was the purpose of this exploration.

What are the next questions calling for investigation? A more extensive and systematic survey will first have to correct the shortcomings inherent in any first pilot study. This means the more systematic collection of data in order to confirm or refute what has been suggested above and in our fuller report. It means the observation of the variations in these forms of communication which may accompany differences in the attributes of scientists, their disciplines, and their positions. It means the examination of the prevalence of various communication processes in diverse settings, including especially those where the opportunities for personal communication with top experts are more limited. Only then will it be possible to draw inferences from the multiple cross-classification and statistical analysis of observations in the traditional manner of survey analysis.

A more detailed illustration follows of some of the questions that must, in our opinion, be answered in order to have a sound basis to guide scientific communications policy. Not all the questions listed below can be answered by interview survey methods; some could, it is hoped, be answered by the judgment of appropriate experts on the basis of materials which an interview survey can provide. The paragraph headings which follow denote the issue of communications policy to which the research questions listed are pertinent.

Detailed Questions on the Furnishings of Answers to Specific Questions

THE following paragraphs refer to the communication function last discussed: enabling scientists to find the available answers to specific questions.

1. *Is Any Action Indicated?*

In the preceding pages have been reviewed some instances where scientists had need for information on a specific point and obtained it through informal and personal channels rather than through the formal and established means of a literature search. Various courses of action are conceivable which would either make feasible the securing of information of this type through the formal channels, or else would make the informal channels more effective and more generally accessible for this purpose. Any such action, however, will recommend itself only insofar as the kind of information now secured outside the formal channels is really essential, and is not now adequately accessible to all who seek it.

If any doubts exist as to these points, some empirical tools are available for resolving them. To examine how essential the information in question is, it may be useful to collect, more systematically and extensively than here, a sample of items of information secured outside of the formal machinery of literature searches. This could be submitted to a panel of experts for judgment as to the importance of the messages. In those cases where the information was to appear in print later, the experts may judge whether the time saved by using personal contacts was essential.

To examine empirically how adequately the information in question is now available to those who seek it, it would be useful to ascertain what is done by scientists who lack access to personal sources of information when they have call for information comparable to that obtained through personal contact by the scientists interviewed here. One would also want to know how often a need for information was felt by a scientist but remained unfulfilled because he knew of no personal way of pursuing it.

2. *Should More Varieties of Information Be Made Securable through the Literature, or Should Informal Channels Be Made Widely Usable?*

The selection of additional questions for research must be a function of the possible courses of action among which a choice is to be made. The possible courses of action fall into two classes: (*a*) those designed to make the printed media carry more messages of a given kind in such a way that they can be located on demand; (*b*) those designed to make useful personal communication more widely accessible.

If the number of potential users of each of the items of information in question is very large, the printed media will have to be emphasized as much as possible, since personal communication imposes an additional cost in terms of source's time for each additional consumer of a message. If, on the other hand, the number of potential users of any one of these messages is moderate, then one may wish to put more emphasis on enhancing the effectiveness of personal communication. The latter is the only course for those types of information which cannot be economically handled in print. (Cf. Item 6 below.)

To form a judgment of the number of potential users of a message goes beyond the scope of survey research. A survey can, however, give a more accurate picture of the *kind* of messages that are under present conditions diffused through personal channels, and on the basis of this picture it may be possible for experts in the sciences to make some judgment of the potential number of their users.

3. *Should Informal Channels Be Made More Widely Usable by Enhancing Their General Accessibility, or by Making the Likely Sources of Particular Information Easier to Identify?*

Policies which may enhance the effectiveness of personal contacts as channels for securing the answers to specific questions again are of two kinds.

a. All the many steps which might be taken to encourage the free and frequent give-and-take among scientists, or, perhaps, among specific kinds of scientists, are relevant here. The possible steps are very diverse and include, for example, exhortations to make mail inquiries, the encouragement of inter-institutional visits, the scheduling of teaching duties so as to leave some days open for travel, the arrangement of small conferences on limited topics, and also such long-range policies as the geographic location of research centers in such a way as to enhance opportunities for personal meetings among scientists from different institutions.

b. An entirely different approach is to enhance the effectiveness of existing personal contacts by enabling scientists to find out speedily to whom best to address a given question. One possible step in this direction might be the regular publication, in a newsletter or in a column in existing periodicals, of very brief announcements of work in progress. (Compare the *Mouse Newsletter* or the *News Bulletin* of the Society for Vertebrate Paleontology.)

As a guide to the emphasis to give to these two approaches, future research should ascertain to what extent the effectiveness of personal communication for reference purposes is now blocked by ignorance of the right source, or by its inaccessibility. This means seeking out the instances where personal communication at present *fails* to perform this function, in order to see to what such failure is due; and if it is due to inaccessibility, what are the main existing blocks.

4. *Should More Information Be Made Securable through the Literature by Having More of It Printed, or by Making That Which Is Printed Easier to Find?*

One can make the printed media more fully perform the reference functions which are now performed by personal communication by (*a*) getting more of the pertinent information into print or (*b*) making such information, once in print, "findable" when it is needed. Research should therefore ascertain to what extent failures to secure information of certain types through the literature are due to its not being published, or to the difficulty of locating it at will even when published.

5. *What Makes Published Information Hard to Locate?*

If the material is indeed hard to locate on demand even when it is published, then further research should examine the reasons for this. Earlier pages have shown that, within the limits of this pilot study, the information which is most frequently secured through informal means concerns practical details on the use of techniques and apparatus. Can such items, even when they are published, be located on demand? They would ordinarily appear as incidental entries in the course of research reports which are titled and placed according to their subject matter and not according to procedures used. Do indexing and abstracting services catalog procedural items which appear incidentally in the course of reports on research findings? Do, in fact, suitable and generally familiar categories for cataloguing such information exist?

Survey research can answer only some of these questions. It can also provide materials which may help other experts answer further questions.

6. *Why Is Information of Certain Types Seldom Published?*

If it is the case that certain classes of essential information are customarily not published at all, the reasons for this should be ascertained. Perhaps these reasons can be inferred from the nature of the information that remains unpublished, once this has been determined. If not, it may be necessary to interrogate the originators of items of information about the reasons for their failure to publish them. Some possible reasons are:

a. The information is felt difficult to verbalize economically. This might be due to the absence of an adequate, standardized, and generally recognized vocabulary. There might then be a task for semanticists. Survey research could tell the semanticists what type of information is at present felt to be ineffable in this sense. No doubt some information would still remain hard to convey without mutual discussion, or without "being shown." Where that is the case, making the personal channels of communication more widely usable is the only recourse.

b. Standards of publication make communicating information of certain kinds through the printed word unduly lengthy and laborious. It is possible that standards of publication which are necessary in the reporting of scientific findings are not appropriate to the reporting of ancillary matter. This again could not be judged by survey methods, but may be judged by experts on the basis of material collected through a survey.

c. Authors or editors believe that few scientists would utilize the information. If a representative survey should show that the kind of information in question is actually in heavy demand, this fact might persuade authors or editors to change their policy accordingly. More importantly, if steps can be taken actually to increase the utility of such information in published form, for example, by making it easier to locate once published (cf. Item 5 above), the incentives for publishing such information would increase correspondingly.

d. The information is felt to be "trivial" in some inherent, almost esthetic sense, regardless of its utility. Perhaps it is felt that publishing some kinds of information is lacking in dignity. A possible solution for this problem would be the creation of a special facility for the exclusive publication of information of the kind in question, e.g., information on details of techniques, procedures, and materials. For it is very likely that information that appears trivial when juxtaposed with items of greater theoretical significance in a regular scientific periodical would lose its felt triviality when published in a special medium. It would then be read by scientists when they are in search of precisely this kind of information; it would no longer constitute an unwelcome interruption of reports on more fundamental topics. A special medium for this purpose could be a special journal, or a special section of existing journals.

Detailed Questions on Keeping Scientists Up to Date

THE following paragraphs refer to the ways in which scientists are kept abreast of current developments in the research areas which are relevant to their own work in one way or another. Some of the issues that seem to face scientific communication policy in this regard and some of the pertinent questions for further investigation will be outlined. Three main problems appear: (1) to assure that news will reach the scientists it should reach; (2) to reduce the labor and time scientists must invest in keeping up; (3) to increase the promptness with which scientists will hear of current developments.

The problem of promptness will not be separately treated here, but many of the problems which it poses for behavior research are identical with questions listed below in other contexts.

A. RESEARCH QUESTIONS CONCERNING THE WIDER DISSEMINATION OF SCIENTIFIC NEWS

In principle, there are three types of strategy for assuring wider dissemination of scientific news: (1) to have more scientific news covered by the literature and other formal information-disseminating facilities; (2) to make the information which is covered in these media reach more scientists; (3) to make the networks of person-to-person dissemination of information about current developments reach more widely. Consequently, the following questions are posed for future research:

1. Does Any Significant Part of Information about Current Scientific Developments Fail to Appear in the Literature?

Information which scientists found relevant to their work after obtaining it through personal contacts should be examined to see whether it had been published, or was shortly to be published. If not, steps to make the scientific literature cover additional items of news about current developments may be indicated. What these steps would be cannot be stated without knowledge of the type of information that may now be slighted by the literature and the reasons for its failure to appear in print. The pertinent research questions would be the same as those indicated for an analogous situation in Item 6 above. (No doubt the judgment will be reached that certain types of scientific news cannot be incorporated into the periodical literature, at least not without obstructing its other functions. This is one reason for the continued importance of informal and personal channels of communication.)

2. *Why Were Specific Published Items of Scientific News Missed by Scientists?*

If information about current scientific developments which scientists obtained "too late" or through personal channels was in print at the time, it should be examined further to see why it had not come to the scientists' attention through the literature. Three kinds of reasons may be suggested:

a. The item fell outside of the scientist's area of attention[5] *as he had defined it.* Where this is frequent, one obvious recommendation is for scientists to define their areas of attention more adequately. This offers little room for action on the part of scientific organizations. They may be able to offer some guidance to scientists if there should be a systematic tendency for the workers in a given field to slight certain types of information.

On the other hand, no scientist will ever be able to define his attention-area so that *all* relevant information will be within its compass. The total area within which news of relevance to him may occur is too vast. Besides, some information becomes "relevant" only in the process of being acquired: a scientist's interests must be molded to a certain extent by what information comes his way. For these reasons, it is not enough to improve scientists' deliberate "gathering" of information; the likelihood that additional information will reach them "accidentally" must be maximized.

b. The item appeared in publications not regularly scanned by the scientist. Such an item of information will be missed by the scientist even though its topic falls squarely within his pre-defined attention area.

Once again, one possible remedy is for the scientist to adjust his information-gathering practices; he needs to scan additional journals or secondary source material. It is possible that certain portions of the

5. In the report from which this paper is excerpted, a scientist's "area of attention" refers to the fields of research with the current developments of which he intends to keep abreast. His "primary fields of attention" are those where he tries to

literature are rather generally slighted in this way by one category of scientists or another.

But, as before, no scientist can be expected to scan all the possible primary sources of news that may be relevant for him. If scientists frequently miss news items that appeared in journals which they do not scan, it may be more practical to improve their coverage in secondary media—indices, abstracts, reading lists—than to adjust the scientists' reading habits, especially with regard to journals that only intermittently carry news of relevance to the scientists concerned.

It would be important to know whether any substantial number of scientists misses the *same* items of information due to their appearance in journals which they do not scan because they carry news of interest to them too infrequently. Such a situation might call for new secondary media which would list, excerpt, or reprint selected material from these journals for the benefit of workers in certain specialties.

c. The item was concealed in the context of an article on another topic. Such an item could be missed even when it does fall within the scientist's pre-defined attention area and appears in a journal which he regularly covers.

Where that is the case, the proper remedial action would seem to lie in the provision of better titles, subtitles, prefatory abstracts, or whatever other cues scientists may use to select papers for reading. It would again be important to know whether there is some general tendency for certain types of information to be frequently missed in this fashion. This would make possible the formulation of concrete recommendations for the more appropriate titling of papers; and if the tendency is of great prevalence, there might even be call for reprinting the relevant portions of these papers in places where they would be more conspicuous.

3. *In What Fields Is Published Information Most Likely to Be Missed in the Course of Scanning?*

Generally speaking, any steps that will lighten the burden and shorten the time required for the scanning of a given portion of the scientific literature will free scientists to scan additional portions. Possible action toward this end is discussed below.

More specific action may be feasible if it is possible to delineate particular areas of knowledge as the likely loci of "missed" information. Scanning is probably least thorough where it is least efficient: with

keep up with current developments in detail. His "secondary fields of attention" are those with which he "also needs to keep up to some extent, but not as much."

respect to fields of research which only occasionally give rise to information of interest to a given scientist. It is in a scientist's secondary fields of attention that he is most likely to miss some relevant information as he scans the current output. Perhaps publications could be created which would selectively reprint, excerpt, abstract, or index those aspects of a given area of research which would be of interest to specialists in another area. This calls for ascertaining what aspects of the information produced in a given field are of relevance to the workers in other specialties.

4. What Are the Forms of Personal Communication Which Bring Relevant Scientific News to Those Who Have Access to Them?

For a variety of reasons inherent in the nature of basic research work, personal communication will, no doubt, have to continue to supply much of the important news to scientists. (In part, it does this by calling published information to their attention.) In order to foster the operation of personal networks of communication with the requisite discrimination, it is necessary to know what sort of personal communication would most profitably link scientists in given specialties and positions.[6]

It is necessary to recognize the most fruitful *occasions* for person-to-person exchanges (conferences, corridor conversations, visits, etc.); the *positions* whose incumbents can become nodal points of information-exchange (consultantships, service on award committees, editorial duties, etc.); and the nature of *relays* through which information may be usefully passed on and shared (through friends at other institutions, through contact with "good correspondents and readers," through departmental colleagues who return from conferences and visits, etc.). Because the usefulness of personal communication differs from discipline to discipline, and possibly from specialty to specialty, the factors which determine its differential utility must be taken into account.

5. What Is the Present Opportunity for Scientists in Varying Positions to Have Access to the Fruitful Forms of Personal Communications?

The scientists among whom this study was carried out probably have easier access to useful personal communication with other scientists than their colleagues in most other institutions. It is important to know how much access to such communication scientists in various institu-

6. Some possible steps to promote personal interchanges among scientists are discussed in the report submitted to the Foundation.

tions, professional positions, and geographic locations now have, if plans are to be made to have more of them "hooked up" in useful ways with the network of informal information flow.

B. RESEARCH QUESTIONS CONCERNING REDUCTIONS IN THE TIME AND LABOR OF SURVEYING CURRENT DEVELOPMENTS

Great amounts of time and effort are required of scientists who wish to keep abreast with current developments. Three features are chiefly responsible for this situation: (1) the number of different publications which must be surveyed to achieve satisfactory coverage is very large; (2) the screening of papers to read is time-consuming, if one does not wish to risk missing too many useful items of information; (3) the assimilation of the content of many papers demands considerable additional time and application.

Any reduction in the difficulty of screening papers must come from editorial policy which would provide appropriate clues in the titles or other features of published articles. The most useful clues are probably those which are present in the papers which scientists find easiest to screen. This leads to the following empirical question.

6. *What Clues Are Lacking from Articles Which Must Be Examined Closely before Their Pertinence Can Be Determined?*

As scientists scan the contents of a journal, they decide to skip some articles on the basis of easily visible clues, and they examine others more closely. Are there many in this latter group which are eventually found not to have been worth reading? If so, what are the clues which are lacking? This may suggest improvements in the provision of titles, subtitles, prefatory abstracts, or other clues, as a matter of editorial policy.

There are several possible strategies for reducing the number of journals a scientist must scan. One is the greater utilization of secondary source publications for scanning purposes. In order to be useful for this purpose, such publications would have to follow original publication much more promptly than the standard abstracting services are able to do. Special abstracts, lists of titles, or even reproductions of the tables of contents of journals may be considered. Most services of this sort which now exist attempt either to cover the entire output in a given line of work, or to report on its most important segments. It might be more practical to have them report to the workers in a given specialty only that part of the output which appears in journals which they find least worth scanning directly.

A second possible way to reduce the number of different journals a scientist must cover would be the creation of new journals which would publish, reprint, or excerpt articles on topics which cut across the classification of topics according to which journals presently specialize.

A third way would be to institute more sharing of the burden of scanning among colleagues. Basic researchers, at least those interviewed, seem not inclined to entrust to anyone but themselves the actual reading of articles; in this they differ from professionals in certain applied fields, e.g., medical doctors. These researchers may, however, be willing to entrust to each other the scanning of journals, even if not the actual reading of the papers selected. Some of those interviewed do this now.

The choice among these strategies and the further specification of any one of them calls for answers to the following questions:

7. *At What Level of Efficiency Is the Scanning of Each Journal or Other Medium Performed?*

How many articles or pages must a scientist scan in a journal for each one that he eventually reads? or for each one that he eventually finds to have been worth reading? This, it should be noted, is not simply a question about each journal, but rather about the relationship of each journal to each specialty. For a journal which is read from cover to cover by the scientists in one field may contain only one useful article in ten for those working in another.

For any one group of scientists, it would be useful to classify the journals which they scan according to the proportion of their content that is eventually found worth reading. One may then concentrate on those journals which scientists in a given specialty scan with the least efficiency. The information they secure in this inefficient way may be capable of being made available to them in a more efficient manner. This will require answering still another question:

8. *What Is the Nature of the Information Which Scientists Secure in the Media Scanned with Least Efficiency?*

Perhaps this information can be identified by its content; for example, it might be information about the handling of materials or organisms. Or, it may have to be identified in terms relative to the specialty pursued by the reading scientists. In either event, the nature of the information which is secured in the media scanned with low efficiency may suggest utilizing one of the specialized media which were mentioned in Item 6. The number of scientists who share identical difficulties will have to determine the proper course of action.

C. RESEARCH QUESTIONS REGARDING SCIENTISTS' MOTIVATION TO BE INFORMED

Nothing has been said so far of a topic to which behavior research has been fruitfully applied in other communication situations: that of motivating the individual to keep himself informed. This is, for example, a serious problem in the case of medical practitioners, who can devote time to following the news only by resisting strong competing pressures from other professional obligations and from private interests as well. With basic researchers like those with whom we dealt here, no such problem seems to exist: being well informed is recognized as one of their chief responsibilities both by themselves and by their institutions. It remains to be seen whether this is true, for example, for scientists engaged in applied work in industry and elsewhere. It is also possible that a problem of motivating the scientists to keep up with certain types of information exists, even though there is no *general* problem. For example, if it should be felt that some scientists do not take interest in a sufficiently broad span of current developments, the wise assignment to them of certain duties may provide a remedy. Many of the scientists interviewed asserted that their teaching duties caused them to keep up with a much broader range of developments than they otherwise would, and that this had important consequences for their research work. Others related how their interests and attention areas had been broadened by editorial duties, by the writing of books, and so forth. This makes the relation between the breadth of a scientist's attention area and his positions and activities a potentially useful question for research.

THE preceding pages have enumerated some questions for further research and have indicated some possible lines of action for improving scientific communication that might recommend themselves, depending on the answers obtained. These examples of possible action are not, of course, to be regarded as recommendations at this point. Additional research questions are listed in the report submitted by the Bureau of Applied Social Research to the National Science Foundation.

The Social Process of Scientific Discovery

5

MUCH has been written about the psychological components in the process of scientific discovery, usually on the basis of uncontrolled personal experience and introspection. Lately, however, as the introduction to Part 4 indicated, some disciplined research on the psychological components is being undertaken, such as the research on creativity by Morris Stein. The six selections in this part of the reader focus on the social components of the process of scientific discovery that affect that process together with the psychological ones. These six selections show that a variety of disciplined research methods can yield interesting and useful knowledge about the social processes of scientific discovery.

Merton's paper is concerned with the fascinating but little analyzed subject of priority disputes in science. This subject is fascinating because it shows how passionately involved, emotionally and morally, scientists actually are in their work. And in part it is just because of this passionate involvement that the fact of the very frequent occurrence of priority disputes has been so little analyzed. It is the beauty of Merton's dispassionate analysis that he shows how these disputes derive directly from the positive norms of science itself. One of the important institutional norms of science is originality. In their pursuit of this important norm, in their quest for the high rewards that go with achievement in terms of this norm, scientists often fall afoul of each other's claims to priority and, consequently, engage in a variety of patterned forms of dispute about who was truly first.

The frequent occurrence of independent multiple discoveries in science is one of the commonest reasons for priority disputes. Merton and Elinor G. Barber have, for example, compiled a list of more than 250 instances of independent multiple discoveries in science, each instance involving from two to fifteen or more scientists, often in several different countries. The phenomenon of independent multiple discoveries is frequent because, at any given time, the structure of scientific ideas contains all the relevant components for a new discovery; and two or more scientists may independently perceive the new way of putting the existing components together to make that discovery. In the selection by Kuhn, the principle of energy conservation, discovered by no fewer than twelve men in the two decades before 1850, is examined as an example of independent multiple discovery. Kuhn shows how the existing structure of scientific ideas, and other parts of the cultural tradition, brought about this case of independent multiple discovery.

The accumulation of scientific ideas has another important effect besides that of producing independent multiple discoveries; it helps to determine the rate of scientific discovery. The larger the accumulation, some observers of science have argued, the greater the rate of discovery and further accumulation. This is the position taken by Price in describing what he calls "the exponential curve of science." Whether the growth pattern of science is indeed exponential has been disputed, but it is clear that the previous accumulation of scientific ideas at any given point is one of the important determinants of the subsequent rate as well as actual substance of the succeeding accumulation.

In earlier writing about the social process of scientific discovery, there was an emphasis on this process as wholly rational and controllable, a process in which the scientist always knew where he was going in his research. More recent writing has brought out that there is also an element of the unforeseen and uncontrollable, of "happy accident," or of what has come to be called "serendipity." In the selection by Barber and Fox, the serendipity element is examined by comparing what happened when two distinguished medical scientists independently experienced the same unusual occurrence, floppiness in rabbits' ears after the injection of the enzyme papain. Through intensive interviews with the two scientists, Barber and Fox were able to isolate some of the general factors that lead to and away from a serendipitous discovery.

The previous selections have analyzed social factors that help to bring about scientific discovery. Barber's paper, "Resistance by Scientists to Scientific Discovery" analyzes the cultural and social factors that blind scientists to discoveries by themselves and others. The data used are entirely from the history of science. In all probability, however, "resistance" is still operating in science.

Our final selection, asking "Which Scientists Win Nobel Prizes?" is an account of some of the social sources of "greatness" in scientific discovery, but it also indicates some of the reasons, those internal to science as well as external reasons, that the Nobel Prize has its limitations as an indicator of greatness in science. Some of the material on rewards in science in Merton's paper on priority disputes supplements Gray's discussion.

28 *Priorities in Scientific Discovery: A Chapter in the Sociology of Science*

ROBERT K. MERTON

WE CAN ONLY GUESS what historians of the future will say about the condition of present-day sociology. But it seems safe to anticipate one of their observations. When the Trevelyans of 2050 come to write that history—as they well might, for this clan of historians promises to go on forever—they will doubtless find it strange that so few sociologists (and historians) of the twentieth century could bring themselves, in their work, to treat science as one of the great social institutions of the time. They will observe that long after the sociology of science became an identifiable field of inquiry,[1] it remained little cultivated in a world where science loomed large enough to present mankind with the choice of destruction or survival. They may even suggest that somewhere in the process by which social scientists take note of the world as it is and as it once was, a sense of values appears to have become badly scrambled.

This spacious area of neglect may therefore have room for a paper which tries to examine science as a social institution, not in the large but in terms of a few of its principal components.

A CALENDAR OF DISPUTES OVER PRIORITY

We begin by noting the great frequency with which the history of science is punctuated by disputes, often by sordid disputes, over

---

Reprinted from the article by Robert K. Merton in American Sociological Review, *vol. 22 (1957), pp. 635–659. Presidential address read at the annual meeting of the American Sociological Society, August, 1957.*

1. The rudiments of a sociology of science can be found in an overview of the subject by Bernard Barber, *Science and the Social Order*, New York: The Free Press, 1952 [paperback edition, Collier Books, 1962]; Bernard Barber, "Sociology of Science: A Trend Report and Bibliography," *Current Sociology*, Vol. 5, No. 2, Paris: UNESCO, 1957.

priority of discovery. During the last three centuries in which modern science developed, numerous scientists, both great and small, have engaged in such acrimonious controversy. Recall only these few: Keenly aware of the importance of his inventions and discoveries, Galileo became a seasoned campaigner as he vigorously defended his rights to priority first, in his *Defense against the Calumnies and Impostures of Baldassar Capar*, where he showed how his invention of the "geometric and military compass" had been taken from him, and then, in *The Assayer*, where he flayed four other would be rivals; Father Horatio Grassi, who tried "to diminish whatever praise there may be in this [invention of the telescope for use in astronomy] which belongs to me"; Christopher Scheiner, who claimed to have been first to observe the sunspots (although, unknown to both Scheiner and Galileo, Johann Fabricius had published such observations before); an unspecified villain (probably the Frenchman Jean Tarde) who "attempted to rob me of that glory which was mine, pretending not to have seen my writings and trying to represent themselves as the original discoverers of these marvels"; and finally, Simon Mayr, who "had the gall to claim that he had observed the Medicean planets which revolve about Jupiter before I had [and used] a sly way of attempting to establish his priority."[2]

The peerless Newton fought several battles with Robert Hooke over priority in optics and celestial mechanics and entered into a long and painful controversy with Leibniz over the invention of the calculus. Hooke,[3] who has been described as the "universal claimant" because "there was scarcely a discovery in his time which he did not conceive himself to claim" (and, it might be added, often justly so, for he was one of the most inventive men in his century of genius), Hooke, in turn, contested priority not only with Newton but with Huygens over the important invention of the spiral-spring balance for regulating watches to eliminate the effect of gravity.

The calendar of disputes was full also in the eighteenth century. Perhaps the most tedious and sectarian of these was the great "Water Controversy" in which that shy, rich, and noble genius of science, Henry Cavendish, was pushed into a three-way tug-of-war with Watt and Lavoisier over the question of which one had first demonstrated the compound nature of water and thereby removed it from its

2. Galileo, *The Assayer*, 1623, translated by Stillman Drake in *Discoveries and Opinions of Galileo*, New York: Doubleday, 1957, pp. 232–233, 245. Galileo thought it crafty of Mayr to date his book as published in 1609 by using the Julian calendar without indicating that, as a Protestant, he had not accepted the Gregorian calendar adopted by "us Catholics" which would have shifted the date of publication to January 1610, when Galileo had reported having made his first observations. Later in this paper, I shall have more to say about the implications of attaching importance to such short intervals separating rival claims to priority.

3. For scholarly reappraisals of Hooke's role in developing the theory of gravita-

millennia-long position as one of the elements. Earthy battles raged also over claims to the first discovery of heavenly bodies, as in the case of the most dramatic astronomical discovery of the century in which the Englishman John Couch Adams and the Frenchman Urban Jean LeVerrier inferred the existence and predicted the position of the planet now known as Neptune, which was found where their independent computations showed it would be. Medicine had its share of conflicts over priority; for example, Jenner believed himself first to demonstrate that vaccination afforded security against smallpox, but the advocates of Pearson and Rabaut believed otherwise.

Throughout the nineteenth century and down to the present, disputes over priority continued to be frequent and intense. Lister knew he had first introduced antisepsis, but others insisted that Lemaire had done so before. The sensitive and modest Faraday was wounded by the claims of others to several of his major discoveries in physics: one among these, the discovery of electro-magnetic rotation, was said to have been made before by Wollaston; Faraday's onetime mentor, Sir Humphrey Davy (who had himself been involved in similar disputes) actually opposed Faraday's election to the Royal Society on the ground that his was not the original discovery.[4] Laplace, several of the Bernoullis, Legendre, Gauss, Cauchy were only a few of the giants among mathematicians embroiled in quarrels over priority.

What is true of physics, chemistry, astronomy, medicine and mathematics is true also of all the other scientific disciplines, not excluding the social and psychological sciences. As we know, sociology was officially born only after a long period of abnormally severe labor. Nor was the postpartum any more tranquil. It was disturbed by violent controversies between the followers of St. Simon and Comte as they quarreled over the delicate question of which of the two was the father of sociology and which merely the obstetrician. And to come to the very recent past, Janet is but one among several to have claimed that they had the essentials of psycho-analysis before Freud.

To extend the list of priority fights would be industrious and, for this occasion, superfluous. For the moment, it is enough to note that these controversies, far from being a rare exception in science, have long been frequent, harsh, and ugly. They have practically become an integral part of the social relations between scientists. Indeed, the pat-

tion, see Louis Diehl Patterson, "Hooke's Gravitation Theory and Its Influence on Newton," *Isis,* 40 (November, 1949), pp. 327–341; 41 (March, 1950), pp. 32–45; and E. N. da C. Andrade, "Robert Hooke," Wilkins Lecture, *Proceedings of the Royal Society,* Series B, Biological Sciences, 137 (24 July, 1950). The recent biography by Margaret 'Espinasse is too uncritical and defensive of Hooke to be satisfactory; *Robert Hooke,* London: Heinemann, 1956.

4. Bence Jones, *The Life and Letters of Faraday,* London: Longmans, Green, 1870, Vol. I, pp. 336–352.

tern is so common that the Germans have characteristically compounded a word for it, *Prioritätsstreit*.

On the face of it, the pattern of conflict over priority can be easily explained. It seems to be merely a consequence of the same discoveries being made simultaneously, or nearly so, a recurrent event in the history of science which has not exactly escaped the notice of sociologists, or of others, at least since the definitive work of William Ogburn and Dorothy Thomas. But on second glance, the matter does not appear quite so simple.

The bunching of similar or identical discoveries in science is only an *occasion*[5] for disputes over priority, not their *cause* or their *grounds*. After all, scientists also know that discoveries are often made independently. (As we shall see, they not only know this but fear it, and this often activates a rush to ensure their priority.) It would therefore seem a simple matter for scientists to acknowledge that their simultaneous discoveries were independent and that the question of priority is consequently beside the point. On occasion, this is just what has happened, as we shall see in that most moving of all cases of *noblesse oblige* in the history of science, when Darwin and Wallace tried to outdo one another in giving credit to the other for what each had separately worked out. Fifty years after the event, Wallace was still insisting upon the contrast between his own hurried work, written within a week after the great idea came to him, and Darwin's work, based on twenty years of collecting evidence. "*I* was then (as often since) the 'young man in a hurry,'" said the reminiscing Wallace; "*he*, the painstaking and patient student seeking ever the full demonstration of the truth he had discovered, rather than to achieve immediate personal fame."[6]

On other occasions, self-denial has gone even further. For example, the incomparable Euler withheld his long "sought solution to the calculus of variations, until the twenty-three-year-old Lagrange, who had developed a new method needed to reach the solution, could put it into print, "'so as not to deprive you,' Euler informed the young man, 'of any part of the glory which is your due.'"[7] Apart from these and many other examples of generosity in the annals of science, there have doubtless been many more that never found their way into the

5. And not always even the occasion. Disputes over priority have occurred when alleged or actual anticipations of an idea have been placed decades or, at times, even centuries or millennia earlier, when they are generally described as "rediscoveries."

6. This remark is taken from Wallace's commentary at the semi-centenary of the joint discovery, a classic of self-abnegation that deserves to be rescued from the near-oblivion into which it has fallen. For a transcript, see James Marchant, *Alfred Russel Wallace: Letters and Reminiscences*, New York: Harper, 1916, pp. 91–96.

7. E. T. Bell, *Men of Mathematics*, New York: Simon and Schuster, 1937, pp. 155–156. And see the comparable act of generosity on the part of the venerable

pages of history. Nevertheless, the recurrent struggles for priority, with all their intensity of affect, far overshadow these cases of *noblesse oblige,* and it still remains necessary to account for them.

ALLEGED SOURCES OF CONFLICTS OVER PRIORITY

One explanation of these disputes would regard them as mere expressions of human nature. On this view, egotism is natural to the species; scientists, being human, will have their due share and will sometimes express their egotism through self-aggrandizing claims to priority. But, of course, this interpretation does not stand up. The history of social thought is strewn with the corpses of those who have tried, in their theory, to make the hazardous leap from human nature to particular forms of social conduct, as has been observed from the time of Montesquieu, through Comte and Durkheim, to the present.[8]

A second explanation derives these conflicts not from the original nature shared by all men, but from propensities toward egotism found among some men. It assumes that, like other occupations, the occupation of science attracts some ego-centered people, and assumes further that it might even attract many such people, who, hungry for fame, elect to enter a profession that promises enduring fame to the successful. Unlike the argument from nature, this one, dealing with processes of self-selection and of social selection, is not defective in principle. It is possible that differing kinds of personalities tend to be recruited by various occupations and, though I happen to doubt it, it is possible that quarrelsome or contentious personalities are especially apt to be attracted to science and recruited into it. The extent to which this is so is a still unanswered question, but developing inquiry into the type of personality characteristic of those entering the various professions may in due course discover how far it is so.[9] In any event, it should not be difficult to find *some* aggressive men of science.

But even should the processes of selection result in the recruitment of contentious men, there are theoretical reasons for believing that this does not adequately account for the great amount of contention over priority that flares up in science. For one thing, these controversies often involve men of ordinarily modest disposition who act in seemingly self-assertive ways only when they come to defend their rights

Legendre toward the mathematical genius, Niels Abel, then in his twenties, *ibid.,* p. 337.

8. Émile Durkheim had traced this basic theme in sociological theory as early as his Latin thesis of 1892, which has fortunately been translated into French for the benefit of some of us later sociologists. See his *Montesquieu et Rousseau: Précurseurs de la Sociologie,* Paris: Marcel Rivière, 1953, esp. Chapter I.

9. Information about this is sparse and unsatisfactory. As a bare beginning, a study of the Thematic Apperception Test protocols of 64 eminent biological, physical, and social scientists found no signs of their being "particularly aggressive." Anne Roe, *The Making of a Scientist,* New York: Dodd, Mead, 1953, p. 192.

to intellectual property. This has often been remarked, and sometimes with great puzzlement. As Sir Humphrey Davy asked at the time of the great Water Controversy between Cavendish and Watt, how does it happen that this conflict over priority should engage such a man as Cavendish, "unambitious, unassuming, with difficulty . . . persuaded to bring forward his important discoveries . . . and . . . fearful of the voice of fame."[10] And the biographer of Cavendish, writing about the same episode, describes it as "a perplexing dilemma. Two unusually modest and unambitious men, universally respected for their integrity, famous for their discoveries and inventions, are suddenly found standing in a hostile position towards each other. . . ."[11] Evidently, ingrained egotism is not required to engage in a fight for priority.

A second strategic fact shows the inadequacy of explaining these many struggles as owing to egotistic personalities. Very often, the principals themselves, the discoverers or inventors, take no part in arguing their claims to priority (or withdraw from the controversy as they find that it places them in the distasteful role of insisting upon their own merits or of deprecating the merits of their rivals). Instead, it is their friends and followers, or other more detached scientists, who commonly see the assignment of priority as a moral issue that must be fought to a conclusion. For example, it was Wollaston's friends, rather than the distinguished scientist himself, who insinuated that the young Faraday had usurped credit for the experiments on electro-magnetic rotation.[12] Similarly, it was Priestley, De Luc and Blagden, "all men eminent in science and of unblemished character," who embroiled the shy Cavendish and the unassertive Watt in the Water Controversy.[13] Finally, it was the quarrelsome, eminent, and justly esteemed scientist François Arago (whom we shall meet again) and a crowd of as-

10. Sir Humphrey Davy, *Collected Works,* VII, p. 128, quoted by George Wilson, *The Life of the Honorable Henry Cavendish,* London, 1851, p. 63.

11. Wilson, *op. cit.,* p. 64. There can be little doubt about the unassuming character of Cavendish, the pathologically shy recluse, whose unpublished notebooks were crowded with discoveries disproving then widely-held theories and anticipating discoveries not made again for a long time to come. He stands as the example *a fortiori,* for even such a man as this was drawn into a controversy over priority.

The history of science evidently has its own brand of chain-reactions. It was the reading of Wilson's *Life of Cavendish* with its report of Cavendish's long-forgotten experiment on the sparking of air over alkalis which led Ramsay (just as the same experiment led Rayleigh) to the discovery of the element argon. Both Rayleigh and Ramsay delicately set out their respective claims to the discovery, claims not easily disentangled since the two had been in such close touch. They finally agreed to joint publication as "the only solution" to the problem of assigning appropriate credit. The episode gave rise to a great controversy over priority in which neither of the discoverers would take part; the debate is continued in the biographies of the two: by the old friend and collaborator of Ramsay, Morris W. Travers, in *A Life of Sir William Ramsay,* London: Edward Arnold Ltd., 1956, pp. 100, 121–122, 292, *passim;* and by the son of Lord Rayleigh, *John William Strutt: Third Baron Rayleigh,* London: Edward Arnold, 1924, Chapter XI.

tronomers, principally in France and England but also in Germany and Russia, rather than "the shy, gentle and unaffected" co-discoverer of Neptune, Adams, who stirred the pot of conflict over priority until it boiled over and then simmered down into general acknowledgment that the planet had been independently discovered by Adams and LeVerrier.[14] And so, in one after another of the historic quarrels over priority in science.

Now these argumentative associates and bystanders stand to gain little or nothing from successfully prosecuting the claims of their candidate, except in the pickwickian sense of having identified themselves with him or with the nation of which they are all a part. Their behavior can scarcely be explained by egotism. They do not suffer from rival claims to precedence. Their personal status is not being threatened. And yet, over and again, they take up the cudgels in the status-battle[15] and, uninhibited by any semblance of indulging in self-praise, express their great moral indignation over the outrage being perpetrated upon their candidate.

This is, I believe, a particularly significant fact. For, as we know from the sociological theory of institutions, the expression of disinterested moral indignation is a signpost announcing the violation of a social norm.[16] Although the indignant bystanders are themselves not injured by what they take to be the misbehavior of the culprit, they respond with hostility and want to see "fair play," to see that behavior conforms to the rules of the game. The very fact of their entering the fray goes to show that science is a social institution with a distinctive body of norms exerting moral authority and that these norms are invoked particularly when it is felt that they are being violated. In this sense, fights over priority, with all their typical vehemence and passion-

12. Jones, *op. cit.*, pp. 351–352; see also the informative book by T. W. Chalmers, *Historic Researches: Chapters in the History of Physical and Chemical Discovery*, New York: Scribner's, 1952, p. 54.

13. This is the contemporary judgment by Wilson, *op. cit.*, pp. 63–64.

14. Sir Harold Spencer Jones, "John Couch Adams and the Discovery of Neptune," reprinted in James R. Newman, *The World of Mathematics*, New York: Simon and Schuster, 1956, II, pp. 822–839. A list of cases in which associates, rather than principals, took the lead in these conflicts is a very long one. I do not include it here.

15. Sometimes, of course, they act as judges and arbitrators rather than advocates, as was true of Lyell and Hooker in the episode involving Darwin and Wallace. But, as we shall see, the same institutional norms are variously called into play in all these cases.

16. For an acute analysis of the theoretical place of moral obligation, and its correlate, moral indignation, in the theory of institutions, particularly as this was developed in the long course of Durkheim's work, see Talcott Parsons, *The Structure of Social Action*, New York: The Free Press, 1949, pp. 368–470; for further formulations and citations of additional literature, see R. K. Merton, *Social Theory and Social Structure*, New York: The Free Press, 1957 (rev. ed.), pp. 361 ff.

ate feelings, are not merely expressions of hot tempers, although these may of course raise the temperature of controversy; basically, they constitute responses to what are taken to be violations of the institutional norms of intellectual property.

Institutional Norms of Science

TO SAY that these frequent conflicts over priority are rooted in the egotism of human nature, then, explains next to nothing; to say that they are rooted in the contentious personalities of those recruited by science may explain part, but not enough; to say, however, that these conflicts are largely a consequence of the institutional norms of science itself comes closer, I think, to the truth. For, as I shall suggest, it is these norms that exert pressure upon scientists to assert their claims, and this goes far toward explaining the seeming paradox that even those meek and unaggressive men, ordinarily slow to press their own claims in other spheres of life, will often do so in their scientific work.

The ways in which the norms of science help produce this result seem clear enough. On every side, the scientist is reminded that it is his role to advance knowledge and his happiest fulfillment of that role, to advance knowledge greatly. This is only to say, of course, that in the institution of science originality is at a premium. For it is through originality, in greater or smaller increments, that knowledge advances. When the institution of science works efficiently, and like other social institutions, it does not always do so, recognition and esteem accrue to those who have best fulfilled their roles, to those who have made genuinely original contributions to the common stock of knowledge. Then are found those happy circumstances in which self-interest and moral obligation coincide and fuse.

Recognition of what one has accomplished is thus largely a motive derived from institutional emphases. Recognition for originality becomes socially validated testimony that one has successfully lived up to the most exacting requirements of one's role as scientist. The self-

17. It is not only the institution of science, of course, that instills and reinforces the concern with recognition; in some degree, all institutions do. This is evident since the time W. I. Thomas included "recognition" as one of what he called "the four wishes" of men. The point is, rather, that with its emphasis on originality, the institution of science greatly reinforces this concern and indirectly leads scientists to vigorous self-assertion of their priority. For Thomas's fullest account of the four wishes, see *The Unadjusted Girl*, Boston: Little, Brown, 1925, Chapter I.

18. In developing this view, I do not mean to imply that scientists, any more than other men, are merely obedient puppets doing exactly what social institutions require of them. But I do mean to say that, like men in other institutional spheres, scientists tend to develop the values and to channel their motivations in directions the institution defines for them. For an extended formulation of the general theory of institu-

image of the individual scientist will also depend greatly on the appraisals by his scientific peers of the extent to which he has lived up to this exacting and crtically important aspect of his role. As Darwin once phrased it, "My love of natural science . . . has been much aided by the ambition to be esteemed by my fellow naturalists."

Interest in recognition,[17] therefore, need not be, though it can readily become, simply a desire for self-aggrandizement or an expression of egotism. It is, rather, the motivational counterpart on the psychological plane to the emphasis upon originality on the institutional plane. It is not necessary that individual scientists begin with a lust for fame; it is enough that science, with its abiding and often functional emphasis on originality and its assigning of large rewards for originality, makes recognition of priority uppermost. Recognition and fame then become symbol and reward for having done one's job well.

This means that long before we know anything about the distinctive personality of this or that scientist, we know that he will be under pressure to make his contributions to knowledge known to other scientists and that they, in turn, will be under pressure to acknowledge his rights to his intellectual property. To be sure, some scientists are more vulnerable to these pressures than others—some are self-effacing, others self-assertive; some generous in granting recognition, others stingy. But the great frequency of struggles over priority does not result merely from these traits of individual scientists but from the institution of science, which defines originality as a supreme value and thereby makes recognition of one's originality a major concern.[18]

When this recognition of priority is either not granted or fades from view, the scientist loses his scientific property. Although this kind of property shares with other types general recognition of the "owner's" rights, it contrasts sharply in all other respects. Once he has made his contribution, the scientist no longer has exclusive rights of access to it. It becomes part of the public domain of science. Nor has he the right of regulating its use by others by withholding it unless it is acknowledged as his. In short, property rights[19] in science become

tionalized motivation, see Talcott Parsons, *Essays in Sociological Theory*, Glencoe: The Free Press, 1954 (rev. ed.), esp. Chapters II and III.

19. That the notion of property is part and parcel of the institution of science can be seen from the language employed by scientists in speaking of their work. Ramsay, for example, asks Rayleigh's "permission to look into atmospheric nitrogen" on which Rayleigh had been working; the young Clerk Maxwell writes William Thomson, "I do not know the Game laws and Patent laws of science . . . but I certainly intend to poach among your electrical images"; Norbert Wiener describes "differential space, the space of the Brownian motion" as "wholly mine in its purely mathematical aspects, whereas I was only a junior partner in the theory of Banach spaces." Borrowing, trespassing, poaching, credit, stealing, a concept which "belongs" to us—these are only a few of the many terms in the lexicon of property adopted by scientists as a matter of course.

whittled down to just this one: the recognition by others of the scientist's distinctive part in having brought the result into being.

It may be that this concentration of the numerous rights ordinarily bound up in other forms of property into the one right of recognition by others helps produce the great concentration of affect that commonly characterizes disputes over priority. Often, the intensity of affect seems disproportionate to the occasion; for example, when a scientist feels he has not been given enough recognition for what is, in truth, a minor contribution to knowledge, he may respond with as much indignation as the truly inventive scientist, or even with more, if he secretly senses that this is the outermost limit of what he can reasonably hope to contribute.[20] This same concentration of property-rights into the one right of recognition may also account for the deep moral indignation expressed by scientists when one of their number has had his rights to priority denied or challenged. Even though they have no personal stake in the particular episode, they feel strongly about the single property-norm and the expression of their hostility serves the latent function of reaffirming the moral validity of this norm.

NATIONAL CLAIMS TO PRIORITY

In a world made up of national states, each with its own share of ethnocentrism, the new discovery redounds to the credit of the discoverer not as an individual only, but also as a national. From at least the seventeenth century, Britons, Frenchmen, Germans, Dutchmen, and Italians have urged their country's claims to priority; a little later, Americans entered the lists to make it clear that they had primacy.

The seventeenth-century English scientist Wallis, for example, writes: "I would very fain that Mr. Hooke and Mr. Newton would

20. Some of this had occurred to Galileo in his counterattack on Sarsi (pseudonym for Grassi): "Only too clearly does Sarsi show his desire to strip me completely of any praise. Not content with having disproved our reasoning set forth to explain the fact that the tails of comets sometimes appear to be bent in an arc, he adds that nothing new was achieved by me in this, as it had all been published long ago, and then refuted, by Johann Kepler. In the mind of the reader who goes no more deeply than Sarsi's account, the idea will remain that I am not only a thief of other men's ideas, but a petty, mean thief at that, who goes about pilfering even what has been refuted. And who knows; perhaps in Sarsi's eyes the pettiness of the theft does not render me more blameworthy than I would be if I had bravely applied myself to greater thefts. If, instead of filching some trifle, I had more nobly set myself to search out books by some reputable author not as well known in these parts, and had then tried to suppress his name and attribute all his labors to myself, perhaps Sarsi would consider such an enterprise as grand and heroic as the other seems to him cowardly and abject." (Galileo, *The Assayer, op. cit.*, pp. 261–262.)

This type of reaction to what I describe as the "professional adumbrationist" (in the unpublished part of this paper) was expressed also by Benjamin Franklin after he had suffered from claims by others that they had first worked out the experiment of the lightning kite. As he said in part (the rest of his observations are almost equally in point), "The smaller your invention is, the more mortification you receive in having set themselves in earnest for promoting the designs about telescopes, that others may not steal from us what our nation invents, only for the neglect to publish them ourselves." So, also, Halley says of his comet that "if it should return according to our prediction about the year 1758 [as of course it did], impartial posterity will not refuse to acknowledge that this was first discovered by an Englishman."[21]

Or to move abruptly to the present, we see the Russians, now that they have taken a powerful place on the world-scene, beginning to insist on the national character of science and on the importance of finding out who first made a discovery. Although the pattern of national claims to priority is old, the formulation of its rationale in a Russian journal deserves quotation if only because it is so vigorously outspoken:

> Marxism-Leninism shatters into bits the cosmopolitan fiction concerning supra-class, non-national, "universal" science, and definitely proves that science, like all culture in modern society, is national in form and class in content. . . . The slightest inattention to questions of priority in science, the slightest neglect of them, must therefore be condemned, for it plays into the hands of our enemies, who cover their ideological aggression with cosmopolitan talk about the supposed non-existence of questions of priority in science, *i.e.*, the questions concerning which peoples [here, be it noted, collectivities displace the individual scientist] made what contribution to the general store of world culture . . . [And summarizing the answers to these questions in compact summary] The Russian people has the richest history. In the course of this history, it has created the richest culture, and all the other countries of the world have drawn upon it and continue to draw upon it to this day.[22]

the credit of it disputed with you by a rival, whom the jealousy and envy of others are ready to support against you, at least so far as to make the point doubtful. It is not in itself of importance enough for a dispute; no one would think your proofs and reasons worth their attention: and yet if you do not dispute the point, and demonstrate your right, you not only lose the credit of being in that instance *ingenious*, but you suffer the disgrace of not being *ingenuous*; not only of being a plagiary but of being a plagiary for trifles. Had the invention been greater it would have disgraced you less; for men have not so contemptible an idea of him that robs for gold on the highway, as of him that can pick pockets for half-pence and farthings." (Quoted in the informed and far-reaching monograph by I. B. Cohen, *Franklin and Newton*, Philadelphia: The American Philosophical Society, 1956, pp. 75–76.)

21. Louis T. More, *Isaac Newton*, New York: Scribner's, 1934, pp. 146–147, and pp. 241; 477–478.

22. An editorial, "Against the Bourgeois Ideology of Cosmopolitanism," *Voprosy filosofi*, 1948, No. 2, as translated in the *Current Digest of the Soviet Press*, February 1, 1949, Vol. 1, No. 1, pp. 9–10, 12. For an informed account, see David Joravsky, "Soviet Views on the History of Science," *Isis*, 46 (March, 1955), pp. 3–13, esp. at pp. 9n. and 11, which treat of changing Russian attitudes toward priority and simultaneous invention; see also Merton, *Social Theory and Social Structure*, *op. cit.*, pp. 556–560.

Against this background of affirmation, one can better appreciate the recent statement by Khrushchev that "we Russians had the H-Bomb before you" and the comment by the New York Times that "the question of priority in the explosion of the hydrogen bomb is . . . a matter of semantics," to be settled only when we know whether the "prototype-bomb" or "full-fledged bomb" is in question.[23]

The recent propensity of the Russians to claim priority in all manner of inventions and scientific discoveries thus energetically reduplicates the earlier, and now less forceful though far from vanished, propensity of other nations to claim like priorities. The restraint often shown by individual scientists in making such claims becomes rather inconspicuous when official or self-constituted representatives of nations put in their claims.

The Reward-System in Science

LIKE other institutions, the institution of science has developed an elaborate system for allocating rewards to those who variously live up to its norms. Of course, this was not always so. The evolution of this system has been the work of centuries, and it is probably far from finished. In the early days of modern science, Francis Bacon could explain and complain all in one by saying that "it is enough to check the growth of science, that efforts and labours in this field go unrewarded. . . . And it is nothing strange if a thing not held in honour does not prosper."[24] And a half-century later, much the same could be said by Thomas Sprat, the Bishop of Rochester, in his official history of the newly-established Royal Society:

> . . . it is not to be wonder'd, if men have not been very zealous about those studies, which have been so farr remov'd, from present benefit, and from the applause of men. For what should incite them, to bestow their time, and Art, in revealing to mankind, those Mysteries for which, it may be, they would be onely despis'd at last? How few must there needs be, who will be willing, to be impoverish'd for the common good? while they shall see, all the rewards, which might give life to their Industry, passing by them, and bestow'd on the deserts of easier studies?[25]

23. *New York Times*, July 27, 1957, p. 3, col. 1.

24. Francis Bacon, *Novum Organum*, trans. by Ellis and Spedding, London: Routledge, n.d. Book I, Aphorism XCI. The ellipsis in the text above was for brevity's sake; it should be filled out here below because of the pertinence of what Bacon went on to say: "For it does not rest with the same persons to cultivate sciences and to reward them. The growth of them comes from great wits, the prizes and rewards of them are in the hands of the people, or of great persons, who are but in very few cases even moderately learned. Moreover this kind of progress is not only unrewarded with prizes and substantial benefits; it has not even the advantage of popular applause.

The echo of these complaints still reverberates in the halls of universities and scientific societies, but chiefly with regard to material rather than honorific rewards. With the growth and professionalization of science, the system of honorific rewards has become diversely elaborated, and apparently at an accelerated rate.

Heading the list of the immensely varied forms of recognition long in use is eponymy,[26] the practice of affixing the name of the scientist to all or part of what he has found, as with the Copernican system, Hooke's law, Planck's constant, or Halley's comet. In this way, scientists leave their signatures indelibly in history; their names enter into all the scientific languages of the world.

At the rugged and thinly populated peak of this system of eponymy are the men who have put their stamp upon the science and thought of their age. Such men are naturally in very short supply, and these few sometimes have an entire epoch named after them, as when we speak of the Newtonian epoch, the Darwinian era, or the Freudian age.

The graduations of eponymy have the character of a Guttman scale in which those men assigned highest rank are also assigned lesser degrees of honorific recognition. Accordingly, these peerless scientists are typically included also in the next highest ranks of eponymy, in which they are credited with having fathered a new science or a new branch of science (at times, according to the heroic theory, through a kind of parthenogenesis for which they apparently needed no collaborators). Of the illustrious Fathers of this or that science (or of this or that specialty), there is an end, but an end not easily reached. Consider only these few, culled from a list many times this length:

Morgagni, the Father of Pathology
Cuvier, the Father of Palaeontology
Faraday, the Father of Electrotechnics
Daniel Bernoulli, the Father of Mathematical Physics
Bichat, the Father of Histology
van Leeuwenhoek, the Father of Protozoology and Bacteriology
Jenner, the Father of Preventive Medicine
Chladni, the Father of Modern Acoustics
Herbart, the Father of Scientific Pedagogy

For it is a greater matter than the generality of men can take in, and is apt to be overwhelmed and extinguished by the gales of popular opinions."

25. Thomas Sprat, *The History of the Royal Society*, London, 1667, p. 27.

26. Galileo begins his "Message from the Stars," announcing his discovery of the satellites of Jupiter, with a paean to the practice of eponymy which opens with these words: "Surely a distinguished public service has been rendered by those who have protected from envy the noble achievements of men who have excelled in virtue, and have thus preserved from oblivion and neglect those names which deserve immortality." (*Op. cit.*, p. 23.) He then proceeds to call the satellites "the Medicean Stars" in honor of the Grand Duke of Tuscany, who soon becomes his patron.

Wundt, the Father of Experimental Psychology
Pearson, the Father of Biometry and, of course,
Comte, the Father of Sociology

In a science as farflung and differentiated as chemistry, there is room for several paternities. If Robert Boyle is the undisputed Father of Chemistry (and, as his Irish epitaph has it, also the Uncle of the Earl of Cork), then Priestley is the Father of Pneumatic Chemistry, Lavoisier the Father of Modern Chemistry, and the nonpareil Willard Gibbs, the Father of Physical Chemistry.

On occasion, the presumed father of a science is called upon, in the persons of his immediate disciples or later adherents, to prove his paternity, as with Johannes Müller and Albrecht von Haller, who are severally regarded as the Father of Experimental Physiology.

Once established, this eponymous pattern is stepped up to extremes. Each new specialty has its own parent, whose identity is often known only to those at work within the specialty. Thus, Manuel Garcia emerges as the Father of Laryngoscopy, Adolphe Brongiart as the Father of Modern Palaeobotany, Timothy Bright as the Father of Modern Shorthand, and Father Johann Dzierson (whose important work may have influenced Mendel) as the Father of Modern Rational Beekeeping.

Sometimes, a particular form of a discipline bears eponymous witness to the man who first gave it shape, as with Hippocratic medicine, Aristotelian logic, Euclidean geometry, Boolian algebra, and Keynesian economics. Most rarely, the same individual acquires a double immortality, both for what he achieved and for what he failed to achieve, as in the cases of Euclidean and non-Euclidean geometries, and Aristotelian and non-Aristotelian logics.

In rough hierarchic order, the next echelon is comprised by thousands of eponymous laws, theories, theorems, hypotheses, instruments, constants, and distributions. No short list can hope to be representative of the wide range of these scientific contributions that have immortalized the men who made them. But a few examples in haphazard array might include the Brownian movement, the Zeeman effect, Rydberg's constant, Moseley's atomic number, and the Lorenz curve or to come closer home, where we refer only to assured contemporary recognition rather than to possibly permanent fame, the Spearman rank-correlation coefficient, the Rorschach ink-blot, the Thurstone scale, the Bogardus social-distance scale, the Bales categories of interaction, the Guttman scalogram, and the Lazarsfeld latent-structure analysis.

27. It has been suggested that, in medicine at least, eponymic titles are given to diseases only so long as they are poorly understood. "Any disease designated by an eponym is a good subject for research." (O. H. Perry Pepper, *Medical Etymology*, Philadelphia: W. B. Saunders Co., 1949, pp. 11–12.)

28. Exercised by the excesses eponymy in natural history had reached, the usually

Each science, or art based on science, evolves its own distinctive patterns of eponymy to honor those who have made it what it is. In the medical sciences, for example, the attention of posterity is assured to the discoverer or first describer of parts of the body (as with the Eustachian tube, the circle of Willis, Graffian follicles, Wharton's duct, and the canal of Nuck) though, oddly enough, Vesalius, commonly described as the Father of Modern Anatomy has been accorded no one part of the body as distinctly his own. In medicine, also, eponymy registers the first diagnostician of a disease (as with Addison's, Bright's, Hodgkin's, Menière's, and Parkinson's diseases); the inventor of diagnostic tests (as with Romberg's sign, the Wassermann reaction, the Calmette test, and the Babinski reflex); and the inventor of instruments used in research or practice (as with the Kelly pad, the Kelly clamp, and the Kelly rectoscope). Yet, however numerous and diversified this array of eponyms in medicine,[27] they are still reserved, of course, to only a small fraction of the many who have labored in the medical vineyard. Eponymy is a prize that, though large in absolute aggregrate, is limited to the relatively few.

Time does not permit, nor does the occasion require, detailed examination of eponymous practices in all the other sciences. Consider, then, only two other patterns: In a special branch of physics, it became the practice to honor great physicists by attaching their names to electrical and magnetic units (as with volt, ohm, ampere, coulomb, farad, joule, watt, henry, maxwell, gauss, gilbert and oersted). In biology, it is the long-standing practice to append the name of the first describer to the name of a species, a custom which greatly agitated Darwin since, as he saw it, this put "a premium on hasty and careless work" as the "species-mongers" among naturalists try to achieve an easy immortality by "miserably describ[ing] a species in two or three lines."[28] (This, I may say, will not be the last occasion for us to see how the system of rewards in science can be stepped up to such lengths as to get out of hand and defeat its original purposes.)

Eponymy is only the most enduring and perhaps most prestigious kind of recognition institutionalized in science. Were the reward-system confined to this, it would not provide for the many other distinguished scientists without whose work the revolutionary discoveries could not have been made. Graded rewards in the coin of the scientific realm—honorific recognition by fellow-scientists—are distributed among the stratified layers of scientific accomplishment. Merely to list some of these other but still considerable forms of recognition will perhaps be enough

mild Darwin repeatedly denounced this "miserable and degrading passion of mere species naming." What is most in point for us is the way in which the pathological exaggeration of eponymizing highlights the normal role of eponymy in providing its share of incentives for serious and sustained work in science. Francis Darwin, ed., *The Life and Letters of Charles Darwin*, New York: Appleton, 1925, Vol. I, pp. 332–344.

to remind us of the complex structure of the reward-system in science.

In recent generations, the Nobel Prize, with nominations for it made by scientists of distinction throughout the world, is perhaps the pre-eminent token of recognized achievement in science.[29] There is also an iconography of fame in science, with medals honoring famous scientists and the recipients of the award alike (as with the Rumford medal and the Arago medal). Beyond these, are memberships in honorary academies and sciences (for example, the Royal Society and the French Academy of Sciences), and fellowships in national and local societies. In those nations that still preserve a titled aristocracy, scientists have been ennobled, as in England since the time when Queen Anne added laurels to her crown by knighting Newton, not, as might be supposed, because of his superb administrative work as Master of the Mint, but for his scientific discoveries. These things move slowly; it required almost two centuries before another Queen of England would, in 1892, confer a peerage of the realm upon a man of science for his work in science, and thus transform the pre-eminent Sir William Thomson into the no less eminent Lord Kelvin.[30] Scientists themselves have distinguished the stars from the supporting cast by issuing directories of "starred men of science" and universities have been known to accord honorary degrees to scientists along with the larger company of philanthropists, industrialists, businessmen, statesmen and politicians.

Recognition is finally allocated by those guardians of posthumous fame, the historians of science. From the most disciplined scholarly works to the vulgarized and sentimentalized accounts designed for the millions, great attention is paid to priority of discovery, to the iteration and reiteration of "firsts." In this way, many historians of science help maintain the prevailing institutional emphasis on the importance of priority. One of the most eminent among them, the late George Sarton, at once expresses and exemplifies the commemorative function of his-

29. On the machinery and results of the Nobel and other prize-awards, see Barber, *Science and the Social Order, op. cit.*, pp. 108 ff.; Leo Moulin, "The Nobel Prizes for the Sciences, 1901–1950," *British Journal of Sociology*, 6 (September, 1955), pp. 246–263.

30. For caustic comment on the lag in according such recognition to men of science, see excerpts from newspapers of the day in Silvanus P. Thompson, *The Life of William Thomson: Baron Kelvin of Largs*, London: Macmillan, 1910, Vol. II, pp. 906–907.

31. George Sarton, *The Study of the History of Science*, Cambridge: Harvard University Press, 1936, pp. 3–4, 35–36. Sarton goes on to observe that this practice of identifying first events "never fails to involve him [the historian] in new difficulties, because creations absolutely *de novo* are very rare, if they occur at all; most novelties are only novel combinations of old elements and the degree of novelty is thus a matter of interpretation, which may vary considerably according to the historian's experience, standpoint, or prejudices. . . . It is always risky, yet when every reasonable precaution has been taken one must be willing to run the risk and make the challenge, for this is the only means of being corrected, if correction be needed." (*Ibid.*, p. 36.) This is a telling sign of the deep-rooted sentiment that recognition for originality in

toriography when he writes that ". . . the first scholar to conceive that subject [the history of science] as an independent discipline and to realize its importance was . . . Auguste Comte." He then goes on to propose that great scholar, Paul Tannery, as most deserving to be called "the father of our studies," and finally states the thesis that ". . . as the historian is expected to determine not only the relative truth of scientific ideas at different chronological states, but also their relative novelty, he is irresistibly led to the fixation of *first* events."[31]

Although scientific knowledge is impersonal, although its claim to truth must be assessed entirely apart from its source, the historian of science is called upon to prevent scientific knowledge from sinking (or rising) into anonymity, to preserve the collective memory of its origins. Anonymous givers have no place in this scheme of things. Eponymity, not anonymity, is the standard. And, as we have seen, outstanding scientists, in turn, labor hard to have their names inscribed in the golden book of firsts.[32]

Seen in composite, from the eponyms enduringly recording the names of scientists in the international language of science to the immense array of parochial and ephemeral prizes, the reward-system of science reinforces and perpetuates the institutional emphasis upon originality. It is in this specific sense that originality can be said to be a major institutional goal of modern science, at times, the paramount one, and recognition for originality a derived, but often as heavily emphasized, goal. In the organized competition to contribute to man's scientific knowledge, the race *is* to the swift, to him who gets there first with his contribution in hand.

INSTITUTIONAL NORM OF HUMILITY

If the institution of science placed great value *only* on originality, scientists would perhaps attach even more importance to recogni-

science must be expressed, that it is an obligation—"the historian is expected . . ." —to search out the "first" to contribute an idea or finding, even though a comprehensive view of the cumulative and interlocking character of scientific inquiry suggests that the attribution of "firsts" is often difficult and sometimes arbitrary. For a further statement on this matter of priority, see George Sarton, *The Study of the History of Mathematics*, Cambridge: Harvard University Press, 1936, pp. 33–36.

I cannot undertake here to examine the attitudes commonly manifested by historians of science toward this emphasis on searching out priorities. It can be said that these too are often ambivalent.

32. This was presumably not always so. As is well known, medieval authors often tried to cloak their writings in anonymity. But this is not the place to examine the complex subject of variations in cultural emphases upon originality and recognition. For some observations on this, see George Sarton, *A Guide to the History of Science*, Waltham, Mass.: Chronica Botanica Co., 1952, p. 23, who reminds us of ancient and medieval practices in which "modest authors would try to pass off their own compositions under the name of an illustrious author of an earlier time," ghost-writing in reverse. See also R. K. Merton, *Science, Technology and Society in Seventeenth Century England*, Bruges, Belgium: Osiris, 1938, pp. 360–632, at p. 528.

tion of priority than they do. But, of course, this value does not stand alone. It is only one of a complex set making up the ethos of science —disinterestedness, universalism, organized scepticism, communism of intellectual property, and humility being some of the others.[33] Among these, the socially enforced value of humility is in most immediate point, serving, as it does, to reduce the misbehavior of scientists below the rate that would occur if importance were assigned only to originality and the establishing of priority.

The value of humility takes diverse expression. One form is the practice of acknowledging the heavy indebtedness to the legacy of knowledge bequeathed by predecessors. This kind of humility is perhaps best expressed in the epigram Newton made his own: "If I have seen farther, it is by standing on the shoulders of giants" (this, incidentally, in a letter to Hooke who was then challenging Newton's priority in the theory of colors).[34] That this tradition has not always been honored in practice can be inferred from the admiration that Darwin, himself lavish in such acknowledgments, expressed to Lyell for "the elaborate honesty with which you quote the words of all living and dead geologists."[35] Exploring the literature of a field of science becomes not only an instrumental practice, designed to learn from the past, but a commemorative practice, designed to pay homage to those who have prepared the way for one's work.

Humility is expected also in the form of the scientist's insisting upon his personal limitations and the limitations of scientific knowledge altogether. Galileo taught himself and his pupils to say, "I do not know." Perhaps another often-quoted image by Newton most fully expresses this kind of humility in the face of what is yet to be known:

> I do not know what I may appear to the world, but to myself I seem to have been only like a boy playing on the seashore, and diverting myself in now and then finding a smoother pebble or a prettier shell than ordinary, whilst the great ocean of truth lay all undiscovered before me.[36]

33. For a review of other values of science, see Barber, *op. cit.*, Chapter IV; Merton, *Social Theory and Social Structure*, *op. cit.*, pp. 552–561; H. A. Shepard, "The Value System of a University Research Group," *American Sociological Review*, 19 (August, 1954), pp. 456–462.

34. Alexander Koyré, "An unpublished letter of Robert Hooke to Isaac Newton," *Isis*, 43 (December, 1952) pp. 312–337, at p. 315.

35. Darwin, *op. cit.*, I, p. 263.

36. David Brewster, *Memoirs of the Life, Writings, and Discoveries of Sir Isaac Newton*, Edinburgh and London, 1855, Volume II, Chapter xxvii. For our purposes, unlike those of the historian, it is a matter of indifference whether Newton actually felt acutely modest or was merely conforming to expectation. In either case, he expresses the norm of personal humility, which is widely held to be appropriate. I. B. Cohen, (*op. cit.*, pp. 47, 58, passim) repeatedly and incisively makes the point that both admirers and critics of Newton have failed to make the indispensable distinction between what he said and what he did.

If this contrast between public image ("what I may appear to the world") and self-image ("but to myself I seem") is fitting for the greatest among scientists, it is presumably not entirely out of place for the rest. The same theme continues unabated. Laplace, the Newton of France, in spite of what has been described as "his desire to shine in the constantly changing spotlight of public esteem," reportedly utters an epigrammatic paraphrase of Newton in his last words, "What we know is not much; what we do not know is immense."[37] Lagrange summarizes his lifetime of discovery in the one phrase, "I do not know." And Lord Kelvin, at the Jubilee celebrating his fifty years as a distinguished scientist in the course of which he was honored by scores of scientific societies and academies, characterizes his lifelong effort to develop a grand and comprehensive theory of the properties of matter by the word, "Failure."[38]

Like all human values, the value of modesty can be vulgarized and run into the ground by excessive and thoughtless repetition. It can become merely conventional, emptied of substance and genuine feeling. There really *can* be too much of a good thing. It is perhaps this excess which led Charles Richet, himself a Nobel laureate, to report the quiet self-appraisal by a celebrated scientist: "I possess every good quality, but the one that distinguishes me above all is modesty."[39] Other scientists, for example, the great Harvard mathematician, George Birkhoff, will have no truck with modesty, whether false, prim, or genuine. Having been told by a Mexican physicist of his hope that the United States would continue "to send us savants of your stature," Birkhoff sturdily replied, "Professor Erro, in the States I *am* the only one of my stature." And as Norbert Wiener is reported to have said in his obituary address for Birkhoff, "He was the first among us and he accepted the fact. He was not modest."[40] Nevertheless, such forthright acknowledgment of one's eminence is not quite the norm among scientists.

It would appear, then, that the institution of science, like other institutions, incorporates potentially incompatible values: among them,

37. Bell, *op. cit.*, p. 172. Bell refers also to "a common and engaging trait of the truly eminent scientist in his frequent confession of how little he knows. . . ." What he describes as a trait of the scientist can also be seen as an expectation on the part of the community of scientists. It is not that many scientists *happen* to be humble men; they are *expected* to be humble. See E. T. Bell, "Mathematics and Speculation," *The Scientific Monthly*, 32 (March, 1931), pp. 193–209, at p. 204.

38. G. F. Fitzgerald, *Lord Kelvin, 1846–99*. Jubilee Commemoration Volume, with an Essay on his Works, 1899; S. P. Thompson, *Life of William Thomson*, Vol. II, Chapter XXIV.

39. See the gallery of trenchant pen-portraits of scientists in Charles Richet, *The Natural History of a Savant*, trans. by Sir Oliver Lodge, New York: Doran, 1927, p. 86.

40. Carlos Graef Fernandez (as transcribed by Samuel Kaplan), "My Tilt with Albert Einstein," *American Scientist*, 44 (April, 1956), pp. 204–211, at p. 204.

the value of originality, which leads scientists to want their priority to be recognized, and the value of humility, which leads them to insist on how little they have been able to accomplish. These values are not real contradictories, of course—" 'tis a poor thing, but my own"—but they do call for opposed kinds of behavior. To blend these potential incompatibles[41] into a single orientation, to reconcile them in practice, is no easy matter. Rather, as we shall now see, the tension between these kindred values—kindred as Cain and Abel were kin—creates an inner conflict among men of science who have internalized both of them and generates a distinct ambivalence toward the claiming of priorities.

Ambivalence toward Priority

THE components of this ambivalence are fairly clear. After all, to insist on one's originality by claiming priority is not exactly humble and to dismiss one's priority by ignoring it is not exactly to affirm the value of orginality.[42] As a result of this conflict, scientists come to despise themselves for wanting that which the institutional values of science have led them to want.

With the rare candor that distinguishes him, Darwin so clearly exhibits this agitated ambivalence in its every detail that this one case can be taken as paradigmatic for many others (which are matters of less detailed and less candid record). In his *Autobiography*, he writes that, even before his historic voyage on the Beagle in 1831, he was "ambitious to take a fair place among scientific men—whether more ambitious or less so than most of my fellow-workers, I can form no opinion."[43] A quarter of a century after this voyage, he is still wrestling with his ambition, exclaiming in a letter that "I wish I could set less value on the bauble fame, either present or posthumous, than I do, but not, I think, to any extreme degree. . . ."[44]

Two years before the traumatizing news from Wallace, reporting his formulation of the theory of evolution, Darwin writes his now-famous letter to Lyell, explaining that he is not quite ready to publish his views, as Lyell had suggested he do in order not to be forestalled, and again

41. For further examination of the problem of blending incompatible norms into stable patterns of behavior, in this case among physicians, see R. K. Merton, "Some Preliminaries to a Sociology of Medical Education," in R. K. Merton, G. C. Reader and P. L. Kendall, eds., *The Student-Physician*, Cambridge: Harvard University Press, 1957, p. 72 ff. As is well known, R. S. Lynd has set forth the general notion that institutional norms are organized as near-incompatibles; see his *Knowledge for What?*, Princeton: Princeton University Press, 1939, Chapter III.

42. Strictly speaking, originality and priority are of course not the same thing. Belated independent rediscoveries of what was long since known may represent great originality on the part of the rediscoverer, as is perhaps best shown in the remarkable case of the self-taught twentieth-century Indian mathematician, Srinivasa Ramanujan, who, all unknowing that it had been done before, re-created much of early nineteenth-century mathematics, and more besides. *Cf.* G. H. Hardy, *Ramanujan: Twelve Lec-*

expressing his uncontrollable ambivalence in these words: "I rather hate the idea of writing for priority, yet I certainly should be vexed if any one were to publish my doctrines before me."[45]

And then, In June 1858, the blow falls. What Lyell warned would happen and what Darwin could not bring himself to believe could happen, as all the world knows, did happen. Here is Darwin writing Lyell of the crushing event:

> [Wallace] has today sent me the enclosed, and asked me to forward it to you. It seems to me well worth reading. Your words have come true with a vengeance—that I should be forestalled. . . . I never saw a more striking coincidence; if Wallace had my MS. sketch written out in 1842, he could not have made a better short abstract! Even his terms now stand as heads of my chapters. . . . So all my originality, whatever it may amount to, will be smashed. . . .[46]

Humility and disinterestedness urge Darwin to give up his claim to priority; the wish for originality and recognition urges him that all need not be lost. At first, with typical magnanimity, but without pretense of equanimity, he makes the desperate decision to step aside altogether. A week later, he is writing Lyell again; perhaps he might publish a short version of his long-standing text, "a dozen pages or so." And yet, he says in his anguished letter, "I cannot persuade myself that I can do so honourably." Torn by his mixed feelings, he concludes his letter, "My good dear friend, forgive me. This is a trumpery letter, influenced by trumpery feelings." And in an effort finally to purge himself of his feelings, he appends a postscript, "I will never trouble you or Hooker on the subject again."[47]

The next day he writes Lyell once more, this time to repudiate the postscript. Again, he registers his ambivalence: "It seems hard on me that I should lose my priority of many years' standing, but I cannot feel at all sure that this alters the justice of the case. First impressions are generally right, and I at first thought it would be dishonourable in me now to publish."[48]

tures Suggested by His Life and Work, Cambridge: Harvard University Press, 1940. Edwin G. Boring, who has long been interested in the subject of priority in science, has, among many other perceptive observations, noted the lack of identity between originality and priority. See, for example, his early paper, "The Problem of Originality in Science," *American Journal of Psychology,* 39 (December, 1927), pp. 70–90, esp. at p. 78.

43. Darwin, *op. cit.*, p. 54.
44. *Ibid.*, p. 452.
45. *Ibid.*, pp. 426–427.
46. *Ibid.*, p. 473.
47. *Ibid.*, pp. 474–475.
48. *Ibid.*, p. 475.

As fate would have it, Darwin is then prostrated by the death of his infant daughter. He manages to respond to the request of his friend Hooker and sends him the Wallace manuscript and his own original sketch of 1844, "solely," he writes, "that you may see by your own handwriting that you did read it. . . . Do not waste much time. It is miserable in me to care at all about priority."[49]

Other members of the scientific community do what the tormented Darwin will not do for himself. Lyell and Hooker take matters in hand and arrange for the momentous session in which both papers are read at the Linnean Society. And as they put it in their letter prefacing the publication of the joint paper of "Messrs. C. Darwin and A. Wallace," "in adopting our present course . . . we have explained to him [Darwin] that we are not solely considering the relative claims to priority of himself and his friend, but the interests of science generally."[50] Despite this disclaimer of interest in priority, be it noted that scientific *knowledge* is not the richer or the poorer for having credit given where credit is due: it is the social *institution* of science and individual men of science that would suffer from repeated failures to allocate credit justly.

This historic and not merely historical episode so plainly exhibits the ambivalence occasioned by the double concern with priority and modesty that it need not be examined further. Had the institutionalized emphasis on originality been alone in point, the claim to priority would have invited neither self-blame nor self-contempt; publication of the long antecedent work would have proclaimed its own originality. But the value of originality was joined with the value of humility and modesty. To insist on priority would be to trumpet one's own excellence, but scientific peers and friends of the discoverers, acting as a third party in accord with the institutional norms, could with full propriety announce the joint claims to originality that the discoverers could not bring themselves to do. Underneath it all lies a deep and agitated ambivalence toward priority.

I have not yet counted the recorded cases of debates about priority in science and the manner of their outcome. Such a count, moreover, will not tell the full story for it will not include the doubtless numerous

49. *Ibid.*, p. 476.

50. "On the Tendency of Species to Form Varieties and on the Perpetuation of Varieties and Species by Natural Means of Selection," by C. Darwin and A. R. Wallace. Communicated by Sir C. Lyell and J. D. Hooker, *Journal of the Linnean Society*, 3 (1859), p. 45. Read July 1, 1858.

51. Merton, *op. cit.*, p. 166. Scientists do not all occupy similar positions in the social structure; there are, consequently, differentials in access to *opportunity* for scientific achievement (and, of course, differences of individual capacity for achievement). The theory of the relations of social structure to anomie requires us to explore differential pressures upon those scientists variously located in the social structure. Contrast only the disputatious Robert Hooke, a socially mobile man whose rise in status resulted wholly from his scientific achievements, and the singularly undisputa-

instances in which independent ideas and discoveries were never announced by those who found their ideas anticipated in print. Nevertheless, I have the strong impression that disputes, even bitter disputes, over priority outnumber the cases of despondent but unreserved admission that the other fellow had made the discovery first.

The institutional values of modesty and humility are apparently not always enough to counteract both the institutional emphasis upon originality and the actual workings of the system of allocating rewards. Originality, as exemplified by the new idea or the new finding, is more readily observable by others in science and is more fully rewarded than the often unobservable kind of humility that keeps an independent discoverer from reporting that he had had the same idea or the same finding. Moreover, after publication by another, it is often difficult, if not impossible, to demonstrate that one had independently arrived at the same result. For these and other reasons, it is generally an unequal contest between the values of recognized originality and of modesty. Great modesty may elicit respect, but great originality promises everlasting fame.

In short, the social organization of science allocates honor in a way that tends to vitiate the institutional emphasis upon modesty. It is this, I believe, which goes far toward explaining why so many scientists, even those who are ordinarily men of the most scrupulous integrity, will go to great lengths to press their claims to priority of discovery. As I have often suggested, perhaps too often, any *extreme* institutional

> . . . emphasis upon achievement—whether this be scientific productivity, accumulation of wealth or, by a small stretch of the imagination, the conquests of a Don Juan—will attenuate conformity to the institutional norms governing behavior designed to achieve the particular form "success," especially among those who are socially disadvantaged in the competitive race.[51]

Or more specifically and more completely, great concern with the goal of recognition for originality can generate a tendency toward sharp practices just inside the rules of the game or sharper practices far out-

tious Henry Cavendish, high-born and very rich (far richer, and, by the canons of Burke's peerage, more elevated even than that other great aristocrat of science, Robert Boyle) who, in the words of Biot, was "*le plus riche de tous les savants; et probablement aussi, le plus savant de tous les riches.*" Or consider what Norbert Wiener has said of himself, "I was competitive beyond the run of younger mathematicians, and I knew equally that this was not a very pretty attitude. However, it was not an attitude which I was free to assume or to reject. I was quite aware that I was an out among ins and I would get no shred of recognition that I did not force." (*I Am a Mathematician,* New York: Doubleday, 1956, p. 87.) But these are only straws in the wind; once again, limitations of space allow me only to identify a problem, not to examine it.

side. That this has been the case with the behavior of scientists who were all-out to have their originality recognized, the rest of this paper will try to show.

Types of Response to Cultural Emphasis on Originality

FRAUD IN SCIENCE

The extreme form of deviant behavior in science would of course be the use of fraud to obtain credit for an original discovery. For reasons to be examined, the annals of science include very few instances of downright fraud although, in the nature of the case, an accurate estimate of frequency is impossible. Darwin, for example, said that he knew of only "three intentionally falsified statements" in science.[52] Yet, some time before, his contemporary, Charles Babbage, the mathematician and inventor of calculating machines (one of which prophetically made use of perforated cards), had angrily taken a classified inventory of fraud in science.[53]

At the extreme are hoaxes and forgery: the concocting of false data in science and learning—or, more accurately, in pseudo-science and anti-scholarship. Literary documents have been forged in abundance, at times, by men of previously unblemished reputation, in order to gain money or fame. Though no one can say with confidence, it appears that love of money was at the root of the forgery of fifty or so rare nineteenth-century pamphlets by that prince of bibliographers, that court of last appeal for the authentication of rare books and manuscripts, Thomas J. Wise. Of quite another stripe was John Payne Collier, the Shakesperian scholar who, unrivalled for his genuine finds in Elizabethan drama and "encouraged by the steadily growing plaudits of his colleagues," could not rest content with his measure of fame and proceeded to forge, with great and knowledgeable skill, a yet-uncounted array of literary papers.[54] But these rogues seem idle alongside the fecund and audacious Vrain-Lucas who, in the space of eight years, created

52. Darwin, *op. cit.*, p. 84.

53. Charles Babbage, *The Decline of Science in England*, London, 1830, pp. 174–183. George Lundberg has independently noted that "a scientist's greed for applause [sometimes] becomes greater than his devotion to truth." [*Social Research*, New York: Longmans Green, 1929, p. 34 (and in less detail, in the second edition, 1946, p. 52).]

54. I have drawn these examples of frauds in anti-scholarship from the zestful and careful account by Richard D. Altick, *The Scholar Adventurers*, New York: Macmillan, 1951, Chapters 2 and 6.

55. The definitive reports on the Vrain-Lucas affair by M. P. Faugère and by Henri Bordier and Mabille are not available to me at this telling; substantial details, including extracts from the court-proceedings, are given by the paleographer, Étienne Charavay, *Affair Vrain-Lucas: Étude Critique*, Paris, 1870; a more accessible summary

more than 27,000 pieces of manuscript, all duly sold to Michel Chasles, perhaps the outstanding French geometer of the mid-nineteenth century, whose credulity stretches our own, inasmuch as this vast collection included letters by Pontius Pilate, Mary Magdalene, the resurrected Lazuras, Ovid, Luther, Dante, Shakspere, Galileo, Pascal, and Newton, all written on paper and in modern French. Most provocative among these documents was the correspondence between Pascal and the then eleven-year-old Newton (all in French, of course, although even at the advanced age of thirty-one Newton could struggle through French only with the aid of a dictionary), for these letters made it plain that Pascal, not Newton, had to the greater glory of France, first discovered the law of gravitation, a momentous correction of history, which for several years excited the interest of the *Académie des Sciences* and usurped many pages of the *Comptes Rendus* until, in 1869, Vrain-Lucas was finally brought to book and sentenced to two years in prison. For our purposes, it is altogether fitting that Vrain-Lucas should have had Pascal address this maxim to the boy Newton: "*Tout homme qui n'aspire pas à se faire un nom n'exécutera jamais rien de grand.*"[55]

Such lavish forgery is unknown to science proper, but the pressure to demonstrate the truth of a theory or to produce a sensational discovery has occasionally led to the faking of scientific evidence. The biologist Paul Kammerer produced specimens of spotted salamanders designed to prove the Lamarckian thesis experimentally; was thereupon offered a chair at the University of Moscow where in 1925 the Lamarckian views of Michurin held reign; and upon proof that the specimens were fakes, attributed the fraud to a research assistant and committed suicide.[56] Most recently, the Piltdown man—that is, the skull and jaw from which his existence was inferred—has been shown, after forty years of uneasy acceptance, to be a carefully contrived hoax.[57]

Excessive concern with "success" in scientific work has on occasion led to the types of fraud Babbage picturesquely described as "trimming" and "cooking." The trimmer clips off "little bits here and there from observations which differ most in excess from the mean, and [sticks]

that does not, however, do full justice to the prodigious inventiveness of Vrain-Lucas is provided by J. A. Farrer, *Literary Forgeries*, London: Longmans Green, 1907, Chapter XII. The biographer of Newton, Sir David Brewster, at the age of 87, did his share to safeguard the integrity of historical scholarship, but this did not prevent Chasles from prizing the three thousand letters of Gallileo which he had acquired from his friend, although they happened to be in French, rather than in the Latin or Italian in which Galileo wrote.

56. Martin Gardner, *In the Name of Science*, New York: G. P. Putnam's Sons, 1952, p. 143; W. S. Beck, *Modern Science and the Nature of Life*, New York: Harcourt, Brace, 1957, pp. 201–202; Conway Zirkle, "The Citation of Fraudulent Data," *Science*, 120 (30 July, 1954), pp. 189–190.

57. William L. Straus, Jr., "The Great Piltdown Hoax," *Science*, 119 (26 February, 1954), pp. 265–269.

. . . them on to those which are too small . . . [for the unallowable purpose of] 'equitable adjustment.'" The cook makes "multitudes of observations" and selects only those which agree with an hypothesis and, as Babbage says, "the cook must be very unlucky if he cannot pick out fifteen or twenty which will do for serving up." This eagerness to demonstrate a thesis can, on occasion, lead even truth to be fed with cooked data, as it did for the neurotic scientist, described by Lawrence Kubie, "who had proved his case, but was so driven by his anxieties that he had to bolster an already proved theorem by falsifying some quite unnecessary additional statistical data."[58]

The great cultural emphasis upon recognition for original discovery can lead by gradations from these rare practices of outright fraud to more frequent practices just beyond the edge of acceptability, sometimes without the scientist's being aware that he has exceeded allowable limits. Scientists may find themselves reporting only "successful experiments or results, so-called, and neglecting to report 'failures.'" Alan Gregg, that informed observer of the world of medical research, practice, and education, reports the case of

> . . . the medical scientist of the greatest distinction who told me that during his graduate fellowship at one of the great English universities he encountered for the first time the idea that in scientific work one should be really honest in reporting the results of his experiments. Before that time he had always been told and had quite naturally assumed that the point was to get his observations and theories accepted by others, and published.[59]

Yet, these deviant practices should be seen in perspective. What evidence there is suggests that they are extremely infrequent, and this temporary focus upon them will surely not be distorted into regarding the exceptional case as the typical. Apart from the moral integrity of scientists themselves and this is, of course, the major basis for honesty in science, there is much in the social organization of science that provides a further compelling basis for honest work. Scientific research is typically, if not always, under the exacting scrutiny of fellow-experts,

58. Lawrence S. Kubie, M.D., "Some Unsolved Problems of the Scientific Career," *American Scientist*, 41 (1953), pp. 596–613; 42 (1954), pp. 104–112 [reprinted as Chapter 14 of this book], at p. 606.

59. Alan Gregg, *Challenges to Contemporary Medicine*, New York: Columbia University Press, 1956, p. 115.

60. As stated by the historian of astronomy, Agnes Mae Clerke, in her article on Laplace in the eleventh edition of the *Encyclopaedia Britannica*. Some of Clerke's further observations are much in point: "In the delicate task of apportioning his own large share of merit, he certainly does not err on the side of modesty; but it would perhaps be as difficult to produce an instance of injustice, as of generosity in his estimate of others. Far more serious blame attaches to his all but total suppression in the body of the work—and the fault pervades the whole of his writings—of the names of his

involving, as it usually though not always does, the verifiability of results by others. Scientific inquiry is in effect subject to rigorous policing, to a degree perhaps unparalleled in any other field of human activity. Personal honesty is supported by the public and testable character of science. As Babbage remarked, "the cook would [at best] procure a temporary reputation . . . at the expense of his permanent fame."

Competition in the realm of science, intensified by the great emphasis on original and significant discoveries, may occasionally generate incentives for eclipsing rivals by illicit or dubious means. But this seldom occurs in the form of preparing fraudulent data; instead, it appears in quite other forms of deviant behavior involving spurious claims to discovery. More concretely, it is an occasional theft rather than forgery, and more often, libel and slander rather than theft that are found on the small seamy side of science.

PLAGIARY: FACT AND SLANDER

Deviant behavior most often takes the form of occasional plagiaries and many slanderous charges or insinuations of plagiary. The historical record shows relatively few cases (and of course the record may be defective) in which one scientist actually pilfered another. We are assured that in the *Mécanique céleste* (until then, outranked only by Newton's *Principia*) "theorems and formulae are appropriated wholesale without acknowledgement" by Laplace.[60] Or, to take a marginal case, Sir Everard Home, the distinguished English surgeon who was appointed custodian of the unpublished papers of his even more distinguished brother-in-law, John Hunter, published 116 papers of uncertain origin in the *Philosophical Transactions* after Hunter's death, and burned Hunter's manuscripts, an action greatly criticized by knowledgeable and suspicious contemporaries.[61] It is true also that Robert Boyle, not impressed by the thought that theft of his ideas might be a high tribute to his talent, was in 1688 driven to the desperate expedient of printing an "Advertisement about the Loss of many of his Writings," later describing the theft of his work and reporting that he

predecessors and contemporaries . . . a production which may be described as the organized result of a century of patient toil presents itself to the world as the offspring of a single brain." And yet, since these matters are seldom all of a piece, "Biot relates that, when he himself was beginning his career, Laplace introduced him at the Institute for the purpose of explaining his supposed discovery of equations of mixed differences, and afterwards showed him, under a strict pledge of secrecy, the papers, then yellow with age, in which he had long before obtained the same results." (Vol. XVI, pp. 201–202.) As we shall see, Gauss, who was meticulous in acknowledging predecessors, treated the young Bolyai as did Laplace the young Biot.

61. Ralph H. Major, *A History of Medicine*, Oxford: Blackwell Scientific Publications, 1954, Vol. II, p. 703.

would from then on write only on loose sheets, in the hope that these would tempt thieves less than "bulky packets" and, going on to say that he was resolved to send his writings to press without extensive revision in order to avoid prolonged delays.[62] But even with such cases of larceny on the grand scale, the aggregate of demonstrable theft in modern science is not large.

What does loom large is the repeated practice of charging others with pilfering scientific ideas. Falsely accused of plagiarizing Harvey in physiology, Snell in optics, and Harriot and Fermat in geometry, Descartes in turn accuses Hobbes and the teen-age Pascal of plagiarizing him.[63] To maintain his property, Descartes implores his friend Mersenne, "I also beg you to tell him [Hobbes] as little as possible about what you know of my unpublished opinions, for if I'm not greatly mistaken, he is a man who is seeking to acquire a reputation at my expense and through shady practices."[64] All unknowing that the serene and unambitious Gauss had long since discovered the method of least squares, Legendre, himself "a man of the highest character and scrupulously fair," practically accuses Gauss of having filched the idea from him and complains that Gauss, already so well-stocked with momentous discoveries, might at least have had the decency not to adopt his brainchild.[65]

At times, the rivalrous concern with priority can go so far as to set, not the Egyptians against the Egyptians, but brother against brother, as in the case of the great eighteenth-century mathematicians, the brothers Jacob and Johannes Bernoulli, who repeatedly and bitterly attacked one another's claims to priority. (Johannes improved on this by throwing his own son out of the house for having won a prize from the French Academy on which he himself had had his eye.)[66]

Or to turn to our own province, Comte, tormented by the suggestion that his law of three stages had really been originated by St.-

62. The account by A. M. Clerke in the article on Boyle in the *Dictionary of National Biography* is somewhat mistaken in attributing charges of plagiary to the published Advertisement. This speaks only of losses of manuscript through "unwelcome accidents" (e.g., the upsetting of corrosive liquors over a file of manuscripts) and at most hints at less impersonal sources of loss. But a later unpublished paper by Boyle, dug up by his biographer Birch, is levelled against the numerous plagiarists of his works. This document, running to three folio pages of print, is a compendium of the ingenious devices for thievery developed by the grand larcenists of seventeenth-century science. See *The Works of the Honourable Robert Boyle*, in six volumes, to which is prefixed The Life of the Author, by J. Birch, London, 1772, Volume I, pp. cxxv–cxxviii, ccxxii–ccxxiv.

63. For the case of Harvey, see A. R. Hall, *The Scientific Revolution*, 1500–1800, London: Longmans, Green 1954, p. 148; for Hobbes, see Descartes, *Oeuvres* (edited by Charles Adam and Paul Tannery), *Correspondance*, Paris, 1899, Vol. III, pp. 283 ff.; for Pascal, see *ibid.*, 1903, Vol. V, p. 366.

64. Descartes, *ibid.*, Vol. III, p. 320.

65. Bell, *op. cit.*, pp. 259–260. Legendre seems to have been particularly sensitive to these matters, perhaps because he was often victimized; note Clerke's remark that

Simon, denounces his one-time master and describes him as a "superficial and depraved charlatan."[67] Again, to take Freud's own paraphrase, Janet claims that "everything good in psychoanalysis repeats, with slight modifications, the views of Janet—everything else in psychoanalysis being bad."[68] Freud refuses to lock horns with Janet in what he describes as "gladiator fights in front of the noble mob," but some years later, his disciple, Ernest Jones, reports that at a London Congress he has "put an end to" Janet's pretensions, and Freud applauds him in a letter that urges him to "strike while the iron is hot," in the interests of "fair play."[69]

So the almost changeless pattern repeats itself. Two or more scientists quietly announce a discovery. Since it is often the case that these are truly independent discoveries, with each scientist having separately exhibited originality of mind, the process is sometimes stabilized at that point, with due credit to both, as in the instance of Darwin and Wallace. But since the situation is often ambiguous with the role of each not easy to demonstrate and since each *knows* that he had himself arrived at the discovery, and since the institutionalized stakes of reputation are high and the joy of discovery immense, this is often not a stable solution. One or another of the discoverers—or frequently, his colleagues or fellow-nationals—suggests that he rather than his rival was really first, and that the independence of the rival is at least unproved. Then begins the familiar deterioration of standards governing conflictful interaction: the other side, grouping their forces, counter with the opinion that plagiary had indeed occurred, that let him whom the shoe fits wear it and furthermore, to make matters quite clear, the shoe is on the other foot. Reinforced by group-loyalties and often by chauvinism, the controversy gains force, mutual recriminations of plagiary abound, and there develops an atmosphere of thoroughgoing hostility and mutual distrust.

between Laplace and Legendre "there was a feeling of 'more than coldness,' owing to his appropriation, with scant acknowledgement, of the other's labors." *Encyclopaedia Britannica*, Vol. XVI, p. 202.

66. Bell, *op. cit.*, p. 134.

67. Frank E. Manuel, *The New World of Henri Saint-Simon*, Cambridge: Harvard University Press, 1956, pp. 340–342; also Richard L. Hawkins, *Auguste Comte and the United States*, Cambridge: Harvard University Press, 1936, pp. 81–82, as cited by Manuel.

68. Sigmund Freud, *History of the Psychoanalytic Movement*, London: Hogarth Press; also, Freud, *An Autobiographical Study*, London: Hogarth Press, 1948, pp. 54–55, where he seeks "to put an end to the glib repetition of the view that whatever is of value in psycho-analysis is merely borrowed from the ideas of Janet . . . historically psycho-analysis is completely independent of Janet's discoveries, just as in its content it diverges from them and goes far beyond them." For Janet's not always delicate insinuations, see his *Psychological Healing*, New York: Macmillan, 1925, I, pp. 601–640.

69. Ernest Jones, *Sigmund Freud: Life and Work*, London: Hogarth Press, 1955, Vol. II, p. 112.

On some occasions, this can lead to outright deceit in order to buttress valid claims, as with Newton in his controversy with Leibniz over the invention of the calculus. When the Royal Society finally established a committee to adjudicate the rival claims, Newton, who was then president of the Royal Society, packed the committee, helped direct its activities, anonymously wrote the preface for the second published report—the draft is in his handwriting—and included in that preface a disarming reference to the old legal maxim that "no one is a proper witness for himself [and that] he would be an iniquitous Judge, and would crush underfoot the laws of all the people, who would admit anyone as a lawful witness in his own cause."[70] We can gauge the immense pressures for self-vindication that must have operated for such a man as Newton to have adopted these means for defense of his valid claims. It was not because Newton was so weak but because the institutionalized values were so strong that he was driven to such lengths.

This interplay of offensive and defensive maneuvers—no doubt, students of the theory of games can recast it more rigorously—thus gives further emphasis to priority. Scientists try to exonerate themselves in advance from possible charges of filching by going to great lengths to establish their priority of discovery. Often, this kind of anticipatory defense produces the very result it was designed to avoid by inviting others to show that prior announcement or publication need not mean there was no plagiary.

The effort to safeguard priority and to have proof of one's integrity has led to a variety of institutional arrangements designed to cope with this strain on the system of rewards. In the seventeenth century, for example, and even as late as the nineteenth, discoveries were sometimes reported in the form of anagrams—as with Galileo's "triple star" of Saturn and Hooke's law of tension—for the double purpose of establising priority of conception and of yet not putting rivals on to one's original ideas, until they had been further worked out.[71] Then, as now,

70. There is a sizable library discussing the Newton-Leibniz controversy. I have drawn chiefly upon More, *op. cit.*, who devotes the whole of Chapter XV to this subject; Auguste de Morgan, *Essays on the Life and Works of Newton*, Chicago: Open Court Pub. Co., 1914, esp. Appendix II; and Brewster, *op. cit.*, Chapter XXII; *cf.* Cohen, *op. cit.*, who is properly critical of the biography by More at various points (e.g., pp. 84–85). On the basis of his examination of the Portsmouth Papers, More concludes that "the principals, and practically all those associated with them wantonly made statements which were false; and not one of them came through with a clean record." (P. 567.) E. N. da C. Andrade has aptly summed up Newton's ambivalence in this judgment: "Evidence can be cited for the view that Newton was modest or most overweening; the truth is that he was a very complex character . . . when not worried or irritated he was modest about his achievements." See also Andrade's *Sir Isaac Newton*, London: Collins, 1954, esp. pp. 131–132.

71. The earlier widespread use of anagrams is well-known. As late as the nineteenth century, the physicists Balfour Stewart and P. G. Tait reintroduced this prac-

complex ideas were quickly published in abstracts, as when Halley urged Newton to do so in order to secure "his invention to himself till such time as he would be at leisure to publish it."[72] There is also the long-standing practice of depositing sealed and dated manuscripts with scientific academies in order to protect both priority and idea.[73] Scientific journals often print the date on which the manuscript of a published article was received, thus serving, even apart from such intent, to register the time it first came to notice. Numerous personal expedients have been developed: for example, letters detailing one's own ideas are sent off to a potential rival, thus disarming him; preliminary and confidential reports are circulated among a chosen few; personal records of research are meticulously dated (as by Kelvin). Finally, it has often been suggested that the functional equivalent of a patent-office be established in science to adjudicate rival claims to priority.[74]

In prolonged and yet overly quick summary, these are some of the forms of deviance invited by the institutional emphasis on priority and some of the institutional expedients devised to reduce the frequency of these deviations. But as we would expect from the theory of alternative responses to excessively emphasized goals, other forms of behavior, verging toward deviance though still well within the law and not as subject to moral disapproval as the foregoing, have also made their appearance.

ALTERNATIVE RESPONSES TO EMPHASIS ON ORIGINALITY

The large majority of scientists, like the large majority of artists, writers, doctors, bankers, and bookkeepers, have little prospect of great and decisive originality. For most of us artisans of research, getting things into print becomes a symbolic equivalent to making a significant discovery. Nor could science advance without the great unending flow of papers reporting careful investigations, even if these are routine rather than distinctly original. The indispensable reporting of research

tice and "to secure priority . . . [took] the unusual step of publishing [their idea] as an anagram in *Nature* some months before the publication of their book." Sir. J. J. Thomson, *Recollections and Reflections,* London: G. Bell, 1936, p. 22.

72. Thomas Birch, *The History of the Royal Society of London,* London, 1756–1757, Vol. IV, p. 437.

73. For a recent instance, see the episode described by Wiener in which the race between Bouligand and Wiener to contribute new concepts "in potential theory" ended in a "dead heat," since Bouligand had submitted his "results to the [French] Academy in a sealed envelope, after a custom sanctioned by centuries of academy tradition." (Wiener, *op. cit.,* p. 92.)

74. J. Hettinger, "Problems of Scientific Property and Its Solution," *Science Progress,* 26 (January, 1932), pp. 449–461; also the paper by Dr. A. L. Soresi, of the New York Academy of Medicine, cited by Bernhard J. Stern, *Social Factors in Medical Progress,* New York: Columbia University Press, 1927, p. 108.

can, however, become converted into an itch to publish that, in turn, becomes aggravated by the tendency, in many academic institutions, to transform the sheer number of publications into a ritualized measure of scientific or scholarly accomplishment.[75]

The urge to publish is given a further push by the moral imperative of science to make one's work known to others; it is the obverse to the culturally repudiated practice of jealously hoarding scientific knowledge for oneself. As Priestley liked to say, "whenever he discovered a new fact in science, he instantly proclaimed it to the world, in order that other minds might be employed upon it besides his own."[76] Indeed, John Aubrey, that seventeenth-century master of the thumbnail biography and member of the Royal Society, could extend the moral imperative for communication of knowledge to justify even plagiary if the original author will not put his ideas into print. In his view it was better to have scientific goods stolen and circulated than to have them lost entirely.[77]

To this point (and I provide comfort by reporting that the end of the paper is in sight), we have examined types of deviant responses to the institutional emphasis on priority that are *active* responses: the fabrication of "data," aggressive self-assertion, the denouncing of rivals, plagiary, and charges of plagiary. Other scientists have responded to the same pressures *passively* or at least by internalizing their aggressions and directing them against themselves.[78] Since these passive responses, unlike the active ones, are private and often not publicly observable, they seldom enter the historical record. This need not mean, of course, that passive withdrawal from the competition for originality in science is infrequent; it might simply mean that the men responding in this fashion do not come to public notice, unless they do so

75. There is not space here to examine the institutional conditions which lead the piling up of publications to become a virtually ritualistic activity.

76. Priestley's remark as paraphrased by his longtime friend, T. L. Hawkes, had reported by George Wilson, *op. cit.*, p. 111. The 17th-century Dutch genius of microscopy, Anton van Leeuwenhoek, also adopted a policy, as he described it, that "whenever I found out anything remarkable, I have thought it my duty to put down my discovery on paper, so that all ingenious people might be informed thereof." (Quoted by Major, *History of Medicine*, Vol. I, p. 531.) The same sentiment was expressed by St.-Simon, among many others. *Cf.* Manuel, *op. cit.*, pp. 63–64.

77. Aubrey could say, irresponsibly and probably without malice, that the mathematician John Wallis "may stand with much glory upon his owne basis, and need not be beholding to any man, for Fame, yet he is so greedy of glorie, that he steales feathers from others to adorn his own cap; e.g. he lies at watch, at Sir Christopher Wren's discours, Mr. Robert Hooke's, &c.; puts down their notions in his note booke, and then prints it, without owneing the authors. This frequently of which they complaine. But though he does an Injury to the Inventors, he does good to Learning, in publishing such curious notions, which the author (especially Sir Christopher Wren) might never have the leisure to write of himselfe." (John Aubrey, *Brief Lives*, ed. by Andrew Clark, Oxford, 1898, Vol. II, pp. 281–282.)

after their accomplishments have qualified them for the pages of history.

Chief among these passive deviant responses is what has been described, on occasion, as *retreatism,* the abandoning of the once-esteemed cultural goal of originality and of practices directed toward reaching that goal. In such instances, the scientist withdraws from the field of inquiry, either by giving up science altogether or by confining himself to some alternative role in it, such as teaching or administration. (This does not say, of course, that teaching and administration do not have their own attractions, or that they are less significant than inquiry; I refer here only to the scientists who reluctantly abandon their research because it does not measure up to their own standards of excellence.)

A few historical instances of such retreatism must stand in place of more. The nineteenth-century physicist Waterston, his classic paper on molecular velocity having been rejected by the Royal Society as "nothing but nonsense," becomes hopelessly discouraged and leaves science altogether.[79] Deeply disappointed by the lack of response to his historic papers on heredity, Mendel refuses to publish the now-permanently lost results of his further research and, after becoming abbot of his monastery, gives up his research on heredity.[80] Robert Mayer, tormented by refusals to grant him priority for the principle of conservation of energy, tries a suicide leap from a third-story window and succeeds only in breaking his legs and being straitjacketed, for a time, in an insane asylum.[81]

Perhaps the most telling instance of retreatism in mathematics is that of Janos Bolyai, inventor of one of the non-Euclidean geometries. The young Bolyai tries to obey his mathematician-father who, out of the bitter fruits of his own experience, warns his son to give up any

78. The distinction between active and passive forms of deviant behavior is drawn from Talcott Parsons, *The Social System,* Glencoe: The Free Press, 1951, pp. 256–267.

79. Murray, *op. cit.,* pp. 346–348; and David L. Watson, *Scientists are Human,* London: Watts & Co., 1938, pp. 58, 80; Baron Rayleigh, *op. cit.,* 169–171. Evidently, Sidney Lee, the editor of the *Dictionary of National Biography* by the time it reached the volume in which Waterston should have had an honored place, could not penetrate the obscurity into which the great discoverer was plunged by the unfounded rejection of his work; there is no biography of Waterston in the *DNB.*

80. Hugo Iltis, *Life of Mendel,* New York: W. W. Norton, 1932, pp. 111–112; and see Mendel's prophetic remark, "My time will come," p. 282.

81. Mayer's having been rejected by his liberal friends who took part in the revolution of 1848, which he as a conservative opposed, may have contributed to his disturbance. For some recent evidence on how Mayer's priority was safeguarded by the lay-sociologist Josef Popper, see Otto Blüh, "The Value of Inspiration: A Study on Julius Robert Mayer and Josef Popper-Lynkeus," *Isis,* 43 (September, 1952), pp. 211–220. Blüh's opinion that claims of priority in science are no longer taken seriously seems exaggerated.

effort to prove the postulate on parallels—or, as his father more picturesquely put it, to "detest it just as much as lewd intercourse; it can deprive you of all your leisure, your health, your rest, and the whole happiness of your life." He dutifully becomes an army officer instead, but his demon does not permit the twenty-one-year-old Bolyai to leave the postulate alone. After years of work, he develops his geometry, sends the manuscript to his father who in turn transmits it to Gauss, the prince of mathematicians, for a magisterial opinion. Gauss sees in the work proof of authentic genius, writes the elder Bolyai so, and adds, in all truth, that he cannot express his enthusiasm as fully as he would like, for "to praise it, would be to praise myself. Indeed, the whole contents of the work, the path taken by your son, the results to which he is led, coincide almost entirely with my meditations, which have occupied my mind partly for the last thirty or thirty-five years . . . I am very glad that it is just the son of my old friend who takes the precedence of me in such a remarkable manner." Delighted by this accolade, the elder Bolyai sends the letter to his son, innocently saying that it is "very satisfactory and redounds to the honor of our country and our nation." Young Bolyai reads the letter, but has no eye for the statements which say that his ideas are sound, that in the judgment of the incomparable Gauss he is blessed with genius. He sees only that Gauss has anticipated him. For a time, he believes that his father must have previously confided his ideas to Gauss who had thereupon made them his own.[82] His priority lost, and, with the further blow, years later, of coming upon Lobachevsky's non-Euclidean geometry, he never again publishes any work in mathematics.[83]

Apart from historical cases of notable scientists retreating from the field after denial of the recognition owing them, there are many contemporary cases that come to the notice of psychiatrists rather than historians. Since Lawrence Kubie is almost alone among psychiatrists to have described these in print, I shall draw upon his pertinent account

82. The principal source on the Bolyais, including the germane correspondence, is Paul Stäckel, *Wolfgang und Johann Bolyai, Geometrische Untersuchungen*, Leipzig: 1913, two vols. which was not available to me at this writing. An excellent short account is provided by Roberto Bonola, *Non-Euclidean Geometry* (trans. by H. S. Carslaw), La Salle, Illinois: Open Court Publishing Company, 1938, 2d rev. ed., pp. 96–113; see also Dirk J. Struik, *A Concise History of Mathematics*, New York: Dover Publications, 1948, Vol. II, p. 251–254; Franz Schmidt, "Lebensgeschichte des Ungarischen Mathematikers Johann Bolyai de Bolya," *Abhandlungen zur Geschichte der Mathematik*, 8 (1898), pp. 135–146.

83. Two letters provide context for Bolyai's great fall from the high peak of exhilaration into the slough of despond. In 1823, he writes his father: ". . . the goal is not yet reached, but I have made such wonderful discoveries that I have been almost overwhelmed by them, and it would be the cause of continual regret if they were lost. When you will see them, you too will recognize it. In the meantime I can say only this: *I have created a new universe from nothing.* All that I have sent you till now is but a house of cards, compared to the tower. I am as fully persuaded that it will bring

of the maladaptations of scientists suffering from an unquenched thirst for original discovery and ensuing praise.

When the scientist's aspirations become too lofty to be realized, the result sometimes is apathy, imbued with fantasy. In Kubie's words,

> . . . the young scientist may dwell for years in secret contemplation of his own unspoken hope of making great scientific discoveries. As time goes on, his silence begins to frighten him; and in the effort to master his fear, he may build up a secret feeling that his very silence is august, and that once he is ready to reveal his theories, they will shake the world. Thus a secret megalomania can hide among the ambitions of the young research worker.[84]

Perhaps most stressful of all is the situation in which the recognition accorded the scientist is not proportioned to his industry or even to the merit of his work. He may find himself serving primarily to remove obstacles to fundamental discoveries by others. His "negative experiments clear the road for the steady advance of science, but at the same time they clear the road for the more glamorous successes of other scientists, who may have used no greater intelligence, skill or devotion; perhaps even less."[85] Like other men, scientists become disturbed by the pan-human problem of evil, in which "the fortunes of men seem to bear practically no relation to their merits and efforts."[86]

Kubie hazards some further observations that read almost as if they were describing the behavior of delinquents in response to a condition of relative anomie.

> Success or failure, whether in specific investigations or in an entire career may be almost accidental, with chance a major factor in determining not *what* is discovered, but when and by whom. . . . Yet young students are not warned that their future success

me honor, as if I had already completed the discovery." And just as, a generation later, Lyell was prophetically to warn Darwin of being forestalled, so does the elder Bolyai warn the younger: "If you have really succeeded in the question, it is right that no time be lost in making it public, for two reasons first, because ideas pass easily from one to another, who can anticipate its publication; and secondly, there is some truth in this, that many things have an epoch, in which they are found at the same time in several places, just as the violets appear on every side in spring. Also every scientific struggle is just a serious war, in which I cannot say when peace will arrive. Thus we ought to conquer when we are able, since the advantage is always to the first comer." (Quoted by Bomola, *op. cit.*, pp. 98, 99.) Small wonder that though young Bolyai continued to work sporadically in mathematics, he never again published the results of his work.

84. Kubie, "Some Unsolved Problems of the Scientific Career," *op. cit.*, p. 110.

85. *Ibid.*

86. Gilbert Murray, quoted in a similar theoretical context by Merton, *op. cit.*, p. 147.

> may be determined by forces which are outside their own creative capacity or their willingness to work hard.[87]

As a result of this, Kubie suspects the emergence of what he calls a

> . . . new psychosocial ailment among scientists which may not be wholly unrelated to the gangster tradition of dead-end kids. Are we witnessing the development of a generation of hardened, cynical, amoral, embittered, disillusioned young scientists?

Lacking the evidence, this had best be left as a rhetorical question. But the import of the question needs comment. There have been diagnoses of the ways in which a culture giving emphasis to aspirations for all, aspirations which cannot be realized by many, exerts a pressure for deviant behavior and for cynicism, for rejection of the reigning moralities and the rules of the game. We see here the possibility that the same pressures may in some degree be at work in the institution of science. But even though the pressures are severe, they need not produce deviant behavior. There are great differences between the social structure of science and other social structures in which deviance is frequent. Among other things, the institution of science continues to have an abiding emphasis on other values that curb the culturally induced tendency toward deviation, an emphasis on the value of truth by whomsoever it is found, and a commitment to the disinterested pursuit of truth. Simply because we have focused on the deviant behavior of scientists, we should not forget how relatively rare this is. Only a few try to gain reputation by means that will lose them repute. Scientists may feel the pressures whose institutional sources I have tried to describe, but we can suppose that most will continue in the future as they manifestly have done in the past to abide by the institutional norms.

87. *Ibid.*, pp. 111–112. This reading of the case is not inconsistent with the facts of multiple independent discoveries and inventions. As the long history of multiple discoveries makes clear, and as W. F. Ogburn and D. S. Thomas among the sociologists have shown, certain discoveries become almost "inevitable" when the cultural base cumulates to a certain level. But this still leaves some indeterminacy in the matter of *who* will *first* make the discovery. Kubie mentions some "near-misses" of discoveries that suggest undoubted merit is not all when it comes to the *first* formulation of a discovery, and this list can be greatly extended. In the nature of the case, moreover, we often do not know of those scientists who have abandoned a line of inquiry that was moving toward a particular discovery when they found it had been made and announced by another. These "personal tragedies" of near-discovery—tragedy in terms of the prevailing cultural belief that all credit is due him who is "first"—are the silent tragedies that leave no mark in the historiography of science.

88. Quoted in Lloyd Stevenson, *Sir Frederick Banting*, London: Heinemann Medical Books, 1947, p. 301. Two hundred years before, John Morgan, the celebrated founder of the first American medical school, had expressed the same conception, but in sociologically more acceptable terms. To his mind, personal motivation for fame

Functions and Dysfunctions of Emphasis on Priority

IT HAS sometimes been said that the emphasis upon recognition of priority has the function of motivating scientists to make discoveries. For example, Sir Frederick Banting, the major figure in the discovery of insulin-therapy for diabetes, was long disturbed by the conviction that the chief of his department had been given too much credit for what he had contributed to the discovery. Time and again, Banting returned to the importance of allocating due credit for a discovery:

> . . . it makes research men [he said]. It stimulates the individuality and develops personality. Our religion, our moral fabric, our very basis of life are centered round the idea of reward. It is not abnormal therefore that the research man should desire the kudos of his own work and his own idea. If this is taken away from him, the greatest stimulant for work is withdrawn.[88]

From this, it would seem that the institutional emphasis is maintained with an eye to its functional utility. But as I have tried to show, the emphasis upon priority is often not confined within functional limits. Once it becomes established, forces of rivalrous interaction lead it to get out of hand. Recognition of priority, operating to reward those who advanced science materially by being the first to make a significant discovery, becomes a sentiment in its own right. Rationalized as a means of providing incentives for original work and as expressing esteem for those who have done much to advance science, it becomes transformed into an end-in-itself. It becomes stepped up to a dysfunctional extreme far beyond the limits of utility.[89] It can even reach the revealing extreme where, for example, the permanent secretary of the French Academy of Sciences, François Arago, could

was linked with the social benefit of the advancement of science. Men of science, he said, "have the highest motives that can animate the pursuits of a generous mind. They consider themselves as under the notice of the public, to which every ingenious person labours to approve himself. A love of fame and a laudable ambition allure him with the most powerful charms. These passions have, in all ages, fired the souls of heroes, of patriots, of lovers of science, have made them renowned in war, eminent in government and peace, justly celebrated for the improvement of polite and useful knowledge." In effect, "other-directedness" can be functional to the society, providing that the criteria of judgment by others are sound. See John Morgan, *A Discourse upon the Institution of Medical Schools in America*, photo-offset reprint of first edition, Philadelphia, 1765, Baltimore: The Johns Hopkins Press, 1937, pp. 59–60.

89. For suggestive observations on the process of "stepping up patterns to unanticipated extremities," a process which he called "perseveration," see W. I. Thomas, *Primitive Behavior*, New York: McGraw-Hill, 1937, p. 9 and passim; see also, Merton, *op. cit.*, pp. 199 ff. As I have tried to show in this paper, science has experienced this stepping-up of functional norms to an extreme at which they become dysfunctional to the workings of the institution.

exclaim (apropos of the controversy involving Cavendish and Watt) that to describe discoveries as having been made "'about the same time' proves nothing; questions as to priority may depend on weeks, on days, on hours, on minutes."[90]

When the criteria of priority become as finely discriminated as this—and Arago only put in words what many others have expressed in behavior—then priority has lost all functional significance. For when two scientists independently make the same discovery months or weeks apart, to say nothing of days or hours, it can scarcely be thought that one has exhibited greater originality than the other or that the short interim that separates them can be used to speed up the rate of scientific achievement.

Conclusion

THE interpretation I have tried to develop here is not, I am happy to say, a new one. Nor do I consider it fully established and beyond debate. After all, neither under the laws of logic nor under the laws of any other realm, must one become permanently wed to an hypothesis simply because one has tentatively embraced it. But the interpretation does seem to account for some of the otherwise puzzling aspects of conflicts over priority in science and it is closely bound to a body of sociological theory.

In short review, the interpretation is this. Like other social institutions, the institution of science has its characteristic values, norms, and organization. Among these, the emphasis on the value of originality has a self-evident rationale, for it is originality that does much to advance science. Like other institutions also, science has its system of allocating rewards for performance of roles. These rewards are largely honorific, since even today, when science is largely professionalized, the pursuit of science is culturally defined as being primarily a disinterested search for truth and only secondarily, a means of earning a livelihood. In line with the value-emphasis, rewards are to be meted out in accord with the measure of accomplishment. When the institution operates effectively, the augmenting of knowledge and the augmenting of personal fame go hand in hand; the institutional goal and the personal reward are tied together. But these institutional values have the defects of their qualities. The institution can get partly out of control, as the emphasis upon originality and its recognition is stepped up. The more thoroughly scientists ascribe an unlimited value to originality, the more they are in this sense dedicated to the ad-

90. M. [F.] Arago, *Historical Éloge of James Watt*, trans. by J. P. Muirhead, London, 1839, p. 106. The whole of this document and Arago's role in the Adams-LeVerrier controversy clearly exemplify the forces producing conflicts over priority.

vancement of knowledge, the greater is their involvement in the successful outcome of inquiry and their emotional vulnerability to failure.

Against this cultural and social background, one can begin to glimpse the sources, other than idiosyncratic ones, of the misbehavior of individual scientists. The culture of science is, in this measure, pathogenic. It can lead scientists to develop an extreme concern with recognition which is in turn the validation by peers of the worth of their work. Contentiousness, self-assertive claims, secretiveness lest one be forestalled, reporting only the data that support an hypothesis, false charges of plagiarism, even the occasional theft of ideas and in rare cases, the fabrication of data—all these have appeared in the history of science and can be thought of as deviant behavior in response to a discrepancy between the enormous emphasis in the culture of science upon original discovery and the actual difficulty many scientists experience in making an original discovery. In this situation of stress, all manner of adaptive behaviors are called into play, some of these being far beyond the mores of science.

All this can be put more generally. We have heard much in recent years about the dangers brought about by emphasis on the relativity of values, about the precarious condition of a society in which men do not believe in values deeply enough and do not feel strongly enough about what they do believe. If there is a lesson to be learned from this review of some consequences of a belief in the absolute importance of originality, perhaps it is the old lesson that unrestricted belief in absolutes has its dangers too. It can produce the kind of fanatic zeal in which anything goes. In this way, the absolutizing of values can be just as damaging as the decay of values to the life of men in society.[91]

91. Limitations of time and space do not allow me to do as I originally intended: to examine patterns of rediscovery, that is, the independent but considerably later discovery of something that had been found before but was since lost to view. These patterns have their own sociological characteristics which have not been considered here. A systematic sociological investigation of priority and rediscovery in science is being planned to test the validity of the interpretations set out in this paper and of other hypotheses mercifully omitted from it.

It is of some interest that just when this paper was in galley proof, all the world came to experience the social, political, and scientific repercussions of a spectacular "first" in science, when Russian scientists put a man-made sphere into space.

29 *Energy Conservation as an Example of Simultaneous Discovery*

THOMAS S. KUHN

BETWEEN 1842 AND 1847, the hypothesis of energy conservation was publicly announced by four widely scattered European scientists —Mayer, Joule, Colding, and Helmholtz—all but the last working in complete ignorance of the others.[1] The coincidence is conspicuous, yet these four announcements are unique only in combining generality of formulation with concrete quantitative applications. Sadi Carnot, before 1832, Marc Séguin in 1839, Karl Holtzmann in 1845, and G. A. Hirn in 1854, all recorded their independent convictions that heat and

---

Reprinted from Critical Problems in the History of Science, *edited by Marshall Clagett (Madison: University of Wisconsin Press, 1959), pp. 321–356.*

1. J. R. Mayer, "Bemerkungen über die Kräfte der unbelebten Natur," *Ann. d. Chem. u. Pharm.*, XLII (1842). I have used the reprint in J. J. Weyrauch's excellent collection, *Die Mechanik der Wärme in gesammelten Schriften von Robert Mayer* (Stuttgart, 1893), pp. 23–30. This volume is cited below as Weyrauch, I. The same author's companion volume, *Kleinere Schriften und Briefe von Robert Mayer* (Stuttgart, 1893), is cited as Weyrauch, II.

James P. Joule, "On the Calorific Effects of Magneto-Electricity, and on the Mechanical Value of Heat," *Phil. Mag.*, XXIII (1843). I have used the version in *The Scientific Papers of James Prescott Joule* (London, 1884), pp. 123–59. This volume is cited below as Joule, *Papers*.

L. A. Colding, "Undersögelse on de almindelige Naturkraefter og deres gjensidige Afhaengighed og isaerdeleshed om den ved visse faste Legemers Gnidning udviklede Varme," *Dansk. Vid. Selsk.*, II (1851), 121–46. I am indebted to Miss Kirsten Emilie Hedebol for preparing a translation of this paper. It is, of course, far fuller than the unpublished original which Colding read to the Royal Society of Denmark in 1843, but it includes much information about that original. See also, L. A. Colding, "On the History of the Principle of the Conservation of Energy," *Phil. Mag.*, XXVII (1864), 56–64.

H. von Helmholtz, *Ueber die Erhaltung der Kraft. Eine physikalische Abhandlung* (Berlin, 1847). I have used the annotated reprint in *Wissenschaftliche Abhandlungen von Hermann Helmholtz* (Leipzig, 1882), I, 12–75. This set is cited below as Helmholtz, *Abhandlungen*.

work are quantitatively interchangeable, and all computed a value for the conversion coefficient or an equivalent.[2] The convertibility of heat and work is, of course, only a special case of energy conservation, but the generality lacking in this second group of announcements occurs elsewhere in the literature of the period. Between 1837 and 1844, C. F. Mohr, William Grove, Faraday, and Liebig, all described the world of phenomena as manifesting but a single "force," one which could appear in electrical, thermal, dynamical, and many other forms, but which

2. Carnot's version of the conservation hypothesis is scattered through a notebook written between the publication of his memoir in 1824 and his death in 1832. The most authoritative version of the notes is E. Picard, *Sadi Carnot, biographie et manuscript* (Paris, 1927); a more convenient source is the appendix to the recent reprint of Carnot's *Réflexions sur la puissance motrice du feu* (Paris, 1953). Notice that Carnot considered the material in these notes quite incompatible with the main thesis of his famous *Réflexions*. In fact, the essentials of his thesis proved to be salvageable, but a change in both its statement and its derivation was required.

Marc Séguin, *De l'influence des chemins de fer et de l'art de les construire* (Paris, 1839), pp. xvi, 380–96.

Karl Holtzmann, *Über die Wärme und Elasticität der Gase und Dämpfe* (Mannheim, 1845). I have used the translation by W. Francis in *Taylor's Scientific Memoirs*, IV (1846), 189–217. Since Holtzmann believed in the caloric theory of heat and used it in his monograph, he is a strange candidate for a list of discoverers of energy conservation. He also believed, however, that the same amount of work spent in compressing a gas isothermally must always produce the same increment of heat in the gas. As a result he made one of the early computations of Joule's coefficient and his work is therefore repeatedly cited by the early writers on thermodynamics as containing an important ingredient of their theory. Holtzmann can scarcely be said to have caught any part of energy conservation as we define that theory today. But for this investigation of simultaneous discovery the judgment of his contemporaries is more relevant than our own. To several of them Holtzmann seemed an active participant in the evolution of the conservation theory.

G. A. Hirn, "Études sur les principaux phénomènes que présentent les frottements médiats, et sur les diverses manières de déterminer la valeur mécanique des matières employées au graissage des machines," *Bulletin de la societé industrielle de Mulhouse*, XXVI (1854), 188–237; and "Notice sur les lois de la production du calorique par les frottements médiats," *ibid.*, pp. 238–77. It is hard to believe that Hirn was completely ignorant of the work of Mayer, Joule, Helmholtz, Clausius, and Kelvin when he wrote the "Études" in 1854. But after reading his paper, I find his claim to independent discovery (presented in the "Notice") entirely convincing. Since none of the standard histories cites these articles or even recognizes the existence of Hirn's claim, it seems appropriate to sketch its basis here.

Hirn's investigation deals with the relative effectiveness of various engine lubricants as a function of pressure at the bearing, applied torque, etc. Quite unexpectedly, or so he says, his measurements led to the conclusion that: "The absolute quantity of caloric developed by mediated friction [e.g., friction between two surfaces separated by a lubricant] is directly and uniquely proportional to the mechanical work absorbed by this friction. And if we express the work in kilograms raised to the height of one meter and the quantity of caloric in calories, we find that the ratio of these two numbers is very nearly 0.0027 [corresponding to 370 kg.m./cal.], whatever the velocity and the temperature and whatever the lubricating material" (p. 202). Until almost 1860 Hirn had doubts about the law's validity for impure lubricants or in the absence of lubrication (see particularly his *Récherches sur l'équivalent mécanique de la chaleur* [Paris, 1858], p. 83.) But despite these doubts, his work obviously displays one of the mid-nineteenth-century routes to an important part of energy conservation.

could never, in all its transformations, be created or destroyed.[3] That so-called force is the one known to later scientists as energy. History of science offers no more striking instance of the phenomenon known as simultaneous discovery.

Already we have named twelve men who, within a short period of time, grasped for themselves essential parts of the concept of energy and its conservation. Their number could be increased, but not fruitfully.[4] The present multiplicity sufficiently suggests that in the two decades before 1850 the climate of European scientific thought included elements able to guide receptive scientists to a significant new view of nature. Isolating these elements within the works of the men affected by them may tell us something of the nature of simultaneous discovery. Conceivably, it may even give substance to those obvious yet totally unexpressive truisms: a scientific discovery must fit the times, or the time must be ripe. The problem is challenging. A preliminary identification of the sources of the phenomenon called simultaneous discovery is therefore the main objective of this paper.

Before proceeding towards that objective, however, we must briefly pause over the phrase simultaneous discovery itself. Does it sufficiently describe the phenomenon we are investigating? In the ideal case of

3. C. F. Mohr, "Ueber die Natur der Wärme," *Zeit. f. Phys.*, V (1837), 419–45; and "Ansichten über der Natur der Wärme," *Ann. d. Chem. u. Pharm.*, XXIV (1837), 141–47.

William R. Grove, *On the Correlation of Physical Forces: being the substance of a course of lectures delivered in the London Institution in the Year* 1843 (London, 1846). Grove states that in this first edition he has introduced no new material since the lectures were delivered. The later and more accessible editions are greatly revised in the light of subsequent work.

Michael Faraday, *Experimental Researches in Electricity* (London, 1844), II, 101–104. The original "Seventeenth Series" of which this is a part was read to the Royal Society in March, 1840.

Justus Liebig, *Chemische Briefe* (Heidelberg, 1844), pp. 114–20. With this work, as with Grove's, one must beware of changes introduced in editions published after the conservation of energy was a recognized scientific law.

4. Since a few of my conclusions depend upon the particular list of names selected for study, a few words about the selection procedure seem essential. I have tried to include all the men who were thought by their contemporaries or immediate successors to have reached independently some significant part of energy conservation. To this group I have added Carnot and Hirn whose work would surely have been so regarded if it had been known. Their lack of actual influence is irrelevant from the viewpoint of this investigation.

This procedure has yielded the present list of twelve names, and I am aware of only four others for whom a place might be claimed. They are von Haller, Roget, Kaufmann, and Rumford. Despite P. S. Epstein's impassioned defense (*Textbook of Thermodynamics* [New York, 1937], pp. 27–34), von Haller has no place on the list. The notion that fluid friction in the arteries and veins contributes to body heat implies no part of the notion of energy conservation. Any theory that accounts for frictional generation of heat can embrace von Haller's conception. A better case can be made for Roget who did use the impossibility of perpetual motion to argue against the contact theory of galvanism (see note 27). I have omitted him only because he seems unaware of the possibility of extending the argument and because his own conceptions are duplicated in the work of Faraday, who did extend them.

Hermann von Kaufmann probably should be included. According to Georg Helm simultaneous discovery two or more men would announce the same thing at the same time and in complete ignorance of each other's work, but nothing remotely like that happened during the development of energy conservation. The violations of simultaneity and mutual influence are secondary. But no two of our men even said the same thing. Until close to the end of the period of discovery, few of their papers have more than fragmentary resemblances retrievable in isolated sentences and paragraphs. Skillful excerpting is, for example, required to make Mohr's defense of the dynamical theory of heat resemble Liebig's discussion of the intrinsic limits of the electric motor. A diagram of the overlapping passages in the papers by the pioneers of energy conservation would resemble an unfinished crossword puzzle.

Fortunately no diagram is needed to grasp the most essential differences. Some pioneers, like Séguin and Carnot, discussed only a special case of energy conservation, and these two used very different approaches. Others, like Mohr and Grove, announced a universal conservation principle, but, as we shall see, their occasional attempts to quantify their imperishable "force" leave its concrete significance in doubt. Only in view of what happened later can we say that all these partial statements even deal with the same aspect of nature.[5] Nor is

his work is identical with Holtzmann's (*Die Energetik nach ihrer geschichtlichen Entwickelung* [Leipzig, 1898], p. 64). But I have been unable to see Kaufmann's writings, and Holtzmann's case is already somewhat doubtful, so that it has seemed better not to overload the list. As to Rumford, whose case is the most difficult of all, I shall point out below that before 1825 the dynamical theory of heat did not lead its adherents to energy conservation. Until the mid-century there was no necessary, or even likely, connection between the two sets of ideas. But Rumford was more than a dynamical theorist. He also said: "It would follow necessarily, from [the dynamical theory] . . . that the sum of the active forces in the universe must always remain constant" (*Complete Works* [London, 1876], III, 172), and this does sound like energy conservation. Perhaps it is. But if so, Rumford seems totally unaware of its significance. I cannot find the remark applied or even repeated elsewhere in his works. My inclination, therefore, is to regard the sentence as an easy echo, appropriate before a French audience, of the eighteenth-century theorem about the conservation of *vis viva*. Both Daniel Bernoulli and Lavoisier and Laplace had applied that theorem to the dynamical theory before (see note 95) without obtaining anything like energy conservation. I know of no reason to suppose that Rumford saw further than they.

5. This may well explain why the pioneers seem to have profited so little from each other's work, even when they read it. Our twelve men were not, in fact, strictly independent. Grove and Helmholtz knew Joule's work and cited it in their papers of 1843 and 1847 (Grove, *Physical Forces*, pp. 39, 52; Helmholtz, *Abhandlungen*, I, 33, 35, 37, 55). Joule, in turn, knew and cited the work of Faraday (*Papers*, p. 189). Liebig, though he did not cite Mohr and Mayer, must have known their work, for it was published in his own journal. (See also G. W. A. Kahlbaum, *Liebig und Friedrich Mohr, Briefe, 1834–1870* [Braunschweig, 1897], for Liebig's knowledge of Mohr's theory.) Very possibly more precise biographical information would disclose other interdependencies as well.

But these interdependencies, at least the identifiable ones, seem unimportant. In 1847 Helmholtz seems to have been unaware both of the generality of Joule's conclusions and of their large-scale overlap with his own. He cites only Joule's experimental findings, and these very selectively and critically. Not until the priority controversies of the second half-century, does Helmholtz seem to have recognized the extent to which he had been anticipated. Much the same holds for the relation between Joule this problem of divergent discoveries restricted to those scientists whose formulations were obviously incomplete. Mayer, Colding, Joule, and Helmholtz were not saying the same things at the dates usually given for their discoveries of energy conservation. More than *amour propre* underlies Joule's subsequent claim that the discovery he had announced in 1843 was different from the one published by Mayer in 1842.[6] In these years their papers have important areas of overlap, but not until Mayer's book of 1845 and Joule's publications of 1844 and 1847 do their theories become substantially coextensive.[7]

In short, though the phrase "simultaneous discovery" points to the central problem of this paper, it does not, if taken at all literally, describe it. Even to the historian acquainted with the concepts of energy conservation, the pioneers do not all communicate the same thing. To each other, at the time, they often communicated nothing at all. What we see in their works is not really the simultaneous discovery of energy conservation. Rather it is the rapid and often disorderly emergence of the experimental and conceptual elements from which that theory was shortly to be compounded. It is these elements that concern us. We know why they were there: Energy *is* conserved; nature behaves that way. But we do not know why these elements suddenly became accessible and recognizable. That is the fundamental problem of this paper. Why, in the years 1830 to 1850, did so many of the experiments and

and Faraday. From the latter Joule took illustrations, but not inspiration. Liebig's case may prove even more revealing. He could have neglected to cite Mohr and Mayer simply because they provided no relevant illustration and did not even seem to be dealing with the same subject matter. Apparently the men whom we call early exponents of energy conservation could occasionally read each other's works without quite recognizing that they were talking about the same things. For that matter, the fact that so many of them wrote from different professional and intellectual backgrounds may account for the infrequency with which they even saw each other's writings.

6. J. P. Joule, "Sur l'équivalent mécanique du calorique," *Comptes rendus*, XXVIII (1849), 132–35. I have used the reprint in Weyrauch, II, 276–80. This is only the first salvo in the priority controversy, but it already shows what the controversy is going to be about. Which of two (and later more than two) different statements is to be equated with *the* conservation of energy?

7. J. R. Mayer, *Die organischc Bewegung in ihrem Zusammenhange mit dem Stoffwechsel* (Heilbronn, 1845) in Weyrouch, I, 45–128. Most of Joule's papers between 1843 and 1847 are relevant, but particularly: "On the Changes of Temperature produced by the Rarefaction and Condensation of Air" (1845) and "On Matter, Living Force, and Heat" (1847) in *Papers*, pp. 172–89, 265–81.

8. This formulation has at least one considerable advantage over the usual version. It does not imply or even permit the question, "Who *really* discovered conservation of energy first?" As a century of fruitless controversy has demonstrated, a suitable extension or restriction in the definition of energy conservation will award the crown to almost any one of the pioneers, an additional indication that they cannot have discovered the same thing.

The present formulation also bars a second impossible question, "Did Faraday [or Séguin, or Mohr, or any one of the other pioneers, at will] really grasp the concept of energy conservation, even intuitively? Does he really belong on the list of pioneers?" Those questions have no conceivable answer, except in terms of the respondent's taste.

concepts required for a full statement of energy conservation lie so close to the surface of scientific consciousness?[8]

This question could easily be taken as a request for a list of all those almost innumerable factors that caused the individual pioneers to make the particular discoveries that they did. Interpreted in this way, it has no answer, at least none that the historian can give. But the historian can attempt another sort of response. A contemplative immersion in the works of the pioneers and their contemporaries may reveal a subgroup of factors which seem more significant than the others, because of their frequent recurrence, their specificity to the period, and their decisive effect upon individual research.[9] The depth of my acquaintance with the literature permits, as yet, no definitive judgments. Nevertheless, I am already quite sure about *two* such factors, and I suspect the relevance of a third. Let me call them the "availability of conversion processes," the "concern with engines," and the "philosophy of nature." I shall consider them in order.

THE availability of conversion processes resulted principally from the stream of discoveries that flowed from Volta's invention of the battery in 1800. According to the theory of galvanism most prevalent, at least, in France and England, the electric current was itself gained at the expense of forces of chemical affinity, and this conversion proved to be only the first step in a chain.[10] Electric current invariably produced

But whatever answer taste may dictate, Faraday (or Séguin, etc.) provides useful evidence about the forces that led to the discovery of energy conservation.

9. These three criteria, particularly the second and third, determine the orientation of this study in a way that may not be immediately apparent. They direct attention away from the *prerequisites* to the discovery of energy conservation and towards what might be called the *trigger-factors* responsible for simultaneous discovery. For example, the following pages will show implicitly that all of the pioneers made significant use of the conceptual and experimental elements of calorimetry and that many of them also depended upon the new chemical conceptions derived from the work of Lavoisier and his contemporaries. These and many other developments within the sciences presumably had to occur before conservation of energy, as we know it, could be discovered. I have not, however, explicitly isolated elements like these below, because they do not seem to distinguish the pioneers from their predecessors. Since both calorimetry and the new chemistry had been the common property of all scientists for some years before the period of simultaneous discovery, they cannot have provided the immediate stimuli that triggered the work of the pioneers. As prerequisites for discovery, these elements have an interest and importance all their own. But their study is unlikely very much to illuminate the problem of simultaneous discovery to which this paper is directed. [This note has been added to the original manuscript in response to points raised during the discussion that followed the oral presentation.]

10. Faraday provides scarce and useful information about the progress of the significant controversy between the exponents of the chemical and contact theories of galvanism (*Experimental Researches,* II, 18–20). According to his account, the chemical theory was dominant in France and England from at least 1825, but the contact theory was still dominant in Germany and Italy when Faraday wrote in 1840. Does the dominance of the contact theory in Germany account for the rather surprising way in which both Mayer and Helmholtz neglect the battery in their accounts of energy transformations?

heat and, under appropriate conditions, light as well. Or, by electrolysis, the current could vanquish forces of chemical affinity, bringing the chain of transformations full circle. These were the first fruits of Volta's work; other more striking conversion discoveries followed during the decade and a half after 1820.[11] In that year Oersted demonstrated the magnetic effects of a current; magnetism, in turn, could produce motion, and motion had long been known to produce electricity through friction. Another chain of conversions was closed. Then, in 1822, Seebeck showed that heat applied to a bimetallic junction would produce a current directly. Twelve years later Peltier reversed this striking example of conversion, demonstrating that the current could, on occasions, absorb heat, producing cold. Induced currents, discovered by Faraday in 1831, were only another, if particularly striking, member of a class of phenomena already characteristic of nineteenth-century science. In the decade after 1827, the progress of photography added yet another example, and Melloni's identification of light with radiant heat confirmed a long-standing suspicion about the fundamental connection between two other apparently disparate aspects of nature.[12]

Some conversion processes had, of course, been available before 1800. Motion had already produced electrostatic charges, and the resulting attractions and repulsions had produced motion. Static generators had occasionally engendered chemical reactions, including dissociations, and chemical reactions produced both light and heat.[13] Harnessed by the steam engine, heat could produce motion, and motion, in turn, engendered heat through friction and percussion. Yet in the eighteenth century these were isolated phenomena; few seemed of central importance to scientific research; and those few were studied by different groups. Only in the decade after 1830, when they were increasingly classified with the many other examples discovered in rapid succession by nineteenth-century scientists, did they begin to look like conversion processes at all.[14] By that time scientists were proceeding inevitably in the laboratory from a variety of chemical, thermal, electrical, magnetic, or dynamical phenomena to phenomena of any of the other types and to optical phenomena as well. Previously separate problems were gaining multiple interrelationships, and that is what Mary Sommerville had in mind when, in 1834, she gave her famous

11. For the following discoveries see Sir Edmund Whittaker, *A History of the Theories of Aether and Electricity*, Vol. 1, *The Classical Theories* (London, 1951), pp. 81–84, 88–89, 170–71, 236–37. For Oersted's discovery see also, R. C. Stauffer, "Persistent Errors Regarding Oersted's Discovery of Electromagnetism," *Isis*, XLIV (1953), 307–10.

12. F. Cajori, *A History of Physics* (New York, 1922), pp. 158, 172–74. Grove makes a particular point of the early photographic processes (*Physical Forces*, pp. 27–32). Mohr gives great emphasis to Melloni's work (*Zeit. f. Phys.*, V [1837], 419).

13. For the chemical effects of static electricity see Whittaker, *Aether and Electricity*, p. 74, n. 2.

popularization of science the title, *On the Connexion of the Physical Sciences.* "The progress of modern science," she said in her preface, "especially within the last five years, has been remarkable for a tendency to . . . unite detached branches [of science, so that today] . . . there exists such a bond of union, that proficiency cannot be attained in any one branch without a knowledge of others."[15] Mrs. Sommerville's remark isolates the "new look" that physical science had acquired between 1800 and 1835. That new look, together with the discoveries that produced it, proved to be a major requisite for the emergence of energy conservation.

Yet, precisely because it produced a "look" rather than a single clearly defined laboratory phenomenon, the availability of conversion processes enters the development of energy conservation in an immense variety of ways. Faraday and Grove achieved an idea very close to conservation from a survey of the whole network of conversion processes taken together. For them conservation was quite literally a rationalization of the phenomenon Mrs. Sommerville described as the new "connexion." C. F. Mohr, on the other hand, took the idea of *conservation* from a quite different source, probably metaphysical.[16] But, as we shall see, it is only because he attempted to elucidate and defend this idea in terms of the new conversion processes that Mohr's initial conception came to look like conservation *of energy.* Mayer and Helmholtz present still another approach. They began by applying their concepts of conservation to well-known older phenomena. But until they extended their theories to embrace the new discoveries, they were not developing the same theory as men like Mohr and Grove. Still another group, consisting of Carnot, Séguin, Holtzmann, and Hirn, ignored the new conversion processes entirely. But they would not be discoverers of energy conservation if men like Joule, Helmholtz, and Colding had not shown that the thermal phenomena with which these steam engineers dealt were integral parts of the new network of conversions.

There is, I think, excellent reason for the complexity and variety of these relationships. In an important sense, though one which will demand later qualification, the conservation of energy is nothing less than the theoretical counterpart of the laboratory conversion processes discovered during the first four decades of the nineteenth century. Each laboratory conversion corresponds in the theory to a transforma-

14. The single exception is significant and is discussed at some length below. During the eighteenth century steam engines were occasionally regarded as conversion devices.

15. Mary Sommerville, *On the Connexion of the Physical Sciences* (London, 1834), unpaginated "Preface."

16. Reasons for distinguishing Mohr's approach from that of Grove and Faraday will be examined below (note 83). The accompanying text will consider possible sources of Mohr's conviction about the conservation of "force."

tion in the form of energy. That is why, as we shall see, Grove and Faraday could derive conservation from the network of laboratory conversions itself. But the very homomorphism between the theory, energy conservation, and the earlier network of laboratory conversion processes indicates that one did not have to start by grasping the network whole. Liebig and Joule, for example, started from a single conversion process and were led by the "connexion" between the sciences through the entire network. Mohr and Colding started with a metaphysical idea and transformed it by application to the network. In short, just because the new nineteenth-century discoveries formed a network of "connexions" between previously distinct parts of science, they could be grasped either individually or whole in a large variety of ways and still lead to the same ultimate result. That, I think, explains why they could enter the pioneers' research in so many different ways. More important, it explains why the researches of the pioneers, despite the variety of their starting points, ultimately converged to a common outcome. What Mrs. Sommerville had called the new "connexions" between the sciences often proved to be the links that joined disparate approaches and enunciations into a single discovery.

The sequence of Joule's researches clearly illustrates the way in which the network of conversion processes actually marked out the experimental ground of energy conservation and thus provided the essential links between the various pioneers. When Joule first wrote in 1838, his exclusive concern with the design of improved electric motors effectively isolates him from all the other pioneers of energy conservation except Liebig. He was simply working on one of the many new problems born from nineteenth-century discovery. By 1840 his systematic evaluations of motors in terms of work and "duty" establishes a link to the researches of the steam engineers, Carnot, Séguin, Hirn, and Holtzmann.[17] But these "connexions" vanished in 1841 and 1842 when Joule's discouragement with motor design forced him to seek instead a fundamental improvement in the batteries that drove them. Now he was concerned with new discoveries in chemistry, and he absorbed entirely Faraday's view of the essential role of chemical processes in galvanism. In addition, his research in these years was concentrated upon what turned out to have been two of the numerous conversion processes selected by Grove and Mohr to illustrate their

17. The first eleven items in Joule's *Papers* (pp. 1–53) are exclusively concerned with improving first motors and then electromagnets, and these items cover the period 1838–41. The systematic evaluations of motors in terms of the engineering concepts, work and "duty," occur on pp. 21–25, 48. For Joule's earliest published use of the concept work or its equivalent, see p. 4.

18. Joule's concern with batteries and more particularly with the electrical production of heat by batteries dominates the five major contributions in *Papers*, pp. 53–

vague metaphysical hypothesis.[18] The "connexions" with the work of other pioneers are steadily increasing in number.

In 1843, prompted by the discovery of an error in his earlier work with batteries, Joule reintroduced the motor and the concept of mechanical work. Now the link to steam engineering is re-established, and simultaneously Joule's papers begin, for the first time, to read like investigations of energy relations.[19] But even in 1843 the resemblance to energy conservation is incomplete. Only as Joule traced still other new "connexions" during the years 1844 to 1847 does his theory really encompass the views of such disparate figures as Faraday, Mayer, and Helmholtz.[20] Starting from an isolated problem, Joule had involuntarily traced much of the connective tissue between the new nineteenth-century discoveries. As he did so, his work was linked increasingly to that of the other pioneers, and only when many such links had appeared did his discovery resemble energy conservation.

Joule's work shows that energy conservation could be discovered by starting from a single conversion process and tracing the network. But, as we have already indicated, that is not the only way in which conversion processes could effect the discovery of energy conservation. C. F. Mohr, for example, probably drew his initial concept of conservation from a source independent of the new conversion processes, but then used the new discoveries to clarify and elaborate his ideas. In 1839, close to the end of a long and often incoherent defense of the dynamical theory of heat, Mohr suddenly burst out: "Besides the known 54 chemical elements, there is, in the nature of things, only one other agent, and that is called force; it can appear under various circumstances as motion, chemical affinity, cohesion, electricity, light, heat, and magnetism, and from any one of these types of phenomena all the others can be called forth."[21] A knowledge of energy conservation makes the import of these sentences clear. But in the absence of such knowledge, they would have been almost meaningless except that Mohr proceeded immediately to two systematic pages of experimental examples. The experiments were, of course, just the new and old conversion processes listed above, the new ones in the lead, and they are essential to Mohr's argument. They alone specify his subject and show its close similarity to Joule's.

Mohr and Joule illustrate two of the ways in which conversion

123. My remark that Joule was led to batteries by his discouragement with motor design is a conjecture, but it seems extremely probable.

19. See note 1. This is the paper in which Joule is usually said to have announced energy conservation.

20. See note 7.

21. *Zeit. f. Phys.*, V (1837), 442.

processes could effect the discoveries of energy conservation. But, as my final example from the works of Faraday and Grove will indicate, these are not the only ways. Though Faraday and Grove reached conclusions much like Mohr's, their route to the conclusions includes none of the same sudden leaps. Unlike Mohr, they seem to have derived energy conservation directly from the experimental conversion processes that they had already studied so fully in their own researches. Because their route is continuous, the homomorphism of energy conservation with the new conversion processes appears most clearly of all in their work.

In 1834, Faraday concluded five lectures on the new discoveries in chemistry and galvanism with a sixth on the "Relations of Chemical Affinity, Electricity, Heat, Magnetism, and other powers of Matter." His notes supply the gist of this last lecture in the words: "We cannot say that any one [of these powers] is the cause of the others, but only that all are connected and due to one common cause." To illustrate "the connection," Faraday then gave nine experimental demonstrations of "the production of any one [power] from another, or the conversion of one into another."[22] Grove's development seems parallel. In 1842 he included a remark almost identical with Faraday's in a lecture with the significant title, "On the Progress of Physical Science."[23] In the following year he expanded this isolated remark into his famous lecture series, *On the Correlation of Physical Forces*. "The position which I seek to establish in this Essay is," he said, "that [any one] of the various imponderable agencies . . . viz., Heat, Light, Electricity, Magnetism, Chemical Affinity, and Motion, . . . may, as a force, produce or be convertible into the other[s]; thus heat may mediately or immediately produce electricity, electricity may produce heat; and so of the rest."[24]

This is the concept of the universal convertibility of natural powers, and it is not, let us be clear, the same as the notion of conservation. But most of the remaining steps proved to be small and rather obvious.[25] All but one, to be discussed below, can be taken by applying to the concept of universal convertibility the perennially serviceable philosophic tags about the equality of cause and effect or the impossibility of perpetual motion. Since any power can produce any other *and be produced by it*, the equality of cause and effect demands a uniform quantitative equivalence between each pair of powers. If there is no such equivalence,

22. Bence Jones, *The Life and Letters of Faraday* (London, 1870), II, 47.

23. *A Lecture on the Progress of Physical Science since the Opening of the London Institution* (London, 1842). Though the title page is dated 1842, the date is immediately followed by "[Not Published]." I do not know when the actual printing took place, but a prefatory remark of the author's indicates that the text itself was written very shortly after the lecture's delivery.

24. *Physical Forces*, p. 8.

25. Reasons for calling the remaining steps "obvious" are given in the closing paragraphs of this paper (see note 92).

then a properly chosen series of conversions will result in the creation of power, that is, in perpetual motion.[26] In all its manifestations and conversions, power must be conserved. This realization came neither all at once, nor fully to all, nor with complete logical rigor. But it did come.

Though he had no general conception of conversion processes, Peter Mark Roget, in 1829, opposed Volta's contact theory of galvanism because it implied a creation of power from nothing.[27] Faraday independently reproduced the argument in 1840 and immediately applied it to conversions in general. "We have," he said, "many processes by which the form of the power may be so changed that an apparent *conversion* of one into another takes place. . . . But in no cases . . . is there a pure creation of force; a production of power without a corresponding exhaustion of something to supply it."[28] In 1842 Grove devised the argument once more in order to prove the impossibility of inducing an electric current from static magnetism, and in the following year he generalized still further.[29] If it were true, he wrote, "that motion [could] be subdivided or changed in character, so as to become heat, electricity, etc.; it ought to follow, that when we collect the dissipated and changed forces, and reconvert them, the initial motion, affecting the same amount of matter with the same velocity, should be reproduced, and so of the change of matter produced by the other forces."[30] In the context of Grove's exhaustive discussion of the known conversion processes, this quotation is a full statement of all but the quantitative components of energy. Furthermore, Grove knew what was missing. "The great problem that remains to be solved, in regard to the correlation of physical forces, is," he wrote, "the establishment of their equivalent of power, or their measureable relation to a given standard,"[31] Conversion phenomena could carry scientists no further towards the enunciation of energy conservation.

Grove's case brings this discussion of conversion processes almost full circle. In his lectures energy conservation appears as the straightforward theoretical counterpart of nineteenth-century laboratory discoveries, and that was the suggestion from which I began. Only two of the pioneers, it is true, actually derived their versions of energy conservation from these new discoveries alone. But because such a derivation

26. Strictly speaking, this derivation is valid only if all the transformations of energy are reversible, which they are not. But that logical shortcoming completely escaped the notice of the pioneers.

27. P. M. Roget, *Treatise on Galvanism* (London,1829). I have seen only the excerpt quoted by Faraday, *Experimental Researches*, II, 103, n. 2.

28. *Experimental Researches*, II, 103.

29. *Progress of Physical Science*, p. 20.

30. *Physical Forces*, p. 47.

31. *Ibid.*, p. 45.

was possible, every one of the pioneers was decisively affected by the availability of conversion processes. Six of them dealt with the new discoveries from the start of their research. Without these discoveries, Joule, Mohr, Faraday, Grove, Liebig, and Colding would not be on our list at all.[32] The other six pioneers show the importance of conversion processes in a subtler but no less important way. Mayer and Helmholtz were late in turning to the new discoveries, but only when they did so, did they become candidates for the same list as the first six. Carnot, Séguin, Hirn, and Holtzmann are the most interesting of all. None of them even mentioned the new conversion processes. But their contributions, being uniformly obscure, would have vanished from history entirely if they had not been gathered into the larger network explored by the men we have already examined.[33] When conversion processes did not govern an individual's work, they often governed that work's reception. If they had not been available, the problem of simultaneous discovery might not exist at all. Certainly it would look very different.

NEVERTHELESS, the view which Grove and Faraday derived from conversion processes is not identical with what scientists now call the conservation of energy, and we must not underestimate the importance of the missing element. Grove's *Physical Forces* contains the layman's view of energy conservation. In an expanded and revised form it proved to be one of the most effective and sought after popularizations of the new scientific law.[34] But this role was achieved only after the work of Joule, Mayer, Helmholtz, and their successors had provided a full quantitative substructure for the conception of force correlation. Anyone who has worked through a mathematical and numerical treatment of energy conservation may well wonder whether, in the absence of such substructure, Grove would have had anything to popularize. The "meas-

32. I am not quite sure that this is true of Colding, particularly since I have not seen his unpublished paper of 1843. The early pages of his 1851 paper (note 1) contain many examples of conversion processes and are thus reminiscent of Mohr's approach. Also, Colding was a protégé of Oersted whose chief renown derived from his discovery of electromagnetic conversions. On the other hand, most of the conversion processes cited explicitly by Golding date from the eighteenth century. In Colding's case, I suspect a prior tie between conversion processes and metaphysics (see note 83 and accompanying text). Very probably neither can be viewed as either logically or psychologically the more fundamental in the development of his thought.

33. Carnot's notes were not published until 1872 and then only because they contained anticipations of an important scientific law. Séguin had to call attention to the relevant passages in his book of 1839. Hirn did not bother to claim credit, but only attached a note denying plagiarism to his 1854 paper. That paper was published in an engineering journal that I have never seen cited by a scientist. Holtzmann's paper is the exception in that it was not obscure. But if other men had not discovered conservation of energy, Holtzmann's memoir would have continued to look like another one of the extensions of Carnot's memoir, for that is basically what it was (see note 2).

urable relation to a given standard" of the various physical forces is an essential ingredient of energy conservation as we know it, and neither Grove, Faraday, Roget, nor Mohr was able even to approach it.

The quantification of energy conservation proved, in fact, insuperably difficult for those pioneers whose principal intellectual equipment consisted of concepts to the new conversion processes. Grove thought he had found the clue to quantification in Dulong and Petit's law relating chemical affinity and heat.[35] Mohr believed he had produced the quantitative relationship when he equated the heat employed to raise the temperature of water 1° with the static force necessary to compress the same water to its original volume.[36] Mayer initially measured force by the momentum which it could produce.[37] These random leads were all totally unproductive, and of this group only Mayer succeeded in transcending them. To do so he had to use concepts belonging to a very different aspect of nineteenth-century science, an aspect to which I previously referred as the concern with engines, and whose existence I shall now take for granted as a well-known by-product of the Industrial Revolution. As we examine this aspect of science, we shall find the main source of the concepts—particularly of mechanical effect or work—required for the quantitative formulation of energy conservation. In addition, we shall find a multitude of experiments and of qualitative conceptions so closely related to energy conservation that they collectively provide something very like a second and independent route to it.

Let me begin by considering the concept of work. Its discussion will provide relevant background as well as opportunity for a few essential remarks on a more usual view about the sources of the quantitative concepts underlying energy conservation. Most histories or pre-histories of the conservation of energy imply that the model for quantifying conversion processes was the dynamical theorem known almost from the begninning of the eighteenth century as the conservation of *vis viva*.[38] That theorem has a distinguished role in the history of dynamics, and it also turns out to have been a special case of energy conservation.

34. Between 1850 and 1875 Grove's book was reprinted at least six times in England, three times in America, twice in France, and once in Germany. The extensions were, of course, numerous, but I am aware of only two essential revisions. In the original discussion of heat (pp. 8–11), Grove suggested that macroscopic motion appears as heat only to the extent that it is *not* transformed to microscopic motion. In addition, of course, Grove's few attempts at quantification were quite off the track (see below).

35. *Physical Forces,* p. 46.

36. *Zeit. f. Phys.,* V (1837), 422–23.

37. Weyrauch, II, 102–105. This is in his first paper, "Ueber die quantitative und qualitative Bestimmung der Kräfte," sent to Poggendorf in 1841 but not published until after Mayer's death. Before he wrote his second paper, the first to be published, Mayer had learned a bit more physics.

38. It would be more precise to say that most pre-histories of energy conservation are principally lists of anticipations, and these occur particularly often in the early literature on *vis viva.*

It could have provided a model. Yet I think the prevalent impression that it did so is misleading. The conservation of *vis viva* was important to Helmholtz's derivation of energy conservation, and a special case (free fall) of the same dynamical theorem was ultimately of great assistance to Mayer. But these men also drew significant elements from a second generally separate tradition—that of water, wind, and steam engineering—and that tradition is all important to the work of the other five pioneers who produced a quantitative version of energy conservation.

There is excellent reason why this should be so. *Vis viva* is mv^2, the product of mass by the square of velocity. But until a late date that quantity appears in the works of none of the pioneers except Carnot, Mayer, and Helmholtz. As a group the pioneers were scarcely interested in energy of motion, much less in using it as a basic quantitative measure. What they did use, at least those who were successful, was *f-s*, the product of force times distance, a quantity known variously under the names mechanical effect, mechanical power, and work. That quantity does not, however, occur as an independent conceptual entity in the dynamical literature. More precisely it scarcely occurs there until 1820 when the French (and only the French) literature was suddenly enriched by a series of theoretical works on such subjects as the theory of machines and of industrial mechanics. These new books did make work a significant independent conceptual entity, and they did relate it explicitly to *vis viva*. But the concept was not invented for these books. On the contrary it was borrowed from a century of engineering practice where its use had usually been quite independent of both *vis viva* and its conservation. That source within the engineering tradition is all that the pioneers of energy conservation required and as much as most of them used.

39. The early eighteenth-century literature contains many general statements about the conservation of *vis viva* regarded as a metaphysical force. These formulations will be discussed briefly below. For the present notice only that none of them is suitable for application to the technical problems of dynamics, and it is with those formulations that we are here concerned. An excellent discussion of both the dynamical and metaphysical formulations is included in A. E. Haas, *Die Entwicklungsgeschichte des Satzes von der Erhaltung der Kraft* (Vienna, 1909), generally the fullest and most reliable pre-history of energy conservation. Other useful details can be found in Hans Schimank, "Die geschichtliche Entwicklung des Kraftbegriffs bis zum Aufkommen der Energetik," in *Robert Mayer und das Energieprinzip, 1842–1942*, ed. H. Schimank and E. Pietsch (Berlin, 1942). I am indebted to Professor Erwin Hiebert for calling these two useful and little known works to my attention.

40. Christian Huyghens, *Horologium oscillatorium* (Paris, 1673). I have used the German edition, *Die Pendcluhr*, ed. A. Heckscher and A. V. Oettingen, Ostwald's Klassiker der Exakten Wissenschaften, No. 192 (Leipzig, 1913), p. 112.

41. D. Bernoulli, *Hydrodynamica, sive de viribus et motibus fluidorum, commentarii* (Basle, 1738), p. 12.

42. J. L. d'Alembert, *Traité de dynamique* (Paris, 1743). I have been able to see only the second edition (Paris, 1758) where the relevant material occurs on pp. 252–

Another paper will be needed to document this conclusion, but let me illustrate the considerations from which it derives. Until 1743 the general dynamical significance of the conservation of *vis viva* must be recaptured from its application to two special sorts of problems: elastic impact and constrained fall.[39] Force times distance has no relevance to the former, since elastic impact numerically conserves *vis viva*. For other applications, e.g., the bachistochrone and isochronous pendulum, vertical displacement rather than force times distance appears in the conservation theorem. Huyghen's statement that the center of gravity of a system of masses can ascend no higher than its initial position of rest is typical.[40] Compare Daniel Bernoulli's famous formulation of 1738: Conservation of *vis viva* is "the equality of actual descent with potential ascent."[41]

The more general formulations, inaugurated by d'Alembert's *Traité* in 1743, suppress even vertical displacement, which might conceivably be called an embryonic conception of work. D'Alembert states that the forces acting on a system of interconnected bodies will increase its *vis viva* by the amount $\Sigma m_i u_i^2$, where the u_i are the velocities that the masses m_i would have acquired if moved freely over the same paths by the same forces.[42] Here, as in Daniel Bernoulli's subsequent version of the general theorem, force times distance enters only in certain particular applications to permit the computation of individual u_i's; it has neither general significance nor a name; *vis viva* is the conceptual parameter.[43] The same parameter dominates the later analytic formulations. Euler's *Mechanica*, Lagrange's *Mécanique analytique*, and Laplace's *Mécanique céleste* give exclusive emphasis to central forces derivable from potential functions.[44] In these works the integral of force times differential path element occurs only in the derivation of the conservation law. The law itself equates *vis viva* with a function of position coördinates.

53. D'Alembert's discussion of the changes introduced since the first edition give no reason to suspect he has altered the original formulation at this point.

43. D. Bernoulli, "Remarques sur le principe de la conservation des forces vives pris dans un sens general," *Hist. Acad. de Berlin* (1748), pp. 356–64.

44. L. Euler, *Mechanica sive motus scientia analytice exposita*, in *Opera omnia* (Leipzig and Berlin, 1911–present), ser. 2, II, 74–77. The first edition was Petersburg, 1736.

J.-L. Lagrange, *Mécanique analytique* (Paris, 1788), pp. 206–9. I cite the first edition because the second, as reprinted in volumes 11 and 12 of Lagrange's *Oeuvres* (Paris, 1867–92), contains a very significant change. In the first edition, the conservation of *vis viva* is formulated only for time-independent constraints and for central or other integrable forces. It then takes the form $\Sigma m_i v_i^2 = 2H + 2\ \Sigma m_i \pi_i$, where H is a constant of integration and the π_i are functions of the position coordinates. In the second edition, Paris, 1811–15 (*Oeuvres*, XI, 306–10), Lagrange repeats the above but restricts it to a particular class of elastic bodies in order to take account of Lazare Carnot's engineering treatise (note 45) which he cites. For a fuller account of the engineering problem treated by Carnot, he refers his readers to his own *Théorie des fonctions analytiques* (Paris, 1797), pp. 399–410, where his version of Carnot's engineering problem is formulated more explicitly. That formulation makes

Not until 1782, in Lazare Carnot's *Essai sur les machines en général*, did force times distance begin to receive a name and a conceptual priority in dynamical theory.[45] Nor was this new dynamical view of the concept work really worked out or propagated until the years 1819 to 1839 when it received full expression in the works of Navier, Coriolis, Poncelet, and others.[46] All these works are concerned with the analysis of machines in motion. As a result, work—the integral of force with respect to distance—is their fundamental conceptual parameter. Among other significant and typical results of this reformulation were: the introduction of the term work and of units for its measure; the redefinition of *vis viva* as $\frac{1}{2}mv^2$ to preserve the conceptual priority of the measure work; and the explicit formulation of the conservation law in terms of the equality of work done and kinetic energy created.[47] Only when thus reformulated did the conservation of *vis viva* provide a convenient conceptual model for the quantification of conversion processes, and then almost none of the pioneers used it. Instead, they returned to the same older engineering tradition in which Lazare Carnot and his French successors had found the concepts needed for their new versions of the dynamical conservation theorem.

the impact of the engineering tradition quite apparent, for the concept work now begins to appear. Lagrange states that the increment of *vis viva* between two dynamical states of the system is $2(P) + 2(Q) + \ldots$, where (P) — Lagrange calls it an "aire"—is $\Sigma_i \int P_i dp_i$, and P_i is the force on the *i*th body in the direction of the position coordinates P_i. These "aires" are, of course, just work.

P. S. Laplace, *Traité de mécanique céleste* (Paris, 1798–1825). The relevant passages are more readily found in *Oeuvres complètes* (Paris, 1878–1825), I, 57–61. Mathematically, this treatment of 1798 actually resembles Lagrange's 1797 formulation rather than the earlier 1788 form. But, as in the pre-engineering formulations, the conservation law which includes a work integral is rapidly passed over in favor of the more restricted statement employing a potential function.

45. L. N. M. Carnot, *Essai sur les machines en général* (Dijon, 1782). I have consulted this work in Carnot's *Oeuvres mathematiques* (Basle, 1797) but rely principally on the expanded and more influential second edition, *Principes fondementaux de l'équilibre et du mouvement* (Paris, 1803). Carnot introduces several terms for what we call work, the most important being, "force vive latent" and "moment d'activité" (*ibid.*, pp. 38–43). Of these he says, "The kind of quantity to which I have given the name *moment of activity* plays a very large role in the theory of machines in motion: for in general it is this quantity which one must economize as much as possible in order to derive from an agent [i.e., a source of power] all the [mechanical] effect of which it is capable" (*ibid.*, p. 257).

46. A useful survey of the early history of this important movement is C. L. M. H. Navier, "Details historiques sur l'emploi du principes des forces vives dans la theorie des machines et sur diverses roues hydraulique," *Ann. Chim. Phys.*, IX (1818), 146–59. I suspect that Navier's edition of B. de F. Belidor's *Architecture hydraulique* (Paris, 1819) contains the first developed presentation of the new engineering physics, but I have not yet seen this work. The standard treatises are: G. Coriolis, *Du calcul de l'effet des machines, ou considérations sur l'emploi des moteurs et sur leur évaluation pour servir d'introduction à l'étude speciale des machines* (Paris, 1829); C. L. M. H. Navier, *Résumé des leçons données a l'école des ponts et chaussées sur l'application de la mécanique à l'établissement des constructions et des machines* (Paris, 1838). Vol. 2; and J.-V. Poncelet, *Introduction à la mécanique industrielle*, ed. Kratz (3rd ed.; Paris, 1870). This work originally appeared in 1829

Sadi Carnot is the single complete exception. His manuscript notes proceed from the assertion that heat is motion to the conviction that it is molecular *vis viva* and that its increment must therefore be equal to work done. These steps imply an immediate command of the relation between work and *vis viva.* Mayer and Helmholtz might also have been exceptions, for both could have made good use of the French reformulation. But neither seems to have known it. Both began by taking work (or rather the product of weight times height) as the measure of "force," and each then rederived something very like the French reformulation for himself.[48] The other six pioneers who reached or came close to the quantification of conversion processes could not even have used the reformulation. Unlike Mayer and Helmholtz, they applied the concept work directly to a problem in which *vis viva* is constant from cycle to cycle and therefore does not enter. Joule and Liebig are typical. Both began by comparing the "duty" of the electric motor with that of the steam engine. How much weight, they both asked, can each of these engines raise through a fixed distance for a given expenditure of coal or zinc? That question is basic to their entire research programs as it is to

(part had appeared in lithoprint in 1827); the much enlarged and now standard edition from which the third is taken appeared in 1830–39.

47. The formal adoption of the term work (*travail*) is often credited to Poncelet (introduction, p. 64), though many others had used it casually before; Poncelet also (pp. 74–75) gives a useful account of the units (*dynamique, dyname, dynamie,* etc.) commonly used to measure this quantity. Coriolis *Du calcul de l'effect des machines,* p. iv) is the first to insist that *vis viva* be $\frac{1}{2}mv^2$, so that it will be numerically equal to the work it can produce; he also makes much use of the term *travail,* which Poncelet may have borrowed from him. The reformulation of the conservation law proceeds gradually from Lazare Carnot through all these later works.

48. As soon as he considers a quantitative problem in his first published paper, Mayer says: "A cause, which effects the raising of a weight, is a force; since this force brings about the fall of a body, we shall call it fall-force [Fallkraft]" (Weyrauch, I, 24). This is the engineering, not the theoretical dynamical, measure. By applying it to the problem of free fall, Mayer immediately derives $\frac{1}{2}mv^2$ (note the fraction) as the measure of energy of motion. The very crudeness of his derivation together with its lack of generality indicates his ignorance of the French engineering texts. The one French text he does mention in his writings (G. Lamé, *Cours de physique de l'école polytechnique* [2nd ed.; Paris, 1840]), does not deal with *vis viva* or its conservation at all.

Helmholtz uses the terms *Arbeitskraft, bewegende Kraft, mechanische Arbeit,* and *Arbeit* for his fundamental measurable force (Helmholtz, *Abhandlungen,* I, 12, 17–18). I have not as yet been able to trace these terms in the earlier German literature, but their parallels in the French and English engineering traditions are obvious. Also, the term *bewegende Kraft* is used by the translator of Clapeyron's version of Sadi Carnot's memoir as equivalent to the French *puissance motrice* (*Pogg. Ann.,* LIX [1843], 446), and Helmholtz cites this translation (p. 17, n. 1). To this extent the tie to the engineering tradition is explicit.

Helmholtz was not, however, aware of the French theoretical engineering tradition. Like Mayer, he derives the factor of ½ in the definition of energy of motion and is unaware of any precedent for it (p. 18). More significant, he fails completely to identify $\int Pdp$ as work or *Arbeitskraft,* and instead calls it the "sum of the tensions" [*Summe der Spannkräfte*] over the space dimension of the motion.

the programs of Carnot, Seguin, Holtzmann, and Hirn. It is not, however, a question drawn from either the new or old dynamics.

But neither, except for its application to the electrical case, is it a novel question. The evaluation of engines in terms of the weight each could raise to a given height is implicit in Savery's engine descriptions of 1702 and explicit in Parent's discussion of water wheels in 1704.[49] Under a variety of names, particularly mechanical effect, weight times height provided the basic measure of engine achievement throughout the engineering works of Desagulier, Smeaton, and Watt.[50] Borda applied the same measure to hydraulic machines and Coulomb to wind and animal power.[51] These examples, drawn from all parts of the eighteenth century, but increasing in density towards its close, could be multiplied almost indefinitely. Yet even these few should prepare the way for a little noted but virtually decisive statistic. Of the nine pioneers who succeeded, partially or completely, in quantifying conversion processes, all but Mayer and Helmholz were either trained as engineers or were working directly on engines when they made their contributions to energy conservation. Of the six who computed independent values of the conversion coefficient, all but Mayer were concerned with engines either in fact or by training.[52] To make the computation they needed the concept work, and the source of that concept was principally the engineering tradition.[53]

The concept work is the most decisive contribution to energy con-

49. The unit implicit in Savery's work is really the horsepower, but this includes weight times height as a part. See H. W. Dickinson and Rhys Jenkins, *James Watt and the Steam Engine* (Oxford, 1927), p. 353–54. Antoine Parent, "Sur le plus grande perfection possible des machines," *Hist. Acad. Roy.* (1704), pp. 323–38.

50. J. T. Desagulier, A *Course of Experimental Philosophy* (3rd ed., 2 vols.; London, 1763), particularly I, 132, and II, 412. This posthumous edition is practically a reprint of the second edition (London, 1749).

John Smeaton, "An Experimental Inquiry concerning the Natural Powers of Water and Wind to turn Mills, and other Machines, depending on a Circular Motion," *Phil. Trans.*, LI (1759), 51. Here the measure is weight times height per unit time. The time dependence is, however, dropped in his "An Experimental Examination of the Quantity and Proportion of Mechanic Power necessary to be employed in giving different degrees of Velocity to Heavy Bodies," *Phil. Trans.* LXVI (1776), 458.

For Watt see Dickinson and Jenkins, *James Watt*, pp. 353–56.

51. J. C. Borda, "Mémoires sur les rues hydrauliques," *Mem. l'Acad. Roy.* (1767), p. 272. Here the measure is weight times vertical speed. Height replaces speed in C. Coulomb, "Observation theorique et experimentales sur l'effet des moulins à vent, et sur la figure de leurs ailes," *ibid.* (1781), p. 68, and "Resultat de plusieurs expériences destinée à determiner la quantité d'action que les hommes peuvent fournir par leur travail journalier, suivant les differentes manières dont ils emploient leurs forces," *Mem. de l'Inst.*, II (1799), 381.

52. Mayer states that he loved to build model water wheels as a boy and that he learned the impossibility of perpetual motion in studying them (Weyrauch, II, p. 390). He could have learned simultaneously the proper measure of the product of machines.

53. Professor Hiebert asks if the concept of mechanical work may not have

servation made by the nineteenth-century concern with engines. That is why I have devoted so much space to it. But the concern with engines contributed to the emergence of energy conservation in a number of other ways besides, and we must consider at least a few of them. For example, long before the discovery of electro-chemical conversion processes, men interested in steam and water engines had occasionally seen them as devices for transforming the force latent in fuel or falling water to the mechanical force that raises weight. "I am persuaded," said Daniel Bernoulli in 1738, "that if all the *vis viva* hidden in a cubic foot of coal were called forth and usefully applied to the motion of a machine, more could be achieved than by the daily labor of eight or ten men."[54] Apparently that remark, made at the height of the controversy over metaphysical *vis viva,* had no later influence. Yet the same perception of engines recurs again and again, most explicitly in the French engineering writers. Lazare Carnot, for example, says that "the problem of turning a mill stone, whether by the impact of water, or by wind, or by animal power . . . is that of consuming the maximum possible [portion] of the work delivered by these agents."[55] With Coriolis, water, wind, steam, and animals are all simply sources of work, and machines become devices for putting this in useful form and transmitting it to the load.[56] Here, engines by themselves lead to a conception of conversion processes very close to that produced by the new discoveries of the nineteenth century. That aspect of the engine problem may well explain

emerged from elementary statics and particularly from the formulation that derives statics from the principle of virtual velocities. The point needs further research, but my present response must be at least equivocally negative. The elements of statics were an important item in the equipment of all eighteenth-century engineers and the principle of virtual velocities, or an equivalent, therefore recurs in eighteenth-century writings on engineering problems. Quite possibly the engineers could not have evolved the concept work without the aid of the pre-existing static principle. But, as the preceding discussion may indicate, if the eighteenth-century concept work did emerge from the far older principle of virtual velocities, it did so only when that principle was firmly embedded in the engineering tradition and only when that tradition turned its attention to the evaluation of power sources such as animals, falling water, wind, and steam. Therefore, reverting to the vocabulary of note 9, I suggest that the principle of virtual velocities may have been a prerequisite for the discovery of energy conservation but that it can scarcely have been a trigger. [This note added to original manuscript in response to points raised during discussion.]

54. *Hydrodynamica,* p. 231.

55. *De l'équilibre et du mouvement,* p. 258. Notice also that as soon as Lagrange turns to Carnot's problem (note 44), he speaks in the same way. In the *Fonctions analytiques,* he says that waterfalls, coal, gunpowder, animals, etc., all "contain a quantity of *vis viva,* which one can harness but which one cannot increase by any mechanical means. One may [therefore] always regard a machine as intended to destroy a given quantity of *vis viva* [in the load] by consuming some other given *vis viva* [from the source]." (*Oeuvres,* IX, 410.)

56. *Du calcul de l'effect des machines,* chap. 1. For Coriolis the conservation theorem applied to a perfect machine becomes the "Principle of the Transmission of Work."

why the steam engineers—Hirn, Holtzmann, Séguin, and Sadi Carnot—were led to the same aspect of nature as men like Grove and Faraday.

The fact that engines could and occasionally did look like conversion devices may also explain something more. Is this not the reason why engineering concepts proved so readily transferable to the more abstract problems of energy conservation? The concept work is only the most important example of such a transfer. Joule and Liebig reached energy conservation by asking an old engineering question, "What is the 'duty'?" about the new conversion processes in the battery driven electric motor. But that question—how much work for how much fuel?—embraces the notion of a conversion process. In retrospect, it even sounds like the request for a conversion coefficient. Joule, at least, finally answered the question by producing one. Or consider the following more surprising transfer of engineering concepts. Though its fundamental conceptions are incompatible with energy conservation, Sadi Carnot's *Réflexion sur la puissance motrice du feu* was cited by both Helmholtz and Colding as the outstanding application of the impossibility of perpetual motion to a non-mechanical conversion process.[57] Helmholtz may well have borrowed from Carnot's memoir the analytic concept of a cyclic process that played so large a role in his own classic paper.[58] Holtzmann derived his value of the conversion coefficient by a minor modification of Carnot's analytic procedures, and Carnot's own discussion of energy conservation repeatedly employs data and concepts from his earlier and fundamentally incompatible memoir. These examples may give at least a hint of the ease and frequency with which engineering concepts were applied in deriving the abstract scientific conservation law.

My final example of the productiveness of the nineteenth-century concern with engines is less directly tied to engines. Yet it underscores the multiplicity and variety of the relationships that made the engineer-

57. Helmholtz, *Abhandlungen*, I, 17. Colding, "Naturkraefter," *Dansk. Vid. Selsk.*, II (1851), 123–24. Particularly interesting evidence about the apparent similarities between the theory of energy conservation and Carnot's incompatible theory of the heat engine is provided by Carlo Matteucci. His paper, "De la relation qui existe entre la quantité de l'action chimique et la quantité de chaleur, d'électricité et de lumière qu'elle produit," *Bibliothèque universelle de Genève, Supplement*, IV (1847), 375–80, is an attack upon several of the early exponents of energy conservation. He describes his opponents as the group of physicists who "have tried to show that Carnot's celebrated principle about the motive force of heat can be applied to the other imponderable fluids."

58. Helmholtz, *Abhandlungen*, I, 18–19, gives Helmholtz initial abstract formulation of the cyclic process.

59. T. S. Kuhn, "The Caloric Theory of Adiabatic Compression," *Isis*, XLIX (1958), 132–40.

60. John Dalton, "Experimental Essays on the Constitution of Mixed Gases; on the Force of Steam or Vapour from Water and other Liquids in different temperatures, both in a Torricellian Vacuum and in Air; on Evaporation; and on the Expansion of Gases by Heat," *Manch. Mem.* V (1802), 535–602. The second essay, though

ing factor bulk so large in this account of simultaneous discovery. I have shown elsewhere that many of the pioneers shared an important interest in the phenomenon known as adiabatic compression.[59] Qualitatively, the phenomenon provided an ideal demonstration of the conversion of work to heat; quantitatively, adiabatic compression yielded the only means of computing a conversion coeffcient with existing data. The discovery of adiabatic compression has, of course, little or nothing to do with the interest in engines, but the nineteenth-century experiments which the pioneers used so heavily often seem related to just this practical concern. Dalton, and Clément and Désormes, who did important early work on adiabatic compression, also contributed early fundamental measurements on steam, and these measurements were used by many of the engineers.[60] Poisson, who developed an early theory of adiabatic compression, applied it, in the same article, to the steam engine, and his example was immediately followed by Sadi Carnot, Coriolis, Navier, and Poncelet.[61] Séguin, though he uses a different sort of data, seems a member of the same group. Dulong, to whose classic memoir on adiabatic compression many of the pioneers referred, was a close collaborator of Petit, and during the period of their collaboration Petit produced a quantitative account of the steam engine that antedates Carnot's by eight years.[62] There is even a hint of government interest. The prize offered by the French *Institut national* and won in 1812 by the classic research on gases of Delaroche and Bérard may well have grown in part from government interest in engines.[63] Certainly Regnault's later work on the same topic did. His famous investigations of the thermal characteristics of gas and steam bear the imposing title, "Experiments undertaken by order of the Minister of Public Works and at the instigation of the Central Commission for Steam Engines, to determine the principal laws and the numerical data which enter into

it grew out of Dalton's meterological interests, was immediately exploited by both British and French engineers.

Clément and Désormes, "Mémoires sur la théorie des machines à feu," *Bulletin des sciences par la societé philomatique,* VI (1819), 115–18; and "Tableau relatif à la théorie general de la puissance mécanique de la vapeur," *ibid.*, XIII (1826), 50–53. The second paper appears in full in Crelle's *Journal für die Baukunst,* VI (1833), 143–64. For the contributions of these men to adiabatic compression, see my paper, note 59.

61. S. D. Poisson, "Sur la chaleur des gaz et des vapeurs," *Ann. Chim. Phys.*, XXIII (1823), 337–55. For Navier, Coriolis, and Poncelet, all of whom devote chapters to steam engine computations, see note 46.

62. A. T. Petit, "Sur l'emploi du principe des forces vives dans le calcul de l'effet des machines," *Ann. Chim. Phys.*, VIII (1818), 287–305.

63. F. Delaroche and J. Bérard, "Mémoire sur la determination de la chaleur specifique des differents gaz," *Ann. Chim. Phys.*, LXXXV (1813), 72–110, 113–82. I know of no direct evidence relating the prize won by this memoir to the problems of steam engineering, but the Academy did offer a prize for improvement in steam engines as early as 1793. See H. Guerlac, "Some Aspects of Science during the French Revolution," *The Scientific Monthly,* LXXX (1955), 96.

steam engine calculations."[64] One suspects that without these ties to the recognized problems of steam engineering, the important data on adiabatic compression would not have been so accessible to the pioneers of energy conservation. In this instance the concern with engines may not have been essential to the work of the pioneers, but it certainly facilitated their discoveries.

BECAUSE the concern with engines and the nineteenth-century conversion discoveries embrace most of the new technical concepts and experiments common to more than a few of the discoverers of energy conservation, this study of simultaneous discovery might well end here. But a last look at the papers of the pioneers generates an uncomfortable feeling that something is still missing, something that is not perhaps a substantive element at all. This feeling would not exist if all the pioneers had, like Carnot and Joule, begun with a straightforward technical problem and proceeded by stages to the concept of energy conservation. But in the cases of Colding, Helmholtz, Liebig, Mayer, Mohr, and Séguin, the notion of an underlying imperishable metaphysical force seems prior to research and almost unrelated to it. Put bluntly, these pioneers seem to have held an idea capable of becoming conservation of energy for some time before they found evidence for it. The factors previously discussed in this paper may explain why they were ultimately able to clothe the idea and thus to make sense of it. But the discussion does not yet sufficiently account for the idea's existence. One or two such cases among the twelve pioneers might not be troublesome. The sources of scientific inspiration are notoriously inscrutable. But the presence of major conceptual lacunae in six of our twelve cases is surprising. Though I cannot entirely resolve the problem it presents, I must at least touch upon it.

We have already noted a few of the lacunae. Mohr jumped without warning from a defense of the dynamical theory of heat to the statement that there is only one force in nature and that it is quantitatively unal-

64. V. Regnault, *Mém. de l'Acad.*, XXI (1847), 1–767. The title of the work is quoted above.

65. See note 21 and accompanying text.

66. *Chemische Briefe*, pp. 115–17.

67. Colding, "History of Conservation," *Phil. Mag.*, XXVII (1864), 57–58.

68. Leo Koenigsberger, *Hermann von Helmholtz*, tr. F. A. Welby (Oxford, 1906), pp. 25–26, 31–33, implies that Helmholtz's ideas about conservation were complete as early as 1843, and he states that by 1845 the attempt at experimental proof motivated all of Helmholtz's research. But Koenigsberger gives no evidence, and he cannot be quite correct. In two articles on physiological heat written during 1845 and 1846 (*Abhandlungen*, I, 8–11; II, 680–725), Helmholtz fails to notice that body heat may be expended in mechanical work (compare the discussion of Mayer, below). In the second of these papers he also gives the usual caloric explanation of adiabatic compression in terms of the change in heat capacity with pressure. In short, his ideas were by no means complete until 1847 or shortly before. But the papers of 1845 and

terable.[65] Liebig made a similar leap from the duty of electric motors to the statement that the chemical equivalents of the elements determine the work retrievable from chemical processes by either electrical or thermal means.[66] Colding tells us that he got the idea of conservation in 1839, while still a student, but withheld announcement until 1843 so that he might gather evidence.[67] The biography of Helmholtz outlines a similar story.[68] Séguin confidently applied his concept of the convertibility of heat and motion to steam computations, even though his single attempt to confirm the idea had been totally fruitless.[69] Mayer's leap has repeatedly been noted, but its full size is not often remarked. From the light color of venous blood in the tropics, it is a small step to the conclusion that less internal oxidation is needed when the body loses less heat to the environment.[70] Crawford had drawn that conclusion from the same evidence in 1778.[71] Laplace and Lavoisier, in the 1780's, had balanced the same equation relating inspired oxygen to the body's heat losses.[72] A continuous line of research relates their work to the biochemical studies of respiration made by Liebig and Helmholtz in the early 1840's.[73] Though Mayer apparently did not know it, his observation of venous blood was simply a rediscovery of evidence for a well known, though controversial, biochemical theory. But that theory was not the one to which Mayer leaped. Instead Mayer insisted that internal oxidation must be balanced against *both* the body's heat loss *and* the manual labor the body performs. To this formulation, the light color of tropical venous blood is largely irrelevant. Mayer's extension of the theory calls for the discovery that lazy men, rather than hot men, have light venous blood.

The persistent occurrence of mental jumps like these suggests that many of the discoverers of energy conservation were deeply predisposed to see a single indestructible force at the root of all natural phenomena. The predisposition has been noted before, and a number of historians have at least implied that it is a residue of a similar metaphysic generated by the eighteenth-century controversy over the conservation of

1846 do show that in these years Helmholtz was concerned to combat vitalism which he thought implied the creation of force from nothing. Also they show that he already knew the work of Clapeyron and of Holtzmann, which he thought relevant. To this extent, at least, Koenigsberger must be right.

69. *Chemins de fer*, p. 383. Séguin had tried unsuccessfully to measure tne difference in the quantities of heat abstracted from the boiler and delivered to the condenser of a steam engine.

70. Weyrauch, I, 12–14.

71. E. Farber, "The Color of Venous Blood," *Isis*, XLV (1954), 3–9.

72. A. Lavoisier and P. S. Laplace, "Mémoire sur la chaleur," *Hist. de l'Acad.* (1780), pp. 355–408.

73. Helmholtz touches on much of this research in his paper of 1845, "Wärme, physiologisch," for the *Encyclopädische Wörterbuch der medicinischen Wissenschaften* (*Abhandlungen*, II, 680–725).

vis viva. Liebniz, Jean and Daniel Bernoulli, Hermann, and du Châtelet, all said things like, "*Vis* [*viva*] never perishes; it may in truth appear lost, but one can always discover it again in its effects if one can see them."[74] There are a multitude of such statements, and their authors do attempt, however crudely, to trace *vis viva* into and out of non-mechanical phenomena. The parallel to men like Mohr and Colding is very close. Yet eighteenth-century metaphysical sentiments of this sort seem an implausible source of the nineteenth-century predisposition we are examining. Though the technical *dynamical* conservation theorem has a continuous history from the early eighteenth century to the present, its metaphysical counterpart found few or no defenders after 1750.[75] To discover the *metaphysical* theorem, the pioneers of energy conservation would have had to return to books at least a century old. Neither their works nor their biographies suggest that they were significantly influenced by this particuar bit of ancient intellectual history.[76]

Statements like those of both the eighteenth-century Leibnizians and the nineteenth-century pioneers of energy conservation can, however, be found repeatedly in the literature of a second philosophical movement, *Naturphilosophie*.[77] Positing organism as the fundamental metaphor of their universal science, the *Naturphilosophen* constantly sought a single unifying principle for all natural phenomena. Schelling, for example, maintained "that magnetic, electrical chemical, and finally even organic phenomena would be interwoven into one great association . . . [which] extends over the whole of nature."[78] Even before the discovery

74. Haas, *Erhaltung*, p. 16, n. Quoted from *Institutions physiques de Madame la Marquise du Chastellet adressés à Mr. son Fils* (Amsterdam, 1742).

75. Haas, *Erhaltung*, p. 17.

76. None of the pioneers mention the eighteenth-century conservation literature in their original papers. Colding, however, says that he got his first glimpse of conservation while reading d'Alembert in 1839 (*Phil. Mag.*, XXVII [1864], 58), and Koenigsberger says that Helmholtz read d'Alembert and Daniel Bernoulli early in 1843 (*von Holmholtz*, p. 26). These two counterexamples do not, however, really modify my thesis. D'Alembert omitted all mention of the metaphysical conservation theorem from the first edition of his *Traité*, and in the second he explicitly disowned the view ([2nd ed.; Paris, 1758], beginning of the "Avertissement" and pp. xvii–xxiv). In fact, d'Alembert was among the first to insist on freeing dynamics from what he considered to be mere metaphysical speculations. To take his ideas from this source Colding would still have required a strong predisposition. Bernoulli's *Hydrodynamica* is a more appropriate source (see, for example, the text which accompanies note 54), but Koenigsberger makes the very plausible point that Helmholtz consulted Bernoulli in order to work out his preexisting conception of conservation.

77. The roots of *Naturphilosophie* can, of course, be traced back through Kant and Wolff to Leibniz, and Leibniz was the author of the metaphysical conservation theorem about which both Kant and Wolff wrote (Haas, *Erhaltung*, pp. 15–18). The two movements are not, therefore, entirely independent.

78. Quoted by R. C. Stauffer, "Speculation and Experiment in the Background of Oersted's Discovery of Electromagnetism," *Isis*, XLVIII (1957), 37, from Schelling's *Einleitung zu seinem Entwurf eines Systems der Naturphilosophie* (1799).

79. Quoted by Haas, *Erhaltung*, p. 45, n. 61, from Schelling's *Erster Entwurf eines Systems der Naturphilosophie* (1799).

of the battery he insisted that "without doubt only a single force in its various guises is manifest in [the phenomena of] light, electricity, and so forth."[79] These quotations point to an aspect of Schelling's thought fully documented by Brehier and more recently by Stauffer.[80] As a *Naturphilosoph,* Schelling constantly sought out conversion and transformation processes in the science of his day. At the beginning of his career chemistry seemed to him the basic physical science; from 1800 on he increasingly found in galvanism "the true border-phenomenon of both [organic and inorganic] natures."[81] Many of Schelling's followers, whose teaching dominated German and many neighboring universities during the first third of the nineteenth century, gave similar emphasis to the new conversion phenomena. Stauffer has shown that Oersted—a *Naturphilosoph* as well as a scientist—persisted in his long search for a relation between electricity and magnetism largely because of his prior philosophical conviction that one must exist. Once the interaction was discovered, electro-magnetism played a major role in Hebart's further elaboration of the scientific substructure of *Naturphilosophie.*[82] In short, many *Naturphilosophen* drew from their philosophy a view of physical processes very close to that which Faraday and Grove seem to have drawn from the new discoveries of the nineteenth century.[83]

Naturphilosophie could, therefore, have provided an appropriate philosophical background for the discovery of energy conservation. Furthermore, several of the pioneers were acquainted with at least its

80. Émile Bréhier, *Schelling* (Paris, 1912). This is the most helpful discussion I have found and should certainly be added to Stauffer's list of useful aids for studying the complex relations of science and *Naturphilosophie* (*Isis,* XLVIII [1957], 37, n. 21).

81. *Ibid.,* p. 36, from Schelling's "Allgemeiner Deduktion des dynamischen Processes oder der Kategorien der Physik" (1800).

82. Haas, *Erhaltung,* p. 41.

83. It is, of course, impossible to distinguish sharply between the influence of *Naturphilosophie* and that of conversion processes. Breheir (*Schelling,* pp. 23–24) and Windelband (*History of Philosophy,* tr. J. H. Tufts [2nd ed.; New York, 1901], pp. 597–98) both emphasize that conversion processes were themselves a significant source of *Naturphilosophie,* so that the two were often grasped together. This fact must qualify some of the dichotomies set up in the first part of this paper, for the distinction between the two sources of the conservation concept is often equally hard to apply to individual pioneers. I have already pointed out the difficulty in Colding's case (note 32). With Mohr and Liebig I am still inclined to give *Naturphilosophie* the psychological priority, because neither had dealt much with the new conversion processes in their own research and because both make such large leaps. Their cases appear in sharp contrast to those of Grove and Faraday who seem to proceed by a continuous path from conversion processes to conservation. But this continuity may be deceptive. Grove (*Physical Forces,* pp. 25–27) mentions Coleridge, and Coleridge was the principal British exponent of *Naturphilosophie.* Since the problem presented by these examples seems to me both real and unresolved, I had better point out that it affects only the organization, not the main thesis, of this paper, Perhaps conversion processes and *Naturphilosophie* should be considered in the same section. Nevertheless, they would both have to be considered.

essentials. Colding was a protegé of Oersted's.[84] Liebig studied for two years with Schelling, and though he afterwards described these years as a waste, he never surrendered the vitalism he had then imbibed.[85] Hirn cited both Ocken and Kant.[86] Mayer did not study *Naturphilosophie*, but he had close student friends who did.[87] Helmholtz's father, an intimate of the younger Fichte's and a minor *Naturphilosoph* in his own right, constantly exhorted his son to desert strict mechanism.[88] Though Helmholtz himself felt forced to excise all philosophical discussion from his classic memoir, he was able by 1881 to recognize important Kantian residues that had escaped his earlier censorship.[89]

Biographical fragments of this sort do not, of course, prove intellectual indebtedness. They may, however, justify strong suspicion, and they surely provide leads for further research. At the moment I shall only insist that this research should be done and that there are excellent reasons to suppose it will be fruitful. Most of those reasons are given above, but the strongest has not yet been noticed. Though Germany in the 1840's had not yet achieved the scientific eminence of either Britain or France, five of our twelve pioneers were Germans, a sixth, Colding, was a Danish disciple of Oersted's, and a seventh, Hirn, was a self-educated Alsatian who read the *Naturphilosophen*.[90] Unless the *Naturphilosophie* indigenous to the educational environment of these seven men had a productive role in the researches of some, it is hard to see why more than fifty per cent of the pioneers should have

84. Povl Vinding, "Colding, Ludwig August," *Dansk Biografisk Leksikon* (Copenhagen, 1933–44), pp. 377–82. I am grateful to Roy and Ann Lawrence for providing me with a précis of this useful biographical sketch.

85. E. von Meyer, *A History of Chemistry*, tr. G. McGowan (3rd ed.; London, 1906), p. 274. J. T. Merz, *European Thought in the Nineteenth Century* (London, 1923–50), I, 178–218, particularly the last page.

86. G. A. Hirn, "Études sur les lois et sur les principes constituants de l'univers," *Revue d'Alsace*, I (1850), 24–41, 127–42, 183–201; *ibid.*, II (1851), 24–45. References to writings related to *Naturphilosophie* occur relatively often, though they are not very favorable. On the other hand, the very title of this piece suggests *Naturphilosophie*, and the title is appropriate to the contents.

87. B. Hell, "Robert Mayer," *Kantstudien*, XIX (1914), 222–48.

88. Koenigsberger, *von Helmholtz*, pp. 3–5, 30.

89. Helmholtz, *Abhandlungen*, I, 68.

90. Much biographical and bibliographical material for the study of Hirn's life and work can be found in the *Bulletin de la société d'histoire naturelle de Colmar*, I (1899), 183–335.

91. Séguin is the sixth, and the source of his idea remains a complete riddle. He attributes it (*Chemins de fer*, p. xvi) to his uncle Montgolfier, about whom I have been able to get no relevant information.

The statistics above are not meant to imply that those exposed to *Naturphilosophie* were invariably affected by it; nor do I mean to argue that those whose work shows no conceptual lacunae were *ipso facto* not influenced by *Naturphilosophie* (see remarks on Grove in note 83). It is the *predominance* rather than the presence of pioneers from the area dominated by German intellectual traditions that constitutes the puzzle.

been drawn from an area barely through its first generation of significant scientific productivity. Nor is this quite all. If proved, the influence of *Naturphilosophie* may also help explain why this particular group of five Germans, a Dane, and an Alsatian includes five of the six pioneers in whose approaches to energy conservation we have previously noted such marked conceptual lacunae.[91]*

THIS preliminary discussion of simultaneous discovery must end here. Comparing it with the sources, primary and secondary, from which it derives, makes apparent its incompleteness. Almost nothing has been said, for example, about either the dynamical theory of heat or the conception of the impossibility of perpetual motion. Both bulk large in standard histories, and both would require discussion in a more extended treatment. But if I am right, these neglected factors and others like them would not enter a fuller discussion of simultaneous discovery with the urgency of the three discussed here. The impossibility of perpetual motion, for example, was an essential intellectual tool for most of the pioneers. The ways in which many of them arrived at the conservation of energy cannot be understood without it. Yet recognizing the intellectual tool scarcely contributes to an understanding of simultaneous discovery because the impossibility of perpetual motion had been endemic in scientific thought since antiquity.[92] Knowing the tool was there, our question has been: why did it suddenly

* [The following paragraph was added to the original manuscript in response to points raised during the discussion.]

Professor Gillispie, in his paper, calls attention to a little known movement in eighteenth-century France that shows striking parallels to *Naturphilosophie*. If this movement had still been prevalent in nineteenth-century France, my contrast between the German scientific tradition and that prevalent elsewhere in Europe would be questionable. But I find nothing resembling *Naturphilosophie* in any of the nineteenth-century French sources I have examined, and Professor Gillispie assures me that, to the best of his knowledge, the movement to which his paper draws attention had disappeared (except perhaps from parts of biology) by the turn of the century. Notice, in addition, that this eighteenth-century movement, which was particularly prevalent among craftsmen and inventors, may provide a clue to the puzzle of Montgolfier (see note 91).

92. E. Mach, *History and Root of the Principle of the Conservation of Energy*, tr. Philip E. B. Jourdain (Chicago, 1911), pp. 19–41; and Haas, *Erhaltung*, chapt. IV. Remember also that in 1775 the French Academy formally resolved to consider no more purported designs of perpetual motion machines. Almost all of our pioneers make use of the impossibility of perpetual motion, and none feels the slightest necessity of arguing about its validity. In contrast, they do find it necessary to argue at length about the validity of the concept of universal conversions. Grove, for example, opens his *Physical Forces* (pp. 1–3) with a plea for a fair hearing of a radical idea. The idea turns out to be the concept of universal conversions developed at great length in the text (pp. 4–44). The impossibility of perpetual motion is casually applied to this idea without argument in the last seven pages (pp. 45–52). It is facts like these that led me to call the steps from universal conversions to an unquantified version of conservation "rather obvious."

acquire a new significance and a new range of application? For us, that is the more significant question.

The same argument applies in part to my second example of neglected factors. Despite Rumford's deserved fame, the dynamical theory of heat had been close to the surface of scientific consciousness almost since the days of Francis Bacon.[93] Even at the end of the eighteenth century, when temporarily eclipsed by the work of Black and Lavoisier, the dynamical theory was often described in scientific discussions of heat, if only for the sake of refutation.[94] To the extent that the conception of heat as motion figured in the work of the pioneers, we must principally understand why that conception gained a significance after 1830 that it had seldom possessed before.[95] Besides, the dynamical theory did not figure very largely. Only Carnot used it as an essential stepping stone. Mohr leaped from the dynamical theory to conservation, but his paper indicates that other stimuli might have served as well. Grove and Joule adhered to the theory but show substantially no

93. For seventeenth-century theories of heat see, M. Boas, "The establishment of the mechanical philosophy," *Osiris*, X (1952), 412–541. Much information about eighteenth-century theories is scattered through: D. McKie and N. H. de V. Heathcote, *The Discovery of Specific and Latent Heat* (London, 1935), and H. Metzger, *Newton, Stahl, Boerhaave et la doctrine chimique* (Paris, 1930). Much other useful information will be found in G. Berthold, *Rumford und die Mechanische Wärmetheorie* (Heidelberg, 1875), though Berthold skips too rapidly from the seventeenth to the nineteenth century.

94. Since the caloric theory was scarcely presented in a developed form before the publication of Lavoisier's *Traité élémentaire de chimie* in 1789, it could hardly have eradicated the dynamical theory in the decade remaining before the publication of Rumford's work. For evidence that even the most pronounced caloricists continued to discuss it, see Armand Séguin, "Observations générales sur le calorique . . . reflexions sur la théorie de MM. Black, Crawford, Lavoisier, & Laplace," *Ann. de Chim.*, III (1789), 148–242, and V (1790), 191–271, particularly, III, 182–90. The material theory of heat has, of course, roots far older than Lavoisier, but Rumford, Davy, *et al.* are really opposing a new theory, not an old one. Their work, particularly Rumford's, may have kept the dynamical theory alive after 1800, but Rumford did not create the theory. It had not died.

95. It is too seldom recognized that until almost the mid-nineteenth century, brilliant scientists could apply the dynamical conservation of *vis viva* to the theory that heat is motion without at all recognizing that heat and work should then be convertible. Consider the following three examples. Daniel Bernoulli, in the often quoted paragraphs from "Section X" of his *Hydrodynamica* equates heat with particulate *vis viva* and derives the gas laws. Then, in paragraph 40, he applies this theory in computing the height from which a given weight must fall to compress a gas to a given fraction of its initial volume. His solution gives the energy of motion abstracted from the falling weight in order to compress the gas, but fails entirely to notice that this energy must be transferred to the gas particles and must therefore raise the gas's temperature. Lavoisier and Laplace, on pp. 357–59 of their classic memoir (note 72), apply the conservation of energy to the dynamical theory in order to show that for all experimental purposes the caloric and dynamical theories are precisely equivalent. J. B. Biot repeats the same argument, in his *Traité de physique expérimentale et mathematique* (Paris, 1816), I, 66–67, and elsewhere in the same chapter. Grove's mistake about heat (note 34) indicates that even the conception of conversion processes was sometimes insufficient to guide scientists away from this virtually universal mistake.

dependence on it.[96] Holtzmann, Mayer, and Séguin opposed it—Mayer vehemently and to the end of his life.[97] The apparently close connections between energy conservation and the dynamical theory are largely retrospective.[98]

Compare these two neglected factors with the three we have discussed. The rash of conversion discoveries dates from 1800. Technical discussions of dynamical engines were scarcely a recurrent ingredient of scientific literature before 1760 and their density increased steadily from that date.[99] *Naturphilosophie* reached its peak in the first two decades of the nineteenth century.[100] Furthermore, all three of these ingredients, except possibly the last, played important roles in the research of at least half the pioneers. That does not mean that these factors explain either the individual or collective discoveries of energy conservation. Many old discoveries and concepts were essential to the work of all the pioneers; many new ones played significant roles in the work of individuals. We have not and shall not reconstruct the causes of all that occurred. But the three factors discussed above may still provide the fundamental constellation, given the question from which we began: Why, in the years 1830 to 1850, did so many of the experiments and concepts required for a full statement of energy conservation lie so close to the surface of scientific consciousness?

96. Grove, *Physical Forces,* pp. 7–8. Joule, *Papers,* pp. 121–23. Perhaps these two would not have developed their theories if they had not tended to regard heat as motion, but their published works indicate no such decisive connections.

97. Holtzmann's memoir is based on the caloric theory. For Mayer see Weyrauch, I, 265–72, and II, 320, n. 2. For Séguin see *Chemins de fer,* p. xvi.

98. The ease and immediacy with which the dynamical theory was identified with energy conservation is indicated by the contemporary misinterpretations of Mayer quoted in Weyrauch, II, pp. 320 and 428. The classic case, however, is Lord Kelvin's. Having employed the caloric theory in his research and writing until 1850, he opens his famous paper "On the Dynamical Theory of Heat" (*Mathematical and Physical Papers* [Cambridge, 1882], pp. 174–75) with a series of remarks on Davy's having "established" the dynamical theory fifty-three years before. Then he continues, "The recent discoveries made by Mayer and Joule . . . afford, *if required,* a perfect confirmation of Sir Humphry Davy's views" (italics mine). But if Davy established the dynamical theory in 1799 and if the rest of conservation follows from it, as Kelvin implies, what had Kelvin himself been doing before 1852?

99. The abstract theories of dynamical engines have no beginning in time. I pick 1760 because of its relation to the important and widely cited works of Smeaton and Borda (notes 50 and 51).

100. Merz, *European Thought,* I, 178, n. 1.

30 *The Exponential Curve of Science*

DERek J. Price

Is it possible to be scientific about science itself? Can one analyse scientifically the present problems of manpower shortage, of specialisation, and of the vital part which science and technology is playing in national and international affairs? With so much at stake and with so many people involved, considerable attention should be paid to the making of any possible intelligent guesses about what is going to happen next to science "in the large." Economists can now describe and explain tendencies in the economic world, and to some extent help those whose job it is to control it. Statisticians can make similar statements about problems of the growth and change of population, but in the study of science there is no parallel. There are only *ex cathedra* statements from those eminent scientists who sense intuitively some of the problems involved. Decisions and policies are laid down by government committees and commissions who must view each case as it arises without any corpus of theory or general principle to guide them. Even in the Soviet Union, where state planning has been particularly successful in technology and science, and in the U.S.A. where these things are now major industries, there has been little attempt seriously to examine the main principles of general development of science, and to draw conclusions about the likely if not inevitable future.

Let us then examine, however crudely and approximately, anything that will tell us about the rate at which the overall size, complexity, and intensity of science is and has been growing. Since it is natural to prefer a quantitative approach, even if inexact, to any purely qualitative analysis, it is necessary to seek any data that can be obtained by

Reprinted from Discovery, *vol.* 17 (1956), *pp.* 240–243.

a process of "head-counting" throughout a series of annual or other periods. So long as there are enough heads to count so as to secure a good statistical sample, one need not worry at first about just what is being measured, but concentrate only on obtaining a good measure.

Head-Counting

CONVENIENT heads for counting may easily be selected, using national statistics, scientific publications, and figures provided by learned societies. For example one may record the number of *Physics Abstracts*, *Chemical Abstracts*, etc., in each year, so taking some sort of measure of the number of papers published in such a selective field during the period. One may prepare similar figures for an even narrower field of study by making use of any of the specialist bibliographies available. It is possible to measure manpower directly from various national statistics, professional registrars, and analyses of the output of the universities and other training institutions, and the annual expenditure in various branches of science and industry are often similarly available. Other measures may be had from the total numbers of scientific patents taken out during the year, and from sundry selective lists of "great scientists" (e.g. Dictionaries of National Biography) or notable scientific advances.

From such collections of data,[1] chosen capriciously for the ease of getting figures rather than for any significance in themselves, three important conclusions can be drawn:

1. Nearly all the curves of growth show the same trends.
2. The growth is (to a surprising accuracy, $\pm$ *ca.* 1%) exponential.
3. The constant of the exponential curve is such as to effect a doubling in size in an interval of the order of 10–15 years.

The first conclusion seems to indicate that the data collected in fact measure, by different means, the same general phenomenon, and that it is reasonably safe to take any one of these sets of data as a provisional measure of the "size" of that sort of science. One may modify this hypothesis later, but it provides a firm foundation for a preliminary investigation (Figure 1).

Exponential Growth

THE second conclusion, that of exponential growth, merely tells us something that we might have guessed beforehand—though, for some reason, planning committees seem peculiarly blind to it—the

1. Price, D. J., "Quantitative Measures of the Development of Science," *Archives Internationales d'Histoire des Sciences*, 1951, vol. 14, pp. 85–93.

growth of most organisms tends to be directly related to their size; the bigger they get, the faster they grow. This law defines exponential increase and governs the growth of a colony of bacteria; it also apparently governs the size of science. Data which can be carried back to the year 1700 (or even before) indicate clearly that in general the "size" of science has been increasing in this way over the whole period since the Scientific Revolution and the time of Newton. In other words,

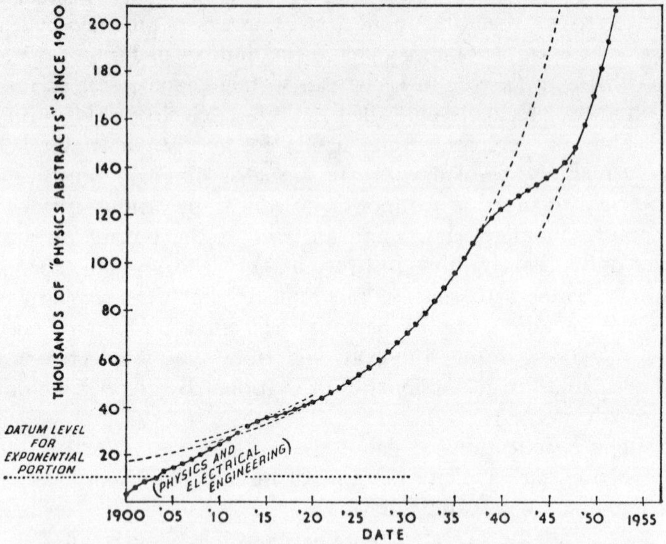

FIGURE 1

Total number of Physics Abstracts published since Jan. 1, 1900. The full curve gives the total, and the broken curve represents the exponential approximation. Parallel curves are drawn to enable the effect of the wars to be illustrated.

one must assume that the *normal* state of growth in science and technology in the past and at present corresponds to a doubling in size every so many years. Any assumption that growth is linear (i.e. it will continue at the present rate) is doomed to be an under-estimate.

The last conclusion, concerning the numerical value of the time constant, is perhaps the most significant. The period of 10–15 years which characterises the growth of science is very considerably shorter than the time corresponding to a generation (say, 25–30 years) and less than that involved in exponential growths associated with non-scientific and non-technological activities of the human race. In point of fact, several other investigations suggest that the time constant for doubling in such fields is about 30–50 years, approximately three times as long at least as that for scientific activities. In a period of half a century, the number of poets, composers, politicians, etc., doubles about once, while the population of scientists doubles at least thrice, multiplying therefore by a factor of eight.

Reduction to Absurdity

A DELIGHTFUL parody of these methods as applied to the growth of the Civil Service and other bodies has been published as "Parkinson's Law" (*Economist,* 1955, vol. 177, p. 365) by my erstwhile colleague and mentor in another clime. Though the burden of his argument is light-hearted he will be shocked to hear that the increase he found, about 5–6% p.a., corresponds exactly to the doubling period found for professional scientists and technologists over the last few centuries.

Those with some mathematical training do not need to be reminded of the "runaway" character of exponential growth. Ten doubling periods correspond to a multiplication in size by a factor of more than 1000, while twenty periods give a factor greater than a million. Now twenty periods each of 10–15 years give a lapse of 200–300 years—sufficient to span the interval from *ca.* A.D. 1700 to the present date. It can readily be seen that this estimate is correct in order of magnitude, for in 1700 the Scientific Revolution was just passing into the Industrial Revolution and the first few of everything had been established, whereas today, estimates of numbers of scientists, journals, etc., run into hundreds of thousands, or even millions, as the factor would predict. An exponential curve is defined by two parameters; one corresponding to the time constant already described, and a second giving the date at which the curve reaches a value of unity. Again, nearly all the curves show agreement in their date of origin, extrapolation indicating it to be *ca.* 1700–50 so far as can be seen, and this is in agreement with what has already been said.

Since the exponential growth of everything scientific is so much more rapid than that of anything in the rest of our civilisation (crude population figures included) it follows that saturation must be reached sooner or later if such growth is maintained even approximately (Figure 2). It is important to note that a state of absurdity may be reached quite quickly, for example, a further 250 years bringing another factor of a million would give about 100 scientists for every man, woman and child in the world today! If one makes reasonable guesses as to the maximum saturation of scientists and technologists in our population, it would seem that the present law of growth would yield that value in something like 50–70 years from now. This is an astonishingly short expected lifetime for our mode of civilisation, and it is therefore worth while to make a crude estimate of the effect of saturation on exponential growth.

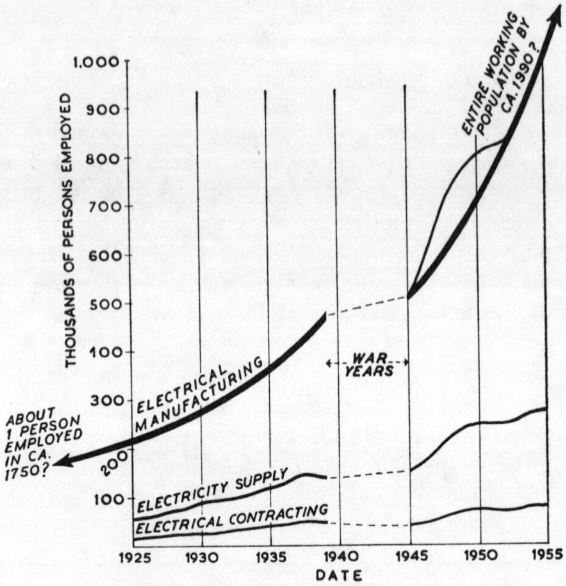

FIGURE 2

Growth of employment in the electrical industry (manufacturing, supply, and contracting from 1925 to 1955). Adapted from figures published by The Manchester Guardian, *of March 20, 1956: "The Electrical Industry Today" by Dr. Willis Jackson, F.R.S.*

Saturated Growth

THE usual physical law associated with saturated growth shows the exponential curve modifying into an S-shaped curve approaching the maximum saturation level (Figure 3). The "half-way house" occurs at a date corresponding to saturation on the purely exponential curve. Approximately two doubling periods before this crucial date is reached, the curve of growth begins to fall sensibly below the exponential figures. Two doubling periods after the crucial date, the growth approaches quite closely to saturation value. Returning to the present estimate it would seem that we may expect two or three decades more before the regular exponential growth will have to fall off markedly as it approaches a "turn-over" near some saturation level. This may not sound so serious, being a gradual change, but one must remember that normal growth has stayed with us for the last two or three centuries so that we have come to regard it as the only healthy manner of evolution. Any curtailment, however small, must react very forcibly on a national scale, having a host of side effects which are bound to be unpleasant and to inspire all conceivable palliative measures. Since the above calculations are only very approximate, giving order of magnitude values only, it may be reasonable to interpret the present manpower shortage, most apparent in a lack of science teachers, as evidence that the growth is following this general pattern and is already dangerously near its critical period. One might alleviate the growing pains by increased training facilities, change of emphasis from Arts faculties to Science, greater incentives to teachers and scientists; but these things cannot change the overall pattern, but will merely postpone the crisis for very few decades.

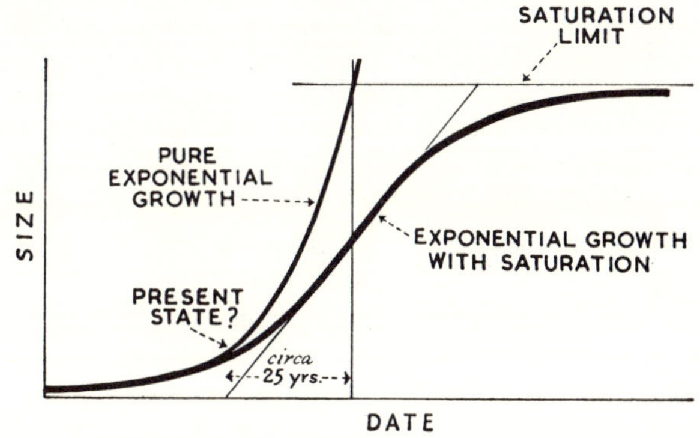

FIGURE 3

Diagrammatic representation of the effect of saturation on growth.

Internal Repressive Effects

THE saturation process just described may be regarded as an external repressive influence in the growth of science. Other, similar analyses can be used to indicate that science carries its own internal repressive effects which seem likely to choke steady development. Every research worker today is familiar with the problem of keeping up with what is affectionately known as "the Literature." There are now so many learned journals that even the finest specialist libraries cannot take them all, and their workers can only read a very small selection of them. Even abstract journals are now becoming so unwieldy that it is already evident that some radically new technique must be evolved if publication is to continue as a useful contribution in the same sense as it was in the past. A quantitative analysis makes the mechanism of the change apparent: scientific journals increase in number as if they had started in about 1700 and doubled every 15 years (actually they started in 1665, but the first 10 years or so constitute a too small, non-statistical group). Abstract journals started about 1830 when there were already about 300 general journals in existence, and these abstract journals have been doubling in number, again every 15 years. The number of scientific journals is now approaching an order of magnitude of 100,000, and the number of abstract journals is now near the critical figure of about 300. That is why we are beginning to wonder about "abstracts of abstracts," new techniques for boiling down and analysing all the work at present being carried out (Figure 4).

Problems of Specialisation

A SIMILAR technique can be applied to some extent to the problems associated with specialisation and with the amount of knowledge that an intending researcher must absorb before he can reach the research front. It is not difficult to show in a general fashion that these are processes exercising a retarding effect on the growth of science, producing narrower and less flexible specialists—the process having a time constant similar to that found elsewhere in science, or perhaps (if one is optimistic) a little longer. Rather interestingly, if one attempts to measure "high-level" advances in science rather than its crude size, it seems that the growth constant is considerably larger, taking about three times as long to double. If this is true, it would mean that to double the usefulness of science involves multiplying by about eight the gross number of workers and the total expenditure of manpower and national income. It also entails the curious, though half-expected law, that the gross size of science goes up as approximately the cube of the attainment of science. To some extent this is a consequence of the distribution law governing the relative numbers of scientists of various degrees of excellence and inferiority. Alternatively one may regard science in the image of a pyramid of theory and experiment. The height represents the attainment, but the bulk of science must be measured by the cubic capacity of the structure.

Some Side Results

OTHER interesting side results of these investigations concern the effect of wars on the progress of science (no observable stimulus after

1938–47), and the tendency for virile, new fields just opening up to grow much faster than average until they settle down to the normal time-constant of growth. Similarly, old fields show the tendency to fall off, growing more slowly than normal. One may well use this technique to make objective judgments on the virility or sterility of certain

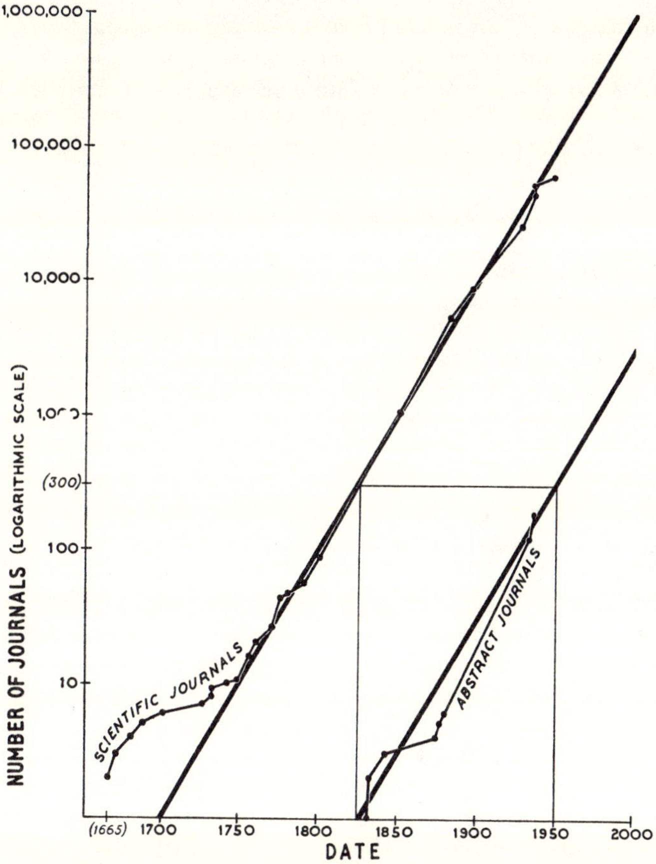

FIGURE 4

Number of scientific journals and abstract journals.

fields of inquiry. Quite clearly there are many other ways in which this sort of analysis might provide useful information. It need not be stressed that any further investigation must also pay attention to increasing the accuracy of the very approximate statistical data already cited, and to a critical examination of the meanings of the quantities being measured. There are also many other econometric laws applicable to science in the large. It is possible to make statistical generalisations about the distribution in quality of scientific work and scientific research workers. It should be possible to develop a calculus that would tell us the optimum ratio between science teachers and scientists, and between pure scientists and technologists. Working scientists are for ever using a sort of poetic imagery which represents a subconscious attempt to get a model of science as a whole—one speaks of "the Research front" and "an isolated field" a "borderline investigation," etc., without any effort to rationalise such terms. If, as seems likely, there is indeed something in the structure of science that makes such terms reasonable, it should be possible to make such loose jargon precise in the form of theories about the general structure of science.

Use of the Technique

UNTIL there is a lot more of this sort of theory and until its limitations and advantages are better understood it is impossible to guess just how useful such techniques might be to scientists who have to organise their own researches, and to national institutions and committees who, however reluctantly, must undertake some sort of planning activity. Already, even the crude theory outlined above is sufficient to indicate that we are very near a crisis which might not otherwise be obvious until it is too late. Theory gives us the grace of a few decades in which to lay plans and to understand better what is happening during the predicted cataclysmic change in the very nature of scientific advance.

It is certainly not sufficient to utter a bald counsel of despair that we are approaching a period of saturation when there cannot be enough scientists to meet all demands, and when those scientists will tend to be stifled by their own flood of literature and intense overspecialisation. One must back up the econometric analysis by the historical analysis which it augments by an added mathematically induced perspective. One must decide what is wanted, as well as what is possible, before policy decisions can be taken. If there is saturation it follows that there must be some sort of rationing, natural or imposed, both in the internal distribution of scientists amongst their fields of inquiry and amongst the needs of the nation and industry who apply the results of scientific inquiry, all competing for the same body of manpower. At present the writing can be seen on the wall but it is no man's professional business to read it and take warning.

31 *The Case of the Floppy-Eared Rabbits: An Instance of Serendipity Gained and Serendipity Lost*

BERNARD BARBER AND RENÉE C. FOX

As WITH SO MANY other basic social processes, the actual process of scientific research and discovery is not well understood.[1] There has been little systematic observation of the research and discovery process as it actually occurs, and even less controlled research. Moreover, the form in which discoveries are reported by scientists to their colleagues in professional journals tends to conceal important aspects of this process. Because of certain norms that are strongly institutionalized in their professional community, scientists are expected to focus their reports on the logical structure of the methods used and the ideas discovered in research in relation to the established conceptual framework of the relevant scientific specialty. The primary function of such reports is conceived to be that of indicating how the new observations and ideas being advanced may require a change—by further generalization or systematization—in the conceptual structure of a given scientific field. All else that has occurred in the actual research process is considered "incidental." Thus scientists are praised for presenting their research in a way that is elegantly bare of anything that does not serve this primary function and are deterred from reporting "irrelevant" social and psychological aspects of the research process, however interesting these matters may be in other contexts. As a result of such norms and practices, the reporting of scientific research may be characterized by what has been called "retrospective falsification." By selecting only those components of the actual re-

Reprinted from American Journal of Sociology, *vol.* 64 (1958), *pp.* 128–136.

1. For an account of what is known see Bernard Barber, *Science and the Social Order* (New York: Free Press, 1952), chap. 9, "The Social Process of Invention and Discovery," pp. 191–206.

search process that serve their primary purpose, scientific papers leave out a great deal, of course, as many scientists have indicated in their memoirs and in their informal talks with one another. Selection, then, unwittingly distorts and, in that special sense, falsifies what has happened in research at it actually goes on in the laboratory and its environs.

Public reports to the community of scientists thus have their own function. Their dysfunctionality for the sociology of scientific discovery, which is concerned with not one but all the components of the research process as a social process, is of no immediate concern to the practicing research scientist. And yet what is lost in "retrospective falsification" may be of no small importance to him, if only indirectly. For it is not unlikely that here, as everywhere else in the world of nature, knowledge is power, in this case power to increase the fruitfulness of scientific research by enlarging our systematic knowledge of it. The sociology of scientific discovery would seem to be an especially desirable area for further theoretical and empirical development.

One component of the actual process of scientific discovery that is left out or concealed in research reports following the practice of "retrospective falsification" is the element of unforeseen development, of happy or lucky chance, of what Robert K. Merton has called "the serendipity pattern."[2] By its very nature, scientific research is a voyage into the unknown by routes that are in some measure unpredictable and unplannable. Chance or luck is therefore as inevitable in scientific research as are logic and what Pasteur called "the prepared mind." Yet little is known systematically about this inevitable serendipity component.

For this reason it seemed to us desirable to take the opportunity recently provided by the reporting of an instance of *serendipity gained* by Dr. Lewis Thomas, now professor and chairman of the Department of Medicine in the College of Medicine of New York University and formerly professor and chairman of the Department of Pathology.[3] Then, shortly after hearing about Dr. Thomas' discovery, we learned from medical research and teaching colleagues of an instance of *serendipity lost* on the very same kind of chance occurrence; unexpected floppiness in rabbits' ears after they had been injected intravenously with the proteolytic enzyme papain. This instance of seren-

2. For discussions of serendipity see Walter B. Cannon, *The Way of an Investigator* (New York: W. W. Norton & Co., 1945), chap. vi, "Gains from Serendipity," pp. 68–78; and Robert K. Merton, *Social Theory and Social Structure* (rev. ed.; New York: Free Press, 1957), pp. 103–8. Our colleagues, Robert K. Merton and Elinor G. Barber, are now engaged in an investigation and clarification of the variety of meanings of "chance" that are lumped under the notion of serendipity by different users of that term.

3. Lewis Thomas, "Reversible Collapse of Rabbit Ears after Intravenous Papain,

dipity lost had occurred in the course of research by Dr. Aaron Kellner, associate professor in the Department of Pathology of Cornell University Medical College and director of its central laboratories. This opportunity for *comparative* study seemed even more promising for our further understanding of the serendipity pattern. Here were two comparable medical scientists, we reasoned, both carrying out investigations in the field of experimental pathology, affiliated with distinguished medical schools, and of approximately the same level of demonstrated research ability (so far as it was in our layman's capacity to judge). In the course of their research both men had had occasion to inject rabbits intravenously with papain, and both had observed the phenomenon of ear collapse following the injection.

In spite of these similarities in their professional backgrounds and although they had both accidentally encountered the same phenomenon, one of these scientists had gone on to make a discovery based on this chance occurrence, whereas the other had not. It seemed to us that a detailed comparison of Dr. Thomas' and Dr. Kellner's experiences with the floppy-eared rabbits offered a quasi-experimental opportunity to identify some of the factors that contribute to a positive experience with serendipity in research and some of the factors conducive to a negative experience with it.

We asked for and were generously granted intensive interviews with Dr. Thomas and Dr. Kellner.[4] Each reported to us that they had experienced both "positive serendipity" and "negative serendipity" in their research. That is, each had made a number of serendipitous discoveries based on chance occurrences in their planned experiments, and on other occasions each had missed the significance of like occurrences that other researchers had later transformed into discoveries. Apparently, both positive and negative serendipity are common experiences for scientific researchers. Indeed, we shall see that one of the chief reasons why Dr. Kellner experienced serendipity lost with respect to the discovery that Dr. Thomas made was that he was experiencing serendipity gained with respect to some other aspects of the very same experimental situation. Conversely, Dr. Thomas had reached a stalemate on some of his other research, and this gave him added incentive to pursue intensively the phenomenon of ear collapse. Partly as a consequence of these experiences, in what were similar experimental situa-

and Prevention of Recovery by Cortisone," *Journal of Experimental Medicine,* CIV (1956), 245–52. This case first came to our attention through a report in the *New York Times.* The pictures printed in Dr. Thomas' original article and in the *Times* will indicate why we have called this "the case of the floppy-eared rabbits."

4. These interviews lasted about two hours each. They are another instance of the "tandem interviewing" described by Harry V. Kincaid and Margaret Bright, "Interviewing the Business Elite," *American Journal of Sociology,* LXIII (1957), 304–11.

tions, the two researchers each saw something and missed something else.

On the basis of our focused interviews with these two scientists, we can describe some of the recurring elements in their experiences with serendipity.[5] We think that these patterns may also be relevant to instances of serendipity experienced by other investigators.

Serendipity Gained

DR. THOMAS

Observing the established norms for reporting scientific research, in his article in the *Journal of Experimental Medicine,* Dr. Thomas did not mention his experience with serendipity. In the manner typical of such reports he began his article with the statement, "For reasons not relevant to the present discussion rabbits were injected intravenously with a solution of crude papain." (By contrast, though not called by this term, serendipity was featured in the accounts of this research that appeared in the *New York Times* and the *New York Herald Tribune.* "An accidental sidelight of one research project had the startling effect of wilting the ears of the rabbit," said the *Times* article. "This bizarre phenomenon, accidentally discovered . . ." was the way the *Herald Tribune* described the same phenomenon. The prominence accorded the "accidental" nature of the discovery in the press is related to the fact that these articles were written by journalists for a lay audience. The kind of interest in scientific research that is characteristic of science reporters and the audience for whom they write and their conceptions of the form in which information about research ought to be communicated differ from those of professional scientists.)[6]

Although Dr. Thomas did not mention serendipity in his article for the *Journal of Experimental Medicine,* in his interview he reported both his general acquaintance with the serendipity pattern ("Serendipity is a familiar term. . . . I first heard about it in Dr. Cannon's class . . .") and his awareness of the chance occurrence of floppy-eared rabbits in his own research. Dr. Thomas first noticed the reversible collapse of rabbit ears after intravenous papain about seven years ago, when he was working on the effects of proteolytic enzymes as a class:

> I was trying to explore the notion that the cardiac and blood vessel lesions in certain hypersensitivity states may be due to re-

5. In this paper we shall concentrate on the instances of serendipity gained by Dr. Thomas and lost by Dr. Kellner and give somewhat less attention to elements of negative serendipity in Dr. Thomas' experiments and elements of positive serendipity in those of Dr. Kellner.

6. Further discussion of this point lies beyond the scope of this paper. But in a society like ours, in which science has become "front-page news," some of the char-

lease of proteolytic enzymes. It's an attractive idea on which there's little evidence. And it's been picked up at some time or another by almost everyone working on hypersensitivity. For this investigation I used trypsin, because it was the most available enzyme around the laboratory, and I got nothing. We also happened to have papain; I don't know where it had come from; but because it was there, I tried it. I also tried a third enzyme, ficin. It comes from figs, and it's commonly used. It has catholic tastes and so it's quite useful in the laboratory. So I had these three enzymes. The other two didn't produce lesions. Nor did papain. But what the papain did was always produce these bizarre cosmetic changes. . . . It was one of the most uniform reactions I'd ever seen in biology. It always happened. And it looked as if something important must have happened to cause this reaction.

Some of the elements of serendipitous discovery are clearly illustrated in this account by Dr. Thomas. The scientific researcher, while in pursuit of some other specific goals, accidentally ("we also happened to have papain . . .") produces an unusual, recurrent, and sometimes striking ("bizarre") effect. Only the element of creative imagination, which is necessary to complete an instance of serendipity by supplying an explanation of the unusual effect, is not yet present. Indeed, the explanation was to elude Dr. Thomas, as it eluded Dr. Kellner, and probably others as well, for several years. This was not for lack of trying by Dr. Thomas. He immediately did seek an explanation:

> I chased it like crazy. But I didn't do the right thing. . . . I did the expected things. I had sections cut, and I had them stained by all the techniques available at the time. And I studied what I believed to be the constituents of a rabbit's ear. I looked at all the sections, but I couldn't see anything the matter. The connective tissue was intact. There was no change in the amount of elastic tissue. There was no inflammation, no tissue damage. I expected to find a great deal, because I thought we had destroyed something.

Dr. Thomas also studied the cartilage of the rabbit's ear, and judged it to be "normal" (". . . The cells were healthy-looking and there were nice nuclei. I decided there was no damage to the cartilage. And that was that . . ."). However, he admitted that at the time his consideration of the cartilage was routine and relatively casual, because

acteristics and special problems of science reporting merit serious study. A recently published work on this topic that has come to our attention is entitled *When Doctors Meet Reporters* (New York: New York University Press, 1957). This is a discussion by science writers and physicians of the controversy between the press and the medical profession, compiled from the record of a series of conferences sponsored by the Josiah Macy, Jr., Foundation.

he did not seriously entertain the idea that the phenomenon of ear collapse might be associated with changes in this tissue:

> I hadn't thought of cartilage. You're not likely to, because it's not considered interesting. . . . I know my own idea has always been that cartilage is a quiet, inactive tissue.

Dr. Thomas' preconceptions about the methods appropriate for studying the ear-collapsing effect of papain, his expectation that it would probably be associated with damage in the connective or elastic tissues, and the conviction he shared with colleagues that cartilage is "inert and relatively uninteresting"—these guided his initial inquiries into this phenomenon. But the same preconceptions, expectations, and convictions also blinded him to the physical and chemical changes in the ear cartilage matrix which, a number of years later, were to seem "obvious" to him as the alterations underlying the collapsing ears. Here again, another general aspect of the research process comes into the clear. Because the methods and assumptions on which a systematic investigation is built selectively focus the researcher's attention, to a certain extent they sometimes constrict his imagination and bias his observations.

Although he was "very chagrined" about his failure, Dr. Thomas finally had to turn away from his floppy-eared rabbits because he was "terribly busy working on another problem at the time," with which he was "making progress." Also, Dr. Thomas reported, "I had already used all the rabbits I could afford. So I was able to persuade myself to abandon this other research." The gratifications of research success elsewhere and the lack of adequate resources to continue with his rabbit experiments combined to make Dr. Thomas accept failure, at least temporarily. As is usually the case in the reporting of scientific research, these experiments and their negative outcome were not written up for professional journals. (There is too much failure of this sort in research to permit of its publication, except occasionally, even though it might be instructive for some other scientists in carrying out their research. Since there is no way of determining what might be instructive failures and since space in professional journals is at a premium, generally only accounts of successful experiments are submitted to such journals and published by them.)

Despite his decision to turn his attention to other, more productive research, Dr. Thomas did not completely forget the floppy-eared rabbits. His interest was kept alive by a number of things. As he explained, the collapse of the rabbit ears and their subsequent reversal "was one of the most uniform reactions I'd ever seen in biology." The "unfailing regularity" with which it occurred is not often observed in scientific research. Thus the apparent invariance of this phenomenon never ceased to intrigue Dr. Thomas, who continued to feel that an important and powerful biological happening might be responsible. The effect of papain on rabbit ears had two additional qualities that helped to sustain Dr. Thomas' interest in it. The spectacle of rabbits with "ears collapsed limply at either side of the head, rather like the ears of spaniels,"[7] was both dramatic and entertaining.

In the intervening years Dr. Thomas described this phenomenon to a number of colleagues in pathology, biochemistry, and clinical investigation, who were equally intrigued and of the opinion that a significant amount of demonstrable tissue damage must be associated with such a striking and uniform reaction. Dr. Thomas also reported that twice he "put the experiment on" for some of his more skeptical colleagues. ("They didn't believe me when I told them what happened. They didn't really believe that you can get that much change and not a trace of anything having happened when you look in the miscroscope.") As so often happens in science, an unsolved puzzle was kept in mind for eventual solution through informal exchanges between scientists, rather than through the formal medium of published communications.

A few years ago Dr. Thomas once again accidentally came upon the floppy-eared rabbits in the course of another investigation:

> I was looking for a way . . . to reduce the level of fibrinogen in the blood of rabbits. I had been studying a form of fibrinoid which occurs inside blood vessels in the generalized Schwartzman reaction and which seems to be derived from fibrinogen. My working hypothesis was that if I depleted the fibrinogen and, as a result, fibrinoid did not occur, this would help. It had been reported that if you inject proteolytic enzyme, this will deplete fibrinogen. So I tried to inhibit the Schwartzman reaction by injecting papain intravenously into the rabbits. It didn't work with respect to fibrinogen. . . . But the same damned thing happened again to the rabbits' ears!

This time, however, Dr. Thomas was to solve the puzzle of the collapsed rabbit ears and realize a complete instance of serendipitous discovery. He describes what subsequently happened:

> I was teaching second-year medical students in pathology. We have these small seminars with them: two-hour sessions in the morning, twice a week, with six to eight students. These are seminars devoted to experimental pathology and the theoretical aspects of the mechanism of disease. The students have a chance to see what we, the faculty, are up to in the laboratory. I happened to have a session with the students at the same time that this thing with the rabbits'

7. Thomas, *op. cit.*, p. 245.

ears happened again. I thought it would be an entertaining thing to show them . . . a spectacular thing. The students were very interested in it. I explained to them that we couldn't really explain what the hell was going on here. I did this experiment on purpose for them, to see what they would think. . . . Besides which, I was in irons on my other experiments. There was not much doing on those. I was not being brilliant on these other problems. . . . Well, this time I did what I didn't do before. I simultaneously cut sections of the ears of rabbits after I'd given them papain *and* sections of normal ears. This is the part of the story I'm most ashamed of. It still makes me writhe to think of it. There was no damage to the tissue in the sense of a lesion. But what had taken place was a quantitative change in the matrix of the cartilage. The only way you could make sense of this change was simultaneously to compare sections taken from the ears of rabbits which had been injected with papain with comparable sections from the ears of rabbits of the same age and size which had not received papain. . . . Before this I had always been so struck by the enormity of the change that when I didn't see something obvious, I concluded there was nothing. . . . Also, I didn't have a lot of rabbits to work with before.

Judging from Dr. Thomas' account, it appears that a number of factors contributed to his reported experimental success. First, his teaching duties played a creative role in this regard. They impelled him to run the experiment with papain again and kept his attention focused on its implications for basic science rather than on its potentialities for practical application. Dr. Thomas said that he used the experiment to "convey to students what experimental pathology is like." Second, because he had reached an impasse in some of his other research, Dr. Thomas had more time and further inclination to study the ear-collapsing effect of papain that he had had a few years earlier, when the progress he was making on other research helped to "persuade" him to "abandon" the problem of the floppy-eared rabbits. Third, Dr. Thomas had more laboratory resources at his command than previously, notably a larger supply of rabbits. (In this regard it is interesting to note that, according to Dr. Thomas' article in the *Journal of Experimental Medicine,* 250 rabbits, all told, were used in the experiments reported.) Finally, the fact that he now had more laboratory animals with which to work and that he wanted to present the phenomenon of reversible ear collapse to students in a way that would make it an effective teaching exercise led Dr. Thomas to modify his method for examining rabbit tissues. In his earlier experiments, Dr. Thomas had compared histological sections made of the ears of rabbits who had received an injection of papain with his own mental image of normal rabbit-ear tissue. This time, however, he actually made sections from the ear tissue of rabbits which did *not* receive papain, as well as from those which did, and simultaneously examined the two. As he reported, this comparison enabled him to see for the first time that "drastic" quantitative changes had occurred in the cartilaginous tissue obtained from the ears of the rabbits injected with papain. In the words of the *Journal* article,

> The ear cartilage showed loss of a major portion of the intercellular matrix, and complete absence of basophilia from the small amount of remaining matrix. The cartilage cells appeared somewhat larger, and rounder than normal, and lay in close contact with each other. . . . (The contrast between the normal ear cartilage and tissue obtained 4 hours after injection is illustrated in Figs. 3*A* and 3*B* of this article.)

Immediately thereafter, Dr. Thomas and his associates found that these changes occur not only in ear cartilage but in all other cartilaginous tissues as well.

How significant or useful Dr. Thomas' serendipitous discovery will be cannot yet be specified. The serendipity pattern characterizes small discoveries as well as great. Dr. Thomas and his associates are currently investigating some of the questions raised by the phenomenon of papain-collapsed ears and the alterations in cartilage now known to underlie it. In addition, Dr. Thomas reported that some of his "biochemist and clinical friends" have become interested enough in certain of his findings to "go to work with papain, too." Two of the major problems under study in Dr. Thomas' laboratory are biochemical: the one concerning the nature of the change in cartilage; the other, the nature of the factor in papain that causes collapse of rabbits' ears and lysis of cartilage matrix in all tissues. Attempts are also being made to identify the antibody that causes rabbits to become immune to the factor responsible for ear collapse after two weeks of injection. The way in which cortisone prolongs the reaction to papain and the possible effect that papain may have on the joints as well as the cartilage are also being considered. Though at the time he was interviewed Dr. Thomas could not predict whether his findings (to date) would prove "important" or not, there was some evidence to suggest that certain basic discoveries about the constituents and properties of cartilaginous tissue might be forthcoming and that the experiments thus far conducted might have "practical usefulness" for studies of the postulated role of cortisone in the metabolism of sulfated mucopolysaccharides and of the relationship between cartilage and the electrolyte imbalance associated with congestive heart failure.

In the research on reversible ear collapse that Dr. Thomas has conducted since his initial serendipitous discovery, the planned and the unplanned, the foreseen and the accidental, the logical and the lucky have continued to interact. For example, Dr. Thomas' discovery that cortisone prevents or greatly delays the "return of papain-collapsed ears to their normal shape and rigidity" came about as a result of a carefully planned experiment that he undertook to test the effect of cortisone on the reaction to papain. On the other hand, his discovery that "repeated injections of papain, over a period of two or three weeks, brings about immunity to the phenomenon of ear collapse" was an unanticipated consequence of the fact that he used the same rabbit to demonstrate the floppy ears to several different groups of medical students:

> I was so completely sold on the uniformity of this thing that I used the same rabbit [for each seminar]. . . . The third time it didn't work. I was appalled by it. The students were there, and the rabbit's ears were still in place. . . . At first I thought that perhaps the technician had given him the wrong stuff. But then when I checked on that and gave the same stuff to the other rabbits and it *did* work I realized that the rabbit had become immune. This is a potentially hot finding. . . .

Serendipity Lost

DR. KELLNER

In our interview with Dr. Thomas we told him that we had heard about another medical scientist who had noticed the reversible collapse of rabbits' ears when he had injected them intravenously with papain. Dr. Thomas was not at all surprised. "That must be Kellner," he said. "He must have seen it. He was doomed to see it." Dr. Thomas was acquainted with the reports that Dr. Kellner and his associates had published on "Selective Necrosis of Cardiac and Skeletal Muscle Induced Experimentally by Means of Proteolytic Enzyme Solutions Given Intravenously" and on "Blood Coagulation Defect Induced in Rabbits by Papain Solutions Injected Intravenously."[8] He took it for granted that, in the course of these reported experiments which had entailed papain solution given intravenously to rabbits, a competent scientist like Dr. Kellner had also seen the resulting collapse of rabbits' ears, with its "unfailing regularity" and its "flamboyant" character. And, indeed, our interview with Dr. Kellner revealed that he had observed the floppiness, apparently at about the same time as Dr. Thomas:

8. See, Aaron Kellner and Theodore Robertson, "Selective Necrosis of Cardiac and Skeletal Muscle Induced Experimentally by Means of Proteolytic Enzyme Solutions Given Intravenously," *Journal of Experimental Medicine*, XCIX (1954), 387–

> We called them the floppy-eared rabbits. . . . Five or six years ago we published our first article on the work we were doing with papain; that was in 1951 and our definitive article was published in 1954. . . . We gave papain to the animals and we had done it thirty or forty times before we noticed these changes in the rabbits' ears.

Thus Dr. Kellner's observation of what he and his colleagues dubbed "the floppy-eared rabbits" represents, when taken together with Dr. Thomas' experience, an instance of independent multiple observation, which often occurs in science and frequently leads to independent multiple invention and discovery.

Once he had noticed the phenomenon of ear collapse, Dr. Kellner did what Dr. Thomas and any research scientist would have done in the presence of such an unexpected and striking regularity: he looked for an answer to the puzzle it represented. "I was a little curious about it at the time, and followed it up to the extent of making sections of the rabbits' ears." However, for one of those trivial reasons that sometimes affect the course of research—the obviously amusing quality of floppiness in rabbits' ears—Dr. Kellner did not take the phenomenon as seriously as he took other aspects of the experimental situation involving the injection of papain.

In effect, Dr. Kellner and his associates closed out their interest in the phenomenon of the reversible collapse of rabbits' ears following intravenous injection of papain by using it as an assay test for the potency and amount of papain to be injected. "Every laboratory technician we've had since 1951," he told us in the interview, "has known about these floppy ears because we've used them to assay papain, to tell us if it's potent and how potent." If the injected rabbit died from the dose of papain he received, the researchers knew that the papain injection was too potent; if there was no change in the rabbit's ears, the papain was not potent enough, but "if the rabbit lived and his ears drooped, it was just right." Although "we knew all about it, and used it that way . . . as a rule of thumb," Dr. Kellner commented, "I didn't write it up." Nor did he ever have "any intention of publishing it as a method of assaying papain." He knew that an applied technological discovery of this sort would not be suitable for publication in the basic science–oriented professional journals to which he and his colleagues submit reports of experimental work.

However, two factors apparently were much more important in leading Dr. Kellner away from investigating this phenomenon. First, like Dr. Thomas, Dr. Kellner thought of cartilage as relatively inert tissue.

404; and Aaron Kellner, Theodore Robertson, and Howard O. Mott, "Blood Coagulation Defect Induced in Rabbits by Papain Solutions Injected Intravenously," abstract in *Federation Proceedings*, Vol. X (1951), No. 1.

Second, because of his pre-established special research interests, Dr. Kellner's attention was predominantly trained on muscle tissue:

> Since I was primarily interested in research questions having to do with the muscles of the heart, I was thinking in terms of muscle. That blinded me, so that changes in the cartilage didn't occur to me as a possibility. I was looking for muscles in the sections, and I never dreamed it was cartilage.

Like Dr. Thomas at the beginning of his research and like all scientists at some stages in their research, Dr. Kellner was "misled" by his preconceptions.

However, as we already know, in keeping with his special research interests, Dr. Kellner noticed and intensively followed up two other serendipitous results that occur when papain is injected intravenously into rabbits: focal necrosis of cardiac and skeletal muscle and a blood coagulation defect, which in certain respects resembles that of hemophilia.[9]

It was the selective necrosis of cardiac and skeletal music that Dr. Kellner studied with the greatest degree of seriousness and interest. Dr. Kellner told us that he is "particularly interested in cardio-vascular disease," and so the lesions in the myocardium was the chance observation that he particularly "chose to follow . . . the one closest to me." Not only did Dr. Kellner himself have a special interest in the necrosis of cardiac muscle, but also his "laboratory and the people associated with me," he said, provided "the physical and intellectual tools to cope with this phenomenon." Dr. Kellner and his colleagues also did a certain amount of "work tracking down the cause of the blood coagulation defect"; but, because this line of inquiry "led [them] far afield" from investigative work in which they were especially interested and competent, they eventually "let that go" as they had let go the phenomenon of floppiness in rabbits' ears. Dr. Kellner indicated in his interview that the potential usefulness of his work with the selective necrosis of cardiac and skeletal muscle cannot yet be precisely ascertained. However, in his article in the *Journal of Experimental Medicine* he suggested that this serendipitous finding "has interesting implications for the pathogenesis of the morphological changes in rheumatic fever, periarteritis nodosa, and other hypersensitivity states."

Thus Dr. Kellner did not have the experience of serendipity gained with respect to the significance of floppiness in rabbits' ears after intravenous injection of papain for a variety of reasons, some trivial apparently, others important. The most important reasons, it seems, were his research preconceptions and the occurrence of other serendipitous phenomena in the same experimental situation.

9. See Kellner and Robertson, *op. cit.*, and Kellner, Robertson, and Mott, *op. cit.*

In summary, although the ultimate outcome of their respective laboratory encounters with floppiness in rabbits' ears was quite different, there are some interesting similarities between the serendipity-gained experience of Dr. Thomas and the serendipity-lost experience of Dr. Kellner. Initially, the attention of both men was caught by the striking uniformity with which the collapse of rabbit ears occurred after intravenous papain and by the "bizarre," entertaining qualities of this cosmetic effect. In their subsequent investigations of this phenomenon, both were to some extent misled by certain of their interests and preconceptions. Lack of progress in accounting for ear collapse, combined with success in other research in which they were engaged at the time, eventually led both Dr. Thomas and Dr. Kellner to discontinue their work with the floppy-eared rabbits.

However, there were also some significant differences in the two experiences. Dr. Thomas seems to have been more impressed with the regularity of this particular phenomenon than Dr. Kellner and somewhat less amused by it. Unlike Dr. Kellner, Dr. Thomas never lost interest in the floppy-eared rabbits. When he came upon this reaction again at a time when he was "blocked" on other research, he began actively to reconsider the problem of what might have caused it. Eventual success was more likely to result from this continuing concern on Dr. Thomas' part. And Dr. Kellner, of course, was drawn off in other research directions by seeing other serendipitous phenomena in the same situation and by his success in following up those other leads.

These differences between Dr. Thomas and Dr. Kellner seem to account at least in part for the serendipity-gained outcome of the case of the floppy-eared rabbits for the one, and the serendipity-lost outcome for the other.

Experiences with both serendipity gained and serendipity lost are probably frequent occurrences for many scientific researchers. For, as Dr. Kellner pointed out in our interview with him, scientific investigations often entail "doing something that no one has done before, [so] you don't always know how to do it or exactly what to do":

> Should you boil or freeze, filter or centrifuge? These are the kinds of crossroads you come to all the time. . . . It's always possible to do four, five, or six things, and you have to choose between them. . . . How do you decide?

In this comparative study of one instance of serendipity gained and serendipity lost, we have tried to make inferences about some of the factors that led one investigator down the path to a successful and potentially important discovery and another to follow a somewhat different, though eventually perhaps a no less fruitful, trail of research. A large enough series of such case studies could suggest how often and in what ways these factors (and others that might prove relevant) influence the paths that open up to investigators in the course of their research, the choices they made between them, and the experimental findings that result from such choices. Case studies of this kind might also contribute a good deal to the detailed, systematic study of "the ways in which scientists actually . . . think, feel and act," which Robert K. Merton says could perhaps teach us more "in a comparatively few years, about the psychology and sociology of science than in all the years that have gone before."[10]

10. See his Foreword to *Science and the Social Order* by Bernard Barber (New York: The Free Press of Glencoe, 1952), p. xxii. [Paperback edition, Collier Books, p. 19.]

32 *Resistance by Scientists to Scientific Discovery*

BERNARD BARBER

IN THE STUDY of the history and sociology of science, there has been a relative lack of attention to one of the interesting aspects of the social process of discovery—the resistance on the part of scientists themselves to scientific discovery. General and specialized histories of science and biographies and autobiographies of scientists, as well as intensive discussions of the processes by which discoveries are made and accepted, all tend to make, at the most, passing reference to this subject. In two systematic analyses of the social process of scientific discovery and invention, for example—analyses which tried to be as inclusive of empirical fact and theoretical problem as possible—there is only passing reference to such resistance in the one instance and none at all in the second.[1] This neglect is all the more notable in view of the close scrutiny that scholars have given the subject of resistance to scientific discovery by social groups other than scientists. There has been a great deal of attention paid to resistance on the part of eco-

---

Reprinted by permission from Science, *vol.* 134, *no.* 3479 (*Sept.* 1, 1961), *pp.* 596–602.

1. S. C. Gilfillan, *The Sociology of Invention* (Follet, Chicago, 1935); B. Barber, *Science and the Social Order* (Free Press, New York, 1952), chap. 9. [Paperback edition, Collier Books, 1962.]

2. P. G. Frank, in *The Validation of Scientific Theories*, P. G. Frank, Ed. (Beacon Press, Boston, 1957); J. Rossman, *The Psychology of the Inventor* (Inventors Publishing Co., Washington, D.C., 1931), chap. 11; R. H. Shryock, *The Development of Modern Medicine* (Univ. of Pennsylvania Press, Philadelphia, 1936), chap. 3; B. J. Stern, in *Technological Trends and National Policy* (Government Printing Office, Washington, D.C., 1937); V. H. Whitney, *Am. J. Sociol.* 56, 247 (1950); J. Stamp, *The Science of Social Adjustment* (Macmillan, London, 1937), pp. 34 ff.; A. C. Ivy, *Science* 108, 1 (1948).

3. T. S. Kuhn, *The Copernican Revolution* (Harvard Univ. Press, Cambridge, Mass., 1957).

nomic, technological, religious, and ideological elements and groups outside science itself.[2, 3] Indeed, the tendency of such elements to resist seems to be emphasized disproportionately as against the support which they also give to science. In the matter of religion, for example, are we not all a little too much aware that religion has resisted scientific discovery, not enough aware of the large support it has given to Western science?[4, 5]

The mere assertion that scientists themselves sometimes resist scientific discovery clashes, of course, with the stereotype of the scientist as "the open-minded man." The norm of open-mindedness is one of the strongest of the scientist's values. As Philipp Frank has recently put it, "Every influence of moral, religious, or political considerations upon the acceptance of a theory is regarded as 'illegitimate' by the so-called 'community of scientists.'" And Robert Oppenheimer emphasizes the "importance" of "the open mind," in a book by that title, as a value not only for science but for society as a whole.[6] But values alone, and especially one value by itself, cannot be a sufficient basis for explaining human behavior. However strong a value is, however large its actual influence on behavior, it usually exerts this influence only in conjunction with a number of other cultural and social elements, which sometimes reinforce it, sometimes give it limits.

This article is an investigation of the elements within science which limit the norm and practice of "open-mindedness." My purpose is to draw a more accurate picture of the actual process of scientific discovery, to see resistance by scientists themselves as a constant phenomenon with specifiable cultural and social sources. This purpose, moreover, implies a practical consequence. For if we learn more about resistance to scientific discovery, we shall know more also about the sources of acceptance, just as we know more about health when we successfully study disease. By knowing more about both resistance and acceptance in scientific discovery, we may be able to reduce the former by a little bit and thereby increase the latter in the same measure.

Helmholtz, Planck, and Lister

ALTHOUGH the resistance by scientists themselves to scientific discovery has been neglected in systematic analysis, it would be surpris-

4. A. N. Whitehead, *Science and the Modern World* (Macmillan, New York, 1947), chap. 1; R. K. Merton, *Osiris* 4, pt. 2 (1938).
5. C. C. Gillispie, *Genesis and Geology* (Harvard Univ. Press, Cambridge, Mass., 1951).
6. R. Oppenheimer, *The Open Mind* (Simon and Schuster, New York, 1955).
7. Quoted from von Helmholtz's *Vorträge und Reden* in R. H. Murray (8).
8. R. H. Murray, *Science and Scientists in the Nineteenth Century* (Sheldon, London, 1825).
9. Lord Kelvin also commented on the "resistance" to Faraday. In his article on "Heat" for the 9th edition of the *Encyclopaedia Britannica* he made a comment on

ing indeed if it had never been noted at all. If nowhere else, we should find it in the writings of those scientists who have suffered from resistance on the part of other scientists. Helmholtz, for example, made aware of such resistance by his own experience, commiserated with Faraday on "the fact that the greatest benefactors of mankind usually do not obtain a full reward during their life-time, and that new ideas need the more time for gaining general assent the more really original they are."[7-9] Max Planck is another who noticed resistance in general because he had experienced it himself, in regard to some new ideas on the second law of thermodynamics which he worked out in his doctoral dissertation submitted to the University of Munich in 1879. Ironically, one of those who resisted the ideas proposed in Planck's paper, according to his account, was Helmholtz:

> None of my professors at the University had any understanding for its contents [says Planck]. I found no interest, let alone approval, even among the very physicists who were closely connected with the topic. Helmholtz probably did not even read my paper at all. Kirchhoff expressly disapproved . . . I did not succeed in reaching Clausius. He did not answer my letters, and I did not find him at home when I tried to see him in person at Bonn. I carried on a correspondence with Carl Neumann, of Leipzig, but it remained totally fruitless.[10]

And Lister, in a graduation address to medical students, warned them all against blindness to new ideas in science, blindness such as he had encountered in advancing his theory of antisepsis.

Scientists Are Also Human

Too often, unfortunately, where resistance by scientists has been noted, it has been merely noted, merely alleged, without detailed substantiation and without attempt at explanation. Sometimes, when explanations are offered, they are notably vague and all-inclusive, thus proving too little by trying to prove too much. One such explanation is contained in the frequently repeated phrase, "After all, scientists are also human beings," a phrase implying that scientists are more human when they err than when they are right.[11] Other such vague explanations can be found in phrases such as "*Zeitgeist*," "human nature,"

the circumstance "that fifty years passed before the scientific world was converted by the experiments of Davy and Rumford to the rational conclusion as to the non-materiality of heat: 'a remarkable instance of the tremendous inefficiency of bad logic in confounding public opinion and obstructing true philosophic thought.' " [S. P. Thompson, *The Life of William Thomson, Baron Kelvin of Largs* (Macmillan, London, 1910)].

10. M. Planck, *Scientific Autobiography*, F. Gaynor, trans. (Philosophical Library, New York, 1949), p. 18.

11. See D. L. Watson, *Scientists Are Human* (Watts, London, 1938).

"lack of progressive spirit," "fear of novelty," and "climate of opinion."

As one of these phrases, "fear of novelty," may indicate, there has also been a tendency, where some explanation of the sources of resistance is offered, to express a psychologistic bias—that is, to attribute resistance exclusively to inherent and ineradicable traits or instincts of the human personality. Thus, Wilfred Trotter, in discussing the response to scientific discovery, asserts that "the mind delights in a static environment," that "change from without . . . seems in its very essence to be repulsive and an object of fear," and that "a little self-examination tells us pretty easily how deeply rooted in the mind is the fear of the new."[12] And Beveridge, in *The Art of Scientific Investigation*, says, "there is in all of us a psychological tendency to resist new ideas."[13] A full understanding of resistance will, of course, have to include the psychological dimension—the factor of individual personality. But it must also include the cultural and social dimensions—those shared and patterned idea-systems and those patterns of social interaction that also contribute to resistance. It is these cultural and social elements that I shall discuss here, but with full awareness that psychological elements are contributory causes of resistance.

Because resistance by scientists has been largely neglected as a subject for systematic investigation, we find that there is sometimes a tendency, when such resistance is noted, to exaggerate the extent to which it occurs. Thus, Murray says that the discoverer must *always* expect to meet with opposition from his fellow scientists. And Trotter goes overboard in the same way:

> . . . the reception of new ideas tends always to be grudging or hostile. . . . Apart from the happy few whose work has already great prestige or lies in fields that are being actively expanded at the moment, discoverers of new truths always find their ideas resisted.[14]

Such exaggerations can be eliminated by more systematic and objective study.

Finally, in the absence of such systematic and objective study, many of those who have noted resistance have been excessively embittered and moralistic. Oliver Heaviside is reported to have exclaimed bitterly, when his important contributions to mathematical physics were ignored for 25 years, "Even men who are not Cambridge mathematicians deserve justice."[15] And Planck's reaction to the resistance he experienced was similar.

> This experience [he said] gave me also an opportunity to learn a new fact—a remarkable one, in my opinion: A new scientific truth

12. W. Trotter, *Collected Papers* (Humphrey Milford, London, 1941).

13. W. I. B. Beveridge, *The Art of Scientific Investigation* (Random House, New York, rev. ed., 1959).

14. Trotter, *op. cit.*, p. 26.

> does not triumph by convincing its opponents and making them see the light, but rather because its opponents eventually die, and a new generation grows up that is familiar with it.[16]

Such bitterness is not tempered by objective understanding of resistance as a constant phenomenon in science, a pattern in which all scientists may sometimes and perhaps often participate, now on the side of the resisters, now on that of the resisted. Instead, such bitterness takes the moralistic view that resistance is due to "human vanities," to "little minds and ignorable minds." Such views impede the objective analysis that is required.

In his discussion of the Idols—idols of the tribe, of the cave, of the market-place, and of the theatre—Francis Bacon long ago suggested that a variety of preconceived ideas, general and particular, affect the thinking of all men, especially in the face of innovation. Similarly, more recent sociological theory has shown that while the variety of idea-systems that make up a given culture are functionally necessary, on the whole, for man to carry on his life in society and in the natural environment, these several idea-systems may also have their dysfunctional or negative effects. Just because the established culture defines the situation for man, usually helpfully, it also, sometimes harmfully, blinds him to other ways of conceiving that situation. Cultural blinders are one of the constant sources of resistance to innovations of all kinds. And scientists, for all the methods they have invented to strip away their distorting idols, or cultural blinders, and for all the training they receive in evading the negative effects of such blinders, are still as other men, though surely in considerably lesser measure because of these methods and this special training. Scientists suffer, along with the rest of us, from the ironies that evil sometimes comes from good, that one noble vision may exclude another, and that good scientific ideas occasionally obstruct the introduction of better ones.

Substantive Concepts

SEVERAL different kinds of cultural resistance to discovery may be distinguished. We may turn first to the way in which the substantive concepts and theories held by scientists at any given time become a source of resistance to new ideas. And our illusations begin with the very origins of modern science. In his magisterial discussion of the Copernican revolution, Kuhn[17] tells us not only about the nonscientific opposition to the heliocentric theory but also about the resistance from the astronomer-scientists of the time. Even after the publication of *De*

15. H. Levy, *Universe of Science* (Century, New York, 1933), p. 197.
16. Planck, *op. cit.*
17. Kuhn, *op. cit.*

Revolutionibus, the belief of most astronomers in the stability of the earth was unshaken. The idea of the earth's motion was either ignored or dismissed as absurd. Even the great astronomer-observer Brahe remained a life-long opponent of Copernicanism; he was unable to break with the traditional patterns of thought about the earth's lack of motion. And his immense prestige helped to postpone the conversion of other astronomers to the new theory. Of course, religious, philosophical, and ideological conceptions were closely interwoven with substantive scientific theories in the culture of the scientists of that time, but it seems clear that the latter as well as the former played their part in the resistance to the Copernican discoveries.

Moving to the early nineteenth century, we learn that the scientists of the day resisted Thomas Young's wave theory of light because they were, as Gillispie says, faithful to a corpuscular model.[18] By the end of the century, when scientists had swung over to the wave theory, the validity of Young's earlier discovery was recognized. Substantive scientific theory was also one of the sources of resistance to Pasteur's discovery of the biological character of fermentation processes. The established theory that these processes are wholly chemical was held to by many scientists, including Liebig, for a long time,[19] The same preconceptions were also the source of the resistance to Lister's germ theory of disease, although in this case, as in that of Pasteur, various other factors were important.

Because it illustrates a variety of sources of scientific resistance to discovery, I shall return several times to the case of Mendel's theory of genetic inheritance. For the present, I mention it only in connection with the source of resistance under discussion, substantive scientific theories themselves. Mendelian theory, it seems clear, was resisted from the time of its announcement, in 1865, until the end of the century, because Mendel's conception of the separate inheritance of characteristics ran counter to the predominant conception of joint and total inheritance of biological characteristics.[20] It was not until botany changed its conceptions and concentrated its research on the separate inheritance of unit characteristics that Mendel's theory and Mendel himself were independently rediscovered by de Vries, a Dutchman, by Carl Correns, working in Tübingen, and by Erich Tschermak, a Viennese, all in the same year, 1900.

New conceptions about the electronic constitution of the atom were also resisted by scientists when fundamental discoveries in this field

18. C. C. Gillispie, *The Edge of Objectivity* (Princeton Univ. Press, Princeton, N.J., 1960).

19. R. Vallery-Radot, *The Life of Pasteur*, R. L. Devonshire, trans. (Garden City Publishing Co., New York, 1926), pp. 175, 215.

20. I. Krumbiegel, *Gregor Mendel und das Schicksal Seiner Entdeckung* (Wissenschaftliche Verlagsgesellschaft, Stuttgart, 1957). H. Iltis, *Life of Mendel*, E. Paul and

were being made at the end of the 19th century. The established scientific notion was that of the absolute physical irreducibility of the atom. When Arrhenius published his theory of electrolytic dissociation, his ideas met with resistance for a time, though eventually, thanks in part to Ostwald, the theory was accepted and Arrhenius was given the Nobel prize for it.[21] Similarly, Lord Kelvin regarded the announcement of Röntgen's discovery of x-rays as a hoax, and as late as 1907 he was still resisting the discovery, by Ramsay and Soddy, that helium could be produced from radium, and resisting Rutherford's theory of the electronic composition of the atom, one of the fundamental discoveries of modern physics. Throughout his long and distinguished life in science Kelvin never discarded the concept that the atom is an indivisible unit.[22]

Let us take one final illustration, from contemporary science. In a recent case history of the role of chance in scientific discovery it was reported that two able scientists, who observed, independently and by chance, the phenomenon of floppiness in rabbits' ears after the injection of the enzyme papain, both missed making a discovery because they shared the established scientific view that cartilage is a relatively inert and uninteresting type of tissue.[23] Eventually one of the scientists did go on to make a discovery which altered the established view of cartilage, but for a long time even he had been blinded by his scientific preconceptions. This case is especially interesting because it shows how resistance occurs not only between two or more scientists but also within an individual scientist. Because of their substantive conceptions and theories, scientists sometimes miss discoveries that are literally right before their eyes.

Methodological Conceptions

THE methodological conceptions scientists entertain at any given time constitute a second cultural source of resistance to scientific discovery and are as important as substantive ideas in determining response to innovations. Some scientists, for example, tend to be antitheoretical, resisting, on that methodological ground, certain discoveries. "In Baconian science," says Gillispie, "the bird-watcher comes into his own while genius, ever theorizing in far places, is suspect. And this is why Bacon would have none of Kepler or Copernicus or Gilbert or anyone who would extend a few ideas or calculations into a system of the

C. Paul, trans. (W. W. Norton, New York, 1932).

21. J. J. Thomson, *Recollections and Reflections* (Bell, London, 1936), p. 390.

22. S. P. Thompson, *The Life of William Thomson: Baron Kelvin of Largs* (Macmillan, London, 1910).

23. B. Barber and R. C. Fox, *Am. J. Sociol.* 64, 128 (1958). [Reprinted in this book.]

world."[24] Goethe too, as Helmholtz pointed out in his discussion of Goethe's scientific researches, was antitheoretical.[25] A more recent discussion of Goethe's scientific work also finds him antianalytical and antiabstract.[26] Perhaps Helmholtz had been made aware of Goethe's antitheoretical bias because his own discovery of the conservation of energy had been resisted as being too theoretical, not sufficiently experimental. German physicists were probably antitheoretical in Helmholtz's day because they feared a revival of the speculations of the Hegelian "nature-philosophy" against which they had fought so long, and eventually successfully.

Viewed in another way, Goethe's antitheoretical bias took the form of a positive preference for scientific work based on intuition and the direct evidence of the senses. "We must look upon his theory of colour as a forlorn hope," says Helmholtz, "as a desperate attempt to rescue from the attacks of science the belief in the direct truth of our sensations."[27] Goethe felt passionately that Newton was wrong in analyzing color into its quantitative components by means of prisms and theories. Color, for him, was a qualitative essence projected onto the physical world by the innate biological character and functioning of the human being.

Later scientists also have resisted discovery because of their preference for the evidence of the senses. Otto Hahn, noted for his discoveries in radioactivity, who received the Nobel prize for his splitting of the uranium atom in 1939, reports the following case:

> Emil Fischer was also one of those who found it difficult to grasp the fact that it is also possible by radioactive methods of measurement to detect, and to recognize from their chemical properties, substances in quantities quite beyond the world of the weighable; as is the case, for example, with the active deposits of radium, thorium, and actinium. At my inaugural lecture in the spring of 1907, Fischer declared that somehow he could not believe those things. For certain substances the most delicate test was afforded by the sense of smell and no more delicate test could be found than that.[28]

Another methodological source of resistance is the tendency of scientists to think in terms of established models, indeed to reject propositions just because they cannot be put in the form of some model. This

24. Gillispie, *The Edge of Objectivity.*
25. H. von Helmholtz, *Popular Scientific Lectures* (Appleton, New York, 1873).
26. Gillispie, *The Edge of Objectivity.*
27. Helmholtz, *op. cit.*
28. O. Hahn, *New Atoms, Progress and Some Memories* (Elsevier, New York, 1950), pp. 154–155.

seems to have been a reason for resistance to discoveries in the theory of electromagnetism during the 19th century. Ampère's theory of magnetic currents, for example, was resisted by Joseph Henry and others because they did not see how it could be fitted into the Newtonian mechanical model.[29] They refused to accept Ampère's view that the atoms of the Newtonian model had electrical properties which caused magnetic phenomena. And Lord Kelvin's resistance to Clerk Maxwell's electromagnetic theory of light was due, says Kelvin's biographer,[30] to the fact that Kelvin found himself unable to translate into a dynamical model the abstract equations of Maxwell's theory. Kelvin himself, in the lectures he had given in Baltimore in 1884, had said, "I never satisfy myself until I can make a mechanical model of a thing. If I can make a mechanical model I can understand it. As long as I cannot make a mechanical model all the way through I cannot understand; and that is why I cannot get the electromagnetic theory."[31] Thus, models, while usually extremely helpful in science, can also be a source of blindness.

Scientists' positions on the usefulness of mathematics is a last methodological source of resistance to discovery. Some scientists are excessively partial to mathematics, others excessively hostile. Thus, when Faraday made his experimental discoveries on electromagnetism, Gillispie tells us, few mathematical physicists gave them any serious attention. The discoveries were regarded with indulgence or a touch of scorn as another example of the mathematical incapacity of the British, their barbarous emphasis on experiment, and their theoretical immaturity.[32] Clerk Maxwell, however, resolved that he "would be Faraday's mathematicus"—that is, put Faraday's experimental discoveries into more mathematical, general, and theoretical a form. Initial resistance was thus overcome. Long ago Augustus De Morgan commented on the antimathematical prejudice of English astronomers of his time. In 1845, he pointed out, the Englishman Adams had, on the basis of mathematical calculations, communicated his discovery of the planet Neptune to his English colleagues. Because they distrusted mathematics, his discovery was not published, and eight months later the Frenchman Leverrier announced and published his simultaneous discovery of the planet, once again on the basis of mathematical calculations. Because the French admired mathematics, Leverrier's discovery was published first, and thus he gained a kind of priority over Adams.[33]

Mendel was another scientist whose ideas were resisted because of

29. T. Coulson, *Joseph Henry: His Life and Work* (Princeton Univ. Press, Princeton, N.J., 1950), p. 36.

30. Thompson, *op. cit.*

31. *Ibid.*

32. Gillispie, *The Edge of Objectivity*.

33. S. E. De Morgan, *Memoir of Augustus De Morgan* (Longmans, Green, London, 1882).

the antimathematical preconceptions of the botany of his time. "It must be admitted, however," says his biographer, Iltis,

> . . . that the attention of most of the hearers [when he read his classic monograph, "Experiments in Plant-Hybridization," before the Brünn Society for the Study of Natural Science in 1865] was inclined to wander when the lecturer was engaged in rather difficult mathematical deductions; and probably not a soul among them really understood what Mendel was driving at. . . . Many of Mendel's auditors must have been repelled by the strange linking of botany with mathematics, which may have reminded some of the less expert among them of the mystical numbers of the Pythagoreans. . . .[34]

Note that the alleged "difficult mathematical deductions" are what we should now consider very simple statistics. And it was not just the audience in Brünn that had no interest in or knowledge of mathematics. Mendel's other biographer Krumbiegel, tells us that even the more sophisticated group of scientists at the Vienna Zoological-Botanical Society would have given Mendel's theory as poor a reception, and for the same reasons.

In some quarters the antimathematical prejudice persisted in biology for a long time after Mendel's discovery, indeed until after he had been rediscovered. In his biography of Galton, Karl Pearson reports that he sent a paper to the Royal Society in October 1900, eventually published in November 1901, containing statistics in application to a biological problem.[35] Before the paper was published, he says, "a resolution of the Council [of the Royal Society] was conveyed to me, requesting that in future papers mathematics should be kept apart from biological applications." As a result of this, Pearson wrote to Galton, "I want to ask your opinion about resigning my fellowship of the Royal Society." Galton advised against resigning, but he did help Pearson to found the journal *Biometrika*, so that there would be a place in which mathematics in biology would be explicitly encouraged. Galton wrote an article for the first issue of the new journal, explaining the need for this new agency of "mutual encouragement and support" for mathematics in biology and saying that "a new science cannot depend on a welcome from the followers of the older ones, and [therefore] . . . it is advisable to establish a special Journal for Biometry."[36] It seems strange to us

34. Iltis, *op. cit.*

35. K. Pearson, *The Life, Letters and Labours of Francis Galton* (Cambridge Univ. Press, Cambridge, England, 1924), vol. 3, pp. 100, 282–283.

36. *Biometrika* 1, 7 (1901–02).

37. Gillispie, *The Edge of Objectivity*.

38. Gillispie, *Genesis and Geology*.

now that prejudice against mathematics should have been a source of resistance to innovation in biology only 60 years ago.

Religious Ideas

ALTHOUGH we have heard more of the way in which religious forces outside science have hindered its progress, the religious ideas of scientists themselves constitute, after substantive and methodological conceptions, a third cultural source of resistance to scientific innovation. Such internal resistance goes back to the beginning of modern science. We have seen that the astronomer colleagues of Copernicus resisted his ideas in part because of their religious beliefs, and we know that Leibniz, for example, criticized Newton "for failing to make providential destiny part of physics."[37] Scientists themselves felt that science should justify God and His world. Gradually, of course, physics and religion were accommodated one to the other, certainly among scientists themselves. But all during the first half of the nineteenth century resistance to discovery in geology persisted among scientists for religious reasons. The difficulty, as Gillispie has put it on the basis of his classic analysis of geology during this period, "appears to be one of religion (in a crude sense) *in* science rather than one of religion *versus* scientists." The most embarrassing obstacles faced by the new sciences were cast up by the curious providential materialism of the scientists themselves.[38] When, in the 1840's, Robert Chambers published his *Vestiges of Creation,* declaring a developmental view of the universe, the theory of development was so at variance with the religious views which all scientists accepted that "they all spoke out: Herschel, Whewell, Forbes, Owen, Prichard, Huxley, Lyell, Sedgwick, Murchison, Buckland, Agassiz, Miller, and others."[39]

Religious resistance continued and was manifested against Darwin, of course, although many of the scientists who had resisted earlier versions of developmentalism accepted Darwin's evolutionary theory, Huxley being not the least among them. In England, Richard Owen offered the greatest resistance on scientific grounds, while in America and, in fact, internationally, Louis Agassiz was the leading critic of Darwinism on religious grounds.[40]

In more recent times, biology, like physics before it, has been successfully accommodated to religious ideas, and religious convictions are no longer a source of resistance to innovation in these fields. Re-

39. *Ibid.,* p. 133. That scientists were religious also, and in the same way, in America can be seen in A. H. Dupree, *Asa Gray* (Harvard Univ. Press, Cambridge, Mass., 1959).

40. Gillispie, *Genesis and Geology;* Dupree, *ibid;* and E. Lurie, *Louis Agassiz: A Life in Science* (Univ. of Chicago Press, Chicago, 1960).

sistance to discoveries in the psychological and social sciences that stems from religious convictions is perhaps another story, but one that does not concern us here.

In addition to shared idea-systems, the patterns of social interaction among scientists also become sources of resistance to discovery. Here again we are dealing with elements that, on the whole, probably serve to advance science but that occasionally produce negative, or dysfunctional, effects.

Professional Standing

THE first of these social sources of resistance is the relative professional standing of the discoverer. In general, higher professional standing in science is achieved by the more competent, those who have demonstrated their capacity for being creative in their own right and for judging the discoveries of others. But sometimes, when discoveries are made by scientists of lower standing, they are resisted by scientists of higher standing partly because of the authority the higher position provides. Huxley commented on this social source of resistance in a letter he wrote in 1852:

> For instance, I know that the paper I have just sent in is very original and of some importance, and I am equally sure that if it is referred to the judgment of my "particular" friend that it will not be published. He won't be able to say a word against it, but he will pooh-pooh it to a dead certainty. You will ask with wonderment, Why? Because for the last twenty years [. . .] has been regarded as the great authority in these matters, and has had no one tread on his heels, until, at last, I think, he has come to look upon the Natural World as his special preserve, and "no poachers allowed." So I must manoeuvre a little to get my poor memoir kept out of his hands.[41]

Niels Henrik Abel, early in the nineteenth century, made important discoveries on a classical mathematical problem, equations of the fifth degree.[42] Not only was Abel himself unknown but there was no one of any considerable professional standing in his own country, Norway (then part of Denmark), to sponsor his work. He sent his paper to various foreign mathematicians, the great Gauss among them. But Gauss merely filed the leaflet away unread, and it was found uncut after his death, among his papers. Ohm was another whose work, in this case experimental, was ignored partly because he was of low professional standing. The researches of an obscure teacher of mathematics at the

41. Murray, *op. cit.*, p. 367.
42. O. Ore, *Niels Henrik Abel: Mathematician Extraordinary* (Univ. of Minnesota Press, Minneapolis, 1957).

Jesuit Gymnasium in Cologne made little impression upon the more noted scientists of the German universities.

Perhaps the classical instance of low professional standing helping to create resistance to a scientist's discoveries is that of Mendel. The notion that Mendel was "obscure," in the sense that his work did not come to the attention of competent and noted professionals in his field, can no longer be accepted. First of all, the proceedings volume of the Brünn society in which his monograph was printed was exchanged with proceedings volumes of more than 120 other societies, universities, and academies at home and abroad. Copies of his monograph went to Vienna and Berlin, to London and Petersburg, to Rome and Upsala.[43] In London, according to Bateson, the monograph was received by the Royal Society and the Linnaean Society.[44] Moreover, we know from the extensive correspondence between them—correspondence which was later published by Mendel's rediscoverer, Correns—that Mendel sent his paper to one of the distinguished botanists of his time, Carl von Nägeli of Munich. Von Nägeli resisted Mendel's theories for a number of reasons: because his own substantive theories about inheritance were different and because he was unsympathetic to Mendel's use of mathematics, but also because he looked down, from his position of authority, upon the unimportant monk from Brünn. Mendel had written deferentially to von Nägeli, in letters that amounted to small monographs. In these letters, Mendel addressed von Nägeli most respectfully, as an acknowledged master of the subject in which they were both interested. But von Nägeli was the victim of his own position as a scientific pundit. Mendel seemed to him a mere amateur expressing fantastic notions, or at least notions contrary to his own. Von Nägeli's letters to Mendel seem unduly critical to present readers, more than a little supercilious. Nevertheless, the modest Mendel was delighted that the great man had even deigned to reply and sent cordial thanks for the gift of von Nägeli's monograph. On both sides, von Nägeli was defined as the great authority, Mendel as the inferior asking for consideration his position did not warrant. Ironically, Mendel took von Nägeli's advice, to change from experiments on peas to work on hawkweed, a plant not at all suitable at that time for the study of inheritance of separate characteristics. The result was that Mendel labored in a blind alley for the rest of his scientific life.

Nor was von Nägeli unique. Others, such as W. O. Focke, Hermann Hoffman, and Kerner von Marilaun, also dismissed Mendel's work because he seemed "an insignificant provincial" to them. Focke did list Mendel's monograph in his own treatise, *Die Pflanzenmischlinge,*

43. Iltis, *op. cit.*
44. R. A. Fisher, *Ann. Sci.* 1, 116 (1933).

but only for the sake of completeness. Focke paid much more attention to those botanists who had produced quantitatively large and apparently more important contributions—men such as Kölreuter, Gärtner, Wichura, and Wiegmann, of higher professional standing.[45] Certainly, in this case, quantity of publication was inadequate as a measure of professional worth. Focke's listing of Mendel served only to bring his work, directly and indirectly, to the attention of Correns, de Vries, and von Tschermak after they had independently rediscovered the Mendelian principle of inheritance.

Mendel met with resistance from the authorities in his field after his discovery was published. But sometimes men of higher professional standing sit in judgment on lesser figures *before* publication and prevent a discovery's getting into print. This can be illustrated by an incident in the life of Lord Rayleigh. For the British Association meeting at Birmingham in 1886, Rayleigh submitted a paper under the title, "An Experiment to show that a Divided Electric Current may be greater in both Branches than in the Mains." "His name," says his son and biographer,

> . . . was either omitted or accidentally detached, and the Committee "turned it down" as the work of one of those curious persons called paradoxers. However, when the authorship was discovered, the paper was found to have merits after all. It would seem that even in the late nineteenth century, and in spite of all that had been written by the apostles of free discussion, authority could prevail when argument had failed![46]

So says the fourth Baron Rayleigh, and we may wonder whether his remark does not still apply, some 75 years later.

Professional Specialization

ANOTHER social source of resistance is the pattern of specialization that prevails in science at any given time. On the whole, of course, as with any social or other type of system, such specialization is efficient for internal and environmental purposes. Specialization concentrates and focuses the requisite knowledge and skill where they are needed. But occasionally the negative aspect of specialization shows itself, and innovative "outsiders" to a field of specialization are resisted by the "insiders." Thus, when Helmholtz announced his theory of the conservation of energy, it met with resistance partly because he was

45. H. F. Roberts, *Plant Hybridization before Mendel* (Princeton Univ. Press, Princeton, N.J., 1929), pp. 210–211.

46. R. J. Strutt, *John William Strutt, Third Baron Rayleigh* (Arnold, London, 1924), p. 228.

47. Murray, *op. cit.*, p. 97.

not a specialist in what we now think of as physics. Referring in the later years of his life to the opposition of the acknowledged experts, Helmholtz said he met with such a remark as this from some of the older men: "This has already been well known to us; what does this young medical man imagine when he thinks it necessary to explain so minutely all this to us?"[47] To be sure, on the other side, medical specialists have a long history of resisting scientific innovations from what they define as "the outside." Pasteur met with violent resistance from the medical men of his time when he advanced his germ theory. He regretted that he was not a medical specialist, for the medical men thought of him as a mere chemist poaching on their scientific preserves, not worthy of their attention. In France, even before Pasteur, Magendie had met with resistance for attempting to introduce chemistry into medicine.[48] If medicine now listens more respectfully to nonmedical science and its discoveries, it is partly because many nonmedical scientists have themselves become experts in a variety of medical-science specialties and so are no longer "outsiders."

Societies, "Schools," and Seniority

SCIENTIFIC organizations, as we may safely infer from their large number and their historical persistence, serve a variety of useful purposes for their members. And of course scientific publications are indispensable for communication in science. But occasionally, when organizations or publications are incompetently staffed and run, they may serve as another social source of resistance to innovation in science. There have been no scholarly investigations into the true history of our scientific organizations and publications, but something is known and points in the direction I have suggested. In the early nineteenth century, for example, even the Royal Society fell on bad days. Lyons tells us that a contemporary, Granville, "severely criticized the shortcomings of the Society" during that period.[49] Granville gave numerous instances in which the selection or rejection of papers by the Committee of Papers was the result of bad judgment. Sometimes the paper had not been read by any Fellow who was an authority on the subject with which it dealt. In other cases, none of the members of the committee who made the judgment could have had any expert opinion in the matter. It was such an incompetent committee, for example, that resisted Waterston's new molecular theory of gases when he submitted

48. J. M. D. Olmstead, *François Magendie, Pioneer in Experimental Physiology and Scientific Medicine in the 19th Century* (Schuman, New York, 1944), pp. 173–175.

49. H. Lyons, *The Royal Society, 1661–1940* (Cambridge Univ. Press, Cambridge, England, 1944), p. 254.

a paper making this contribution. The referee of the Royal Society who rejected the paper wrote on it, "The paper is nothing but nonsense." As a result, Waterston's work lay in utter oblivion until rescued by Rayleigh some 45 years later.[50] Many present-day misjudgments of this kind probably occur, although the multiplicity of publication outlets now provides more than one chance for a significant paper ignored by the incompetent to appear in print.

The rivalries of what are called "schools" are frequently alleged to be another social source of resistance in science. Huxley, for example, is reported to have said, two years before his death, "'Authorities,' 'disciples,' and 'schools' are the curse of science; and do more to interfere with the work of the scientific spirit than all its enemies."[51] Murray suggests that the supposed warfare between science and theology is equaled only by the warfare among rival schools in each of the scientific specialties. Unfortunately, just what the term *school* means is usually left unclear, and no empirical evidence of anything but the most meager and unsystematic character is ever offered by way of illustration.[52] No doubt some harmful resistance to discovery, as well as some useful competition, comes out of the rivalry of "schools" in science, but until the concept itself is clarified, with definite indicators specified, and until research is carried out on this more adequate basis, we can only feel that "there is something there" that deserves a scholarly treatment it has not yet received.

That the older resist the younger in science is another pattern that has often been noted by scientists themselves and by those who study science as a social phenomenon. "I do not," said Lavoisier in the closing sentences of his memoir *Reflections on Phlogiston* (read before the Academy of Sciences in 1785),

> . . . expect my ideas to be adopted all at once. The human mind gets creased into a way of seeing things. Those who have envisaged nature according to a certain point of view during much of their career, rise only with difficulty to new ideas. It is the passage of time, therefore, which must confirm or destroy the opinions I have presented. Meanwhile, I observe with great satisfaction that the young people are beginning to study the science without prejudice. . . .[53]

Or again, Hans Zinsser remarks in his autobiography,

> That academies and learned societies—commonly dominated by the older foofoos of any profession—are slow to react to new

50. Trotter, *op. cit.*, p. 26.

51. C. Bibby, *T. H. Huxley: Scientist, Humanist, and Educator* (Horizon, New York, 1959), p. 18.

52. For the best available sociological essay, see F. Znaniecki, *The Social Role of*

ideas is in the nature of things. For, as Bacon says, *scientia inflat,* and the dignitaries who hold high honors for past accomplishment do not usually like to see the current of progress rush too rapidly out of their reach.[54]

Now of course the older workers in science do not always resist the younger in their innovations, nor can it be physical aging in itself that is the source of such resistance as does occur. If we scrutinize carefully the two comments I have just quoted and examine other, similar ones with equal care, we can see that *aging* is an omnibus term which actually covers a variety of cultural and social sources of resistance. Indeed, we may put it this way, that as scientists get older they are more likely to be subject to one or another of the several cultural and social sources of resistance I have analyzed here. As a scientist gets older he is more likely to be restricted in his response to innovation by his substantive and methodological preconceptions and by his other cultural accumulations; he is more likely to have high professional standing, to have specialized interests, to be a member or official of an established organization, and to be associated with a "school." The likelihood of all these things increases with the passage of time, and so the older scientist, just by living longer, is more likely to acquire a cultural and social incubus. But this is not always so, and the older workers in science are often the most ardent champions of innovation.

After this long recital of the cultural and social sources of resistance, by scientists, to scientific discovery, I need to emphasize a point I have already made. That some resistance occurs, that it has specifiable sources in culture and social interaction, that it may be in some measure inevitable, is not proof either that there is more resistance than acceptance in science or that scientists are no more open-minded than other men. On the contrary, the powerful norm of open-mindedness in science, the objective tests by which concepts and theories often can be validated, and the social mechanisms for ensuring competition among ideas new and old—all these make up a social system in which objectivity is greater than it is in other social areas, resistance less. The development of modern science demonstrates this ever so clearly. Nevertheless, some resistance remains, and it is this we seek to understand and thus perhaps to reduce. If "the edge of objectivity" in science, as Charles Gillispie has recently pointed out, requires us to take physical and biological nature as it is, without projecting our wishes upon it, so also we have to take man's social nature, or his behavior in society, as it is. As men in society, scientists

the Man of Knowledge (Columbia Univ. Press, New York, 1940), chap. 3.

53. Gillispie, *The Edge of Objectivity*.

54. H. Zinsser, *As I Remember Him: The Biography of* R. S. Little, Brown, Boston, 1940), p. 105.

are sometimes the agents, sometimes the objects, of resistance to their own discoveries.[55]

55. For invaluable aid in the preparation of this article I am indebted to Dr. Elinor G. Barber. The Council for Atomic Age Studies of Columbia University assisted with a grant for typing expenses.

33 *Which Scientists Win Nobel Prizes?*

GEORGE W. GRAY

HALF THE TIME AND TEMPER spent in arguing the question of America's so-called "lag" in science might be spared if we had a reliable yardstick of excellence. Actually, one useful, if limited, measure of comparative national prestige exists in the sixty-year record of the Nobel Prize awards in science. To the extent that the nation shares the glory of its individual stars, the United States actually is not doing badly. In 1960, in fact, this country went over the top in the accumulated total of science prizes awarded to various nations. Another great year for the U.S. was 1946, when we took all the science awards and the Peace Prize as well. But we came late to this achievement.

Thousands of scientists have engaged in physical, chemical, and biological research over the past sixty years, but only 215 of them won recognition from the Nobel Prize juries in Stockholm. Many were little known to the public at the time. The German physician Werner Forssmann, who received an award in 1956, said he felt like a village priest suddenly raised to the cardinalate. C. J. Davisson, a staff physicist at the Bell Telephone Laboratories, was astonished to find himself "transformed overnight from an exceedingly private person to something in the nature of a semipublic institution." At airports, convention halls, and other public places he was embarrassingly aware of being pointed out as "that Nobel Prize-winner," and everywhere he was met by an attitude of deference. Some of the chosen were already known—the Curies, Rutherford, Einstein—but they too gained the added aura that impressed Forssmann and Davisson.

This power to impart prestige seems to be peculiarly a property of the three awards in science. I know of no instance in which receipt of the Nobel Prize for Literature transformed a little-known writer to "something in the nature of a semipublic institution"—although some of the choices have been exceedingly minor authors. Years ago the Oxford classicist Gilbert Murray, weighing the Nobel-winners in literature against those in science, noted that "this age is clearly better at science than at poetry." Regarding the Peace Prize, what impressed Murray most was "the number of times in which it could not be awarded at all." Last December it was not awarded—the seventeenth year when no one meriting the Prize for Peace was found—but the Prizes for Physics, for Chemistry, and for Physiology and Medicine were given, and two of them went to Americans. This event marked the arrival, at last, of American scientists at the top in terms of absolute winnings.

Alfred Nobel's will became operative in 1900 and the first awards were made in 1901. The Physics Prize of that year went to a German (Röntgen), the Chemistry Prize to a Dutchman (van't Hoff), the Physiology and Medicine Prize to another German (von Behring), the Literature Prize to a Frenchman (Sully-Prudhomme), and the Peace Prize was divided between a Swiss (Dunant) and a Frenchman (Passy). And so it went for several consecutive years. By 1906 German scientists held six Prizes, British four, French two, and Dutch two. America did win the Nobel Peace Committee's attention that year and it gave President Theodore Roosevelt the 1906 Prize for promoting the conference which ended the Japanese-Russian War. But honors had a way of seeking the colorful T. R., and to the author of "speak softly and carry a big stick" the award was just another in his collection of trophies. People took it for granted that the President of the United States would be the object of attention, and there was no great stir over his Nobel Prize for Peace.

Michelson Takes First

THERE was a stir among scientists the following year when it became known that Albert Abraham Michelson was to receive the 1907 Prize for Physics. The news came quietly, with no advance fanfare. It occupied only a paragraph in the *New York Times*, datelined Washington, November 29:

> The State Department has been advised by American Minister Graves at Stockholm that Prof. A. A. Michelson of Chicago is to be awarded the Nobel Prize for physicists. Dr. Michelson is the discoverer of a new method of determining the velocity of light.

The *Times* decided there was enough national interest in the Chicago man's success to put the tiny dispatch on the front page. Few who read it, except his colleagues in the physical sciences, realized what lay back of the award. Michelson was inventor of the interferometer, a wondrously exact instrument of optical measurement, and he had devised important improvements in the spectroscope. His most famous work was the attempt to measure the flight of the Earth through the ether of space. This was in the 1880s, when he was professor of physics at the Case School of Applied Science in Cleveland and had as a near neighbor Edwin W. Morley, professor of chemistry at Western Reserve College. The two collaborated in the investigation, which became known as the Michelson-Morley Experiment. Their finding was negative—they were not able to detect any relative motion—but the result nevertheless was epochal. For it cast doubt on the theory of the ether, was one of the results that set Einstein on his trail of relativity, and played a key part in the revolution of physics.

In 1890 Michelson transferred to Clark University and two years later to the University of Chicago, then in process of reorganization with Rockefeller millions. The conversion of this little Baptist college into the many departmented university had already provided the local newspapers with several field days of excitement, and now—in 1907—the Nobel Prize! The Chicago *Tribune* wrote a four-deck heading over its story and printed the piece prominently with a two-column photograph of Michelson. For the Midwestern metropolis and its rising young university the Nobel Prize was another triumph.

Michelson went to Stockholm for the presentation. This is usually a brilliant ritual, with the king presiding and notables of the realm attending, held on December 10. But on the ninth the nation had been thrown into mourning by the death of King Oscar. The program planned for the Stockholm Concert Hall was called off, and the conferring of awards took place almost privately in a room of the Swedish Royal Academy of Science. Germany, France, and Great Britain shared the honors with the United States—the Chemistry Prize going to Eduard Buchner of the University of Berlin, the Physiology and Medicine Prize to Charles Laveran of the University of Paris, and the Literature Prize to Rudyard Kipling. The Peace Prize, which is administered separately by a Norwegian Nobel Committee, was presented to E. T. Moretta of Italy at a ceremony held the same day in Oslo. Michelson brought home a check for 139,000 Swedish crowns (good for about $40,000 in the U.S.), a diploma citing his achievements that had won the award, and a gold medal bearing the image of Alfred Nobel.

The Stars Don't Count

THOUGH Michelson was first to win, he was not the first American scientist to be considered. There were twelve nominees for the Physics Prize in 1901, and among them was the American astrophysicist W. W. Campbell, director of the Lick Observatory. The proposal of Campbell's name at least performed the useful service of determining whether or not the physics of the stars is physics under the terms of Nobel's will. The Swedes decided "No," and ever since a firm thumbs-down has been the official attitude toward students of starlight—though George Ellery Hale, Sir Arthur Eddington, and other astrophysical worthies have been proposed over the years.

In 1902 Michelson's former neighbor of Cleveland days, his collaborator in the Michelson-Morley Experiment, was nominated for the Chemistry Prize. According to his biographer, Morley stood second in the voting and lost only to the great Emil Fischer of Berlin. Morley was an adept measurer of infinitesimal qualities and would have adorned the Nobel roll, but the nod from Stockholm was withheld in 1902 to go five years later to his younger colleague.

It was another five years before Michelson had a companion. This was Alexis Carrel of the Rockefeller Institute, awarded the Physiology and Medicine Prize in 1912. Two years later Theodore W. Richards of Harvard received the Chemistry Prize. Workers in the United States now held one science prize in each of the three categories, but meanwhile workers in Europe had increased their leads. Another nine years passed before America scored again, with the award of the 1923 Physics Prize to Robert A. Millikan. This gave the United States a total of four—but by then Germany had 19½, Great Britain 9½, France 7½, and The Netherlands 4. The pace quickened. In the next decade-and-a-half U.S. scientists' winnings were 6½ prizes—while Germans gained nearly double that. Deutschland was still *über Alles*.

The tide turned in 1939. Ernest O. Lawrence of the University of California was named for the Physics Prize, Adolf Butenandt of the Kaiser Wilhelm Institute for Biochemistry and Leonard Ruzicka of the Zurich Technical High School jointly for the Chemistry Prize, and Gerhard Domagk, another German, for the Physiology and Medicine Prize. On the face of it, Germany still seemed to be riding at the head of the procession. But Hitler, enraged by an earlier award of the Peace Prize to one of his political prisoners, ordered Butenandt and Domagk to renounce the proffered honors—and they obeyed.

Since then, scientists working in Germany institutions have won only two full prizes, one half-prize, and one third-of-a-prize. During the same period scientists in the United States have won twenty-three full prizes, one half-prize, and one third-of-a-prize. Except for 1940 to 1942, when awarding was suspended on account of the war, there has been only one year that brought no Nobel recognition to the United States. Several times Americans won two science prizes in a year, and once they were invited to Stockholm to receive all three. This was the memorable 1946. P. W. Bridgman (of Harvard) brought back the

Physics Prize, H. J. Muller (of Indiana University) the Physiology and Medicine Prize, and J. B. Sumner (of Cornell) and J. H. Northrop and W. M. Stanley (both of the Rockefeller Institute) among them the Chemistry Prize. In addition, the Norwegian Nobel Committee selected two Americans, Emily G. Balch and John R. Mott, to receive the Peace Prize.

Despite their postwar decline in winnings, the Germans had accumulated such a backlog before the war that they held place at the top of the Nobel roster right up to 1960. It was last December's awards to two professors of the University of California—the Physics Prize to Donald A. Glaser and the Chemistry Prize to Willard F. Libby—that put the United States in the lead with a total of 37⅓ prizes. The score by nations is shown in Table 1.

TABLE 1. The Score by Nations (1901–60) Nobel Prize-winnings by Scientists

	Scientists	*Prizes*
United States	61	37⅓
Germany	42	35⅚
Great Britain	36	25½
France	16	11½
Sweden	8	7½
Netherlands	8	6½
Switzerland	8	5⅚
Denmark	6	5½
Austria	6	4½
Italy	4	3
Russia	5	2½
Canada	3	2
Belgium	2	2
Czechoslovakia	1	1
Finland	1	1
Hungary	1	1
India	1	1
Japan	1	1
Argentina	1	½
Australia	1	½
Ireland	1	½
Portugal	1	½
Spain	1	½
Total	215	157

The Place at the Time

MANY of the winners are not native to the lands in which they lived at the time they received their awards. Michelson was born in

Posen (under Prussian rule) and came to the United States as a boy. Carrel, an *émigré* from France, was educated there and began work as a surgeon at the University of Lyons. Karl Landsteiner was not only born, educated, and launched upon his career in Austria, but he made his discovery of blood types at the University of Vienna. This was two decades before he came to the United States. Here, as a member of the Rockefeller Institute, he continued his study of blood groups and greatly enlarged the findings which, eight years later, won him a Nobel Prize. So, in listing the laureates, we identify Landsteiner, Carrel, and many other foreign-born and foreign-educated winners as of the United States. The criterion is *the place* in which the recipient is working *at the time* he is awarded the prize. Under this rule Britain has credit for the prize-winning of Max Born, Boris Chain, and Hans Krebs, although all were born and began their research careers in Germany; and France has credit for the prize-winning of the Polish-born Marie Curie and the Russian-born Elie Metchnikoff.

Every institution with which a laureate has had association shares to some degree in his honor. His birthplace and school, the college he attended as an undergraduate, the town he lived in—each shines by reflected glory. Last December the newspapers carried a story about Leonia, New Jersey, as having been the home town of three Nobel Prize men: Professors Urey, Fermi, and Libby lived in this suburb during their years of service with Columbia University. College publications sometimes list as their laureates former students who have won the award, faculty members to whom the award has come, and holders of the Nobel Prize who have joined their faculty since receiving the award elsewhere. This wholesale operation leads to confusion for the statistician, though there is a justification for each claim. On the whole, it seems best to apply the rule of *the place at the time* to institutions as well as to nations, although it occasionally appears to work a hardship.

Thus, Robert A. Millikan was a member of the physics department of the University of Chicago for more than a score of years. During this connection, working in Chicago laboratories with Chicago equipment, he performed the delicate oil-drop experiment with which he measured the electric charge on the electron. Later he resigned from Chicago, transferred to the California Institute of Technology, and had been in Pasadena only two years when Stockholm awarded the Physics Prize for his finding achieved at Chicago. Under the rule of *the place at the time,* Caltech got the distinction. There have been many occurrences of this kind. Indeed, the rule operated to the University of Chicago's advantage on one occasion. This was in 1927 when a share in the Physics Prize of that year was awarded to Chicago's Professor Arthur H. Compton for his discovery of the energy transformation now known as the Compton Effect, a finding he had made several years earlier during his professorship at Washington University, St. Louis.

Having both Michelson and Compton, Chicago was the first American university with two Nobel-winners on its staff. But this unique status did not last, and the palm for bringing Nobel awards to the United States passed to others. The 37⅓ prizes that have been received are distributed among twenty-four institutions, as shown in Table 2.

TABLE 2. Nobel Prizes Won by U.S. Institutions

	Scientists	*Prizes*
University of California	8	6
Harvard University	8	4⅓
California Institute of Technology	5	3¾
Columbia University	6	3⅔
Rockefeller Institute	6	3¼
University of Chicago	2	1½
Stanford University	3	1½
Cornell University	2	1½
Institute for Advanced Study	2	1½
Carnegie Institute of Technology	1	1
Washington University, St. Louis	3	1
Indiana University	1	1
General Electric Laboratories	1	1
Rockefeller Foundation	1	1
Rutgers University	1	1
Bell Telephone Laboratories	2	⅚
Mayo Clinic	2	⅔
New York University	1	½
University of Wisconsin	1	½
St. Louis University	1	½
Beckman Instruments, Inc.	1	⅓
University of Illinois	1	⅓
University of Rochester	1	⅓
Western Reserve University	1	⅓
Total	61	37⅓

Until last December Harvard headed this list, but California's gain of both the Physics and the Chemistry awards in 1960 was such a giant step that it gave the West Coast university the lead by a wide margin. Never before have two full prizes come to an institution in a single year.

It is also true that never before has there been a university so large as California, with a staff of 5,639 teachers and researchers distributed among five campuses widely spaced over the state. Analysis of California's winnings shows that five of the six prizes went to one campus (Berkeley); and of these, four were received by members of one department, the Radiation Laboratory. Except for the chemistry department on the Berkeley campus and the chemistry department on the Los Angeles campus, other divisions of the immense university have had no luck with the Nobel electors—and California has yet to win its first Prize for Physiology and Medicine.

The per-capita test is applicable also to nations, of course. If the reckoning is made on that basis, Denmark (with a population ranging from 2,450,000 in 1900 to about 4,500,000 now) rates highest with its total of 5½ Nobel Prizes in science. When winnings are calculated thus, relative to the total population of the country, the United States drops to eighth place, following this sequence:

Denmark	one prize per	632,000	people
Switzerland		754,000	
Sweden		837,000	
Netherlands		1,246,000	
Austria		1,503,000	
Germany		1,781.000	
Great Britain		1,822,000	
United States		3,415,000	

The figures above are derived from averages, whereas a strict accounting would require a calculation for each year in which a Nobel Prize was received by the country in question—but I doubt if this would change the order.

Out of Luck

AS ONE notes the winners, it is impossible not to be reminded of the losers. When the Nobel Prizes were instituted, Yale had on its faculty the most creative thinker in chemistry that America has yet produced—J. Willard Gibbs. He has been called "the Newton of chemistry." In the book *Nobel, the Man and His Prizes* (edited by the Nobel Foundation and published in 1950) there is a chapter on the chemistry awards written by Arne Westgren, then chairman of the Swedish Royal Academy's Nobel Committee on Chemistry. "Gibbs unquestionably deserved to be awarded the Nobel Prize for his work on thermodynamics," Westgren writes. "But he was never nominated. The time was clearly not yet ripe for a true evaluation of his work. This is greatly to be regretted, as the name of J. W. Gibbs as the first or possibly the second on the list of Nobel Prize-winners for Chemistry would undoubtedly have been an honorable addition."

Yale lost again in 1917. Its professor of comparative anatomy, Ross G. Harrison, was among those nominated for the Physiology and Medicine Prize. Selection for this award is the responsibility of the Caroline Medical Institute in Stockholm, and its Nobel Committee recommended Harrison because of his successful use of tissue culture in the study of embryonic development. Then, strangely, the Caroline Institute decided to make no award for 1917. To be sure, that was a war year, but the Swedish Royal Academy of Science awarded the Physics Prize, the Swedish Academy (of Letters) the Literature Prize, and the Norwegian Nobel Committee the Peace Prize. Sixteen years later Harrison again was nominated, but this time the committee voted for Thomas Hunt Morgan of Caltech.

Although the prizes are restricted to three fields, the competition yearly grows more intense. Luck undoubtedly plays a part, as it does in discovery itself. The instances in which a discoverer or inventor stumbled on his finding by chance are legion, and whether or not he wins the vote of a Nobel Prize jury depends (1) on being nominated and (2) on the judgment of the very human men who make the selections. After all, it is a staggering task that Alfred Nobel bequeathed to his countrymen: to select from the laboratories and research chambers of the world the one achievement in physics, the one in chemistry, and the one in physiology or medicine which "shall have conferred the greatest benefit on mankind." It is understandable why so often a prize has been divided and conferred jointly on two and even three winners.

Whatever the role of luck may be—and regardless of the circumstances that astronomers, geologists, mathematicians, oceanographers, and specialists in dozens of other sciences are ineligible—the Nobel Prize remains the supreme accolade. It is, a university president declared in congratulating one of his faculty on winning, "the highest scholastic honor that the world has to offer a man or woman for intellectual attainments."

In 1959 Stanford University rejoiced in the receipt of an award by Arthur Kornberg, its professor of biochemistry. Kornberg is an alumnus of the College of the City of New York, and last year his fellow alumni had a medal struck bearing his image and the inscription "First Nobel Laureate of the City College." They presented the medal to Kornberg at a ceremony in December. It's the first time, to my knowledge, that a prize has been conferred to celebrate the winning of another prize.

The Social Responsibilities of Science

6

AS science has grown in the modern world, it has increasingly and ever more powerfully had direct and indirect influences on many different parts of society. These influences, for good and ill alike, have made both scientists and nonscientists aware that they must constantly ask the question, what are the social responsibilities of science? The answer, or better, answers to this question involve an understanding of the many reciprocal relationships between science and society. Hence, all that has gone before in this reader is at least indirectly relevant to the question and its answers. But in this part of the reader the question is confronted directly in five selections, the authors of which include sociologists, a political scientist, and a philosopher. They do not say everything we need to know about the social responsibilities of science, but they give us some of the fundamentals.

The moving, at times poetic, statement by Max Weber, "Science as a Vocation," tells us what the values of science are and how science as an objective discipline is related to other social values. Parsons carries this analysis a little farther and suggests how social science is an indispensable asset in using our science and realizing our values as fully as possible.

As the American government has used the services of its scientists more and more since World War II, and as scientists have more and more volunteered for that service in areas where they knew they saw farther or more deeply than the government, some dilemmas have arisen for both parties that have not always been seen or wisely dealt with. Sayre, applying the point of view of a political scientist, clarifies these dilemmas and suggests ways of dealing with them that recognize both the interests of science and the values of democracy in appropriate measure.

As scientists have undertaken their social responsibilities more fully since World War II, one of the key problems in their relations with the government has been the conflict between their need for autonomy and the government's need for security and secrecy. Shils' essay, part of a larger book on this subject, is a valuable analysis of the sources of the conflict and of desirable patterns for its resolution. Fortunately, it would seem, considerable progress has been made in the last few years toward some of these more satisfactory patterns. The McCarthy era, which directly provoked Shils's concern and analysis, is now behind us and seems unlikely to return.

Finally, in the last selection, Nagel combines analysis with moral exhortation to indicate the responsibility modern man has for the maintenance and strengthening of reason and what is its best human instrument, science in all its forms. Nagel shows the inadequacy of values which attack science and criticizes the grounds on which such attacks have been made.

34 *Science as a Vocation*

MAX WEBER

THIS MUCH I deem necessary to say about the external conditions of the academic man's vocation. But I believe that actually you wish to hear of something else, namely, of the *inward* calling for science. In our time, the internal situation, in contrast to the organization of science as a vocation, is first of all conditioned by the facts that science has entered a phase of specialization previously unknown and that this will forever remain the case. Not only externally, but inwardly, matters stand at a point where the individual can acquire the sure consciousness of achieving something truly perfect in the field of science only in case he is a strict specialist.

All work that overlaps neighboring fields, such as we occasionally undertake and which the sociologists must necessarily undertake again and again, is burdened with the resigned realization that at best one provides the specialist with useful questions upon which he would not so easily hit from his own specialized point of view. One's own work must inevitably remain highly imperfect. Only by strict specialization can the scientific worker become fully conscious, for once and perhaps never again in his lifetime, that he has achieved something that will endure. A really definitive and good accomplishment is today always a specialized accomplishment. And whoever lacks the capacity to put on blinders, so to speak, and to come up to the idea that the fate of his soul depends upon whether or not he makes the correct conjecture at this passage of this manuscript may as well stay away from science. He will never have what one may call the "personal experience" of science. Without this strange intoxication, ridiculed by every outsider; without this passion, this "thousands of years must pass before you enter into life and thousands more wait in silence"—according to whether or not you succeed in making this conjecture; without this, you have *no* calling for science and you should do something else. For nothing is worthy of man as man unless he can pursue it with passionate devotion.

Yet it is a fact that no amount of such enthusiasm, however sincere and profound it may be, can compel a problem to yield scientific results. Certainly enthusiasm is a prerequisite of the "inspiration" which is decisive. Nowadays in circles of youth there is a widespread notion that science has become a problem in calculation, fabricated in laboratories or statistical filing systems just as "in a factory," a calculation involving only the cool intellect and not one's "heart and soul." First of all one must say that such comments lack all clarity about what goes on in a factory or in a laboratory. In both some idea has to occur to someone's mind, and it has to be a correct idea, if one is to accomplish anything worthwhile. And such intuition cannot be forced. It has nothing to do with any cold calculation. Certainly calculation is also an indispensable prerequisite. No sociologist, for instance, should think himself too good, even in his old age, to make tens of thousands of quite trivial computations in his head and perhaps for months at a time. One cannot with impunity try to transfer this task entirely to mechanical assistants if one wishes to figure something, even though the final result is often small indeed. But if no "idea" occurs to his mind about the direction of his computations and, during his computations, about the bearing of the emergent single results, then even this small result will not be yielded.

Normally such an "idea" is prepared only on the soil of very hard work, but certainly this is not always the case. Scientifically, a dilettante's idea may have the very same or even a greater bearing for science than that of a specialist. Many of our very best hypotheses and insights are due precisely to dilettantes. The dilettante differs from the expert, as Helmholtz has said of Robert Mayer, only in that he lacks a firm and reliable work procedure. Consequently he is usually not in the position to control, to estimate, or to exploit the idea in its bearings. The idea is not a substitute for work; and work, in turn, cannot substitute for or compel an idea, just as little as enthusiasm can. Both, enthusiasm and work, and above all both of them *jointly*, can entice the idea.

Ideas occur to us when they please, not when it pleases us. The best ideas do indeed occur to one's mind in the way in which Ihering describes it: when smoking a cigar on the sofa; or as Helmholtz states of himself with scientific exactitude: when taking a walk on a slowly ascending street; or in a similar way. In any case, ideas come when we do not expect them, and not when we are brooding and searching at our desks. Yet ideas would certainly not come to mind had we not brooded at our desks and searched for answers with passionate devotion.

However this may be, the scientific worker has to take into his bargain the risk that enters into all scientific work: Does an "idea" occur or does it not? He may be an excellent worker and yet never have had any valuable idea of his own. It is a grave error to believe that this is so only in science, and that things for instance in a business office are different from a laboratory. A merchant or a big industrialist without "business imagination," that is, without ideas or ideal intuitions, will for all his life remain a man who would better have remained a clerk or a technical official. He will never be truly creative in organization. Inspiration in the field of science by no means plays any greater role, as academic conceit fancies, than it does in the field of mastering problems of practical life by a modern entrepreneur. On the other hand, and this also is often misconstrued, inspiration plays no less a role in science than it does in the realm of art. It is a childish notion to think that a mathematician attains any scientifically valuable results by sitting at his desk with a ruler, calculating machines or other mechanical means. The mathematical imagination of a Weierstrass is naturally quite differently oriented in meaning and result than is the imagination of an artist, and differs basically in quality. But the psychological processes do not differ. Both are frenzy (in the sense of Plato's "mania") and "inspiration."

Now, whether we have scientific inspiration depends upon destinies that are hidden from us, and besides upon "gifts." Last but not least, because of this indubitable truth, a very understandable attitude has become popular, especially among youth, and has put them in the service of idols whose cult today occupies a broad place on all street corners and in all periodicals. These idols are "personality" and "personal experience." Both are intimately connected, the notion prevails that the latter constitutes the former and belongs to it. People belabor themselves in trying to "experience" life—for that befits a personality, conscious of its rank and station. And if we do not succeed in "experiencing" life, we must at least pretend to have this gift of grace. Formerly we called this "experience," in plain German, "sensation"; and I believe that we then had a more adequate idea of what personality is and what it signifies.

Ladies and gentlemen. In the field of science only he who is devoted *solely* to the work at hand has "personality." And this holds not only for the field of science; we know of no great artist who has ever done anything but serve his work and only his work. As far as his art is concerned, even with a personality of Goethe's rank, it has been detrimental to take the liberty of trying to make his "life" into a work of art. And even if one doubts this, one has to be a Goethe in order to dare permit oneself such liberty. Everybody will admit at least this much: that even with a man like Goethe, who appears once in a thousand years, this liberty did not go unpaid for. In politics matters are not different, but we shall not discuss that today. In the field of science, however, the man who makes himself the impresario of the subject to which he should be devoted, and steps upon the stage and seeks to legitimate himself through "experience," asking: How can I prove that I am something other than a mere "specialist" and how can I manage to say something in form or in content that nobody else has ever said?—such a man is no "personality." Today such conduct is a crowd phenomenon, and it always makes a petty impression and debases the one who is thus concerned. Instead of this, an inner devotion to the task, and that alone, should lift the scientist to the height and dignity of the subject he pretends to serve. And in this it is not different with the artist.

In contrast with these preconditions which scientific work shares with art, science has a fate that profoundly distinguishes it from artistic work. Scientific work is chained to the course of progress; whereas in the realm of art there is no progress in the same sense. It is not true that the work of art of a period that has worked out new technical means, or, for instance, the laws of perspective, stands therefore artistically higher than a work of art devoid of all knowledge of those means and laws—if its form does justice to the material, that is, if its object has been chosen and formed so that it could be artistically mastered without applying those conditions and means. A work of art which is genuine "fulfilment" is never surpassed; it will never be antiquated. Individuals may differ in appreciating the personal significance of works of art, but no one will ever be able to say of such a work that it is "outstripped" by another work which is also "fulfilment."

In science, each of us knows that what he has accomplished will be antiquated in ten, twenty, fifty years. That is the fate to which science is subjected; it is the very *meaning* of scientific work, to which it is devoted in a quite specific sense, as compared with other spheres of culture for which in general the same holds. Every scientific "fulfilment" raises new "questions"; it *asks* to be "surpassed" and outdated. Whoever wishes to serve science has to resign himself to this fact. Scientific works certainly can last as "gratifications" because of their artistic quality, or they may remain important as a means of training. Yet they will be surpassed scientifically—let that be repeated—for it is our common fate and, more, our common goal. We cannot work without hoping that others will advance further than we have. In principle, this progress goes on *ad infinitum*. And with this we come to inquire into the *meaning* of science. For, after all, it is not self-evident that something subordinate to such a law is sensible and meaningful in itself. Why does one engage in doing something that in reality never comes, and never can come, to an end?

One does it, first, for purely practical, in the broader sense of the word, for technical, purposes: in order to be able to orient our practical activities to the expectations that scientific experience places at our disposal. Good. Yet this has meaning only to practitioners. What is the attitude of the academic man towards his vocation—that is, if he is at all in quest of such a personal attitude? He maintains that he engages in "science for science's sake" and not merely because others, by exploiting science, bring about commercial or technical success and can better feed, dress, illuminate, and govern. But what does he who allows himself to be integrated into this specialized organization, running on *ad infinitum,* hope to accomplish that is significant in these productions that are always destined to be outdated? This question requires a few general considerations.

Scientific progress is a fraction, the most important fraction, of the process of intellectualization which we have been undergoing for thousands of years and which nowadays is usually judged in such an extremely negative way. Let us first clarify what this intellectualist rationalization, created by science and by scientifically oriented technology, means practically.

Does it mean that we, today, for instance, everyone sitting in this hall, have a greater knowledge of the conditions of life under which we exist than has an American Indian or a Hottentot? Hardly. Unless he is a physicist, one who rides on the streetcar has no idea how the car happened to get into motion. And he does not need to know. He is satisfied that he may "count" on the behavior of the streetcar, and he orients his conduct according to this expectation; but he knows nothing about what it takes to produce such a car so that it can move. The savage knows incomparably more about his tools. When we spend money today I bet that even if there are colleagues of political economy here in the hall, almost every one of them will hold a different answer in readiness to the question: How does it happen that one can buy something for money—sometimes more and sometimes less? The savage knows what he does in order to get his daily food and which institutions serve him in this pursuit. The increasing intellectualization and rationalization do *not,* therefore, indicate an increased and general knowledge of the conditions under which one lives.

It means something else, namely, the knowledge or belief that if one but wished one *could* learn it at any time. Hence, it means that principally there are no mysterious incalculable forces that come into play, but rather that one can, in principle, master all things by calculation. This means that the world is disenchanted. One need no longer have recourse to magical means in order to master or implore the spirits, as did the savage, for whom such mysterious powers existed. Technical means and calculations perform the service. This above all is what intellectualization means.

Now, this process of disenchantment, which has continued to exist in Occidental culture for millennia, and, in general, this "progress," to which science belongs as a link and motive force, do they have any meanings that go beyond the purely practical and technical? You will find this question raised in the most principled form in the works of Leo Tolstoi. He came to raise the question in a peculiar way. All his broodings increasingly revolved around the problem of whether or not death is a meaningful phenomenon. And his answer was: for civilized man death has no meaning. It has none because the individual life of civilized man, placed into an infinite "progress," according to its own imminent meaning should never come to an end; for there is always a further step ahead of one who stands in the march of progress. And no man who comes to die stands upon the peak which lies in infinity. Abraham, or some peasant of the past, died "old and satiated with life" because he stood in the organic cycle of life; because his life, in terms of its meaning and on the eve of his days, had given to him what life had to offer; because for him there remained no puzzles he might wish to solve; and therefore he could have had "enough" of life. Whereas civilized man, placed in the midst of the continuous enrichment of culture by ideas, knowledge, and problems, may become "tired of life" but not "satiated with life." He catches only the most minute part of what the life of the spirit brings forth ever anew, and what he seizes is always something provisional and not definitive, and therefore death for him is a meaningless occurrence. And because death is meaningless, civilized life as such is meaningless; by its very "progressiveness" it gives death the imprint of meaninglessness. Throughout his late novels one meets with this thought as the keynote of the Tolstoyan art.

What stand should one take? Has "progress" as such a recognizable meaning that goes beyond the technical, so that to serve it is a meaningful vocation? The question must be raised. But this is no longer merely the question of man's calling *for* science, hence, the problem of what science as a vocation means to its devoted disciples. To raise this question is to ask for the vocation of science within the total life of humanity. What is the value of science?

Here the contrast between the past and the present is tremendous. You will recall the wonderful image at the beginning of the seventh book of Plato's *Republic:* those enchained cavemen whose faces are turned toward the stone wall before them. Behind them lies the source of the light which they cannot see. They are concerned only with the shadowy images that this light throws upon the wall, and they seek to fathom their interrelations. Finally one of them succeeds in shattering his fetters, turns around, and sees the sun. Blinded, he gropes about and stammers of what he saw. The others say he is raving. But gradually he learns to behold the light, and then his task is to descend to the cavemen and to lead them to the light. He is the philosopher; the sun, however, is the truth of science, which alone seizes not upon illusions and shadows but upon the true being.

Well, who today views science in such a manner? Today youth feels rather the reverse: the intellectual constructions of science constitute an unreal realm of artificial abstractions, which with their bony hands seek to grasp the blood-and-the-sap of true life without ever catching up with it. But here in life, in what for Plato was the play of shadows on the walls of the cave, genuine reality is pulsating; and the rest are derivatives of life, lifeless ghosts, and nothing else. How did this change come about?

Plato's passionate enthusiasm in *The Republic* must, in the last analysis, be explained by the fact that for the first time the *concept,* one of the great tools of all scientific knowledge, had been consciously discovered. Socrates had discovered it in its bearing. He was not the only man in the world to discover it. In India one finds the beginnings of a logic that is quite similar to that of Aristotle's. But nowhere else do we find this realization of the significance of the concept. In Greece, for the first time, appeared a handy means by which one could put the logical screws upon somebody so that he could not come out without admitting either that he knew nothing or that this and nothing else was truth, the *eternal* truth that never would vanish as the doings of the blind men vanish. That was the tremendous experience which dawned upon the disciples of Socrates. And from this it seemed to follow that if one only found the right concept of the beautiful, the good, or, for instance, of bravery, of the soul—or whatever—that then one could also grasp its true being. And this, in turn, seemed to open the way for knowing and for teaching how to act rightly in life and, above all, how to act as a citizen of the state; for this question was everything to the Hellenic man, whose thinking was political throughout. And for these reasons one engaged in science.

The second great tool of scientific work, the rational experiment, made its appearance at the side of this discovery of the Hellenic spirit during the Renaissance period. The experiment is a means of reliably controlling experience. Without it, present-day empirical science would be impossible. There were experiments earlier; for instance, in India physiological experiments were made in the service of ascetic yoga technique; in Hellenic antiquity, mathematical experiments were made for purposes of war technology; and in the Middle Ages, for purposes of mining. But to raise the experiment to a principle of research was the achievement of the Renaissance. They were the great innovators in *art*, who were the pioneers of experiment. Leonardo and his like and, above all, the sixteenth-century experimenters in music with their experimental pianos were characteristic. From these circles the experiment entered science, especially through Galileo, and it entered theory through Bacon; and then it was taken over by the various exact disciplines of the continental universities, first of all those of Italy and then those of the Netherlands.

What did science mean to these men who stood at the threshold of modern times? To artistic experimenters of the type of Leonardo and the musical innovators, science meant the path to *true* art, and that meant for them the path to true *nature*. Art was to be raised to the rank of a science, and this meant at the same time and above all to raise the artist to the rank of the doctor, socially and with reference to the meaning of his life. This is the ambition on which, for instance, Leonardo's sketch book was based. And today? "Science as the way to nature" would sound like blasphemy to youth. Today, youth proclaims the opposite: redemption from the intellectualism of science in order to return to one's own nature and therewith to nature in general. Science as a way to art? Here no criticism is even needed.

But during the period of the rise of the exact sciences one expected a great deal more. If you recall Swammerdam's statement, "Here I bring you the proof of God's providence in the anatomy of a louse," you will see what the scientific worker, influenced (indirectly) by Protestantism and Puritanism, conceived to be his task: to show the path to God. People no longer found this path among the philosophers, with their concepts and deductions. All pietist theology of the time, above all Spener, knew that God was not to be found along the road by which the Middle Ages had sought him. God is hidden, His ways are not our ways, His thoughts are not our thoughts. In the exact sciences, however, where one could physically grasp His works, one hoped to come upon the traces of what He planned for the world. And today? Who—aside from certain big children who are indeed found in the natural sciences—still believes that the findings of astronomy, biology, physics, or chemistry could teach us anything about the *meaning* of the world? If there is any such "meaning," along what road could one come upon its tracks? If these natural sciences lead to anything in this way, they are apt to make the belief that there is such a thing as the "meaning" of the universe die out at its very roots.

And finally, science as a way "to God"? Science, this specifically irreligious power? That science today is irreligious no one will doubt in his innermost being, even if he will not admit it to himself. Redemption from the rationalism and intellectualism of science is the fundamental presupposition of living in union with the divine. This, or something similar in meaning, is one of the fundamental watchwords one hears among German youth, whose feelings are attuned to religion or who crave religious experiences. They crave not only religious experience but experience as such. The only thing that is strange is the method that is now followed: the spheres of the irrational, the only spheres that intellectualism has not yet touched, are now raised into consciousness and put under its lens. For in practice this is where the modern intellectualist form of romantic irrationalism leads. This method of emancipation from intellectualism may well bring about the very opposite of what those who take to it conceive as its goal.

After Nietzsche's devastating criticism of those "last men" who "invented happiness," I may leave aside altogether the naive optimism in which science—that is, the technique of mastering life which rests upon science—has been celebrated as the way to happiness. Who believes in this?—aside from a few big children in university chairs or editorial offices. Let us resume our argument.

Under these internal presuppositions, what is the meaning of science as a vocation, now after all these former illusions, the "way to true being," the "way to true art," the "way to true nature," the "way to true God," the "way to true happiness," have been dispelled? Tolstoi has given the simplest answer, with the words: "Science is meaningless because it gives no answer to our question, the only question important for us: 'What shall we do and how shall we live?' " That science does not give an answer to this is indisputable. The only question that remains is the sense in which science gives "no" answer, and whether or not science might yet be of some use to the one who puts the question correctly.

Today one usually speaks of science as "free from presuppositions." Is there such a thing? It depends upon what one understands thereby. All scientific work presupposes that the rules of logic and method are valid; these are the general foundations of our orientation in the world; and, at least for our special question, these presuppositions are the least problematic aspect of science. Science further presupposes that what is yielded by scientific work is important in the sense that it is "worth being known." In this, obviously, are contained all our problems. For this presupposition cannot be proved by scientific means. It can only be *interpreted* with reference to its ultimate meaning, which we must reject or accept according to our ultimate position towards life.

Furthermore, the nature of the relationship of scientific work and its presuppositions varies widely according to their structure. The natural sciences, for instance, physics, chemistry, and astronomy, presuppose as self-evident that it is worth while to know the ultimate laws of cosmic events as far as science can construe them. This is the case not only because with such knowledge one can attain technical results but for its own sake, if the quest for such knowledge is to be a "vocation." Yet this presupposition can by no means be proved. And still less can it be proved that the existence of the world which these sciences describe is worth while, that it has any "meaning," or that it makes sense to live in such a world. Science does not ask for the answers to such questions.

Consider modern medicine, a practical technology which is highly developed scientifically. The general "presupposition" of the medical enterprise is stated trivially in the assertion that medical science has the task of maintaining life as such and of diminishing suffering as such to the greatest possible degree. Yet this is problematical. By his means the medical man preserves the life of the mortally ill man, even if the patient implores us to relieve him of life, even if his relatives, to whom his life is worthless and to whom the costs of maintaining his worthless life grow unbearable, grant his redemption from suffering. Perhaps a poor lunatic is involved, whose relatives, whether they admit it or not, wish and must wish for his death. Yet the presuppositions of medicine, and the penal code, prevent the physician from relinquishing his therapeutic efforts. Whether life is worth while living and when—this question is not asked by medicine. Natural science gives us an answer to the question of what we must do if we wish to master life technically. It leaves quite aside, or assumes for its purposes, whether we should and do wish to master life technically and whether it ultimately makes sense to do so.

Consider a discipline such as aesthetics. The fact that there are works of art is given for aesthetics. It seeks to find out under what conditions this fact exists, but it does not raise the question whether or not the realm of art is perhaps a realm of diabolical grandeur, a realm of this world, and therefore, in its core, hostile to God and, in its innermost and aristocratic spirit, hostile to the brotherhood of man. Hence, aesthetics does not ask whether there *should* be works of art.

Consider jurisprudence. It establishes what is valid according to the rules of juristic thought, which is partly bound by logically compelling and partly by conventionally given schemata. Juridical thought holds when certain legal rules and certain methods of interpretations are recognized as binding. Whether there should be law and whether one should establish just these rules—such questions jurisprudence does not answer. It can only state: If one wishes this result, according to the norms of our legal thought, this legal rule is the appropriate means of attaining it.

Consider the historical and cultural sciences. They teach us how to understand and interpret political, artistic, literary, and social phenomena in terms of their origins. But they give us no answer to the question, whether the existence of these cultural phenomena have been and are *worth while.* And they do not answer the further question, whether it is worth the effort required to know them. They presuppose that there is an interest in partaking, through this procedure, of the community of "civilized men." But they cannot prove "scientifically" that this is the case; and that they presuppose this interest by no means proves that it goes wthout saying. In fact it is not at all self-evident.

Finally, let us consider the disciplines close to me: sociology, history, economics, political science, and those types of cultural philosophy that make it their task to interpret these sciences. It is said, and I agree, that politics is out of place in the lecture-room. It does not belong there on the part of the students. If, for instance, in the lecture-room of my former colleague Dietrich Schäfer in Berlin, pacifist students were to surround his desk and make an uproar, I should deplore it just as much as I should deplore the uproar which anti-pacifist students are said to have made against Professor Förster, whose views in many ways are as remote as could be from mine. Neither does politics, however, belong in the lecture-room on the part of the docents, and when the docent is scientifically concerned with politics, it belongs there least of all.

To take a practical political stand is one thing, and to analyze political structures and party positions is another. When speaking in a political meeting about democracy, one does not hide one's personal standpoint; indeed, to come out clearly and take a stand is one's damned duty. The words one uses in such a meeting are not means of scientific analysis but means of canvassing votes and winning over others. They are not plowshares to loosen the soil of contemplative thought; they are swords against the enemies: such words are weapons. It would be an outrage, however, to use words in this fashion in a lecture or in the lecture-room. If, for instance, "democracy" is under discussion, one considers its various forms, analyzes them in the way they function, determines what results for the conditions of life the one form has as compared with the other. Then one confronts the forms of democracy with non-democratic forms of political order and endeavors to come to a position where the student may find the point from which, in terms of his ultimate ideals, he can take a stand. But the true teacher will beware of imposing from the platform any political position upon the student, whether it is expressed or suggested. "To let the facts speak for themselves" is the most unfair way of putting over a political position to the student.

Why should we abstain from doing this? I state in advance that some highly esteemed colleagues are of the opinion that it is not possible to carry through this self-restraint and that, even if it were possible, it would be a whim to avoid declaring oneself. Now one cannot demonstrate scientifically what the duty of an academic teacher is. One can only demand of the teacher that he have the intellectual integrity to see that it is one thing to state facts, to determine mathematical or logical relations or the internal structure of cultural values, while it is another thing to answer questions of the *value* of culture and its individual contents and the question of how one should act in the cultural community and in political associations. These are quite heterogeneous problems. If he asks further why he should not deal with both types of problems in the lecture-room, the answer is: because the prophet and the demagogue do not belong on the academic platform.

To the prophet and the demagogue, it is said: "Go your ways out into the streets and speak openly to the world," that is, speak where criticism is possible. In the lecture-room we stand opposite our audience, and it has to remain silent. I deem it irresponsible to exploit the circumstance that for the sake of their career the students have to attend a teacher's course while there is nobody present to oppose him with criticism. The task of the teacher is to serve the students with his knowledge and scientific experience and not to imprint upon them his personal political views. It is certainly possible that the individual teacher will not entirely succeed in eliminating his personal sympathies. He is then exposed to the sharpest criticism in the forum of his own conscience. And this deficiency does not prove anything; other errors are also possible, for instance, erroneous statements of fact, and yet they prove nothing against the duty of searching for the truth. I also reject this in the very interest of science. I am ready to prove from the works of our historians that whenever the man of science introduces his personal value judgment, a full understanding of the facts *ceases*. But this goes beyond tonight's topic and would require lengthy elucidation.

I ask only: How should a devout Catholic, on the one hand, and a Freemason, on the other, in a course on the forms of church and state or on religious history ever be brought to evaluate these subjects alike? This is out of the question. And yet the academic teacher must desire and must demand of himself to serve the one as well as the other by his knowledge and methods. Now you will rightly say that the devout Catholic will never accept the view of the factors operative in bringing about Christianity which a teacher who is free of his dogmatic presuppositions presents to him. Certainly! The difference, however, lies in the following: Science "free from presuppositions," in the sense of a rejection of religious bonds, does not know of the "miracle" and the "revelation." If it did, science would be unfaithful to its own "presuppositions." The believer knows both, miracle and revelation. And science "free from presuppositions" expects from him no less—and no more—than acknowledgment that *if* the process can be explained without those supernatural interventions, which an empirical explanation has to eliminate as causal factors, the process has to be explained the way science attempts to do. And the believer can do this without being disloyal to his faith.

But has the contribution of science no meaning at all for a man who does not care to know facts as such and to whom only the practical standpoint matters? Perhaps science nevertheless contributes something.

The primary task of a useful teacher is to teach his students to recognize "inconvenient" facts—I mean facts that are inconvenient for their party opinions. And for every party opinion there are facts that are extremely inconvenient, for my own opinion no less than for others. I believe the teacher accomplishes more than a mere intellectual task if he compels his audience to accustom itself to the existence of such facts. I would be so immodest as even to apply the expression "moral achievement," though perhaps this may sound too grandiose for something that should go without saying.

Thus far I have spoken only of practical reasons for avoiding the imposition of a personal point of view. But these are not the only reasons. The impossibility of "scientifically" pleading for practical and interested stands—except in discussing the means for a firmly given and presupposed end—rests upon reasons that lie far deeper.

"Scientific" pleading is meaningless in principle because the various value spheres of the world stand in irreconcilable conflict with each other. The elder Mill, whose philosophy I will not praise otherwise, was on this point right when he said: If one proceeds from pure experience, one arrives at polytheism. This is shallow in formulation and sounds paradoxical, and yet there is truth in it. If anything, we realize again today that something can be sacred not only in spite of its not being beautiful, but rather because and in so far as it is not beautiful. You will find this documented in the fifty-third chapter of the book of Isaiah and in the twenty-first Psalm. And, since Nietzsche, we realize that something can be beautiful, not only in spite of the aspect in which it is not good, but rather in that very aspect. You will find this expressed earlier in the *Fleurs du mal,* as Baudelaire named his volume of poems. It is commonplace to observe that something may be true although it is not beautiful and not holy and not good. Indeed it may be true in precisely those aspects. But all these are only the most elementary cases of the struggle that the gods of the various orders and values are engaged in. I do not know how one might wish to decide "scientifically" the value of French and German culture; for here, too, different gods struggle with one another, now and for all times to come.

We live as did the ancients when their world was not yet disenchanted of its gods and demons, only we live in a different sense. As Hellenic man at times sacrificed to Aphrodite and at other times to Apollo, and, above all, as everybody sacrificed to the gods of his city, so do we still nowadays, only the bearing of man has been disenchanted and denuded of its mystical but inwardly genuine plasticity. Fate, and certainly not "science," holds sway over these gods and their struggles. One can only understand what the godhead is for the one order or for the other, or better, what godhead is in the one or in the other order. With this understanding, however, the matter has reached its limit so far as it can be discussed in a lecture-room and by a professor. Yet the great and vital problem that is contained therein is, of course, very far from being concluded. But forces other than university chairs have their say in this matter.

What man will take upon himself the attempt to "refute scientifically" the ethic of the Sermon on the Mount? For instance, the sentence, "resist no evil," or the image of turning the other cheek? And yet it is clear, in mundane perspective, that this is an ethic of undignified conduct; one has to choose between the religious dignity which this ethic confers and the dignity of manly conduct which preaches something quite different; "resist evil—lest you be co-responsible for an overpowering evil." According to our ultimate standpoint, the one is the devil and the other the God, and the individual has to decide which is God for him and which is the devil. And so it goes throughout all the orders of life.

The grandiose rationalism of an ethical and methodical conduct of life which flows from every religious prophecy has dethroned this polytheism in favor of the "one thing that is needful." Faced with the realities of outer and inner life, Christianity has deemed it necessary to make those compromises and relative judgments, which we all know from its history. Today the routines of everyday life challenge religion. Many old gods ascend from their graves; they are disenchanted and hence take the form of impersonal forces. They strive to gain power over our lives and again they resume their eternal struggle with one another. What is hard for modern man, and especially for the younger generation, is to measure up to *workaday* existence. The ubiquitous chase for "experience" stems from this weakness; for it is weakness not to be able to countenance the stern seriousness of our fateful times.

Our civilization destines us to realize more clearly these struggles again, after our eyes have been blinded for a thousand years—blinded by the allegedly or presumably exclusive orientation towards the grandiose moral fervor of Christian ethics.

But enough of these questions which lead far away. Those of our youth are in error who react to all this by saying, "Yes, but we happen to come to lectures in order to experience something more than mere analyses and statements of fact." The error is that they seek in the professor something different from what stands before them. They crave a leader and not a teacher. But we are placed upon the platform solely as teachers. And these are two different things, as one can readily see. Permit me to take you once more to America, because there one can often observe such matters in their most massive and original shape.

The American boy learns unspeakably less than the German boy. In spite of an incredible number of examinations, his school life has not had the significance of turning him into an absolute creature of examinations, such as the German. For in America, bureaucracy, which presupposes the examination diploma as a ticket of admission to the realm of office prebends, is only in its beginnings. The young American has no respect for anything or anybody, for tradition or for public office—unless it is for the personal achievement of individual men. This is what the American calls "democracy." This is the meaning of democracy, however distorted its intent may in reality be, and this intent is what matters here. The American's conception of the teacher who faces him is: he sells me his knowledge and his methods for my father's money, just as the greengrocer sells my mother cabbage. And that is all. To be sure, if the teacher happens to be a football coach, then, in this field, he is a leader. But if he is not this (or something similar in a different field of sports), he is simply a teacher and nothing more. And no young American would think of having the teacher sell him a *Weltanschauung* or a code of conduct. Now, when formulated in this manner, we should reject this. But the question is whether there is not a grain of salt contained in this feeling, which I have deliberately stated in extreme with some exaggeration.

Fellow students! You come to our lectures and demand from us the qualities of leadership, and you fail to realize in advance that of a hundred professors at least ninety-nine do not and must not claim to be football masters in the vital problems of life, or even to be "leaders" in matters of conduct. Please, consider that a man's value does not depend on whether or not he has leadership qualities. And in any case, the qualities that make a man an excellent scholar and academic teacher are not the qualities that make him a leader to give directions in practical life or, more specifically, in politics. It is pure accident if a teacher also possesses this quality, and it is a critical situation if every teacher on the platform feels himself confronted with the students' expectation that the teacher should claim this quality. It is still more critical if it is left to every academic teacher to set himself up as a leader in the lecture-room. For those who most frequently think of themselves as leaders often qualify least as leaders. But irrespective of whether they are or are not, the platform situation simply offers no possibility of *proving* themselves to be leaders. The professor who feels called upon to act as a counselor of youth and enjoys their trust may prove himself a man in personal human relations with them. And if he feels called upon to intervene in the struggles of world views and party opinions, he may do so outside, in the market place, in the press, in meetings, in associations, wherever he wishes. But after all, it is somewhat too convenient to demonstrate one's courage in taking a stand where the audience and possible opponents are condemned to silence.

Finally, you will put the queston: "If this is so, what then does science actually and positively contribute to practical and personal life?" Therewith we are back again at the problem of science as a "vocation."

First, of course, science contributes to the technology of controlling life by calculating external objects as well as man's activities. Well, you will say, that, after all, amounts to no more than the greengrocer of the American boy. I fully agree.

Second, science can contribute something that the greengrocer cannot: methods of thinking, the tools and the training for thought. Perhaps you will say: well, that is no vegetable, but it amounts to no more than the means for procuring vegetables. Well and good, let us leave it at that for today.

Fortunately, however, the contribution of science does not reach its limit with this. We are in a position to help you to a third objective: to gain *clarity*. Of course, it is presupposed that we ourselves possess clarity. As far as this is the case, we can make clear to you the following:

In practice, you can take this or that position when concerned with a problem of value—for simplicity's sake, please think of social phenomena as examples. If you take such and such a stand, then, according to scientific experience, you have to use such and such a *means* in order to carry out your conviction practically. Now, these means are perhaps such that you believe you must reject them. Then you simply must choose between the end and the inevitable means. Does the end "justify" the means? Or does it not? The teacher can confront you with the necessity of this choice. He cannot do more, so long as he wishes to remain a teacher and not to become a demagogue. He can, of course, also tell you that if you want such and such an end, then you must take into the bargain the subsidiary consequences which according to all experience will occur. Again we find ourselves in the same situation as before. These are still problems that can also emerge for the technician, who in numerous instances has to make decisions according to the principle of the lesser evil or of the relatively best. Only to him one thing, the main thing, is usually given, namely, the end. But as soon as truly "ultimate" problems are at stake for us this is not the case. With this, at long last, we come to the final service that science as such can render to the aim of clarity, and at the same time we come to the limits of science.

Besides we can and we should state: In terms of its meaning, such and such a practical stand can be derived with inner consistency, and hence integrity, from this or that ultimate *weltanschauliche* position. Perhaps it can only be derived from one such fundamental position, or maybe from several, but it cannot be derived from these or those other positions. Figuratively speaking, you serve this god and you offend the other god when you decide to adhere to this position. And if you remain faithful to yourself, you will necessarily come to certain final conclusions that subjectively make sense. This much, in principle at least, can be accomplished. Philosophy, as a special discipline, and the essentially philosophical discussions of principles in the other sciences attempt to achieve this. Thus, if we are competent in our pursuit (which must be presupposed here) we can force the individual, or at least we can help him, to give himself an *account of the ultimate meaning of his own conduct.* This appears to me as not so trifling a thing to do, even for one's own personal life. Again, I am tempted to say of a teacher who succeeds in this: he stands in the service of "moral" forces; he fulfils the duty of bringing about self-clarification and a sense of responsibility. And I believe he will be the more able to accomplish this, the more conscientiously he avoids the desire personally to impose upon or suggest to his audience his own stand.

This proposition, which I present here, always takes its point of departure from the one fundamental fact, that so long as life remains immanent and is interpreted in its own terms, it knows only of an unceasing struggle of these gods with one another. Or speaking directly, the ultimately possible attitudes toward life are irreconcilable, and hence their struggle can never be brought to a final conclusion. Thus it is necessary to make a decisive choice. Whether, under such conditions, science is a worth while "vocation" for somebody, and whether science itself has an objectively valuable "vocation" are again value judgments about which nothing can be said in the lecture-room. To affirm the value of science is a presupposition for teaching there. I personally by my very work answer in the affirmative, and I also do so from precisely the standpoint that hates intellectualism as the worst devil, as youth does today, or usually only fancies it does. In that case the word holds for these youths: "Mind you, the devil is old; grow old to understand him." This does not mean age in the sense of the birth certificate. It means that if one wishes to settle with this devil, one must not take to flight before him as so many like to do nowadays. First of all, one has to see the devil's ways to the end in order to realize his power and his limitations.

Science today is a "vocation" organized in special disciplines in the service of self-clarification and knowledge of interrelated facts. It is not the gift of grace of seers and prophets dispensing sacred values and revelations, nor does it partake of the contemplation of sages and philosophers about the meaning of the universe. This, to be sure, is the inescapable condition of our historical situation. We cannot evade it so long as we remain true to ourselves. And if Tolstoi's question recurs to you: as science does not, who is to answer the question: "What shall we do, and, how shall we arrange our lives?" or, in the words used here tonight: "Which of the warring gods should we serve? Or should we serve perhaps an entirely different god, and who is he?" then one can say that only a prophet or a savior can give the answers. If there is no such man, or if his message is no longer believed in, then you will certainly not compel him to appear on this earth by having thousands of professors, as privileged hirelings of the state, attempt as petty prophets in their lecture-rooms to take over his role. All they will accomplish is to show that they are unaware of the decisive state of affairs: the prophet for whom so many of our younger generation yearn simply does not exist. But this knowledge in its forceful significance has never become vital for them. The inward interest of a truly religiously "musical" man can never be served by veiling to him and to others the fundamental fact that he is destined to live in a godless and prophetless time by giving him the *ersatz* of armchair prophecy. The integrity of his religious organ, it seems to me, must rebel against this.

Now you will be inclined to say: Which stand does one take towards the factual existence of "theology" and its claims to be a "science"? Let us not flinch and evade the answer. To be sure, "theology" and "dogmas" do not exist universally, but neither do they exist for Christianity alone. Rather (going backward in time), they exist in highly developed form also in Islam, in Manicheanism, in Gnosticism, in Orphism, in Parsism, in Buddhism, in the Hindu sects, in Taoism, and in the Upanishads, and, of course, in Judaism. To be sure their systematic development varies greatly. It is no accident that Occidental Christianity—in contrast to the theological possessions of Jewry—has expanded and elaborated theology more systematically, or strives to do so. In the Occident the development of theology has had by far the greatest historical significance. This is the product of the Hellenic spirit, and all theology of the West goes back to it, as (obviously) all theology of the East goes back to Indian thought. All theology represents an intellectual *rationalization* of the possession of sacred values. No science is absolutely free from presuppositions, and no science can prove its fundamental value to the man who rejects these presuppositions. Every theology, however, adds a few specific presuppositions for its work and thus for the justification of its existence. Their meaning and scope vary. Every theology, including for instance Hinduist theology, presupposes that the world must have a *meaning,* and the question is how to interpret this meaning so that it is intellectually conceivable.

It is the same as with Kant's epistemology. He took for his point of departure the presupposition: "Scientific truth exists and it is valid," and then asked: "Under which presuppositions of thought is truth possible and meaningful?" The modern aestheticians (actually or expressly, as for instance, G. v. Lukacs) proceed from the presupposition that "works of art exist," and then ask: "How is their existence meaningful and possible?"

As a rule, theologies, however, do not content themselves with this (essentially religious and philosophical) presupposition. They regularly proceed from the further presupposition that certain "revelations" are facts relevant for salvation and as such make possible a meaningful conduct of life. Hence, these revelations must be believed in. Moreover, theologies presuppose that certain subjective states and acts possess the quality of holiness, that is, they constitute a way of life, or at least elements of one, that is religiously meaningful. Then the question of theology is: How can these presuppositions, which must simply be accepted be meaningfully interpreted in a view of the universe? For theology, these presuppositions as such lie beyond the limits of "science." They do not represent "knowledge," in the usual sense, but rather a "possession." Whoever does not "possess" faith, or the other holy states, cannot have theology as a substitute for them, least of all any other science. On the contrary, in every "positive" theology, the devout reaches the point where the Augustinian sentence holds: *credo non quod, sed quia absurdum est.*

The capacity for the accomplishment of religious virtuosos—the "intellectual sacrifice"—is the decisive characteristic of the positively religious man. That this is so is shown by the fact that in spite (or rather in consequence) of theology (which unveils it) the tension between the value-spheres of "science" and the sphere of "the holy" is unbridgeable. Legitimately, only the disciple offers the "intellectual sacrifice" to the prophet, the believer to the church. Never as yet has a new prophecy emerged (and I repeat here deliberately this image which has offended some) by way of the need of some modern intellectuals to furnish their souls with, so to speak, guaranteed genuine antiques. In doing so, they happen to remember that religion has belonged among such antiques, and of all things religion is what they do not possess. By way of substitute, however, they play at decorating a sort of domestic chapel with small sacred images from all over the world, or they produce surrogates through all sorts of psychic experiences to which they ascribe the dignity of mystic holiness, which they peddle in the book market. This is plain humbug or self-deception. It is, however, no humbug but rather something very sincere and genuine if some of the youth groups who during recent years have quietly grown together give their human community the interpretation or a religious, cosmic, or mystical relation, although occasionally perhaps such interpretation rests on misunderstanding of self. True as it is that every act of genuine brotherliness may be linked with the awareness that it contributes something imperishable to a super-personal realm, it seems to me dubious whether the dignity of purely human and communal relations is enhanced by these religious interpretations. But that is no longer our theme.

The fate of our times is characterized by rationalization and intellectualization and, above all, by the "disenchantment of the world." Precisely the ultimate and most sublime values have retreated from public life either into the transcendental realm of mystic life or into the brotherliness of direct and personal human relations. It is not accidental that our greatest art is intimate and not monumental, nor is it accidental that today only within the smallest and intimate circles, in personal human situations, in *pianissimo*, that something is pulsating that corresponds to the prophetic *pneuma*, which in former times swept through the great communities like a firebrand, welding them together. If we attempt to force and to "invent" a monumental style in art, such miserable monstrosities are produced as the many monuments of the last twenty years. If one tries intellectually to construe new religions without a new and genuine prophecy, then, in an inner sense, something similar will result, but with still worse effects. And academic prophecy, finally, will create only fanatical sects but never a genuine community.

To the person who cannot bear the fate of the times like a man, one must say: may he rather return silently, without the usual publicity build-up of renegades, but simply and plainly. The arms of the old churches are opened widely and compassionately for him. After all, they do not make it hard for him. One way or another he has to bring his "intellectual sacrifice"—that is inevitable. If he can really do it, we shall not rebuke him. For such an intellectual sacrifice in favor of an unconditional religious devotion is ethically quite a different matter than the evasion of the plain duty of intellectual integrity, which sets in if one lacks the courage to clarify one's own ultimate standpoint and rather facilitates this duty by feeble relative judgments. In my eyes, such religious return stands higher than the academic prophecy, which does not clearly realize that in the lecture-rooms of the university no other virtue holds but plain intellectual integrity. Integrity, however, compels us to state that for the many who today tarry for new prophets and saviors, the situation is the same as resounds in the beautiful Edomite watchman's song of the period of exile that has been included among Isaiah's oracles:

> He calleth to me out of Seir, Watchman, what of the night? The watchman said, The morning cometh, and also the night: if ye will enquire, enquire ye: return, come.

The people to whom this was said has enquired and tarried for more than two millennia, and we are shaken when we realize its fate. From this we want to draw the lesson that nothing is gained by yearning and tarrying alone, and we shall act differently. We shall set to work and meet the "demands of the day," in human relations as well as in our vocation. This, however, is plain and simple, if each finds and obeys the demon who holds the fibers of his very life.

35 *Some Aspects of the Relation between Social Science and Ethics*

TALCOTT PARSONS

THE QUESTION posed for this discussion is whether science can determine the ends for which its discoveries are to be used. The general position I shall take is that though science can play an important part in this determination, it is not by itself adequate. However, science at the same time shares with all rational disciplines certain fundamental intellectual methods, and these intellectual methods are as much a part of any rational attempt to determine values as they are of science itself. Therefore there is much in common between the methods of science and the methods by which ends must be determined.

It seems to me that the distinction between science and ethics must, on the level on which I wish to discuss it at least, be considered fundamental. I should like to start by recognizing fully how relative any such distinction must be. I shall not attempt to go into the deeper ontological questions which may be involved from the philosophical point of view. When I speak of science, I mean science as it has traditionally existed in western civilization and as it is likely to develop in the future, which is of immediate concern to scientists now living. In relation to this level, social science must, of course, play a particularly important part since the determination of ends is a matter of human action, and among scientific disciplines it is the social sciences which are directly concerned with the study of human action as such. It is on this level of relativity to a given cultural situation that I should like to take the position that science, including social science, is not by itself

Reprinted from Social Science, *vol. 22 (1947), pp. 213–217.*
This paper was read at a joint meeting of Sections K and L at the Annual Convention of the American Association for the Advancement of Science held at Boston in December, 1946.

competent to determine the most fundamental judgments of value which must underlie the selection of ends.

This, however, is by no means to say that science and ethics should be considered as independent in the sense that they have nothing to do with each other. They are not segregated, unrelated disciplines, but are interdependent parts of the same fundamental system of rational orientation to the world. The great system of positivistic thought which has played such an important role in the last two centuries attempted to maintain the position that science could be self-subsistent as the sole rational discipline in man's cognitive orientation to his world. It is my contention[1] for scientific rather than philosophical reasons that, in the field of the analysis of human social behavior, this positivistic system has broken down. It has proved itself fundamentally inadequate to cope with the scientific problems of that area. The fundamental reason for and consequence of this breakdown is that in order to understand action scientifically on the present levels of knowledge and method, it has proved to be essential to introduce structured systems of values which cannot be grounded only in knowledge of the facts of the situation in which people act, and which alone are accessible to scientific analysis as such. The basic methodological postulate of the ruling conceptual scheme of social science for this purpose is not that social science implies any one particular system of values, but that without recognizing the positive role of such values and of man's attempt to give them a rational grounding and foundation it is not possible to perform its own scientific task.

On the level of present and prospective social science, the close interdependence of science and rational ethics has certain fundamental consequences. If both are to be acknowledged as rational disciplines, it follows in the first place that applied ethics, that is judgments about any actual concrete courses of action, must be directly and fully integrated with the current scientific knowledge that is relevant to the particular concrete situations in question. There is no such thing as a rational solution to the concrete problems of conduct without the best available analysis of the actual facts of the situation and without prediction of the probable empirical consequences of alternative possible measures. Such analysis and prediction is possible on a rational level only as applied social science. This consideration is of fundamental importance since in the actual thinking of our times, social science and philosophical ethics have tended to be seriously out of touch with one another. To a serious extent ethical thought has uncritically assumed the conceptual schemes of a previous phase of social science, and has been guilty of biases on this account. If the ethical

1. Developed and documented in *The Structure of Social Action*, New York, 1937.

theorist, therefore, is to do more than discuss the abstract nature of conduct, if he is to turn to prescription for actual conduct, whether of individuals or the policy of a collectivity, he must do so in terms of adequate and verified social science knowledge of the situations he is dealing with. The relation here is essentially the same as that involved in the engineering knowledge necessary to justify a technological decision, or again in medicine. In the sense of medicine as applied science, whether or not a life is worth saving cannot be a technical decision of medical science, but, if it is decided that it is worth the attempt to save such a life, then no prescription as to what measures should be taken can dispense with the best available scientific knowledge.

The converse is also true. No social science can be adequate unless it has the best available understanding of the role of values in general and of particular systems of values in social life. This understanding must in the nature of the case include a clear analysis of the limits to the scientific guidance of practice in the given state of knowledge and in the given situation. People who are inclined to naive rationalism must be warned that this is an exceedingly complicated and technical problem. The subtle interrelationships on unconscious levels between factors which must be attributed to biological needs, for example, and, on the other hand, value sentiments, are of such a degree of complexity that nothing short of elaborate technical analysis can hope to unravel them. Nevertheless, it is essential to attempt to do this if an adequate understanding of dynamic problems of social action is to be achieved.

It is a further implication of this basic conceptional scheme for the analysis of social action that scientific investigation must itself be regarded as a process of social action. It is not somehow a process of emanation by which eternal truth is miraculously implanted in human minds. Understanding of truth, whether it be the truth of the physical world or of the social, is an achievement—an achievement to be analyzed in the same fundamental terms as the achievement of any other type of goal. If scientific investigation is a process of action, it follows that it, in the same sense as any other action, is governed by moral values and ethical standards. Once having recognized this fundamental fact, social science is in a position to develop certain implications from it which are essential to the treatment of the question before us. Of these, two appear to be particularly significant.

The first is that the values which govern the scientist in his most technical procedures of scientific investigation are not isolated, but are part of a total system of values which govern his action as whole, and which, depending on the degree of rationality of his total system of conduct, is more or less integrated as a system. Secondly, the values held by the scientist could not in the nature of the case be peculiar to this small minority of the total society in which he lives. They must, if science is to be a stably functioning part of the social system, be well integrated with the larger value system which governs the action of all the principal elements in the society; of farmers, industrial workers, businessmen.

It follows from these facts that in a given society in which science has had an important historical role the range of values which could determine the ends for which scientific discoveries will be used is relatively narrow. This is because these values must in the nature of social integration be such as are compatible with a high evaluation of science itself no matter how restricted the minority to whom scientific investigation is a dominant personal goal. The development of science and the position it has attained in a large scale society is unquestionably unique to modern western civilization. It would be incautious to suggest that no society which differed radically from that in which we live could develop science to a comparable degree. It is, however, perfectly safe to assert that no other radically different society known to history has done so, and this fact is quite clearly not fortuitous. It is, as for instance Max Weber has shown, deeply involved with the fundamentally different value orientations which other great civilizations, as for instance those of the Orient, have had. It may, therefore, be argued that if we are true to the great tradition of western science we will have to use the discoveries of science in terms of a system of ends which has deep roots in the broader cultural tradition of the western world. Time will permit only the suggestion of one or two directions in which these close relationships between the ethic of science and other aspects of our value system may be followed out.

From one point of view the dominant standard of science is that of objective impartial truth. We are familiar with the fact that on the part of the scientist this implies, on the one hand, the moral values of intellectual honesty, the impossibility of admitting even to oneself what cannot within the relevant fields be objectively demonstrated. At the same time, it involves a certain humility, a willingness to be guided by the facts regardless of their conflict with personal sentiments or wishes. It cannot be anything but striking that these values in the ethic of science are most intimately associated with those of the freedom of the individual. If there is one statement which can be generally asserted, it is that truth cannot be attained by coercion. Where truth is in conflict with authority the ethic of science must take a perfectly clear stand. Authority cannot prevail when it stands in conflict with truth. Freedom of investigation is the most elementary need of the scientist if science is to flourish, but freedom of investigation cannot be isolated from many other aspects of the freedom of the individual so that the basic "libertarian" values of our society would certainly seem to be most intimately associated with the values of science.

The association of science with democracy is perhaps somewhat less obvious. Nevertheless, there seems to be a fundamental connection.

One of the most fundamental aspects of the values of science is universalism. Truth in the nature of the case cannot be parochial. It cannot be confined in its validity to one section of a population. It is, therefore, as a value, likely to be integrated with other universalistic values, such as the valuation of achievement in terms of performance rather than, for instance, the valuation of status in terms of origin by birth. The kind of ethics which is congenial to science therefore is likely to be extremely uncongenial to the drawing of distinctions in terms of privilege which cannot be justified in universalistic terms. The most obvious pattern, therefore, which is compatible with these values is one which refuses to draw distinctions between men which are not founded in such universalistic values as ability and the like. This is apt to result in a doctrine of equality of opportunity, a criticism of privilege as such, and various things of this sort.

There is no intention to deny the possibility that science will in fact be used, and obviously has been, for ends incompatible with these values. We have seen a very large scale example of this in the case of Nazi Germany which was pre-eminent in science applied to war. It may, nevertheless, still be held that science could not have flourished over a long period under a Nazi regime. As a social scientist, I should be quite willing to defend the thesis that had the Nazi regime continued it would in a relatively short period have dried up the wellsprings of science. Equally I should incline to the view that the problems of the conflict between the values of freedom associated with science and the authoritarianism of the regime constitute a very deep dilemma for the Soviet Union, a dilemma comparable in many respects to the American dilemma, the treatment of the Negro, which Myrdal has analyzed so illuminatingly.

One may sum up, therefore, by saying that while science itself cannot from its own resources determine the ends for which it will be used, where science is an integral part of a great cultural tradition in a major civilization it is in fact integrated with a value system in such a way that many of the basic values with which it is associated, and which will determine these ends, are not arbitrary but are inherent in the cultural tradition itself. Abandonment of them would almost certainly lead to radical changes in the society and culture as a whole, and in all probability to a major diminution in the role of science itself.

Insofar as this value system may be regarded as an integral part of western civilization, the very fact that science is so highly valued in that civilization may, I think, be considered to be an encouraging fact about the relative firmness and definiteness of our ethical orientations. Many of the interpreters of the social and political crisis of our time are inclined to see in it fundamentally a moral crisis—to feel that men in the western world do not have a stable value orientation, and that their social and political confusions and conflicts may be traced primarily to this fact. I should not like to deny radically that there may be important validity in this view. I should, however, like to suggest at the same time that another aspect of the situation may be seriously neglected by people who hold it. There is little doubt that the situation into which millions of people are placed under modern industrial and technological conditions is of a far higher degree of complexity and difficulty than that in the simpler traditional societies. This fact has subjected people to far greater strains than their ancestors knew, and we know that beyond certain points people cannot handle such strains without reacting irrationally. Insofar as this type of factor is involved, the undoubted crisis of our time may be interpreted not as resulting from the moral crisis but from the sheer difficulty of carrying out a moral attitude in practice. Insofar as this is the case, the direction in which help must be sought is clear. It is greater knowledge of the factors determining human action with a view to finding ways of controlling these factors; of easing the burden on people or making them stronger and better capable of bearing it. The prophets of the spiritual crisis are apt to be defeatists in the practical world—they are apt to say there is nothing that can be done; we must await the regenerating influence of a new religion, of a new system of morals, or we must return to the traditional patterns of morality of earlier times. Insofar as the other type of factor is involved, however, a more hopeful position can be taken. It is possible through science to learn better how to control not only the world of nature but of man and his actions. Can anyone doubt that it is at least worthwhile to make the most of this possibility?

36 *Scientists and American Science Policy*

WALLACE S. SAYRE

THE PHRASE "scientists and American science policy" suggests other comparable formulations: soldiers and American military policy, diplomats and American foreign policy, farmers and American farm policy, businessmen and American business policy, educators and American education policy, labor and American labor policy, and a host of other variations. These parallels serve to remind us sharply of the limitations which a democratic order places upon the role of experts as well as upon special interests in the shaping of public policy. If it can be said, for example, that war is too important to be entrusted to the generals and peace too important to be left to the diplomats, then it may be asked whether science policy is not too important to be delegated wholly to the scientists. In a democratic order all policies of significance must secure a wide range of consent, not merely from the general public but also among the many organized groups and institutions that see their interests importantly involved. Scientists do have a special involvement in science policy, but under the rules of a democratic society they have no monopoly in its development or maintenance, nor have they inherently any greater legitimacy or relevance as participants than all the other claimants who aspire to influence the content of science policy.

Scientists, we may assume, aspire to be influential as a group in the determination of public policy over a wide range, especially those elements of public policy which may be described as "science policy." To exercise such influence the scientists must enter the political arena. Scientists in politics encounter the questions posed by the political

Reprinted by permission from Science, *vol.* 133, *no.* 3456 (*March* 24, 1961), *pp.* 859–864.

process to all those who enter: who are they? who speaks for them? what are their goals? what are their strategies?

Who Are the Scientists?

IF SCIENTISTS are to be influential participants in constructing an American science policy, they will need to be self-conscious participants—that is, they must have a visible and concrete identity. That identity is now vague and elusive—to many scientists as well as to the other groups involved in the policy process. "The scientific community," a phrase often submitted as an identification, is a world of uncertain boundaries.

Who are the members of the scientific community? Is it an open community, hospitable to all who desire to enter, or is it open only to those who meet severe tests of eligibility? More specifically, are there "hard scientists," whose membership is taken for granted, and "soft scientists," whose credentials are dubious? Are physicists and chemists members of the scientific community by right, while other natural scientists must submit additional claims for admission? Do all engineers qualify, or only certain types of engineers? Do doctors of medicine have entry, or only research scientists in medicine? Are social scientists full members of the scientific community? The answer of the moment appears to be that the natural scientists are the most fully accredited members of the science community but that the life scientists and the social scientists regard this as a transient condition of affairs.

The difficulties raised by these questions suggest that "the scientific community" is most often used as a strategic phrase, intended by the user to imply a large number of experts where only a few may in fact exist, or to imply unity of view where disagreement may in fact prevail. The phrase may thus belong in that class of invocations, so familiar to the political process, which summon up numbers and legitimacy for a point of view by asserting that "the American people," or "the public," or "all informed observers," or "the experts" demand this or reject that. There is nothing especially astonishing about this, since all participants in the political process indulge in the stratagem, and each participant learns to discount the claims of others, but there may be ground for mild surprise that the code of science permits its extensive use by scientists either as deliberate strategy or in genuine innocence.

If scientists are themselves uncertain as to who all their fellow scientists are, then some ambiguities attend their relationship to American science policy. Are they a small elite group (for example, the approximately 96,000 named in *American Men of Science* for the physical and biological sciences), or do they number several million (as they do if the engineers, the social scientists, and the medical profession are included)? If scientists want to be among the shapers of American science policy rather than simply the objects of that policy, then they must expect these and similar questions from the other participants in the making of science policy. The spokesmen of science will be asked: For whom do you speak? The scientists themselves confront a prior question: Who are to be the accredited spokesmen for the scientists?

Who Speaks for the Scientists?

THE notion of an American science policy, a policy with which the scientists are to be influentially identified, requires the scientists to have leaders who can act as their representatives in that bargaining with public officials and other groups which accompanies the policy-making process. Not every scientist can participate directly in this process; there is not room enough, nor time enough, for a town meeting of all the scientists with all the other groups that have equally legitimate claims to be present. Some few scientists must be selected to speak for the many, but the scientists may choose these few in many different ways. They may let the science spokesmen nominate themselves; they may let non-scientists select the leaders of science; they may develop nominating and electoral devices for choosing their leaders through the votes of all scientists in a single scientist constituency; they may choose their leaders in numerous specialized constituencies; or they may combine these methods in various ways, or invent still other methods.

Tradition and recent practice have already provided some important patterns of choice. The history of American science is rich with examples of the articulate, self-directing, individual scientist of high prestige who felt it his obligation to speak often and boldly in behalf of science and the scientists. Few scientists, and fewer nonscientists, have been inclined to question his representative role, although his peers in prestige and self-confidence have often publicly challenged his advice. Another pattern has been provided by the habit which high-ranking government science officials have of speaking, from their position of special eminence and authority, for the interests of science as they perceive them; this would seem to be, for example, the primary function of some government science advisers and advisory committees. If these advisers are the spokesmen of the scientists, it is relevant to ask: What role did which scientists have in choosing them? Still another pattern has been demonstrated by the role of the National Academy of Sciences since 1863. This quasi-governmental body of scientists, its membership small and its new members elected on the basis of scientific eminence by those who are already members, has for many decades acted upon the assumption that it could and should speak for the scientists in the realm of public policy. The scientists who are not members of the academy have not invariably agreed that the academy spoke for them, or that its silence was to be taken as neutrality on their part on contemporary issues of science policy.

These patterns of individuals and small, elite groups, some self-nominated and some the designees of government officials, speaking for the scientists have been accompanied by several efforts to establish more comprehensive scientist constituencies from which spokesmen might be chosen. The American Association for the Advancement of Science is the most durable of these constituencies. Its own membership is large, and its affiliated societies enlarge its base. Its officials, and especially its committees and its journal *Science*, often speak eloquently for the values and the priorities of science and the scientists. One of the most dramatic assertions of its representative role as spokesman for the scientists was its 1958 Parliament of Science, assembled in Washington, to consider the proposal for a government department of science and other issues. Some privacy surrounded the identity of the delegates to this "parliament," the method of their selection as representatives of the scientists, the record of their deliberations, and the extent of their agreement upon the terms of the published report of the "parliament." The sense in which the AAAS and its "parliament" are authentic spokesmen for the scientists as a scientific community thus cannot be easily determined, either by scientists or nonscientists. The Federation of American Scientists provides still another variation—an association of scientists quite explicitly committed to participation in the political process.

But the most prevalent pattern for choosing the spokesmen of the scientists is provided by the specialized associations of scientists. The officers and committees and journals of the American Physical Society, the American Chemical Society, the American Institute of Biological Sciences, the Federation of American Scientists, the Engineers' Council, the Association of American Geologists—and perhaps a thousand other specialized societies—provide the scientists with hundreds of spokesmen in their specialized areas of interest. These spokesmen do not often speak with one voice upon a given aspect of science policy, nor do they often if ever concern themselves with the elements of a comprehensive science policy. Instead, the leaders of each specialized society tend to express their views upon that segment of science policy which touches significantly the interests of the society's own members. As spokesmen for the scientists, their voices are often competitive, emphasizing separate priorities, asserting specialized rather than general goals. In this characteristic the associations of the scientists share the pluralistic, fragmented, and internally competitive attributes of the other group participants in the American political process—whether political parties, business, labor, agriculture, the professions, nationality groups, or the governmental bureaucracies.

The leaders of still other groups often speak confidently in policy discussions as surrogates for scientists. The Association of Land Grant Colleges and Universities, the American Association of University Presidents, science laboratories and institutes, and the science communication media are prominent among these groups. Do they, too, hold a watching brief for scientists by the scientists' own choice?

Who, then, speaks for the scientists? The answer would seem to lie somewhere in a broad zone of ambiguity. Only the scientists themselves can identify their authentic spokesmen. If they have already done so, it would seem to have been done privately and to have been kept confidential. When and if the scientists undertake an explicit identification of their spokesmen, it is not improbable that they will conclude that no one can speak for all of them, and that in a democratic society we will all, perforce, continue to be confronted by numerous, competing spokesmen for science, each often claiming to speak for more of the scientific community than he in fact represents.

An American Science Policy

UNCERTAINTY thus surrounds the questions: Who are the scientists and who speaks for them? Ambiguity also characterizes the phrase "American science policy." It is not difficult to cite examples of particular science policies; these exist in abundance—from the patents clause of the Constitution to yesterday's progress reports of the National Science Foundation. But the unity and comprehensiveness implied by the phrase "American science policy" are not achieved by merely consolidating and codifying all these separate items of science policy. Something more than this is quite clearly implied and evidently desired by many of those who speak for the scientists. It may be assumed, then, that an American science policy is something aspired to but not yet achieved by the scientists: a unified comprehensive, coherent, rational statement of goals and methods for science in the United States, accepted by and binding upon all the participants in the policy process, and including agreement upon the rules by which the policy may be changed.

The main elements of such a policy might include the following.

1. A preamble, asserting the values of science to society and the nation; a statement defining the boundary line between the governmental and private sectors in science.

2. A statement of the priorities for science in each of these sectors.

3. A ranking of the competing claims of science education, basic research, and applied research, as well as an assignment of priorities among the fields of science—chemistry, engineering, physics, biology, psychology, economics, and perhaps a score of others.

4. A statement of agreement and action upon the structure, location, and assignments of the science agencies in the Executive branch—for example, a unified science department (or, alternatively, decentralized science agencies) and the relation of such a department or such agencies to the President and the Congress.

5. Explicit statements of governmental procedures intended to reflect the values of scientists in such matters as secrecy, personnel loyalty and security, government contracts and grants for research, definitions of "basic" research, and provisions concerning the "chain-of-command" in science activities, including protection for the autonomy of individual scientists.

If such a body of public policy existed, accepted by the scientists and legitimatized by the President and the Congress in a statute, thus attesting the consent of the nation, then an American science policy in the fullest implications of that phrase would have been established.

Is such a unified and comprehensive policy a feasible goal for American scientists? Do they in fact desire it?

If a 1961 Town Meeting of Science were to be assembled, despite the problems of deciding which scientists were eligible to attend, agreement could no doubt be reached on the preamble to an American science policy. Preambles, like political party platforms, are usually triumphs in ambiguity. A viable consensus could probably also be reached on item 5—the "working conditions" for scientists—although ambiguity would overshadow precision here, too. But beyond these two items the available evidence suggests that there are no other major elements of an American science policy upon which one could expect unanimity, or even a clear majority agreement, among the scientists themselves. The document which might emerge from the work of such an assembly of science would most likely be an unstable mixture of vague agreement and sharp minority dissents, a testament to the pluralism of science and the scientists. And if the scientists are not likely to agree upon a unified science policy, the prospects that the nonscientist participants might develop such a policy are even less convincing. There are no apparent powerful incentives for any other great interest group in the American society to develop a unified, comprehensive science policy. And even if agreement were possible among the scientists, there is no persuasive evidence that they could win consent without major concessions to the competing claims of all the other interests that must, in a democratic order, agree to such a significant allocation of social values and resources.

Unity and comprehensiveness are thus not likely to be the hallmarks of American science policy. Talk of a single, comprehensive "American science policy" has an essentially fictitious quality. There will be many science policies, rather than a master science policy. Diversity, inconsistency, compromise, experimentation, pulling and hauling, competition, and continuous revision in science policies are more predictable continuing characteristics than their antonyms. This has been the history of American science policies and this describes their present state of affairs as deplorable. But to live with diversity and accommodations of policy, and yet to be impatient of them, may be the process by which a democratic society achieves progress in science as well as in other fields. In any event, the future seems to offer American scientists more dilemmas than unequivocal answers in science policy.

Persisting Dilemmas for Scientists

SOME of these dilemmas may be illustrated by a brief exploration of a few of the choices concerning governmental arrangements for science—choices which some scientists have helped to make in the past, or which nonscientists have made for them, and still other choices which must yet be made.

SCIENCE ADVISERS

One of these choices involves the aspirations of scientists to give advice to officials at the highest levels of the national government—advice not simply in an area defined as "science policy" but also concerning those elements of foreign policy, defense policy, and domestic policy to which many scientists believe their specialized knowledge is relevant. These aspirations of scientists were reflected in the establishment of the National Academy of Sciences a hundred years ago and in the creation of the National Research Council almost fifty years ago. They are reflected today in the existence of the office of Special Assistant to the President for Science and Technology, the President's Science Advisory Committee, the office of Science Adviser to the Secretary of State, and the General Advisory Committee of the Atomic Energy Commission. The Council of Economic Advisers is still another example.

Attempts to define the role of these advisory institutions raise several important questions. Is their primary responsibility to advise the officials as an autonomous voice of the scientists, or are they, as agents or colleagues of the officials, to explain official policies to the scientists, or are they to participate in working out those accommodations in policy which will build a bridge of collaboration between scientists and officialdom? The history of these institutions of advice reveals the tensions, as well as the temporary adjustments, between these inherently competitive conceptions of the advisory role. For the scientists the dilemma remains unresolved: an autonomous science adviser is soon at the periphery rather than at the center of policy making; an involved adviser is soon the advocate of all official policy rather than its critic, an ambassador from the officials to the scientists rather than the scientists' spokesman, or at best a broker between the scientists and the officials. The scientists who are dismayed by these hard choices may perhaps find some consolation in knowing that all other groups in a democratic order confront similar frustrations.

A DEPARTMENT OF SCIENCE

Another choice involves the recommendation for a unified department of science, or for a department of science and technology. This proposal to concentrate most of the talents and other resources of the scientists in a single agency, and "to give Science a voice at the Cabinet table," is a strategy supported by the precedents of comparable aspirations in agriculture, business, and labor. But the proposal encounters today, as it has since John Wesley Powell advocated it before the Allison Commission in the 1880's, the stubborn pluralism of the scientists themselves, the uncertainties of the scientists about the boundaries of their interests, and the opposition of government scientists more willing to endure their existing, familiar organizational environment than to risk the unknowns of a new and untested arrangement. With the scientists thus embattled among themselves, neither the nonscientist interest groups nor the public officials now seem likely to take a determined initiative on a question the scientists, as such, cannot decide. A department of science, then, waits upon the unlikely event that the scientists will soon be able, and will find it desirable, to decide who they are, who their accredited spokesmen are, and what their common goals are, and, most important, able to conclude that they are sufficiently unified to risk their separate interests to the leadership and fortunes of a single government institution.

AN AUTONOMOUS SCIENCE AGENCY

An alternative choice—the creation of an autonomous science agency, but with a limited assignment—has been at least temporarily decided upon. The National Science Foundation has completed its first decade; its durability now appears convincingly demonstrated. The independence of the agency from the supervision of officialdom is not as great as was hoped for by those spokesmen for the scientists who piloted the proposal through the hearings, the amendments, the debates, and the votes of the 79th and 80th Congresses, past the shock of a Presidential veto indicting excessive autonomy, to the eventual compromise enacted by the 81st Congress. Some of the form, and more than a little of the substance, of autonomy was lost along the way. Annual budgets and annual appropriations are continuing reminders that autonomy is limited, even in decisions about kinds and amount of basic research, and even after sputniks gave the agency higher priorities and the scientists greater authority.

A close look at the composition of the National Science Board since 1950 also raises the question of whether the agency does not more nearly reflect the autonomous voice of university and other administrators of science, alumni from the ranks of scientists though they be, than it does the voice of scientists in the classrooms and laboratories. The task of representing the scientists on the Board has apparently, with the passage of time and with the entry of Presidential and other preferences, been entrusted more to surrogates for scientists than was the expressed expectation of the sponsors and the officials in the discussions accompanying the passage of the National Science Foundation Act of 1950. Surrogates perhaps provide "virtual" or "existential" representation for the scientists; other groups in American society must often accept a similar settlement.

SPECIALIZED SCIENCE AGENCIES

Most existing government science organizations represent a different kind of choice for scientists. These science agencies are immersed in the political system of a large department or "independent" agency, the degree of autonomy of the science unit in that system varying widely. The life scientists, for example, occupy many special units in Agriculture, in Health, Education, and Welfare, and in Interior; the nuclear scientists are found in the Atomic Energy Commission, and other physicists and chemists, in the Bureau of Standards; meteorologists staff the Weather Bureau; scientists of many varieties inhabit Defense Department units; while the geologists have their sanctuary in the Geological Survey, the space scientists have theirs in the National Aeronautics and Space Administration, and the economists have theirs in the Council of Economic Advisers. The other social sciences are less visibly accommodated, but they do staff numerous units in Agriculture, in Commerce, in Health, Education, and Welfare, and in Labor.

The leaders of all these science units have links, strong or attenuated as the case may be, to the associations and institutions of scientists outside the government, but inside the departmental or agency system they share the powers of decision and compete for priorities with other members of the executive hierarchy, and they report to congressional committees whose concerns are not confined to questions of science or the preferences of scientists. In these many science enterprises the scientists are partners with nonscientists rather than autonomous decision makers. They may employ the *mystique* and the *expertise* of science as strategies to maximize their autonomous role, but they cannot realistically expect to be more than senior partners. Most frequently they will be compelled to accept the status of equal partner with nonscientist officials; not infrequently they will find they are actually junior partners. Their hopes for autonomy are, in practice, curbed not only by nonscientist officials in the executive hierarchy and by congressional committees but also by the activities of the interest-group associations in the science bureau's own special constituency. Thus, the Bureau of Mines must listen attentively to the American Mining Congress and the United Mine Workers; the Bureau of Standards, to many industry associations; the Weather Bureau, to the Air Transport Association and the Farm Bureau Federation; the Public Health Service, to the American Medical Association and the American Cancer Society; NASA, to the aviation industry associations; the Atomic Energy Commission, to the electric power associations and many contractor groups; and agricultural research bureaus, to the Cotton Council and numerous other commodity associations. Rare is the science bureau which is not required by its political environment to bargain continuously with, and accommodate its aims and its priorities to, the interest groups in its constituency.

ADVICE TO CONGRESS

Science agencies in the Executive branch have occupied most of the attention of scientists. If they are to pursue their aspirations for a more distinctive and influential role in science policy, the scientists will find it necessary to formulate a general strategy concerning advice to Congress from scientists. No Congressional committee is now organized and staffed to give exclusive and comprehensive attention to science policy and to listen continuously to scientists, although the House Committee on Science and Astronautics takes a broad view of its science role, and its Senate counterpart may follow suit. Most scientists must pursue their congressional interests across almost the whole range of committees and subcommittees in both Senate and House. If unity and comprehensiveness in congressional action on science are desired—unity such as is sometimes proposed for science in the Executive branch—scientists will be required to choose among several apparent alternatives: they can propose a joint committee on science and technology, with a wide-ranging jurisdiction over all the concerns of scientists; or they can propose a comprehensive committee on science and technology in each House, rather than a joint committee; or they can aim at the creation of a joint committee on science policy with a more limited assignment, or of such a committee on science policy in each House. If changes like these were to be made in congressional science committees (an event to be anticipated only after long and determined effort), the scientists still could not expect to enjoy a monopoly of attention from the new committees. Those other groups who now share power with the scientists' spokesmen in the numerous specialized committees and subcommittees would follow the scientists into the new arenas of influence. The scientists might, however, hope to have, at least for a time, higher status and legitimacy as spokesmen before such new committees, and they might also hope that their competitors in the new setting might compete with each other as well as with the scientists. The question which would soon confront the scientists, however, would be, could they establish and maintain their own unity of goals and priorities before the new committees? The odds in favor of an affirmative answer do not seem to be high.

Politics Inescapable

SCIENTISTS influential in the creation, maintenance, and modification of American science policy are scientists in politics. The spokesmen for the scientists need not be party officials nor candidates for, or occupants of, elective public office, but they will have to be active participants in other phases of the political process—as high government science officials, as science advisers to executive officials, as spokesmen for science policies before committees of Congress, as organizers of opinion through the communication media, as officials and leaders of science associations and institutions. The leaders of the scientists cannot escape politics and remain leaders in science; since their leaders cannot escape politics, the scientists as a whole are in politics too—even their silence is interpreted as acquiescence.

Leading American scientists have long entered the political arena with boldness and success. Convincing examples are provided by the zeal and skill with which the Scientific Lazzaroni piloted the National Academy of Sciences through the Congress and secured Lincoln's signature in 1863; by the subtlety and determination with which Powell secured the establishment of the Geological Survey in 1879 through an appropriation bill rider (a technique that is of the essence of politics); by the frequency with which the Cosmos Club has served as the meeting place of an informal caucus of scientists planning the strategy for a scientists' *coup d'état* in the public interest; by the magisterial role of Vannevar Bush in national science policy; by the sophistication of the Federation of the Atomic Scientists in their 1946 attack upon the May-Johnson bill and their shaping of the terms of the McMahon Act.

Scientists in politics share the problems of other participants in the political process. No special dispensation exempts the scientists from the hard choices and continuing difficulties which the political process imposes upon all those who aspire to shape public policy. One course is to seek to maximize the unity of all scientists and to establish legitimacy for the spokesmen of a unified science community. An alternative is to accept diversity and competitive priorities among scientists and to establish the identity of the separate groups of scientists, establishing the legitimacy of their respective spokesmen. Whichever of these two main roads is chosen, the united or the separated scientists will face the necessity of recruiting allies from among organized groups of nonscientists; the scientists cannot exercise a unilateral dominance in the making of science policy. Alliances are created and maintained at a price; the price takes the form of mutually acceptable accommodations in policy or priorities. Scientists in politics meet with varying fortunes in the process of bargaining with allies and opponents: in the Bureau of Agricultural Economics they find an environment too severe for survival; in the National Institutes of Health, an embarrassment of riches; in the Weather Bureau, high-velocity crosswinds of pressure; in the Geological Survey, an atmosphere of quiet and modest benevolence which has existed for a half century, since the belligerent initial decades; in the Bureau of Standards, a favorable equilibrium of forces, but in the Public Health Service, an unsteady equilibrium. Such variations are the common experiences of most participants in the political process. The conditions which determine the range of variations are best understood, anticipated, and managed by those who are politicians—that is, by those who are expert in the political process.

The 1958 Parliament of Science states the scientists' hopes and fears in persuasive terms.

> This scientific revolution will totally dwarf the Industrial Revolution and the other historical instances of great social change. It will be more compelling, and will pose more urgent problems, because of both the pace and the magnitude of the changes which now impend.
>
> What faces man is not, in any restricted sense, a scientific problem. The problem is one of the relation of science to public policy. Scientific issues are vitally and almost universally involved. The special knowledge of the scientist is necessary, to be sure; but that knowledge would be powerless or dangerous if it did not include

all areas of science and if it were not effectively pooled with the contributions of humanists, statesmen, and philosophers and brought to the service of all segments of society.

What is to be done? Scientists certainly have no arrogant illusion that they have the answers. But they do want to help. They are, moreover, convinced that the time is overripe for a more understanding collaboration between their special profession and the rest of society.

The scientists are now inescapably committed to politics if they hope to exercise influence in the shaping of public policy, including science policies. The leaders of the scientists, then, are perforce politicians. As politicians in a democratic order, they are effective in the degree to which they understand the political process, accept its rules, and play their part in the process with more candor than piety, accepting gladly the fact that they are in the battle rather than above it. The spokesmen for science have occasionally lectured the nonscientists, sometimes sternly, upon their obligation to understand science. Perhaps the advice may be reversed: the scientist has an obligation to understand, and to play his significant role forthrightly in, the polity.

NOTE: In the preparation of this article I have drawn upon the bibliography given below and upon data developed through an extensive series of interviews with science officials in Washington during 1959 and 1960 and from a study of pertinent official and unofficial documents—a project supported in part by the Council for Atomic Age Studies, Columbia University.

BIBLIOGRAPHY

AAAS Committee on Science in the Promotion of Human Welfare, *Science*, 132, 68 (1960).

Ann. Am. Acad. Polit. Soc. Sci., 327 (January, 1960).

B. Barber, *Science and the Social Order* (Free Press, New York, 1952). [Paperback edition, Collier Books, 1962.]

W. R. Brode, *Science*, 131, 9 (1960).

A. H. Dupree, *Science in the Federal Government* (Harvard Univ. Press, Cambridge, 1957).

President's Science Advisory Committee, *Strengthening American Science* (1958), and subsequent reports.

President's Scientific Research Board, *Science and Public Policy* (Government Printing Office, Washington, D.C., vols. 1–3.

D. K. Price, *Government and Science* (New York Univ. Press, New York, 1954).

Science, 127, 852 (1958).

U.S. Congress, Interstate and Foreign Commerce Committees, hearings on National Science Foundation Act, 79th, 80th, 81st, 83rd, 85th, and 86th Congresses.

U.S. Congress, Senate Committee on Government Operations, hearings and reports on Science and Technology Act of 1958, 85th Congress; hearings on a Department of Science and Technology, 86th Congress.
A. T. Waterman, *Science*, 131, 1341 (1960).
D. Wolfle, *Science*, 131, 1407 (1960).

37 *The Autonomy of Science*

EDWARD A. SHILS

I. *The Tradition*

THERE IS AN INNER AFFINITY between science and the pluralistic society. The conduct of scientific research requires a pattern of relationships among scientists which is the prototype of the free society. In microcosm, the scientific community mirrors the larger free society. The internal freedom of the scientific community requires also a freedom from external control. The internal freedom of the scientific community requires the autonomy of science, and whatever may be the relationship among the other spheres of a society, their relationship to the scientific community must be pluralistic.

Quite independently of modern individualistic liberalism, the tradition of the free community of science has grown up. The chief elements in this tradition are internal publicity of the results and procedures of scientific research and free admission to the profession of science on the basis of the qualifications necessary for the conduct of scientific research.

The community of science is built around the free communication of ideas among a relatively small number of intellectually interested and qualified persons whose judgment is recognized to be a measure of validity, and whose approbation gives confidence in the truthfulness of discoveries and in the fruitfulness of the paths traversed. This tradition is part of the fundamental constitution of science. Without it science could not exist. Science is not a collection of results of individual investigators who happen to have been working on the

Reprinted from The Torment of Secrecy: The Background and Consequences of American Security Policies, *by* Edward A. Shils *(New York: The Free Press of Glencoe, 1956), pp. 176–191.*

same subjects at more or less the same time. Science is the product of a very informal community of many scientists working on similar or related problems—matching their own results with one another's or using them as the point of departure for their own investigations. The communication of scientists takes place through publication in scientific journals, through the distribution of offprints, through private correspondence and conversation. This has been harshly misunderstood by the custodians of loyalty and security.

This system of communication is not a tradition in the sense that it is an outworn vestige to which people are irrationally attached just because they have been attached to it in the past. It is a tradition which has grown out of the work itself, as a necessary part of it and as an indispensable condition of its continuation. Without free discussion and the possibility of contact with ideas different from his own, a scientist can seldom discover new implications in his own ideas. It is not that someone else tells him what his ideas imply—although this often happens, and with great benefit—nor does he simply add the other scientist's insight to his own. Rather, contact with the other scientist's train of thought sets his own going in directions in which it might not have gone by itself. In this way discoveries are made, science grows, and mankind gains in stature and in welfare.

The decisive fact is that in the autonomous scientific community, decisions as to truth and falsity, as to feasibility, and appropriateness, are made by scientists in the light of standards which their own best judgment recommends. These standards are inherited in the scientific tradition. They are the product of the reflection, observation, and genius of the best minds of several centuries; they undergo continual and gradual revision, clarification and improvement in accordance with the judgment of scientists who respect their own tradition and yet move freely, critically and creatively within it.

Fundamentally, the autonomy of science or of any field of activity is infringed on when its practitioners are required to conform with an ideal of ideological perfection, to an ideal which requires in the individual's past, present, and probable future conduct and associations a complete conformity with an ideal of the perfect American—such as has been implicit in many decisions of American security officers.

The standard of truth in science has nothing to do with the criteria of political success or of political loyalty. A scientific proposition is true or false in accordance with the standards which are appropriate to scientific judgment. Whether a scientific proposition is true or false depends not at all on whether the person who asserts it is a loyal American, a loyal Russian, a disloyal American, or a politically indifferent Frenchman or Pole. A member of the Communist Party might be a poor scientist, but the determination as to whether he is a poor scientist can be made only by qualified scientists who would not consider his Communist affiliation in arriving at their judgment.

The autonomy of science is infringed on when scientists who are qualified by their training, personal qualities and intellectual gifts, as assessed by their peers and seniors, are prevented for extra-scientific reasons from working on problems on which research is possible and for which resources are available. It is infringed on when scientists are unable to discuss, publish, or circulate their work to other scientists interested in the same or related problems. It is infringed on when scientists are unable to leave their country or to enter another country to attend a scientific congress because the government in the country from which they come or to which they wish to go is concerned about their ideological adequacy. It is infringed on when talented young scientists are refused grants which are otherwise available and for which they are otherwise qualified, or when older and well-established scientists are refused research grants for which their achievements and reputation qualify them because their ideological disposition is adjudged to be unsatisfactory.

The introduction of loyalty criteria for the Fellows of the Atomic Energy Commission or for recipients of grants of the U.S. Public Health Service to conduct research on subjects having nothing to do with weapons or military security has been such an infringement. It has likewise been an infringement when a few universities have been reluctant to employ scientists who have been unable to obtain security clearance for work on classified subjects because the universities have hoped to obtain subsequently government contracts which require that all scientists on the projects have security clearance. (The development of similar policies in industrial firms employing scientists and hoping to obtain defense contracts also constitute infringements of the autonomy of economic life.)

The autonomy of science is further infringed by the McCarran-Walter Act and the rigid application of the visa provisions by consular officials and the Visa Division of the State Department when they deny visas to scientists who wish to enter the United States to attend public scientific congresses or to do unclassified research at American universities. The fact that these scientists in some of the many instances, but by no means all, had once been connected or were still peripherally connected with groups now classified as Communist or Communist-front organizations in the United States has not rendered

1. Nowhere is this tendency more visible than in the social sciences. Either out of a desire to obtain funds not otherwise available or because the value of their enquiries in the discovery of truth is less appealing to them, social scientists greatly overstate the immediate practical value of their studies and do less than justice to their

the infringement of autonomy any less real or effective. The withholding of passports from American scientists by the State Department is no less an infringement on the autonomy of science, although fortunately it has been less frequent than the refusal of visas.

These are simple, superficial instances of the intrusion. It is no less an infringement of the autonomy of science when political criteria are self-imposed. When those who are charged with the maintenance of the autonomy of their sphere adopt criteria which emanate from another sphere, they are to that extent renouncing their independence.

When the American Council of Learned Societies, which has in its charge the furtherance of humanistic studies, is forced into a position where it defends itself by pointing to its work on dictionaries, which if they could obtain additional financial support, would allow the United States to compete more effectively with the Soviet Union, the professors of humanistic studies have ceased to see their subjects as intrinsically, autonomously valuable and have taken over the standards of another sphere. Throughout American academic life, there has emerged a tendency to legitimate scientific and scholarly activities on utilitarian grounds—on grounds of aiding the conduct of the Cold War, of contributing to material well-being, etc. It is not that these goals are unworthy or that scientific and scholarly studies should not aid in their realization. The principal of partial autonomy requires consideration for the goals of the other spheres and for the maintenance of the entire society as well. But when there is never an argument given regarding the intrinsic value of the activity there is something wrong. Either the administrators of the learned lack the necessary self-esteem, or they feel so unappreciated by their society that they think they cannot aver their real sentiments about the true value of their intellectual activities.[1]

The tradition of an autonomous community of scientists, regulating their own culture, determining what is true and what is false by observation and analysis freely made and freely communicated, entails a system of publicity far from the populistic tradition of publicity, but in America it has been able, under favorable circumstances, to coexist with and even gain encouragement and sustenance from it. Persons brought up in the culture in which publicity is accepted have felt no strain of contradiction with the world of science. They came indeed to think that the complete freedom of criticism of all and everything and the lack of respect for authority made America more harmonious with science than other, more hierarchical countries. To

potential value in the disinterested search for truth. This cannot be attributed to the preoccupation with secrecy and conspiracy, but it is closely related to the more basic populistic strain in American culture.

those who grew up in the culture of the scientific community, the widespread practice of publicity seemed—insofar as they concerned themselves at all with what went on outside science—to be no more than right.

There was, however, a point at which the two traditions diverged. Ever since the last decade the nineteenth century there have been explosions of populistic hostility against big business, or against government, or against the intellectuals. It has not been by any means continuous or universal, but in each of the past seven decades there has been a flurry of excitement against the people's enemies. The battle was fought "on behalf of the people" by politicians. It was there that the autonomy of science, necessary to its existence, was interpreted as aloofness and a feeling of superiority, the interest in pure science derided as self-indulgence and a sign of inadequate appreciation of the needs of the people. The ivory tower was assailed as morally suspect.

Science, particularly science which bears fruit in technology and in public health, is much appreciated in the United States, and even the pure scientist has rarely been regarded as worse than a harmless eccentric and at best as a new priesthood and the parent of technology and material well-being. But wherever there is a residue of traditionalism there is also a lurking distress about science as the critic of inherited beliefs. Fundamentalism came into existence in America as a theological movement which sough to repulse the movement of the scientific attitude towards traditional religion. Its major engagement with modern science was at Dayton, Tennessee, and it cannot be said that the Scopes trial had any significant consequences for the development of American science. American science was carried on in the great universities with sufficient self-esteem and attachment to the traditions of intellectual freedom, and the physical and biological research encountered no obstacles.[2] The Scopes trial was of interest only as a notification to the larger world of a tense undercurrent of distrust towards science and scientists.

II. *Science, Politics, and Security*

FOR most of the first half of the present century, science lived as it always had in the past, largely unto itself. The increasing industrial application of science did not cause any worry because industrial science was on the whole applied science and the small amount of industrial secrecy involved was of no consequence for the life of

2. The path of the social sciences, too, was quite unobstructed in the great universities, although those subjects have not ceased to be under intermittent attack from the outside, during the present century. It should be noted that the attackers almost always alleged that they acted in defense of traditional institutions and moral

the scientific community and for the ethos of scientists. Publicity within the community of science was unhampered. It was only when governments decided that science was useful and even entirely crucial to military strength and to national survival that the trouble began. On the one hand, science began to benefit from a largesse which hitherto the greatest private patrons, local and state government bodies, and philanthropic and educational foundations could never afford. On the other hand, it began to find itself working under rather embarrassing and constricting circumstances.

The Second World War brought scientists into a new realm, bounded and criss-crossed by secrecy and compartmentalization. During the war scientists chafed at the restraints, complained to one another, and accumulated much distress about the restrictions imposed by the security system. They undoubtedly infringed on the lesser security rules of compartmentalization, not because they were gossips or busybodies, but because they had to do so in order to get their work done. They did not challenge the principle of security against the passage of information into the hands of agents of enemy or potentially hostile powers, and they even respected this principle when the foreign power was a military ally at the time. Dr. Oppenheimer's action regarding Mr. Chevalier is probably only one of many instances in which scientists, although not regarding the subordinate details of the security system as sacred, respected the fundamental principle and refused to disclose classified information to an outsider. As far as we know, the war ended without a serious breach of security by any American scientist. It also ended with military nerves frayed by the wilfulness of scientists, their slipshodness with respect to the details of military security, their individuality and waywardness.

The end of the war brought no remission for American science. The rapid discovery of the expansionist tendency of Soviet policy forced the United States government to remain as active in the field of science as it had been during the war. The government has remained far and away the chief source of funds for scientific research. Normal caution would have dictated an alert program of information and personnel security since a substantial fraction of the research was directly relevant to military technology, and a security system alone would have been difficult enough. The situation was sorely aggravated, however, by the attraction which science and scientists exercised for persons preoccupied with *maximal loyalty* and obsessed simultaneously by secrecy and publicity.

beliefs, that research was scarcely attacked at all, teaching somewhat more and especially in the lesser colleges and state universities, and that the external political activity of social scientists was the most frequent target.

The fundamentalist hostility towards science, the populist preoccupation with conspiracy and its dislike of rarified intellectual interests, had hitherto been without great weight in the life of the scientific community. They became more weighty only when events over which they had no influence presented secrecy and conspiracy as a more realistic problem. Their own imaginations became inflamed and they communicated some of their own agitation to those who had to deal with secrecy and conspiracy as practical problems.

The atomic bomb and the subsequent discussion of the secret as the source of American strength made some sectors of American opinion feel that their survival lay in the hands of university scientists, a species which had not previously ranked very high in their esteem. The arrest of Dr. Alan Nunn May and the disclosure of the atomic energy espionage ring in Canada under the control of Soviet diplomatic officials, the testimony of Miss Bentley and Mr. Chambers, the Hiss and Coplon trials, Fuchs and then Pontecorvo, and the Rosenbergs, were evidence of the vigor of Soviet espionage and of the dangers which it brought for the retention of secret scientific knowledge on which our national security in part depended. It also encouraged a belief that scientists were not reliable. What was more natural than for those already inclined to be suspicious to begin with to suspect the scientists of faltering loyalty and of inadequate reliability in matters of security? After all, a few eminent scientists had, before the war, lent their names to the proclamations of Communist fellow-travelling organizations; a small number of younger scientists had been members of the Communist Party or had expressed sympathy with it before, during, and even after the war; and one scientist was known to have spent an evening with a Soviet consular official.

Thus, among those figures in the legislative branch who arrogated to themselves the protection of the security system, scientists tended to fall under a blanket of suspicion. Their dislike for the security system arising from their devotion to the free tradition of science was mistaken for a hostility towards the security system arising from political considerations and, more particularly, from attachment to subversive political beliefs and movements. Political sympathies of the prewar period, often the sympathies of youthful enthusiasm and of the naiveté of a first discovery of the world's problems, were taken as valid evidence of present unreliability in security matters.

Scientists have been expected to prove themselves beyond others because they bear a mark of original sin. They are intellectuals and very different from the ordinary back-slapping, vigorous, go-ahead sorts of men. A very important administrator of scientists, who is himself no scientist, said not many years ago about a very important scientist X who had annoyed him greatly, something like: "X is abnormal.

If only he had been interested in sports when he was a boy, just like Y." (Y too is an extremely eminent scientist who since this general exculpation has lost his security clearance.)

This might explain the distrust. The demand for perfection in conduct, associations, and loyalty has perhaps another source. The scientist, like the bureaucrat, is a challenge to the legislator while nominally his subordinate. His recondite knowledge gives him the upper hand. The scientists are able to obtain immense sums of money to discover ways of protecting the country. The politican cannot reasonably be expected to possess the knowledge necessary for adequate supervision of the scientist's work, and yet he feels a responsibility for doing so. Where the results of their work cannot be checked in detail, and where indeed the executive branch discourages it, there is a corresponding increase in the fervor to check on their qualities, on their attitudes, and above all on their loyalty.

Partly because what they do is so important to the national military security, partly, too, because although they have an enormous responsibility conferred on them, they are not trusted, scientists (and higher civil servants) have come to bear the brunt of the loyalty-security measures. There can be no doubt that scientists are more scrutinized, picked over, and controlled by the loyalty-security system than any other profession or occupation in the United States.

It is practically impossible to discover with any accuracy just how many scientists who are subject to the system have fallen afoul of it. An official of the Federation of American Scientists, given to moderation in his judgments, estimates very tentatively that somewhere in the neighborhood of a thousand qualified scientists have encountered security difficulties. Few of these have been in the same class of scientific eminence as Dr. Oppenheimer, Dr. Pauling, Dr. Condon, or Dr. Peters, but they have been highly trained and qualified scientists, well capable of contributing much to knowledge and the national welfare.

At least one hundred and probably several hundred foreign scientists, most of them of high distinction and many of the greatest eminence, have been prevented from coming into the United States to participate in discussions with their American colleagues, to present papers at scientific congresses and to conduct non-secret research in American laboratories.

Like most Americans brought up in the tradition of a loose pluralistic society and not given to worrying much about the symbols of the whole society, American scientists rallied without reservations to the war effort. The great achievements of the scientists on behalf of the nation are very well known, especially to those who most harry them. No single scientist has been shown to be a spy. None of the famous scientist spies—Fuchs, Nunn May and Pontecorvo—were American; none of the American spies—Greenglass, Sobel, Gold or Rosenberg—were higher than technicians or engineers. The few American minor scientists who were alleged to be involved in espionage have never been shown to have been spies.

No professionals have been asked to sacrifice so much of their own tradition as the scientists. The freedom of publication and discussion is absolutely central to the tradition of science. Yet for about a dozen years, large numbers of American scientists have been required to forego both these necessities on behalf of "compartmentalization" and "classification." They have suffered from misunderstanding, often well-intentioned, but sometimes suspicious, by their military supervisors and their security officers, who have not always understood their nature and their needs. They have done many things which have been foreign to the tradition of free science. On the whole, they have done so without more than a complaint which has accepted the postulates of the system about which they complained.

No profession in a position so invested with crucial knowledge has maintained its trust so well. No other profession has been so suspected by the fanatics of secrecy and those who fall victim to their agitation;[3] none has been more insulted by the ideologists of hyperpatriotism than scientists.

III. *The Impact of Science*

WHAT has been the effect of the security-loyalty measures? The effect on science as such is difficult to assess. It is impossible even for a scientist of genius to assess a lost opportunity of which he was unaware. The freedom of science is so important to its progress just because it allows the accidental to occur. Discovery of scientific truth is too unpredictable to allow anyone to say exactly what has been lost to American science through the unnecessary restriction of communication and the distraction and harassment generated by a security system made more extreme than realism requires, out of an obsession with loyalty. A small number of senior scientists and some outstanding younger scientists have apparently refused to work in fields dominated by the security procedures. A large number have been prevented from working in these fields, and an even larger group, consisting mainly of younger men of high quality, who are both cleared and willing to work under security procedures, feel constricted in working under what they consider conditions of unjustifiable discrimination, diminution of their freedom, and lack of esteem for the dignity of science.

3. A high official of the United States embassy in London once shouted in reply to a request for help in obtaining a visa for a famous British scientist and philosopher,

There has never been any likelihood of anything remotely approximating a wholesale exodus or a general strike or a widespread boycott of scientists against the government. There is, however, a vague and growing feeling among some of the most distinguished scientists of the older generation, and more commonly in the younger generation, that they are serving masters who have no sense of the tasks scientists and the country as a whole are facing, and who are often utterly inaccessible to reason. There is a growing feeling that life in the university is better than life in a government laboratory and that even within a university, work on unclassified projects is better than work on classified subjects. Scientists appreciate the financial generosity of government departments and they appreciate, too, the efforts of the administrators and military men with whom they have to deal to facilitate their work and to spare them some of the ignominies which otherwise await them. It is not so much the technically necessary secrecy which disturbs them—there are very few scientists and none of serious stature who deny the need for some system of security—as it is the atmosphere which has been generated. One of the most respected American scientists, Professor Hans Bethe, no radical and no enemy of security, has said,

> Perhaps the greatest impediment to the scientist . . . is the political climate of the country. . . . We sense a distrust of scientists and of intellectuals in general, in Washington and some newspapers and among many radio commentators. Concern about the future development of this trend is perhaps the greatest outside influence which hinders the effective work of scientists.

Nor is this view confined to scientists. Administrators of great practical experience have asserted that if the postwar security provisions had obtained during the war, it would not have been possible to have accomplished anything approximating the great successes of wartime scientific research. Some of the consequences of security for science will take years to be uncovered. Only after many years, if ever, will it be known which crucial bits of knowledge lay hidden by classification and had to wait for a long time to be rediscovered by someone else who then made use of them for some important discovery. That such things happen in the history of science is certain even when communication is free. They are even more likely when it is unfree.

It was recently pointed out that if security provisions such as now exist in the United States had existed in Germany early in the present century the discovery of nuclear energy might not have occurred. Professor Einstein's postulation of the equivalence of mass and energy through the concept of relativity was possible only on the basis of

well known for his anti-Communism. "Oh, that's it! He's a scientist. We've had some experience with them. We've been bitten by them."

Max Planck's work on high temperature radiation. If present security regulations existed at that time, Einstein might never have known of Planck's work since it would have been classified as relevant to military security.

A scientist, under our present system, might formulate an idea of military importance, but can be excluded from developing it by reason of his not having a "need to know" (in the security sense of the term). Even though he could contribute greatly to its development, compartmentalization within the security system might prevent him from either contributing to, or learning from, subsequent development. Or the idea might be classified at birth, and other scientists who might otherwise have taken it up and been stimulated by it might never learn about it even though they have been "cleared." The idea, however potentially useful, must then wait for a duplicate discovery. We are not referring here to the financial wastefulness of duplicate discoveries such as arise in consequence of our security policies in forcing British work to duplicate American work with no gain to either country. We refer to the slowing down of scientific progress, and of the progress of military technology which arises from scientific progress.

Older men, more mature or more resigned, or more satisfied to be indifferent if they are left alone to work as they wish, might not feel as strongly as the younger men to whom science and the pursuit of truth is more than a job and to whom the idea of the scientific community is very precious. Such young scientists, perfectly cleared, might still be pushed away and discouraged from doing the research most needed by the country. They feel restricted and frustrated, and as an eminent scientist with great and continuing responsibility in the administration of scientific research has recently said, "There is a limit to the frustration experienced by the really creative scientist, beyond which his creativeness is destroyed irrespective of his willingness to serve." Thus to the extent that militarily relevant science requires for its prosperity the preservation of the integrity of the scientific community, the damage done to it by the excesses and indiscriminateness of the present loyalty-security system hurts both national defense and the spirit and atmosphere which sustains pure science.

Science is, however, far from being only an instrument to be operated for increasing military power or for contributing to economic well-being. It is one of the highest forms of the expression of man's nature and his freedom. Scientific activity is the activity of free men—not of all free men, but of those who have special gifts and qualifications—and its community is the epitome of the free society. Within itself science provides the model of a free society of reasonable men coordinating themselves voluntarily in the light of a transpersonal standard, their own individual intelligence and the judgment of their qualified peers. No one can be coerced in science into beliefs which are contrary to his convictions, and yet his observance of the trans-personal standards of truth and the judgment of his peers provides the framework which facilitates creativity and yet restrains arbitrariness and mere eccentricity.

The scientific community must not only be free internally—here it is obstructed by compartmentalization and overclassification—but it must also be as free as possible in its external relations. Criteria of admission to the scientific community other than evidence of intellectual quality and the moral standards necessary for scientific work harm that community and prevent it from doing its proper work. It was a demand for loyalty which excluded Professors Szilard, Wigner, von Neumann, Franck, Weisskopf, Meitner, and the late Enrico Fermi from the scientific communities in their own countries. The effects of this exclusion and its benefits for science in America are known to all. Yet, in a more certainly conscientious and ostensibly more rational way, the United States government has confined the boundaries of the scientific community, refusing research grants, suspending scientists and denying fellowships on criteria which, although far less brutal, are almost as unrealistic and as irrelevant to truth, national security, or welfare as the Nazi and Fascist criteria.

While on the one hand, the National Science Foundation, the Atomic Energy Commission, and the U.S. Public Health Service have exerted themselves to attract the best young minds to science and to provide them with the means for prosecuting their research, the security-loyalty policies of the government are unwittingly undoing the government's own work by making the scientific career less attractive and more worrisome and distracting than it must be if creativity in scientific discovery is to be fostered.

The elementary requirements of a sound security program would infringe, to some extent, on the autonomy of science. Some work would have to be classified, and the principle of publicity within the field of science would be somewhat restricted. Some scientists would have to be checked for their reliability in the handling of secret information, and this would introduce a restriction of the autonomy of science in regulating its own membership and its allocation to the various scientific tasks to be undertaken. The instrusions of a realistic security program into the autonomous domain of science would on the whole be marginal and of no great significance to the central concerns of science.

As long, however, as the security program mixes considerations of maximal loyalty with those of security, the area of science over which autonomy obtains is much more restricted, and without gain to science or to the nation. Until the preoccupation with loyalty and particularly with maximal loyalty is eliminated from the field of security, it will continue to be so, despite the quiescence of demagogues and the conscientiousness of security officers. The demagogues, although silent for the time being and perhaps for even longer, may well rest. Their influence is more lasting than the uproar they caused.

38 *Malicious Philosophies of Science*

ERNEST NAGEL

THERE IS NO SUBSTANTIAL EVIDENCE for the widely held view that changes in the content and standards of theoretical inquiry are uniquely determined by changes in the economic and political structure of society. To be sure, scientific inquiries are often initiated and subsidized by those concerned with problems of commerce and technology, and the manner in which scientific discoveries are assimilated by a society depends on its economic and political organization. But once a department of inquiry establishes its traditions of workmanship, so the history of science seems to indicate, the course of subsequent developments in it is determined by the materials explored, by the talents and skills available, and by the logic of theoretical investigation.

In almost every age, however, the attitudes which men assume toward personal and social issues have been often justified by them in terms of their understanding of the methods and latest findings of science. Professional scientists have frequently used their specialized knowledge to buttress or criticize the institutions of their day; but publicists, religious leaders, and philosophers have usually played a more prominent role in this task of evaluating the general social import of scientific methods and scientific theories. Such evaluations do not, in most cases, flow from the specific character of scientific methods or their technical achievements; they issue from the social and religious commitments of those who make them, and are symptomatic of the stresses and strains in the social scene.

In the midst of actual and impending disaster, men are inclined to

---

Reprinted from Sovereign Reason, and Other Studies in the Philosophy of Science, *by Ernest Nagel (New York: The Free Press of Glencoe, 1954), pp. 17–35.*

listen to any voice speaking with sufficient authority; and during periods of social crisis, when rational methods of inquiry supply no immediate solutions for pressing problems, spokesmen for institutional and philosophic theologies find a ready audience for a systematic disparagement of the achievements of empirical science. Ideas which the advance of knowledge had partially driven underground during periods of fair social weather, are then insolently proclaimed as panaceas for public and private ills. The assured methods of scientific control and understanding, because they effect no wholesale resolution of problems and because they yield no conclusions beyond the possibility of error and correction, are then declared to be unsuitable guides for rational living.

The mounting economic and political tensions of our own age have not failed to produce a literature of this type. From various quarters—from men of science, historians of ideas, as well as outspoken representatives of theological systems—there has come a flood of criticism of modern science and of the secular naturalism which has accompanied its growth. The criticism has been neither uniform nor consistent. But the common objective of much of it has been the limitation of the authority of science, and the institution of methods other than those of controlled experimentation for discovering the natures and values of things. Many recent evaluations of science have thus had an obviously malicious intent; and the present essay will seek to determine briefly to what extent, in the case of several influential types of philosophies of science, good sense has been sacrificed to such malice.

1

UNDOUBTEDLY the most solidly intrenched intellectual basis for the current disparagement of science is a well-known but nonetheless questionable theory of knowledge upon which experimental method is made to rest. This type of critique starts with the familiar fact that in many theories of the natural sciences, especially physics, the various sensory qualities (such as colors and sounds) receive no *explicit* mention, and that it is only the quantifiable traits of things (such as mass and length) which are noticed. The immediate conclusion which is then drawn is that sensory qualities are not properties of objects in their own right, but are dependent on the activity of an *immaterial mind*. The remainder of the argument may then proceed along either of two

1. Two citations from recent writers will help convey the flavor of the alternative methods which have been proposed.

"Imagination is more adequate to reality than reason, for reality is not rational; therefore poetry and religion are better adapted to the real than the sciences. The real is not abstract and general. It is always concrete and individual; that is the reason why imagination alone can grasp it, whereas the intellect cannot fully conceive it. A theory of imagination . . . is urgently needed as a foundation for ethics, esthetics, philosophy of history and of religion, and even for metaphysics." (Richard Kroner, in a lines of interpretation. According to one of them, more common in earlier centuries than in our own, the traits studied by the natural sciences are the only genuinely real things, while the directly experienced qualities are only a passing appearance. The sights and sounds and smells of the human scene are thus taken to have only a "subjective" existence and to be the otiose by-products of the true executive order of nature. According to the second interpretation, currently highly fashionable, the qualities apprehended in daily experience are the concrete and exclusive reality. It is these qualities which are held to constitute the intrinsic natures of things; and since these qualities are allegedly psychic products, it is they which are regarded as the intelligible substance of the world. In putting to one side the qualitative character of existence the natural sciences are consequently preoccupied with shadowy abstractions, which have at best only a mean practical value; and the laws which are the outcome of scientific inquiry, far from expressing the true nature of things, fail to grasp and convey the dynamic reality of existence.

On either interpretation, therefore, the world is split into two discontinuous realms. One of them, the proper domain of natural science, is a "mysterious universe" forever foreign and essentially unintelligible to the common experience of mankind; the other, the locus of enjoyments and values, is the theater of the mind's activities and creations, and is the only reality in which mind can feel confidently at home. On either alternative, the human scene is endowed with character so distinctive that the procedures of the empirical sciences can provide no guide to it. For the controlled methods of experimentation are held to be relevant only to the realm of abstract quantity, so that the entire field of valuation, of deliberation and moral choice, is exempted from the norms of experimental inquiry. Qualitative reality, which by hypothesis has an inherent connection with mind and consciousness, must therefore be explored by techniques different from those employed in the positive sciences; and in this realm, claims to truth must be subjected to canons of a radically different kind. Imagination, intuition, introspection, and modes of emotional experience, are some of the ways which have been recommended for grasping genuine reality and for understanding human affairs.[1]

paper read before the Third Conference on Science, Philosophy and Religion, as quoted by *The New York Times,* August 29, 1942.)

". . . It is an axiom of sound method that any experience is, in some manner and to some degree, *intrinsically* cognitive. An experience of love . . . is at the same time an insight into the loveable nature of what is loved; an experience of moral urgency . . . is an insight into the rightness of the action to be performed; an experience of reverence . . . is an insight into the divinity of what is reverenced. Every such experience is a growth in wisdom and the wisdom is not testable by scientific techniques. . . ." Philip Wheelwright, "Religion and Social Grammar," *Kenyon Review,* vol. 4 (1942), pp. 203–204.

Nevertheless, the actual character of scientific method offers warrant neither for such attempts to limit its authority, nor for the radical dualism of the qualitative and the quantitative, the mental and the physical, upon which those attempts thrive. A brief mention of some obvious features of experimental procedure will be sufficient to show how inadequate are the analyses on whose basis the authority of scientific method is impugned.

In the first place, however "abstract" scientific theories may be, those theories can be neither understood nor used except in contexts of familiar qualitative discriminations. These contexts are tacitly taken for granted in the explicit formulation of theories, and are neither ignored nor contemned by the practicing scientists. Consider, for example, some of the operations involved in even so elementary a process as the measurement of spatial magnitude: standards of magnitude must be constructed, requiring the use of familiar bodies of daily experience; the relative constancy of the standards must be established, thus necessitating the noting of qualitative changes such as temperature; and the mutual relations of bodies must be discriminated, thus involving the identification and distinction of bodies on the basis of such qualities as color and the texture of surfaces. In general, no metaphysical opposition between the qualitative and the quantitative is forced upon us in this process, since the institution of quantitative standards is simply the ordering and the discrimination of qualitative continua. The view that the subject-matter and the data of the physicist are opaque pointer-readings is clearly a falsification of the scientist's procedure—a falsification which becomes more evident the more thorough is our examination of the full spectrum of science operations.

In the second place, there can be no doubt that the colors, sounds, and other characteristics we perceive in our every-day affairs owe their existence not only to the objects they are commonly believed to qualify, but also to complicated mechanisms (including physiological ones) of which common-sense is frequently unaware. But it does not follow that these qualities are therefore constituted out of some "mental stuff," or that a "mind" (in the sense of a disembodied, experimentally unidentifiable agent) is required for bringing them into existence. Whatever the conditions may be for the occurrence of colors, for example, experience shows extended *surfaces* to be colored; and if it is held that colors are "mental," traits (such as that of being extended) must be attributed to the mind which are the presumptive distinguishing marks of physical objects. In that case, however, what becomes of the notion of mind as a disembodied entity? But dialectic aside, there is no shred of evidence that in addition to complicated physico-organic processes any other "agents" are required to produce the qualitative manifolds of experience. The postulation of an additional agent (held to be something distinctively "mental") is on par with the caprice of endowing the planets with souls in order to account for their motions. The actual procedures of the natural sciences thus offer no ground for the alleged dualism between the mental and the physical; and accordingly, even the semblance of a reason disappears for limiting the scope of experimental techniques.

And in the third place, the "abstractness" with which natural science is charged as a fatal weakness, is in fact a trait of all cognition. All cognition involves the making of distinctions and the recognition of some things as relevant and others as not; in this sense, therefore, all cognition abstracts from its subject-matter and prescinds those features from it which bear on the problems at issue. To *know* the course of planets is not to engage in periodic journeys around the sun; to *know* the factors and conditions of a human transaction is not the same as to participate in its joys and sorrows. More generally, it is not the function of knowledge to reproduce its own subject-matter; and to refuse to make abstractions of any sort is to abandon knowledge in favor of uninformed feeling and blind experience. Accordingly, just what is the quarrel of those critics who find fault with the abstractness of science? Do they seriously claim that the theories of science are not relevant to their subject-matters? Do they maintain that there are ways essentially different from and superior to those employed in the natural science for ascertaining the conditions for the occurrence of things and events? Or do they disdain science simply because it does not supply what no knowledge worthy of the name can offer—an unanalyzed reduplication of its own subject-matter? In either event, their discontent flows from a willful romanticism and a disregard of the historical achievements of the natural sciences; it provides no valid ground for excluding the operations of experimental inquiry from the domain of human affairs.

2

A SECOND widespread critique of scientific methods is at bottom a variant of the one already considered. It does not *explicitly* disown the authority of science in human affairs; but it does recommend the adoption of such vague and irresponsible canons of experimental control that it in effect argues for the exclusion of the logical methods employed in the positive sciences from the study of social problems. This view rests its case on two major claims; that in the past the natural sciences have mistakenly tried to "reduce" all features of the world to "mechanical" or "materialistic" properties; and that recent advances in our knowledge have demonstrated the breakdown of the "mechanical" categories of classical science. The present view, like the previous one, maintains that the human scene is so discontinuous with the "lower levels" of nature that a common logic of inquiry cannot be adequate to all of them; it therefore concludes that problems affecting human destiny must be investigated on the basis of canons of validity and intelligibility which differ radically from those used in the natural sciences.

Let us examine these contentions. And first, what are we to understand by the terms "mechanical" and "materialistic"? When practicing physicists characterize an explanation as "mechanical," they mean a theory which, like the one developed by Galileo and Newton, explains a class of changes *entirely* on the basis of the masses and the spatial and temporal relations of bodies. In this quite precise sense of the word, Maxwell's electro-magnetic theory is not a mechanical theory, and with its advent in the 19th century the earlier hope that the science of mechanics would become the universal science of nature was gradually abandoned. But those who accuse classical science of being "mechanical" are anything but so definite as to the real point of their charge. For according to them the Darwinian theory of organic evolution as well as Maxwell's theory of electro-magnetism are mechanical, although neither of these theories satisfies the physicist's definition of "mechanical." The only clear meaning which can be given to the accusation is that classical science is *deterministic*—in the sense that it attempts to discover the precise conditions for the occurrence of phenomena, without benefit of final causes and without invoking experimentally unidentifiable causal agents.

But if this is the meaning of the charge, the claim that modern science no longer operates with mechanical categories is singularly ill-founded. As already noted, even classical physics recognized that mechanical theories (in the technical sense of the word) are not universally adequate; and recent researches into atomic phenomena have only fortified this conclusion. There is, however, nothing in modern research which requires the abandonment of the *generic* ideal of classical science: to find the determining conditions for the occurrence of phenomena, expressible in terms capable of overt empirical control. Thus, even modern quantum-theory—although it employs technical modes of specifying the character of physical systems which are different from those used in classical mechanics—is deterministic or mechanical (in the loose sense) in so far as it rigorously specifies the unique physical conditions under which certain types of changes will occur. Similarly, modern genetics is no less deterministic than the Darwinian theory, since the former even more completely than the latter has succeeded in disclosing the mechanisms or structures involved in the transmission of characteristics from one generation to another. It is therefore simply not true that recent advances in knowledge have demonstrated the untenability of the logical canons of classical science.

The claim that the world-picture according to natural science "reduces" everything to blind, undifferentiated collocations of material particles, and thus fails to do justice to the distinctive traits of human behavior, likewise rests on a misconception as to what the sciences in fact accomplish. Consider, for example, the following comments of the late Lord Balfour:

> What are we to say about a universe reduced without remainder to collections of electric charges radiating throughout a hypothetical ether? . . . We can certainly act on our environment, and as certainly our actions can never be adequately explained in terms of entities which neither think, nor feel, nor purpose, nor know. It constitutes a spiritual invasion of the physical world:—it is a miracle. . . . We are spiritual beings, and must take account of spiritual values. The story of a man is something more than a mere continuation of the story of matter. It is different in kind. . . .[2]

That this represents a caricature of what the achievements of physics imply, will be evident if we recall that the sciences seek to determine the precise *conditions* under which events come into being and continue to exist. For in ascertaining those conditions the sciences do not *thereby deny* the existence of any traits found in the nature, whether in the human scene or elsewhere. In particular, physics has assumed the task of finding the most general and pervasive constituents and circumstances of existing things; it does not legislate away as unreal or non-existent—and could not do so without contradiction—the things and events into whose conditions of existence it inquires. The explanations which physics offers for the traits and changes it studies, consist of careful specifications of the conditions under which those traits and changes occur; and no other sense of "explanation" is relevant in discussing its findings. Whether these explanations can be stated entirely in terms of a special class of entities and their relations (for example, in terms of the distribution of electrically charged particles), is a specific empirical issue which can be resolved only by a detailed empirical inquiry; it cannot be settled by dialectic, or by an *a priori* fiat such as that the living cannot be explained in terms of non-living.

Criticisms of natural science such as the following are therefore altogether pointless, since they operate with mythological conceptions as to the character of its explanations:

2. Arthur J. Balfour, in an essay contributed to *Science, Religion and Reality* (edited by Joseph Needham), pp. 15–17.

> With the faintest and simplest element of consciousness, natural science meets something for which it has no pigeon-hole anywhere in its system. . . . Mind at its best is autonomous. Granting that it is connected mysteriously and intimately with physical processes that natural science claims as its own, it cannot be reduced to those processes, nor can it be explained by the laws of those processes.[3]

Explanations of "mind" in terms of physical processes do not wipe out the distinction between the behavior of inorganic masses and the distinctive activities of men; nor do they pretend to deduce somehow the direction of those activities from physical laws containing no mention of purposive behavior. Such explanations simply state the generalized conditions for the occurrence of "mind." Accordingly, the only form of "reduction" with which the natural sciences may rightly be charged, is the form which consists in ascertaining the structures under which specfic traits are manifested; and it is clear that if those sciences failed in effecting such a reduction, they would fail in achieving the objective of knowledge. The conclusion seems unavoidable that those who would exclude the logical methods of the natural sciences from fields of social inquiry, on the score that these methods commit their users to the "reductive fallacy," are in effect recommending the abandonment of the quest for the causal determinants of human affairs.

One final claim, associated with the charge that the logic of natural science is "reductive," remains to be considered—the claim that human traits are "emergent properties" on a "higher level" of existence than are those with which physics and chemistry are concerned, so that the methods employed by these disciplines cannot be adequate to the study of the higher emergent qualities. Some examples will make clear the chief features of the theory of emergence. However much we may know about the interaction of hydrogen and oxygen with *other* elements, so it is said, it is impossible to infer from such knowledge the fact that they combine to form water; and in particular, the qualities which emerge when water is formed could never be predicted from those data. Similarly, no amount of knowledge concerning the physics and physiology of the human body makes it possible to deduce the "spiritual" characteristics of the organism as a thinking, purposive creature. Nature is thus conceived as a system of levels of emergence, each level requiring a peculiar mode of study; and *a fortiori*, the distinctive qualities of human beings can be satisfactorily explored and understood only when inquiry into them is conducted on the basis of a logic specific to "spiritual" subject-matter.

3. Brand Blanshard, "Fact, Value, and Science," in *Science and Man* (edited by Ruth Nanda Anshen), pp. 189, 203.

But the following brief remarks will be sufficient to blunt whatever force the argument from the alleged facts of emergence may be supposed to have. It is indeed not possible to *deduce* the properties of water (for example, its transparency or its ability to quench thirst) from those of hydrogen and oxygen, *if* the former properties do not enter into the premises from which the deduction is attempted. For in general, no statement containing a given term "P" can be deduced from a class of statements unless the latter also contain that term. In one sense, therefore, the main contention of emergent evolution is simply a logical truism. In the second place, although the occurrence of certain traits may be left unexplained by one theory, a different theory (perhaps a revised form of the first one) may supply a satisfactory explanation. For example, the theories of physics which were accepted at the beginning of the 19th century were unable to account for any chemical facts, although present-day physics is in the position to explain the occurrence of many chemical reactions. Accordingly, whether a quality is to be regarded as an "emergent" or not is relative to a specific theory, and is not an *inherent* fact about that quality. It also follows that since no theory of science can be regarded as necessarily final, traits which at one time are taken to fall into the province of one specialized discipline, may at some later date be explained on the basis of theories developed in a different branch of science. This has certainly been the history of such sciences as chemistry, biology, and even psychology, in their relation to physics. And finally, if the doctrine of emergence is seen in this light, no clear reasons remain why the logic of experimental inquiry—as conducted in the natural sciences—has no authority over investigations into human affairs. Indeed, as the natural sciences have become more comprehensive they have provided an enriched understanding of human traits. No theoretical limits can be set to such a progressively widening scope of the sciences of nature. And what is no less to the point, these fresh achievements have involved no surrender, on the part of the natural sciences, of the procedural principles under whose guidance they obtained their historical successes.

3

THE views which have been noticed thus far attempt to limit the scope of scientific methods on the basis of considerations that are at least nominally scientific in character. The criticisms of science to which attention must next be directed do not even pretend to adduce scientific grounds for their claims, and are frankly based upon explicit theological and metaphysical commitments for which no experimental evidence is invoked. The chief burden of their complaints is that science offers no "ultimate explanation" for the facts of existence; and their chief recommendation is the cultivation of "ontological wisdom" as the sole method for making "ultimately intelligible" both the order of the cosmos and the nature of the good life.

Some citations from recent writers will exhibit more clearly than would a paraphrase the unique mixture of pontifical dogmatism, oracular wisdom, and condescending obscurantism which seems to be the indispensable intellectual apparatus of this school of criticism. Professor Gilson characterizes the plight of science as follows:

> This world of ours is a world of change; physics, chemistry, biology, can teach us the laws according to which change actually happens to it; what these sciences cannot teach us is why this world, taken together with its laws, its order, and its intelligibility is, or exists. . . . Scientists never ask themselves *why* things happen, but *how* they happen. Now as soon as you substitute the positivist's notion of relation for the metaphysical notion of cause, you at once lose all right to wonder *why* things are, and why they are what they are. . . . Why anything at all is, or exists, science knows not, precisely because it cannot even ask the question. To this supreme question the only answer is that each and every particular existential energy, and each and every particular existing thing depends for its existence upon a pure Act of existence. In order to be the ultimate answer to all existential problems, this supreme cause has to be absolute existence. Being absolute, such a cause is self-sufficient; if it creates, its creative act must be free. Since it creates not only being but order, it must be something which at least eminently contains the only principle of order known to us in experience, namely, thought.[4]

And Professor Maritain, building on the alleged subordination of science to metaphysics, indicates some of the immediate consequences of this hierarchical arrangement:

> Science . . . is distinguished from wisdom in this, that science aims at the detail of some special field of knowing and deals with

4. Étienne Gilson, *God and Philosophy*, pp. 72, 140. Although Whitehead's manner of arriving at his speculative cosmology is radically different from that cultivated by neo-Thomists, his evaluations of the limitations of natural science are frequently not dissimilar. He comments as follows on "the grand doctrine of Nature as a self-sufficient, meaningless complex of facts": "Newton left for empirical investigation the determination of the particular stresses now existing. In this determination he made a magnificent beginning by isolating the stresses indicated by his law of gravitation. But he left no hint, *why in the nature of things there should be any stresses at all*. The arbitrary motion of the bodies were thus explained by the arbitrary stresses between material bodies, conjoined with their spatiality, their mass, and their initial states of motion. By introducing stresses—in particular the law of gravitation—instead

> the secondary, proximate or apparent causes, while wisdom aims at some universal knowing and deals with prime and deepest causes, with the highest sources of being. . . . Wisdom is not only distinct from but also superior to science, in the sense that its object is more universal and more deeply immersed in the mystery of things, and in the sense that the function of defending the first principles of knowledge and of discovering the fundamental structure and organization thereof belongs to wisdom, not to science. . . . Science puts means in man's hands, and teaches men how to apply these means for the happiest outcome, not for him who acts, but for the work to be done. Wisdom deals with ends in man's heart, and teaches man how to use means and apply science for the real goodness and happiness of him who acts, of the person himself. . . . Science is like art in this that though both are good in themselves man can put them to bad uses and bad purposes: while in so far as man uses wisdom . . . he can only use it for good purposes.
>
> The paleontologist does not step out of his sphere when he establishes the hypothesis of evolution and applies it to the origin of the human being. But the philosopher must warn him that he is out of his field when he tries to deny for that reason that the human soul is a spiritual soul which cannot emanate from matter, so that if once upon a time the human organism was produced by a mutation of an animal organism, it was because of the infusion of a soul created by God.[5]

Although criticism of a position is futile when those who hold it make a virtue of its mysteries and when they regard themselves as superior to the usual canons of scientific intelligibility, those who are not so fortunately placed may find the following observations not irrelevant. In the first place, there is a perfectly clear sense in which science does supply answers as to "why" things happen and are what they are. Thus, if we ask why the moon becomes eclipsed at certain times, the answer is that at those times the moon moves into the earth's shadow; if we ask why the moon behaves in this way, the

of the welter of detailed transformations of motion, he greatly increased the systematic aspect of nature. But he left all the factors of the system—more particularly, mass and stress—in the position of detached facts *devoid of reason* for their compresence. He thus illuminated a great philosophic truth, that a dead nature can give no reasons. All the ultimate reasons are in terms of aim at value. A dead nature aims at nothing." A. N. Whitehead, *Modes of Thought*, pp. 183–4, italics not in the text.

5. The first paragraph is taken from the essay "Science and Wisdom," contained in *Science and Man*, pp. 66–7, 72, 94. The second paragraph is from the essay "Science, Philosophy and Faith," contained in the volume *Science, Philosophy and Religion*, the proceedings of the First Conference on Science, Philosophy and Religion, p. 181.

answer is given in part by the theory of gravitation; if we ask why bodies behave in the manner predicted by this theory, the answer is supplied by the general theory of relativity. On the other hand, if we repeat this question concerning relativity theory, no further answer is at present forthcoming, so that for the present at least this theory is an "ultimate" or "brute" fact. Furthermore, if some day relativity theory should become absorbed into a unified field-theory embracing both macroscopic and microscopic phenomena, the unified field-theory would explain why the equations of relativity theory hold, but at the same time it would become the (perhaps only temporary) "ultimate" structural fact. In science the answer to the question "why" is therefore always a theory, from which the specific fact at issue may be deduced when suitable initial conditions are introduced. The point of these familiar remarks is that no matter how far the question "why" is pressed—and it may be pressed indefinitely—it must terminate in a theory which is itself not logically demonstrable. For no theory which explains why things happen as they do and not otherwise can be a logically necessary truth. It follows that those who seek to discern the laws of nature to be necessary, as well as those who "hope to see that it is necessary that there should be an order of nature," are violating an elementary canon of discursive thought.

In the second place, it is obvious that anyone who invokes an "absolute cause" (or God) to explain "why" the world exists, merely postpones settling his accounts with the logic of his question: for the Being who has been postulated as the Creator of the world is simply one more being into the reasons of whose existence it is possible to inquire. If those who invoke such a Being declare that such questions about His existence are not legitimate, they surmount a difficulty only by dogmatically cutting-short a discussion when the intellectual current runs against them. If, on the other hand, the question is answered with the assertion that God is his own cause, the question is resolved only by falling back upon another mystery; and at best, such a "reason" is simply an unclear statement of the grounds upon which scientists regard as unintelligible the *initial* "why" as to the world's existence. But a mystery is no answer if the question to which it is a reply has a definite meaning; and in the end, nothing is gained in the way of intellectual illumination when the discussion terminates in such a manner.

In the third place, the postulation of an "absolute cause" or an "ultimate reason" for the world and its structure provides no answer to any *specific* question which may be asked concerning any particular objects or events in the world. On the contrary, no matter what the world were like, no matter what the course of events might be, the same Ultimate Cause is offered as an "explanation." This is admitted in so many words by Professor Gilson:

> The existence or non-existence of God . . . is a proposition whose negation or affirmation determines no change whatever in the structure of our scientific explanation of the world and is wholly independent of the contents of science as such. Supposing, for instance, there be design in the world, the existence of God cannot be posited as a *scientific* explanation for the presence of a design in the world; it is a *metaphysical* one.[6]

But just what does an "explanation" explain when it explains nothing in particular? What understanding of our world does a metaphysics provide which is compatible both with a design in the processes of nature as well as with its absence, with the existence of specific goods as well as with their non-existence, with one pervasive pattern of causal interactions as well as with another? A high price in unintelligibility must be paid when the canons of scientific discourse and inquiry are abandoned.

And finally, the assumption that there is a superior and more direct way of grasping the secrets of the universe than the painfully slow road of science has been so repeatedly shown to be a romantic illusion, that only those who are unable to profit from the history of the human intellect can seriously maintain it. Certainly, whatever enlightenment we possess about ourselves and the world has been achieved only after the illusion of a "metaphysical wisdom" superior to "mere science" had been abandoned. The methods of science do not guarantee that its conclusions are final and incorrigible by further inquiry; but it is by dropping the pretense of a spurious finality and recognizing the fallibility of its self-corrective procedures that science has won its victories. It may be a comfort to some to learn that in so far as man uses "wisdom" he can aim only at the good; since the most diverse kinds of action—kindly as well as brutal, beneficent as well as costly in human life—are undertaken in the name of wisdom, such a testimonial will doubtless enable everyone engaged in such an undertaking to redouble his zeal without counting the costs. But it is not wisdom but a mark of immaturity to recommend that we simply examine our hearts if we wish to discover the good life; for it is just because men rely so completely and unreflectively on their intuitive insights and passionate impulses that needless sufferings and conflicts occur among them. The point is clear: claims as to what is required by wisdom need to be adjudicated if such claims are to be warranted; and accordingly, objective methods must be instituted, on the basis of which the

6. *Op. cit.*, p. 141.

conditions, the consequences, and the mutual compatibility of different courses of action may be established. But if such methods are introduced, we leave the miasmal swamps of supra-scientific wisdom, and are brought back again to the firm soil of scientific knowledge.

4

THE final variety of current criticism of science to be considered rests its case on the alleged facts of history. The development of science, it is admitted, has brought with it an increase in material power, a broadening of the average span of human life, and a wider distribution of innumerable goods than was possible in earlier days. Nevertheless, so the criticism runs, human happiness has not increased and the quality of life has not improved. On the contrary, increased power over material nature has generated a deadening monotony and uniformity in men's lives, has produced ghastly brutalities, cataclysmic wars, and fierce superstitions, and has undermined personal and social security. Science deals with instrumentalities and is incapable of determining values; and with the spread of secular naturalism and the consequent decline of religious influences, men have grown insensitive to the distinction between good and evil, and have identified material success with ethical excellence. Intellectual historians join hands with preachers and publicists in placing the blame for contemporary Fascism upon the demoralizing effects of positivistic philosophy. And in language solemn and threatening Professor Maritain warns his readers of the dreadful consequences which allegedly flow from scientific naturalism:

> Let us not delude ourselves; an education in which the sciences of phenomena and the corresponding techniques take precedence over philosophical and theological knowledge is already, potentially, a Fascist education; an education in which biology, hygiene and eugenics provide the supreme criteria of morality is already, potentially, a Fascist education.[7]

Whatever may be the validity of the causal imputations contained in such criticism, it cannot be denied, unfortunately, that many of its characterizations of modern society are well founded; it is certainly not the intent of the following comments to dispute them. It must nevertheless be noted that the implied judgment, according to which

7. In the essay "Science, Philosophy and Faith," *op. cit.*, p. 182. Neo-Thomists have no monopoly in the making of such causal imputations. Thus, in his essay "Fact and Value in Social Science," Professor Frank Knight writes as follows: "In the field of social policy, the pernicious notion of instrumentalism, resting on the claim or assumption of a parallelism between social and natural science is actually one of the

the quality of modern life is inferior to that of earlier societies unburdened by an institutionalized natural science, is based on a definite set of preferences or values in terms of which human history is surveyed. But while it is clear that there is nothing reprehensible in employing definite standards of evaluation (for example, such as are involved in Catholic Christianity), such standards need to be made explicit and should not be assumed as self-evident and above criticism. For it is sheer dogmatism to assume that only one conception of spiritual excellence is valid; and it is the height of discourtesy and parochialism to damn a society as immoral simply because its standards of excellence differ from one's own. Moreover, comparative judgments as to the happiness of men are notoriously untrustworthy, unless they are based upon objective measures of well-being. And if the material conditions of life are discounted as indications of "true" happiness, the critic's evaluations of different cultures are a better guide to his own preferences and loyalties than to the ostensible subject-matter of his judgment.

Let us turn to the causal imputations contained in the criticism under consideration. Almost no argument is required to show that if the growth of science may validly be held responsible for the ills of modern society, then the fact that men marry may no less validly be declared the cause of the evils of divorce. For surely, divorce would be impossible unless men first married, just as our present social distresses would not exist unless the advance of scientific knowledge had first made possible our present institutional structures. But to convert marriage into the cause of divorce, and the advance of secular knowledge into the cause of social ills, is to convert the *context* in which problems arise into an *agent* responsible for our inability to master them. As well argue that in order to eliminate the evils of divorce men must stop getting married, as recommend the de-secularization of modern society as a solution for its difficulties; in either case the conversion of context into cause is an unintelligent performance. The development of science has brought with it new opportunities for the exercise of human energies, and has helped set the stage for the emergence of new problems. How many of these problems have remained unsolved because vested interests and the cake of custom have prevented the application to them of the methods of controlled inquiry which the natural sciences use so successfully, it is difficult to judge. But in any event, the indictment of scientific intelligence as

most serious of the sources of danger which threaten destruction to the values of what we have called civilization. . . . It is a serious reflection that the unsatisfactory state of affairs in social science has largely resulted from the very progress of science. . . ." *Science and Man* (edited by Ruth Nanda Anshen), pp. 325–6.

solely responsible for our present difficulties not only involves an arbitrary selection of one factor from a complex of others distinguishable in the social scene; it arbitrarily rejects the one instrument from which a resolution of these difficulties may reasonably be expected.

Consider, finally, the charge that science "cannot determine values," and that therefore the apprehension of the elements of a good life must be obtained through some form of emotional experience. Now whatever be one's views as to the nature of values—whether they are regarded as relative or absolute, dependent on human preferences or not—it must be admitted that a science (such as astronomy) which does not concern itself with values and which does not contain value-terms in its vocabulary, is incompetent to establish value judgments. The thesis that *some* sciences cannot determine values is thus trivially true. On the other hand, every rational appraisal of values must take cognizance of the findings of the natural and social sciences; for if the existential conditions and consequences of the realization of values are not noted, acceptance of a scheme of values is a species of undisciplined romanticism. Accordingly, unless values are to be affirmed on the basis of uncontrolled intuition and impulse, all the elements of scientific analysis—observation, imaginative reconstruction, dialectic elaboration of hypotheses, and experimental verification—must be employed. Knowledge of biology and hygiene are indeed not sufficient for an adequate conception of the moral life; but if one may judge from the historical functions of some philosophic and theologic ideas in perpetuating economic inequality and human slavery, and in sanctioning the brutal shedding of human blood, neither is a knowledge of philosophy and theology.

It is often urged that what is good for man lies outside the province of scientific method, because the determination of human goods requires a sympathetic understanding of the human heart and a sensitive, individualized perception of the qualities of human personality; and the exercise of such powers, it is maintained, has no place in the procedures of science. But this objection rests, at bottom, on a failure to distinguish between the psychological and sociological conditions under which ideas originate, and the validity of those ideas. Thus, it is reported of Schiller that he used to place a rotting apple on his desk for the stimulus the odor of the fruit provided to his writing; but while this is an interesting item about the conditions under which Schiller obtained his inspirations, it has no bearing on the issue as to quality of his poetry. Similarly, the unusual circumstances—whether personal or social—under which many seers and religious prophets obtained their visions are not relevant in a consideration of the soundness of their moral exhortations. More generally, the psychology and the sociology of research are not identical with its logic. Those who disparage the application of scientific methods to the evaluation of human goods, on the ground that those methods exclude the exercise of a sympathetic imagination, are not only mistaken in their factual allegations; they are also well on the road to identifying the sheer vividness and the emotional overtones of ideas with their validity.

Selected Bibliography

This bibliography covers the period from 1952 through 1961. An extensive selected bibliography for the sociology of science prior to 1952 can be found in either Bernard Barber, *Science and the Social Order*, Free Press, 1952 (paperback edition, Collier Books, 1962), or Bernard Barber, "Sociology of Science: A Trend Report and Bibliography," *Current Sociology*, V, No. 2, 1956, UNESCO.

I. *The Social Nature of Science and the Scientific Role*

Carl B. Boyer, "Mathematical Inutility and the Advance of Science," *Science*, 130 (July 3, 1955), 22–25.

J. Bronowski, *The Common Sense of Science*, New York, Modern Library Paperbacks, no date.

Charles Coulston Gillispie, *The Edge of Objectivity: An Essay in the History of Scientific Ideas*, Princeton University Press, 1960.

Ernest Nagel, *Logic without Metaphysics, and Other Studies in the Philosophy of Science*, The Free Press, 1956.

———, *The Structure of Science: Problems in the Logic of Scientific Explanation*, Harcourt, Brace & World, 1961.

Herbert A. Shepard, "Basic Research and the Social System of Pure Science," *Philosophy of Science*, 23 (1956), 48–57.

Evon Z. Vogt and Ray Hyman, *Water Witching, U.S.A.*, University of Chicago Press, 1959.

II. *Science and Society*

Bernard Barber, "An Ignorance We Take for Granted," *Teachers College Record*, 60 (1959), 297–305.

———, "From Utopia to Reality: The Role of Public Opinion in the Understanding and Control of Science," unpublished paper delivered at

Meetings of American Association for Public Opinion Research, May, 1959, Lake George, N.Y.
George Z. F. Bereday, "American and Soviet Scientific Potential," *Social Problems*, 4 (1957), 208–219.
David M. Blank and G. V. Stigler, *The Demand and Supply of Scientific Personnel*, New York, National Bureau of Economic Research, 1957.
Giorgio de Santillana, *The Crime of Galileo*, University of Chicago Press, 1955.
Nicholas DeWitt, *Soviet Professional Manpower: Its Education, Training, and Supply*, Washington, D.C., National Science Foundation, 1955.
———, *Education and Professional Employment in the U.S.S.R.*, National Science Foundation, Washington, D.C., 1961.
———, "Soviet Science: The Institutional Debate," *Bulletin of the Atomic Scientists*, 16 (1960), 208–211.
———, "Reorganization of Science and Research in the U.S.S.R.," *Science*, 133 (June 23, 1961), 1891–1991.
J. Stefan Dupré and Sanford A. Lakoff, *Science and the Nation: Policy and Politics*, Prentice-Hall (Spectrum Books), 1962.
A. Hunter Dupree, *Asa Gray*, Harvard University Press, 1959.
———, *Science in the Federal Government: A History of Policies and Activities to 1940*, Harvard University Press, 1957.
Burton Edelson, "Mutual Obligations: Science and the Military," *Bulletin of the Atomic Scientists*, 14 (1958), 169–172.
Gerald W. Elbers and Paul Duncan, eds., *The Scientific Revolution: Challenge and Promise*, Washington, D.C., Public Affairs Press, 1959.
Renée Fox, "Some Social and Cultural Factors in American Society Conducive to Medical Research on Human Subjects," *Clinical Pharmacology and Therapeutics*, 1 (1960), 423–443.
Sidney H. Gould, ed., *Sciences in Communist China*, AAAS Symposium, vol. 68, Washington, 1961.
Zvi Griliches, "Research Costs and Social Returns: Hybrid Corn and Recent Innovations," *Journal of Political Economy*, 66 (1958), 419–431.
Henry E. Guerlac, "Science and French National Strength," in E. M. Earle, ed., *Modern France*, Princeton University Press, 1951.
Leopold H. Haimson, "Three Generations of the Soviet Intelligentsia," in Howard W. Winger, ed., *Iron Curtains and Scholarship*, University of Chicago Press, 1958, pp. 28–42.
F. G. Hill, "Formative Relations of American Enterprise, Government, and Science," *Political Science Quarterly*, 75 (1960), 400–419.
Brooke Hindle, *The Pursuit of Science in Revolutionary America, 1735–1789*, University of North Carolina Press, 1956.
Walter Hirsch, "The Autonomy of Science in Totalitarian Societies," *Social Forces*, 40 (1961), 15–22.
John Jewkes, "How Much Science?," *Economic Journal*, 70 (1960), 1–16.
David Joravsky, "Soviet Marxism and Biology before Lysenko," *Journal of the History of Ideas*, 20 (1959), 85–104.
Arthur Koestler, *The Sleepwalkers: A History of Man's Changing Vision of the Universe*, Macmillan, 1959.
Alexander Korol, *Soviet Education for Science and Technology*, Technology Press and John Wiley, 1957.
Hillier Krieghbaum, *Science, the News, and the Public*, New York University Press, 1958.

Louis Kriesberg and Larry Rosenberg, "The Impact of Federal Grants on Medical Education: An Essay and an Annotated Bibliography," unpublished, University of Chicago, National Opinion Research Center.

P. Lenard, "Der Arische and der Jüdische Geist" and "Einstein, Heisenberg and Schrödinger," in Leon Poliakov and J. Wulf, *Das Dritte Reich und seine Denker*, Berlin, Arani Verlags-GMBH, 1959, pp. 297, 300.

Edward Lurie, *Louis Agassiz: A Life in Science*, University of Chicago Press, 1960.

Lewis C. Mainzer, "A Public Place for American Science," *Virginia Quarterly Review*, 37 (1961), 398–413.

Leo A. Orleans, *Professional Manpower and Education in Communist China*, Washington, D.C., National Science Foundation, 1961.

Talcott Parsons, "Professional Training and the Role of the Professions in American Society," *Scientific Manpower*, 1958, National Science Foundation, 1959.

Helmuth Plessner, ed., *Untersuchungen zur Lage der deutschen Hochschullehrer*, Göttingen, Vonderbroeck & Ruprecht, 1956, vol. 1, pp. 283–287.

Derek J. deSolla Price, *Science since Babylon*, Yale University Press, 1961.

Don K. Price, *Government and Science: Their Dynamic Relation in American Democracy*, New York University Press, 1954.

Eugene Rabinowitch, "A Survey of Russian Science," in Alex Inkeles and Kent Geiger, eds., *Soviet Society: A Book of Readings*, Houghton, Mifflin, 1961. [Originally published as "Soviet Science—A Survey," in *Problems of Communism*, 8 (1958), United States Information Survey.]

Edward Shils, *The Intellectual between Tradition and Modernity: The Indian Situation*, Comparative Studies in Society and History, Supplement I, Mouton & Co., The Hague.

Richard H. Shryock, "The Interplay of Social and Internal Factors in the History of Modern Medicine," *Scientific Monthly*, 76 (1953), 221–230.

———, "Medicine and Society: An Historical Perspective," *Journal of the Mt. Sinai Hospital*, 19 (1953), 699–715.

Soviet Science, a symposium presented on Dec. 27, 1951, at the Philadelphia meeting of the American Association for the Advancement of Science, 1952.

Walter Sullivan, *Assault on the Unknown*, McGraw-Hill, 1961.

Survey Research Center, Institute for Social Research, University of Michigan, *Satellites, Science, and the Public*, 1959.

———, *The Public Impact of Science in the Mass Media*, 1958.

F. X. Sutton, "Problems and Strategy of Research in Less Developed Countries," a paper delivered to the North American Conference on the Social Implications of Industrialization and Technological Change, UNESCO and the University of Chicago, September, 1960.

Robert C. Tucker, "Stalin and the Uses of Psychology," *World Politics*, 8 (1956), 455–483.

Alexander Vucinich, *The Soviet Academy of Sciences*, Hoover Institute Studies, Series E: Institutions, No. 3, Stanford University Press, 1956.

Helmut Wagner, "The Cultural Sovietization of East Germany," *Social Research*, 24 (1957), 395–426.

Norman Wengert, ed., *Perspectives on Government and Science, The Annals*, 327 (1960).

Gustav Wetter, S.J., "Ideology and Science in the Soviet Union: Recent Developments," in Richard Pipes, ed., *The Russian Intelligentsia,* Columbia University Press, 1961, pp. 141–163. (This is the same essay that appeared earlier in *Daedalus,* Summer, 1960.)

Dael Wolfle, *America's Resources of Specialized Talent,* Harper and Bros., 1954.

III. *The Social Image of the Scientist and His Self-Conceptions*

Howard S. Becker, "An Analytical Model for Studies of the Recruitment of Scientific Manpower," *Scientific Manpower,* 1958, National Science Foundation, 1959.

——— and James W. Carper, "The Development of Identification with an Occupation," *American Journal of Sociology,* 61 (1956), 289–298.

James W. Carper and Howard S. Becker, "Adjustments to Conflicting Expectations in the Development of Identification with an Occupation," *Social Forces,* 36 (1957), 51–56.

Lindsey R. Harmon, "High School Backgrounds of Science Doctorates," *Science,* 133 (March 10, 1961), 679–688.

Bernard Mausner and Judith Mausner, "A Study of the Anti-Scientific Attitude," *Scientific American,* February, 1955.

Opinion Research Corporation, "The Scientific Mind and Management Mind," unpublished report, Princeton, N.J., 1959.

Donald L. Thistlethwaite, "College Environment and the Development of Talent," *Science,* 130 (July 10, 1960), 71–80.

Martin Trow, "Some Implications of the Social Origins of Engineers," *Scientific Manpower,* 1958, National Science Foundation, 1959.

S. Stewart West, "The Ideology of Academic Scientists," *IRE Transactions of Engineering Management,* June, 1960, 54–62.

IV. *The Organization of Scientific Work and Communication among Scientists*

Sir Eric Ashby, *Technology and the Academics,* London, Macmillan (New York, St. Martin's Press), 1958.

Louis B. Barnes, *Organizational Systems and Engineering Groups: A Comparative Study of Two Technical Groups in Industry,* Boston, Division of Research, Harvard Business School, 1960.

Ralph H. Bates, *Scientific Societies in the U.S.,* Columbia University Press, 1958.

Howard M. Baumgartel, "Leadership Style as a Variable in Research Administration," *Administrative Science Quarterly,* 2 (1957), 344–360.

"The Behavioral Sciences at Harvard," report by a faculty committee, June, 1954.

Warren G. Bennis, "The Social Science Research Organization: A Study of the Institutional Practices and Values of Interdisciplinary Research," unpublished doctoral dissertation, Massachusetts Institute of Technology, 1955.

———, "The Social Scientist as a Research Entrepreneur: A Case Study," *Social Problems,* 3 (1955), 44–49.

———, "Some Barriers to Teamwork in Social Research," *Social Problems*, 3 (1956), 223–235.
———, "Values and Organization in a University Social Research Group," *American Sociological Review*, 21 (1956), 555–563.
———, "The Effect of Academic Goods on Their Market," *American Journal of Sociology*, 62 (1956), 28–33.
Bernard Berelson, *Graduate Education in the United States*, McGraw-Hill, 1960.
Paula Brown, "Bureaucracy in a Government Laboratory," *Social Forces*, 32 (1954), 259–268.
——— and Clovis Shepherd, "Factionalism and Organizational Change in a Research Laboratory," *Social Problems*, 3 (1956), 235–243.
Helen L. Brownson, "Research on Handling Scientific Information," *Science*, 132 (Dec. 30, 1960), 1922–1931.
George P. Bush, *Bibliography on Research Administration—Annotated*, Washington, D.C., 1954.
——— and Lowell H. Hattery, "Teamwork and Creativity in Research," *Administrative Science Quarterly*, 1 (1956–1957), 361–372.
Theodore Caplow and Reece J. McGee, *The Academic Marketplace*, Basic Books, 1958.
D. S. L. Cardwell, *The Organisation of Science in England*, Heinemann, 1957.
William V. Consolazio, "Dilemma of Academic Biology in Europe," *Science*, 133 (June 16, 1961), 1892–1896.
Stevan Dedijer, "Why Did Daedalus Leave?," *Science*, 133 (June 30, 1961), 2047–2052.
William E. Dick, "Science and the Press," *Impact of Science on Society*, 5 (1954), 143–173.
J. C. Fisher, "Basic Research in Industry," *Science*, 129 (June 19, 1959), 1653–1657.
David M. Gates, "Basic Research in Europe," *Science*, 128 (Aug. 1, 1958), 227–235.
Barney G. Glaser, "The Resolution Effect of Recognition on Research"; "The Performance Process in a Research Organization"; and "The Impact of Differential Promotion Systems on Careers," three unpublished papers based on a 1961 Columbia University Ph.D. dissertation.
Bentley Glass, "The Academic Scientist, 1940–1960," *AAUP Bulletin*, 46 (1960) 149–155.
Norman Kaplan, "The Role of the Research Administrator," *Administrative Science Quarterly*, 4 (1959), 20–42.
———, "Research Overhead and the Universities: Should Government Support of Basic Research in Private Universities Cover Overhead as Well as Direct Costs?," *Science*, 132 (Aug. 12, 1960), 400–404.
John L. Kennedy and G. H. Putt, "Administration of Research in a Research Corporation," *Administrative Science Quarterly*, 1 (1956–1957), 326–339.
Earl W. Lindveit, *Scientists in Government*, Washington, D.C., Public Affairs Press, 1960.
James L. McCamy, *Science and Public Administration*, University of Alabama Press, 1960.
Edward McCrensky, *Scientific Manpower in Europe: A Comparative Study of Scientific Manpower in the Public Service of Great Britain and Selected European Countries*, New York, Pergamon Press, 1958.

Jack Alan MacWatt, "Improving Scientific Communication," *Science*, 134, Aug. 4, 1961, 313–316.

Simon Marcson, *The Scientist in American Industry: Some Organizational Determinants in Manpower Utilization*, Industrial Relations Section, Princeton University, 1960.

Leo Meltzer, "Scientific Productivity and Organizational Settings," *Journal of Social Issues*, 12 (1956), 32–40.

Herbert Menzel, *The Flow of Information among Scientists: Problems, Opportunities, and Research Questions*, mimeo., Bureau of Applied Social Research, May, 1958.

———, *Review of Studies in the Flow of Information among Scientists*, 2 vols., mimeo., Bureau of Applied Social Research, Columbia University, 1960.

Egon Orowan, "Our Universities and Scientific Creativity," *Bulletin of the Atomic Scientists*, 15 (1959), 236–239.

George Payne, *Britain's Scientific and Technological Manpower*, Stanford, 1960.

Donald C. Pelz, "Interaction and Attitudes between Scientists and the Auxiliary Staff: I. Viewpoint of Staff; II. Viewpoint of Scientists," *Administrative Science Quarterly*, 4 (1959–1960), 321–336, 410–425.

Allan C. Rankin, "The Administrative Processes of Contract and Grant Research," *Administrative Science Quarterly*, 1 (1956–1957), 275–294.

Curt P. Richter, "Free Research vs. Design Research," *Science*, 118 (July 24, 1953), 91–93.

Herbert A. Shepard, "Value System of a University Research Group," *American Sociological Review*, 19 (1954), 456–462.

Clovis Shepherd and Paula Brown, "Status, Prestige, and Esteem in a Research Organization," *Administrative Science Quarterly*, 1 (1956–1957), 340–360.

Richard H. Shryock, "The Academic Profession in the United States," *Bulletin of the American Association of University Professors*, 38 (1952), 32–70.

Gideon Sjoberg, "Science and Changing Publication Patterns," *Philosophy of Science*, 23 (1956), 90–96.

Ileen E. Stewart and V. W. McGurl, "Dues and Membership in Scientific Societies," *Science*, 132 (October, 1960), 939–42.

Paul Weiss, "Knowledge: A Growth Process," *Science*, 131 (June 10, 1960), 1716–19.

v. *The Social Process of Scientific Discovery*

Ernest M. Allen, "Why Are Research Grant Applications Disapproved?," *Science*, 132 (Nov. 25, 1960), 1532–1534.

Joseph Ben-David, "Roles and Innovation in Medicine," *American Journal of Sociology*, 65 (1960), 557–568.

W. I. B. Beveridge, *The Art of Scientific Investigation*, Modern Library Paperbacks, Random House, 1957 (first published in United States in 1950).

I. Bernard Cohen, "Orthodoxy and Scientific Progress," *Proceedings of American Philosophical Society*, 96 (1952), 505–512.

Wayne Dennis, "The Age Decrement in Outstanding Scientific Contributions," *American Psychologist*, 13 (1958), 457–460.

Gerald Holton, "On the Duality and the Growth of Physical Science." *American Scientist,* 41 (1953), 89–99.
John Jewkes, David Sawers, and Richard Stillerman, *The Sources of Invention,* London, Macmillan and Co., 1958.
Thomas S. Kuhn, *The Copernican Revolution: Planetary Astronomy in the Development of Western Thought,* Harvard University Press, 1957.
E. Manniche and G. Falk, "Age and the Nobel Prize," *Behavioral Science,* 2 (1957), 301–307.
Richard R. Nelson, "The Economics of Invention: A Survey of the Literature," *Journal of Business* (University of Chicago), 32 (1959), 101–127.
Michael Polanyi, "Passion and Controversy in Science," *Bulletin of the Atomic Scientists,* 13 (1957), 114–119.
F. Reif, "The Competitive World of the Pure Scientist," *Science,* 134 (1961), 1957–1962.
Morris I. Stein and Shirley J. Heinze, *Creativity and the Individual: Summaries of Selected Literature in Psychology and Psychiatry,* The Free Press, 1960.
UNESCO, "The Right to Scientific Property," *Impact of Science on Society,* 5 (1954), 47–68.

VI. *The Social Responsibility of Science*

AAAS, Committee on Human Welfare, "Science in the Promotion of Human Welfare," *Science,* 132 (July 8, 1960), 68–73.
Arthur S. Barron, "Why Do Scientists Read Science Fiction?," *Bulletin of the Atomic Scientists,* 13 (1957), 62–65, 70.
J. Bronowski, *Science and Human Values,* Julian Messner, 1958.
Clifford Grobstein, "The Social Conscience of United States Science: Sketch of a Decade," *Bulletin of the Atomic Scientists,* 12 (1956), 241–246.
Pendleton Herring, "On Science and the Polity," *Items* (Social Science Research Council Newsletter), 15 (1961), 1–6.
Gerald Holton, "Modern Science and the Intellectual Tradition" in G. de Huszar, ed., *The Intellectuals,* Free Press, 1960, pp. 180–191.
Robert Jungk, *Brighter than a Thousand Suns,* Harcourt, Brace, and World, 1958.
Joan W. Moore and Burton M. Moore, "The Role of the Scientific Elite in the Decision to Use the Atomic Bomb," *Social Problems,* 6 (1958), 78–85.
J. Robert Oppenheimer, *The Open Mind,* Simon and Schuster, 1955.
Eugene Rabinowitch, "Responsibilities of Scientists in the Atomic Age," *Bulletin of the Atomic Scientists,* 15 (1959), 2–7.
Edward Shils, "Freedom and Influence: Observations on the Scientists' Movement in the United States," *Bulletin of the Atomic Scientists,* 13 (1957), 13–18.
C. P. Snow, *Science and Government,* Harvard University Press, 1961.
———, *The Two Cultures and the Scientific Revolution,* The Rede Lecture, 1959, Cambridge University Press, 1959.
Neal Wood, "Utopians of Science," Chapter V in his *Communism and British Intellectuals,* Columbia University Press, 1959.

Name Index

Subject Index